Progress in Probability

Volume 39

Stochastic Modelling in Physical Oceanography

Robert J. Adler
Peter Müller
Boris Rozovskii

Editors

1996

Birkhäuser
Boston • Basel • Berlin

Robert J. Adler
Faculty of Industrial Engineering
and Management
Technion University
Haifa, Israel 32000

Peter Müller
Department of Oceanography
University of Hawaii
Honolulu, HI 96822

B.L. Rozovskii
Center for Applied Mathematical Sciences
University of Southern California
Los Angeles, CA 90089-1113

Library of Congress Cataloging-in-Publication Data

Stochastic modelling in physical oceanography / Robert J. Adler, Peter Müller, Boris Rozovskii, editors.
p. cm. -- (Progress in probability ; v. 39)
Includes bibliographical references.
ISBN 0-8176-3798-2 (hardcover : acid-free paper). -- ISBN 3-7643-3798-2 (hardcover : acid-free paper)
1. Oceanography--Mathematical models. 2. Stochastic processes. I. Adler, Robert J. II. Müller, Peter, 1944- . III. Rozovskii, B. L. (Boris L'vovich) IV. Series: Progress in probability; 39.
GC10.4.M36S76 1996 95-48417
551.46'001'5118--dc20 CIP

Printed on acid-free paper

Birkhäuser

ISBN 0-8176-3798-2
ISBN 3-7643-3798-2
Typeset by the editors in LATEX and AMS-TEX.
Printed and bound by Maple-Vail, York, PA.
Printed in the U.S.A.

9 8 7 6 5 4 3 2 1

Contents

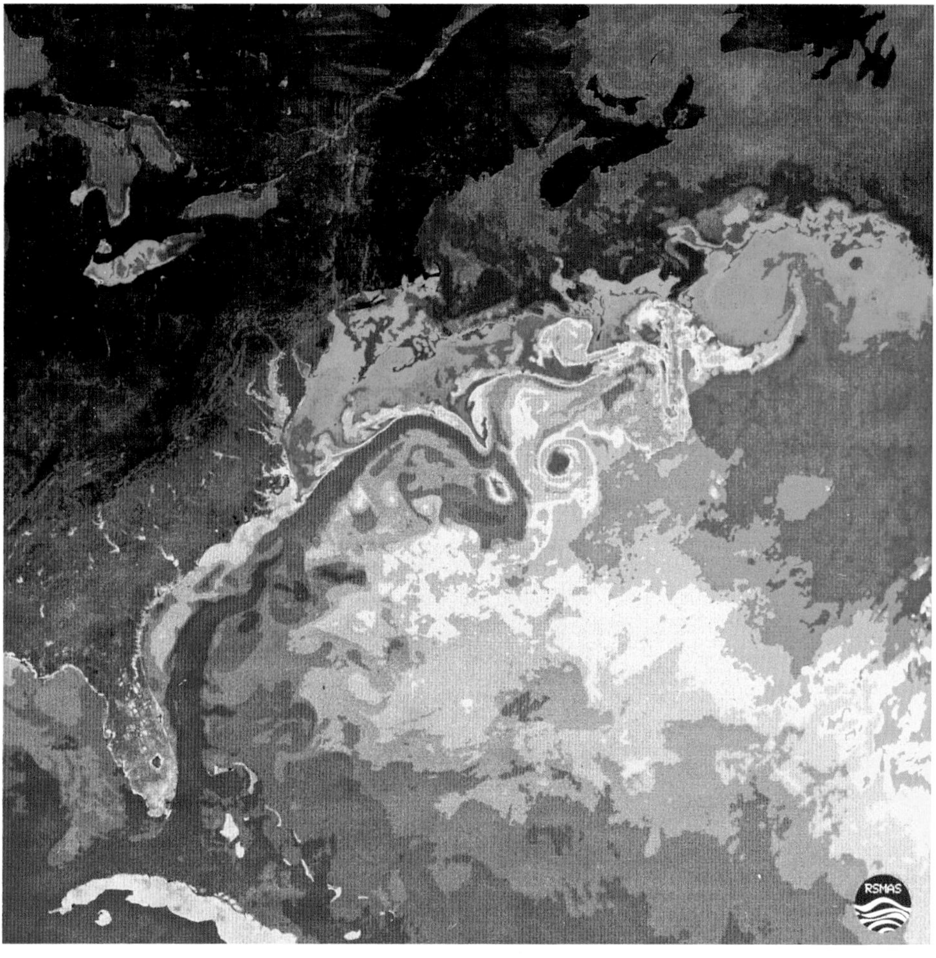

A composite of Advanced Very High Resolution Radiometer (AVHRR) images for an area encompassing the Gulf Stream. Red colors indicate warm waters and the purple colors indicate cooler waters. Reprinted with permission of the Rosenstiel School of Marine and Atmospheric Science Remote Sensing Group at the University of Miami.

Preface

The study of the ocean is almost as old as the history of mankind itself. When the first seafarers set out in their primitive ships they had to understand, as best they could, tides and currents, eddies and vortices, for lack of understanding often led to loss of live.

These primitive oceanographers were, of course, primarily statisticians. They collected what empirical data they could, and passed it down, initially by word of mouth, to their descendants. Data collection continued throughout the millenia, and although data bases became larger, more reliable, and better codified, it was not really until surprisingly recently that mankind began to try to understand the physics behind these data, and, shortly afterwards, to attempt to model it.

The basic modelling tool of physical oceanography is, today, the partial differential equation. Somehow, we all 'know" that if only we could find the right set of equations, with the right initial and boundary conditions, then we could solve the mysteries of ocean dynamics once and for all.

This utopian situation is, however, far away from us. Even if there is such a set of equations, their study poses a fascinating but formidable challenge. In fact, it may well be that the only current consensus on the nature of the equations of physical oceanography is that they must be of a highly nonlinear nature. As well as this, they must couple physical processes across various scales of motion, from the scale of a planetary radius to the scale of molecular diffusion, and it is more than feasible that, because of nonlinearities, microscale processes contribute substantially to macroscale properties of geophysical flows. In addition to all this, there is the practical impossibility of determining the "correct" initial and boundary conditions for such equations (such as the exact shape of the continents or of the ocean floor).

Thus, for the foreseeable future, even the ever increasing power of modern computers provides little hope of reaching oceanographic utopia.

To circumvent these seemingly insurmountable obstacles it is only natural to work with approximations, involving simpler equations that neglect, for example, microscale processes, and model our lack of knowledge as a random noise of some kind. In some sense, this last sentence describes what we mean by "stochastic models in physical oceanography".

The amazing heterogeneity of oceanographic phenomena and structures, from the micro to macro scales, from sound waves to the majestic ocean currents, means that not all models will involve partial differential equations, and that the tools needed to study the different models will be of many varied kinds.

Sometimes, one is interested in what happens at a particular point in the ocean only as a function of time, and then a one dimensional time series is an appropriate model. Sometimes one is interested in data smoothing problems, problems themselves so involved that thinking of the ocean as a locally stationary random field is a sufficiently sophisticated model. Mathematical analyses range from basic model building to parameter estimation and computational algorithms. Sometimes the focus is on the motion of large bodies of water, other times on the dispersion of small tracers in a turbulent flow field.

In this collection of papers, we have attempted to collect a variety of papers from scientists – either mathematicians or more applied physical oceanographers – that indicate what are the main topics of interest in this fascinating subject, and where breakthroughs are being made today. More than breakthroughs, however, we tried to collect here papers that would help to bring closer the two communities that are represented here, by asking our authors to set out challenges for the future.

We believe that both disciplines have much to gain from a fruitful interchange of ideas and problems. While physical oceanography adopted the serious use of stochastic models some two decades ago, many of the models used there remain overly simplisitic, and are far from exploiting the impressive developments in stochastic process theory during the same period. There is no doubt that the exploitation of the more modern theories and the more powerful tools now available will lead to more sophisticated and, more importantly, more realistic models of ocean dynamics. On the other hand, mathematicians have much to gain by concentrating their interest on, and being motivated by, models and problems that are generated by the greatest of all scientific innovators, Mother Nature.

Thus, the papers in this collection can, roughly, be characterised into three groups: Mathematical analysis of models that arise from, or are somehow related to, physical oceanography; expositions by oceanographers of phenomena and models that they are working with today or would like to see developed in the future; and the true interdisciplinary papers, where the borders between the two disciplines and approaches become blurred.

Rather than attempt to classify now into which of these three ill-defined groupings each paper fits, and thus give each paper a label that will almost definitely be out of date in another five years as the barriers between disciplines move, we shall leave it to the reader to now go through them, and make his own classification. All told, these papers give an excellent feel for what is happenning in the stochastic modelling of physical oceanography today, and how it is likely to advance in the near future.

Finally, as editors, we have two sets of acknowledgements to make. The first is to our authors and referees, most of whom worked hard to prepare papers that members of the "other" school could read. Writing for another school is harder than one might imagine at first, and many of the papers went back and forth a few times between author and referee until convergence to an acceptable language and style was found.

The second is to the Office of Naval Research. All three of the editors, and many of the contributors to this volume, have been supported by ONR funds over the years, primarily via the programs in Probability and Statistics and Physical Oceanography. However, our acknowledgment and appreciation goes beyond the usual "thank you for all the money". On this occassion, we wish to thank ONR and its officers for their foresight in developing the interdisciplinary program in *Random Fields in Oceanographic Modelling*, and for funding the associated conferences that brought two initially quite distinct communities together and, ultimately, led to the present collection of papers.

Robert J. Adler
Peter Müller
Boris Rozovskii

November, 1995.

Particle Displacements in Inhomogeneous Turbulence

Andrew F. Bennett

Contents

1. Introduction

Buoys, drifting freely on or beneath the ocean surface, are providing new insights into ocean circulation (Davis, 1991). To the extent that buoys faithfully behave like the fluid which they displace, they belong to the Lagrangian reference frame, rather than the Eulerian. That is, they map fluid properties such as velocity by launch position and time from launch. Moored current meters, on the other hand, map velocity by mooring location and time from installation. Fluid velocity is but one aspect of the ever-changing and intricate circulation of the oceans. It is conventional to conceive of this complexity in terms of random fields, although the approach has yet to be vindicated. That is, the turbulence problem remains unresolved. Nevertheless, in the Lagrangian frame, we seek the probability distribution function (pdf) for the position of a marked fluid particle at any time, given the position and time of launch. This pdf also yields the mean concentration field for a dissolved, nondiffusing substance injected where and when the particle was launched. Higher moments of concentration are also of interest, especially as fluctuations can be as large as means (Yee et al., 1992). However, from the climatological perspective, concentration means are of first concern. They are quite poorly understood, and will be the sole concern here.

The standard 'semiempirical' theory of inhomogeneous turbulent diffusion is beautifully described in Chapter 5 of the encyclopedic text by Monin and Yaglom (1971). The theory consists of a diffusion equation for the displacement pdf, or for the mean concentration. The drift velocity and the diffusivity tensor are functions of field position and time. That is, the diffusion equation has variable coefficients, but they are not dependent on the mean concentration. In its standard form the diffusion equation has, as the coefficient of the gradient of mean concentration, the Eulerian mean velocity. That is, the coefficient is the mean velocity observed by a moored current meter. The diffusion terms are grouped as a divergence of a flux proportional to the mean concentration gradient. The coefficient of proportionality is the so-called diffusivity tensor. This diffusion equation can be derived using simple Markovian-like approximations; incompressibility of the Eulerian velocity field seems to be an essential assumption.

The goals of this chapter are twofold. First, an extremely simple derivation of the semiempirical equation is given in §4 (Closure Theories), following preliminaries in §2 (Kinematics and Tracers), and in §3 (Displacement Statistics). Variants of the standard theory are given in §4 for compressible flow, for isentropic coordinates, and for surface floats. A scale analysis is made of errors in the simple derivation of the standard theory. The second goal of this chapter is also contained in §4, which begins by deriving in an equally simple way an alternative diffusion equation. It has a drift velocity and diffusivity tensor independent of the field position. These alternative coefficients do, however, depend upon the position of the launch of the particle or injection of concentrate. Unlike the semiempirical equation, this alternative equation has Gaussian solutions. Stochastic differential equations for the standard and alternative diffusion equations are given in §5 (Stochastic Processes). Gaussianity (§6) is of course a consequence of central limit theorems such as that owing to Cocke (1972). His theorem seems appropriate here since it addresses sums of nonstationary, dependent vector steps such as would approximate the integral of a particle's velocity in inhomogeneous turbulence. Theoretical probabilists are encouraged to show that the technical conditions assumed by Cocke are in fact consequences of the Navier Stokes equations. A numerical attempt to show Gaussianity for particle displacements, using Monte Carlo evaluation of functional path integrals, is described in §7 and §8. The issue of boundary conditions for turbulent diffusion equations is raised in §9; some oceanic observations of turbulent diffusion are reviewed in §10.

2. Kinematics and tracers

Let $\boldsymbol{u}(\boldsymbol{x}, t)$ be the two or three-dimensional vector value of a fluid velocity field, measured at position $\boldsymbol{x}$ and time t. This is the Eulerian description of fluid motion. The paths of fluid particles (the Lagrangian description) may be constructed from the Eulerian description. Consider a particle known to occupy the position $\boldsymbol{X}$ at time s. It will also pass through position $\boldsymbol{x}$ at time t, provided $\boldsymbol{x} = \boldsymbol{A}(\boldsymbol{X}, s|t)$ where $\boldsymbol{A}$ satisfies the ordinary differential equation

$$\frac{\partial \boldsymbol{A}}{\partial t} = \boldsymbol{u}(\boldsymbol{A}, t) \tag{2.1}$$

subject to

$$\boldsymbol{A}(\boldsymbol{X}, s|s) = \boldsymbol{X}. \tag{2.2}$$

See Fig. 1. The integral representation of (2.1) and (2.2) is

$$\boldsymbol{A}(\boldsymbol{X}, s|t) = \boldsymbol{X} + \int_s^t \boldsymbol{u}(\boldsymbol{X}, s|r)dr, \tag{2.3}$$

where

$$\boldsymbol{u}(\boldsymbol{X}, s|r) \equiv \boldsymbol{u}(\boldsymbol{A}(\boldsymbol{X}, s|r), r) \tag{2.4}$$

is the Lagrangian velocity at time r of the particle passing through $\boldsymbol{X}$ at time s.

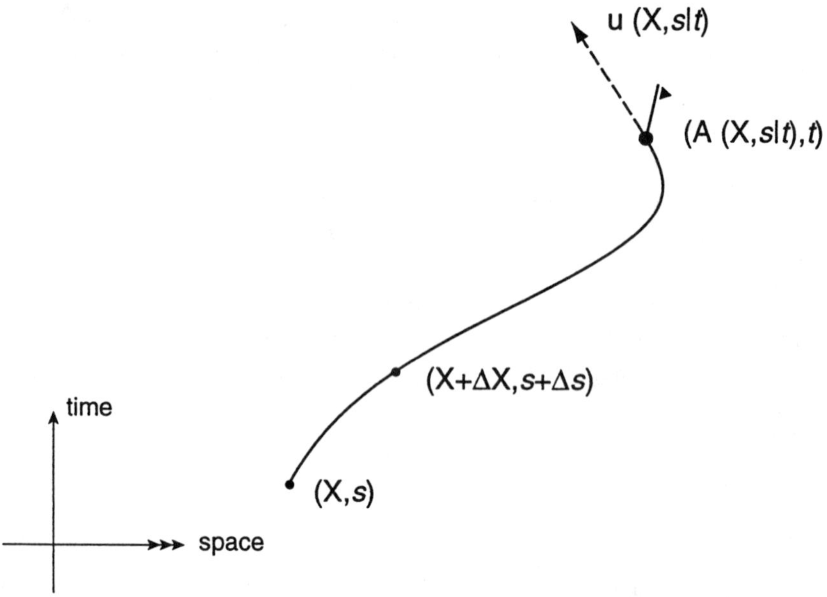

Figure 1. Path of a particle released from position $\boldsymbol{X}$ at time s. At time t the position of the particle is $\boldsymbol{A}(\boldsymbol{X}, s|t$, satisfying (2.1) and (2.2). The velocity of the particle is $\boldsymbol{u}(\boldsymbol{X}, s|t)$, defined by (2.4). The drifting buoy denotes that the velocity is Lagrangian. The point $(\boldsymbol{X} + \Delta\boldsymbol{X}, s + \Delta s)$ defines the same Lagrangian velocity if (2.6) holds, implying (2.5).

It is clear (see Fig. 1) that

$$\boldsymbol{u}(\boldsymbol{X} + \Delta\boldsymbol{X}, s + \Delta s|t) = \boldsymbol{u}(\boldsymbol{X}, s|t), \tag{2.5}$$

provided that

$$\Delta\boldsymbol{X} = \int_s^{s+\Delta s} \boldsymbol{u}(\boldsymbol{X}, s|r)dr; \tag{2.6}$$

thus (Kraichnan, 1965)

$$\left(\frac{\partial}{\partial s} + \boldsymbol{u}(\boldsymbol{X}, s) \cdot r\right) \boldsymbol{u}(\boldsymbol{X}, s|t) = 0. \tag{2.7}$$

The appropriate initial condition for (2.7) is

$$\boldsymbol{u}(\boldsymbol{X}, t|t) = \boldsymbol{u}(\boldsymbol{X}, t). \tag{2.8}$$

In (2.7), r denotes $\left(\frac{\partial}{\partial X_i}\right)$. Similarly it follows that

$$\left(\frac{\partial}{\partial s} + \boldsymbol{u}(\boldsymbol{X}, s) \cdot \mathrm{r}\right) \boldsymbol{A}(\boldsymbol{X}, s|t) = 0. \tag{2.9}$$

That is, a fluid particle retains its identity following the fluid motion. Several useful results will be derived subsequently from (2.7) and (2.9).

Consider now a passive tracer. This is material which does not influence the fluid motion, but which follows the flow. Examples are dissolved matter, or neutrally buoyant particles including plankton. If the concentration per unit mass is $C = C(\boldsymbol{x}, t)$, then a conserved tracer obeys

$$\frac{\partial}{\partial t}(\rho C) + \mathrm{r} \cdot (\rho \boldsymbol{u} C) = 0, \tag{2.10}$$

where $\rho = \rho(\boldsymbol{x}, t)$ is the fluid density and r denotes $\left(\frac{\partial}{\partial x_i}\right)$. Now the total mass per unit volume is also conserved, so

$$\frac{\partial \rho}{\partial t} + \mathrm{r} \cdot (\rho \boldsymbol{u}) = 0, \tag{2.11}$$

and hence

$$\frac{\partial C}{\partial t} + \boldsymbol{u} \cdot \mathrm{r} C = 0. \tag{2.12}$$

Simple generalizations include linear growth rates, distributed sources, and various nonlinear transformations of C: see Bennett and Denman, 1985.

Subject to the initial condition

$$C(\boldsymbol{x}, 0) = C_0(\boldsymbol{x}), \tag{2.13}$$

the solution of (2.12) is, by virtue of (2.9),

$$C(\boldsymbol{x}, t) = C_0(\boldsymbol{A}(\boldsymbol{x}, t|0)). \tag{2.14}$$

Thus the stirring of a conserved passive tracer has a Lagrangian description.

3. Displacement Statistics

Let $\boldsymbol{u}(\boldsymbol{x}, t)$ be a realization of a random velocity field, and let $\langle\ \rangle$ denote the ensemble average. Departures from the ensemble average will be denoted with a superscript prime; for example,

$$\boldsymbol{u}(\boldsymbol{x},t)' = \boldsymbol{u}(\boldsymbol{x},t) - \langle \boldsymbol{u}(\boldsymbol{x},t)\rangle . \tag{3.1}$$

The simplest displacement moment is the time rate of change of the mean displacement $\langle \boldsymbol{A}\rangle - \boldsymbol{X}$, given by the mean Lagrangian velocity:

$$\frac{\partial}{\partial t}\langle \boldsymbol{A}(\boldsymbol{X},s|t)\rangle = \langle \boldsymbol{u}(\boldsymbol{X},s|t)\rangle . \tag{3.2}$$

The time rate of change of the second order displacement moment is the Taylor diffusivity tensor (Taylor, 1921):

$$\begin{aligned}\frac{\partial}{\partial t}\langle A_i(\boldsymbol{X},s|t)'A_j(\boldsymbol{X},s|t)'\rangle = \int_s^t \Big\{ &\langle u_i(\boldsymbol{X},s|t)'u_j(\boldsymbol{X},s|r)'\rangle \\ &+ \langle u_i(\boldsymbol{X},s|r)'u_j(\boldsymbol{X},s|t)'\rangle \Big\} dr.\end{aligned} \tag{3.3}$$

If the first integral in (3.3) is denoted by the tensor $\mathsf{K}_{ij}(\boldsymbol{X},s|t)$, then the right hand side is twice the symmetric part of that tensor: $2\mathsf{K}^S_{ij} \equiv \mathsf{K}_{ij} + \mathsf{K}_{ji}$. If the Lagrangian mean velocity vanishes then K^A_{ij}, the antisymmetric part of K_{ij}, is proportional to the mean angular momentum of the fluid particle about the point $\boldsymbol{X}$ (Rhines, 1977).

There are some simple consequences of assuming that the turbulent velocity field $\boldsymbol{u}$ is *incompressible* and *homogeneous*, that is, if $r \cdot \boldsymbol{u} \equiv 0$ and the probability distribution of $\boldsymbol{u}$ is independent of position $\boldsymbol{x}$. For example, it follows readily from (2.7) and (2.8) that

$$\langle \boldsymbol{u}(\boldsymbol{X},s|t)\rangle = \langle \boldsymbol{u}(\boldsymbol{X},t)\rangle . \tag{3.4}$$

Equation (3.4) states the equality of the Lagrangian and Eulerian mean velocity fields at time t, irrespective of the Lagrangian conditioning time s. Both means are independent of position $\boldsymbol{X}$, by virtue of the assumed homogeneity. It is also readily shown that all (Lagrangian) moments of $\boldsymbol{u}(\boldsymbol{X},s|t)$ agree with the corresponding (Eulerian) moments of $\boldsymbol{u}(\boldsymbol{X},t)$. This implies that the two velocity fields have the same one-point probability distribution functions. The original proof of this result owes to Lumley (1962); see also Monin and Yaglom (1971, p. 573). Lumley's proof assumes ergodicity of $\boldsymbol{u}$. For a rigorous proof, see Port and Stone (1976). Another delicate result may be derived easily from (2.7) and (2.8), again for homogeneous incompressible turbulence:

$$\mathsf{D}(\boldsymbol{X},t|s) = \mathsf{K}(\boldsymbol{X},s|t) \tag{3.5}$$

where

$$\text{(3.6)} \qquad \mathsf{D}_{ij}(\boldsymbol{X}, t|s) = \int_s^t \langle u_i(\boldsymbol{X}, t)' u_j(\boldsymbol{X}, t|r)' \rangle dr,$$

while

$$\text{(3.7)} \qquad \mathsf{K}_{ij}(\boldsymbol{X}, s|t) = \int_s^t \langle u_i(\boldsymbol{X}, s|t)' u_j(\boldsymbol{X}, s|r)' \rangle dr$$

as already mentioned. Note that both **D** and **K** are actually independent of $\boldsymbol{X}$ since the turbulence is assumed to be homogeneous. The "events" represented by (3.5) and (3.6) are shown in Fig. 2.

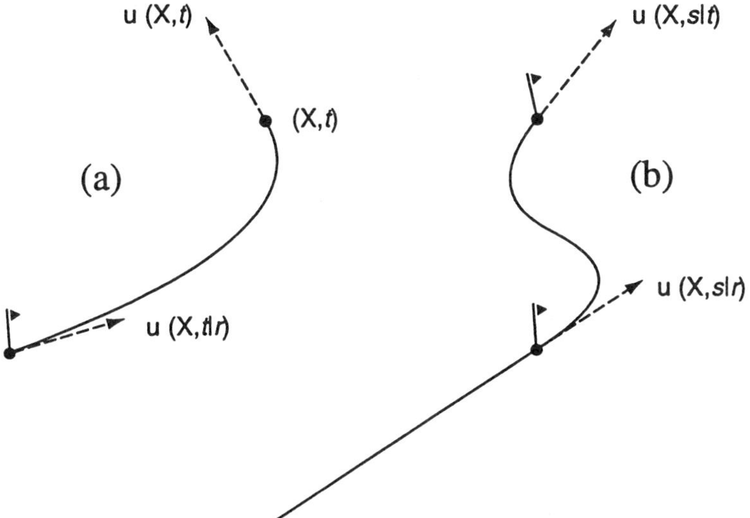

Figure 2. (a) The diffusivity $D_{ij}(\boldsymbol{X}, t|s)$ is determined by the covariance of Eulerian velocity $\boldsymbol{u}(\boldsymbol{X}, t)$, denoted by a rotor, with the Lagrangian velocity $\boldsymbol{u}(\boldsymbol{X}, r)$ for $s \leq r \leq t$. See (3.6). The conditioning parameters are $(\boldsymbol{X}, t)$.

For *incompressible* but *inhomogeneous* turbulence, the following relation between Lagrangian and Eulerian mean velocities is a direct consequence of (2.7) and (2.8):

$$\text{(3.8)} \qquad \langle u_i(\boldsymbol{X}, s|t) \rangle = \langle u_i(\boldsymbol{X}, t) \rangle + \frac{\partial}{\partial X_j} \int_s^t \langle u_j(\boldsymbol{X}, r) u_i(\boldsymbol{X}, r|t) \rangle dr.$$

Note the use of the summation convention in (3.8) and subsequently. The "events" represented by (3.8) are shown in Fig. 3. Rhines (1977) proposed a similar formula involving the Taylor diffusivity, as an approximation for weak velocities

$$(3.9) \qquad \langle u_i(\boldsymbol{X}, s|t)\rangle \cong \langle u_i(\boldsymbol{X}, t)\rangle + \frac{\partial}{\partial X_j} K_{ij}(\boldsymbol{X}, s|t).$$

Comparing (3.8) and (3.9) shows that the former, exact relation requires significantly more information for the evaluation of the Lagrangian mean. Particles must be released at $\boldsymbol{X}$ throughout the time interval $s < r < t$. The latter, approximate relation only requires that particles be released at time s. Haidvogel and Rhines (1983) made analytical and numerical estimates of K_{ij} in idealized fields of Rossby waves.

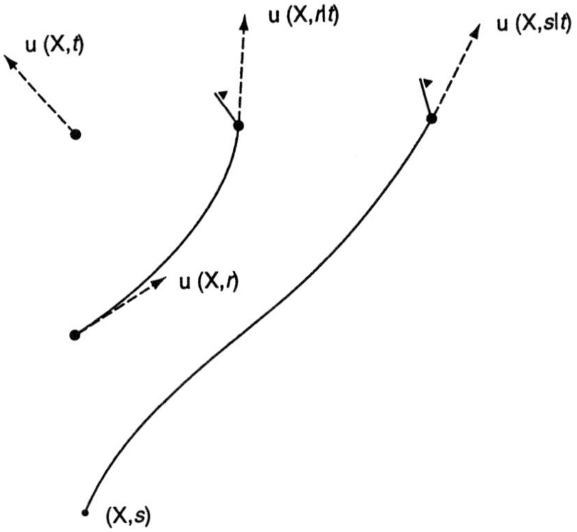

Figure 3. The Stokes drift $\langle \boldsymbol{u}(\boldsymbol{X}, s|t)\rangle - \langle \boldsymbol{u}(\boldsymbol{X}, t)\rangle$ is determined by the gradient of the mean of the (tensor) product of the Eulerian velocity $\boldsymbol{u}(\boldsymbol{X}, r)$ with the Lagrangian velocity $\boldsymbol{u}(\boldsymbol{X}, r|t)$, for $s \le r \le t$. See (3.8). This would be an expensive experiment.

Consider again the stirring of a conserved, passive tracer. Its mean concentration (averaging over the ensemble of flows, for a single initial concentration) is determined by the initial concentration and by the distribution of random displacements:

$$(3.10) \qquad \langle C(\boldsymbol{x}, t)\rangle = \langle C_0\left(\boldsymbol{A}(\boldsymbol{x}, t|0)\right)\rangle$$

$$(3.11) \qquad = \int C_0(\boldsymbol{X}) P(\boldsymbol{X}, 0|\boldsymbol{x}, t)\, d\boldsymbol{X}$$

where $P(\boldsymbol{X}, 0|\boldsymbol{x}, t) d\boldsymbol{X}$ is the probability that a particle passing through $\boldsymbol{x}$ at time t also passed within a $d\boldsymbol{X}$-neighborhood of $\boldsymbol{X}$ at time zero. In general, this probability is not reversible:

$$P(\boldsymbol{X}, 0|\boldsymbol{x}, t) \neq P(\boldsymbol{x}, t|\boldsymbol{X}, 0). \tag{3.12}$$

This is obvious, after a moment's consideration of flow from a simple source.

However, P must be given by

$$P(\boldsymbol{X}, 0|\boldsymbol{x}, t) = \langle \delta\left(\boldsymbol{X} - \boldsymbol{A}(\boldsymbol{x}, t|0)\right)\rangle, \tag{3.13}$$

hence

$$P(\boldsymbol{x}, t|\boldsymbol{X}, 0) = \langle \delta\left(\boldsymbol{x} - \boldsymbol{A}(\boldsymbol{X}, 0|t)\right)\rangle \tag{3.14}$$

$$= \langle \delta(\boldsymbol{X} - \boldsymbol{A}(\boldsymbol{x}, t|0)) \left|\frac{\partial(\boldsymbol{A})}{\partial(\boldsymbol{x})}\right|_{s=0} \rangle, \tag{3.15}$$

where the last symbol denotes the Jacobian determinant of the transformation $\boldsymbol{x} \to \boldsymbol{A}(\boldsymbol{x}, t|s)$. It is readily shown that this determinant is identically equal to unity for incompressible flow, in which case (Lundgren, 1981)

$$P(\boldsymbol{X}, 0|\boldsymbol{x}, t) = P(\boldsymbol{x}, t|\boldsymbol{X}, 0). \tag{3.16}$$

That is, displacement statistics are reversible, for incompressible flow, and we may infer that

$$\langle C(\boldsymbol{x}, t)\rangle = \int C_0(\boldsymbol{X}) P(\boldsymbol{x}, t|\boldsymbol{X}, 0)\, d\boldsymbol{X}. \tag{3.17}$$

For compressible flow, consider mean concentration per unit volume, that is, consider a density-weighted average. It may be shown that if $\rho_0(\boldsymbol{x})$ is the initial density, then

$$\rho(\boldsymbol{x}, t) C(\boldsymbol{x}, t) = \rho_0\left(\boldsymbol{A}(\boldsymbol{x}, t|0)\right) \left|\frac{\partial(\boldsymbol{A})}{\partial(\boldsymbol{x})}\right|_{s=0} C_0\left(\boldsymbol{A}(\boldsymbol{x}, t|0)\right) \tag{3.18}$$

as a consequence of mass conservation expressed in Lagrangian coordinates:

$$\frac{\partial}{\partial s}\left\{\rho(\boldsymbol{x}, t|s) \left|\frac{\partial(\boldsymbol{A})}{\partial(\boldsymbol{x})}\right| (\boldsymbol{x}, t|s)\right\} = 0. \tag{3.19}$$

It then follows as before (R. de Szoeke, personal communication) that

$$\langle \rho(\boldsymbol{x}, t) C(\boldsymbol{x}, t)\rangle = \int \rho_0(\boldsymbol{X}) C_0(\boldsymbol{X}) P(\boldsymbol{x}, t|\boldsymbol{X}, 0)\, d\boldsymbol{X}, \tag{3.20}$$

in spite of (3.12).

4. Closure Theories

The distribution P of fluid particle displacements determines the mean concentration, $\langle C \rangle$, of a passive scalar. The semiempirical equation of turbulent diffusion is an approximate equation for the evolution of P or of $\langle C \rangle$, and has been derived in a number of ways, all of which use quasi-Markovian arguments (e.g., Kraichnan, 1965). The case of interest here is the general one, that is, a nonGaussian, inhomogeneous, nonstationary turbulent velocity field. A particularly simple derivation is described below in §4.2. Variants for compressible flow, for thermodynamic coordinates and for surface floats are given in §4.3, 4.4 and 4.5 respectively. Higher corrections to these second-order closure approximations are examined in §4.6. This section begins, however, with an equally simple derivation (§4.1) of a turbulent diffusion equation which is significantly different from the standard, semiempirical theory.

4.1 Forward Lagrangian Closure. Recall that, for a given realization of the velocity field, it is certainly the case that a particle released from $\boldsymbol{X}$ at s will pass through $\boldsymbol{x}$ at t if, and only if, $\boldsymbol{x} = \boldsymbol{A}(\boldsymbol{X}, s|t)$. That is, the "microscopic" pdf is

$$p(\boldsymbol{x}, t|\boldsymbol{X}, s) = \delta(\boldsymbol{x} - \boldsymbol{A}(\boldsymbol{X}, s|t)). \tag{4.1}$$

The "macroscopic" pdf discussed previously is

$$P(\boldsymbol{x}, t|\boldsymbol{X}, s) = \langle p(\boldsymbol{x}, t|\boldsymbol{X}, s)\rangle, \tag{4.2}$$

where $\boldsymbol{X}$ and s are fixed. It is easily seen that p satisfies a Liouville equation:

$$\frac{\partial p}{\partial t} + u_i(\boldsymbol{X}, s|t)\frac{\partial p}{\partial x_i} = 0, \tag{4.3}$$

and so P satisfies

$$\frac{\partial P}{\partial t} + \langle u_i(\boldsymbol{X}, s|t)\rangle\frac{\partial P}{\partial x_i} = -\langle u_i(\boldsymbol{X}, s|t)' \left(\frac{\partial p}{\partial x_i}\right)' \rangle. \tag{4.4}$$

These equations, (4.3) and (4.4), hold for compressible flow. In order to "close" (4.4), it is necessary to estimate the right hand side (the "mean source") entirely in terms of P and moments of $\boldsymbol{u}$. To this end, consider the microscopic pdf perturbation p', which satisfies

$$\frac{\partial p'}{\partial t} + u_i(\boldsymbol{X}, s|t)\frac{\partial p'}{\partial x_i} = T_1(\boldsymbol{x}, t|\boldsymbol{X}, s) + T_2(\boldsymbol{x}, t|\boldsymbol{X}, s) \tag{4.5}$$

where

$$(4.6)\qquad T_1(\boldsymbol{x},t|\boldsymbol{X},s) \equiv \langle u_i(\boldsymbol{X},s|t)'\frac{\partial p}{\partial x_i}(\boldsymbol{x},t|\boldsymbol{X},s)'\rangle,$$

$$(4.7)\qquad T_2(\boldsymbol{x},t|\boldsymbol{X},s) \equiv -u_i(\boldsymbol{X},s|t)'\frac{\partial P}{\partial x_i}(\boldsymbol{x},t|\boldsymbol{X},s).$$

The solution of (4.5), subject to the appropriate initial condition

$$(4.8)\qquad p(\boldsymbol{x},s|\boldsymbol{X},s)' \equiv 0,$$

is

$$(4.9)\qquad p(\boldsymbol{x},t|\boldsymbol{X},s)' = \sum_{j=1}^{2}\int_s^t T_j\left(\boldsymbol{x}-\int_r^t \boldsymbol{u}(\boldsymbol{X},s|w)\,dw, r\,\middle|\,\boldsymbol{X},s\right)dr$$

which may be used to express the mean source $-\langle u_i'\partial p'/\partial x_i\rangle$ appearing in (4.4). *So far this analysis is exact.* Two approximations will now be made.

(i) The random Lagrangian velocity $\boldsymbol{u}(\boldsymbol{X},s|w)$ appearing in the argument of T_1 is replaced by its mean value $\langle u(\boldsymbol{X},s|w)\rangle$. Consequently the contribution to the mean source involving T_1 is proportional to $\langle u_i(\boldsymbol{X},s|t)'\rangle$, which vanishes. This approximation will be discussed in §4.6.

(ii) The term T_2 contributes a mean source

$$(4.10)\qquad -\left\langle u_i(\boldsymbol{X},s|t)'\left(\frac{\partial p}{\partial x_i}\right)'\right\rangle = \int_s^t \langle u_i(\boldsymbol{X},s|t)'u_j(\boldsymbol{X},s|r)' \times \frac{\partial^2 P}{\partial x_i\partial x_j}\left(\boldsymbol{x}-\int_r^t \boldsymbol{u}(\boldsymbol{X},s|w)\,dw, r\,\middle|\,\boldsymbol{X},s\right)dr\rangle.$$

There is a standard quasi-Markovian approximation (see for example Lundgren (1981), Drummond (1982) and their references) which must inevitably be made in order to obtain a second order closure. In applying the approximation to (4.10), the assumed decorrelation of the Lagrangian velocities for large $t-s$ is employed as a crude justification for replacing r by t in the arguments of P. This too will be discussed in Section 4.6. Substituting the resulting approximation into (4.4) yields the following evolution equation for $P(\boldsymbol{x},t|\boldsymbol{X},s)$:

$$\text{(4.11)} \qquad \frac{\partial P}{\partial t}(\boldsymbol{x},t|\boldsymbol{X},s) + \langle u_j(\boldsymbol{X},s|t)\rangle \frac{\partial P}{\partial x_j}(\boldsymbol{x},t|\boldsymbol{X},s) = \mathsf{K}_{ij}(\boldsymbol{X},s|t)\frac{\partial^2 P}{\partial x_i \partial x_j}(\boldsymbol{x},t|\boldsymbol{X},s).$$

It is easily seen that (4.11) is "rigorously" correct if the Lagrangian velocity $\boldsymbol{u}(\boldsymbol{X},s|t)$ is delta-correlated in time (the argument $\boldsymbol{x}$ must, however, be lagged by the mean displacement).

The coefficients in (4.11) are the Lagrangian mean velocity of a particle known to pass through $\boldsymbol{X}$ at s, and the Taylor diffusivity of that particle. In particular the coefficients are independent of the observer position or field position $\boldsymbol{x}$. Note that K_{ij} may be replaced with K_{ij}^S, assuming that mixed second partial derivatives of P are symmetric.

The initial condition for (4.11) is

$$\text{(4.12)} \qquad P(\boldsymbol{x},s|\boldsymbol{X},s) = \delta(\boldsymbol{x}-\boldsymbol{X}).$$

Thus the solution for (4.11), (4.12) in an unbounded domain is a Gaussian pdf with moments satisfying

$$\text{(4.13)} \qquad \frac{d}{dt}\langle\boldsymbol{x}\rangle = \langle\boldsymbol{u}(\boldsymbol{X},s|t)\rangle$$

subject to

$$\text{(4.14)} \qquad \langle\boldsymbol{x}\rangle = \boldsymbol{X}, \ \text{at} t = s,$$

and

$$\text{(4.15)} \qquad \frac{d}{dt}\{\langle x_i x_j\rangle - \langle x_i\rangle\langle x_j\rangle\} = 2\mathsf{K}_{ij}^S(\boldsymbol{X},s|t)$$

These are no more than identities.

4.2 Backward Lagrangian Closure. Starting again with the microscopic pdf (4.1), and using the "identity conservation" law (2.9), it follows that $p(\boldsymbol{X},s|\boldsymbol{x},t)$ satisfies

$$\text{(4.16)} \qquad \frac{\partial p}{\partial t}(\boldsymbol{X},s|\boldsymbol{x},t) + u_i(\boldsymbol{x},t)\frac{\partial p}{\partial x_i}(\boldsymbol{X},s|\boldsymbol{x},t) = 0.$$

and hence

$$(4.17)\qquad \frac{\partial P}{\partial t}(\boldsymbol{X}, s|\boldsymbol{x}, t) + \langle u_i(\boldsymbol{x}, t)\rangle \frac{\partial P}{\partial x_i}(\boldsymbol{X}, s|\boldsymbol{x}, t) = -\langle u_i(\boldsymbol{x}, t)' \left(\frac{\partial p}{\partial x_i}\right)' \rangle.$$

A closure analogous to that in (a) yields

$$(4.18)\qquad \frac{\partial P}{\partial t}(\boldsymbol{X}, s|\boldsymbol{x}, t) + \langle u_i(\boldsymbol{x}, t)\rangle \frac{\partial P}{\partial x_i}(\boldsymbol{X}, s|\boldsymbol{x}, t)\rangle = \int_s^t \langle u_i(\boldsymbol{x}, t)' \frac{\partial}{\partial x_i} \left\{ u_j(\boldsymbol{x}, t|r)''\rangle\, dr \frac{\partial P}{\partial x_j}(\boldsymbol{X}, s|\boldsymbol{x}, t) \right\}.$$

The right hand side of (4.18) needs careful inspection. Note the positions of the ket "⟩", the infinitesimal "dr", and the braces "{ }." Also notice that

$$(4.19)\qquad u_j(\boldsymbol{x}, t|r)'' \equiv u_j(\boldsymbol{\xi}, t) - \langle u_j\rangle(\boldsymbol{\xi}, t) \text{ where } \boldsymbol{\xi} = \boldsymbol{A}(\boldsymbol{x}, t|r),$$

is a fluctuating Eulerian velocity, evaluated at a random position. It differs from the Lagrangian velocity fluctuation

$$(4.20)\qquad u_j(\boldsymbol{x}, t|r)' = u_j(\boldsymbol{A}(\boldsymbol{x}, t|r), r) - \langle u_j(\boldsymbol{A}(\boldsymbol{x}, t|r), r)\rangle.$$

For incompressible flow, (4.18) becomes

$$(4.21)\qquad \frac{\partial P}{\partial t}(\boldsymbol{X}, s|\boldsymbol{x}, t) + \langle u_i(\boldsymbol{x}, t)\rangle \frac{\partial P}{\partial x_i}(\boldsymbol{X}, s|\boldsymbol{x}, t) = \frac{\partial}{\partial x_i} \left\{ \int_s^t \langle u_i(\boldsymbol{x}, t)' u_j(\boldsymbol{x}, t|r)''\rangle dr \frac{\partial P}{\partial x_j}(\boldsymbol{X}, s|\boldsymbol{x}, t) \right\}.$$

Neglecting the difference between $u_j(\boldsymbol{x}, t|r)'$ and $u_j(\boldsymbol{x}, t|r)''$, that is, neglecting a triple correlation, (4.21) becomes

$$(4.22)\qquad \frac{\partial P}{\partial t}(\boldsymbol{X}, s|\boldsymbol{x}, t) + \langle u_i(\boldsymbol{x}, t)\rangle \frac{\partial P}{\partial x_i}(\boldsymbol{X}, s|\boldsymbol{x}, t) = \frac{\partial}{\partial x_i} \left\{ \mathsf{D}_{ij}(\boldsymbol{x}, t|s) \frac{\partial P}{\partial x_j}(\boldsymbol{X}, s|\boldsymbol{x}, t) \right\}.$$

Combining the backward equation (4.22) with (3.10) yields an evolution equation for mean concentration $\langle C(\boldsymbol{x}, t)\rangle$ which for incompressible flow simplifies to the standard, "semiempirical" turbulent diffusion equation

$$(4.23)\qquad \frac{\partial \langle C\rangle}{\partial t} + \langle u_i(x, t)\rangle \frac{\partial \langle C\rangle}{\partial x_i} = \frac{\partial}{\partial x_i} \left\{ \mathsf{D}_{ij}(x, t|0) \frac{\partial \langle C\rangle}{\partial x_j} \right\}.$$

This equation can be derived "rigorously" if the Eulerian velocity field $\boldsymbol{u}(\boldsymbol{x}, t)$ is delta-correlated in time: see Klyatskin et al. (this volume), Lundgren (1981), etc.

Recalling again the assumption of incompressible flow, (4.22) yields, by virtue of the reversibility of displacement probabilities, the forward equation

$$\frac{\partial P}{\partial t}(\boldsymbol{x}, t|\boldsymbol{X}, s) + \langle u_i(\boldsymbol{x}, t)\rangle \frac{\partial P}{\partial x_i}(\boldsymbol{x}, t|\boldsymbol{X}, s) = \frac{\partial}{\partial x_i}\left\{ \mathsf{D}_{ij}(\boldsymbol{x}, t|s) \frac{\partial P}{\partial x_j}(\boldsymbol{x}, t|\boldsymbol{X}, s) \right\}. \tag{4.24}$$

Compare (4.11) and (4.24). Making a closure approximation to the *exact* equation (4.4) yields (4.11). A closely analogous approximate to the *exact* equation (4.17) yields (4.24). The dangers of rearranging and then partially summing series, which are almost certainly divergent, emerge as always.

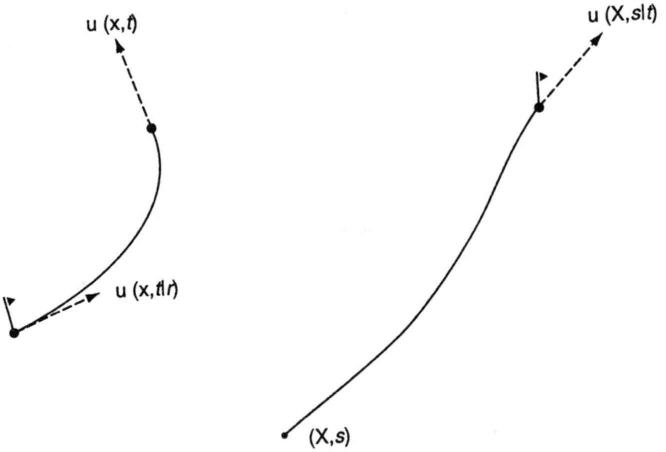

Figure 4. According to (4.25) the Lagrangian mean $\langle \boldsymbol{u}(\boldsymbol{X}, t)\rangle$ may be approximated by a knowledge of $\langle \boldsymbol{u}(\boldsymbol{x}, t)\rangle$, $\mathsf{D}(\boldsymbol{x}, t|s)$ and $P(\boldsymbol{x}, t|\boldsymbol{X}, s)$ for all $\boldsymbol{x}$. The path on the left indicates measurements required for D; see (3.6) and Fig. 2(a).

Note that the first moments of (4.24) satisfy

$$\frac{d}{dt}\langle x_i \rangle = \int \left\{ \langle u_i(\boldsymbol{x}, t)\rangle + \frac{\partial}{\partial x_k} \mathsf{D}_{ik}(\boldsymbol{x}, t|s) \right\} P(\boldsymbol{x}, t|\boldsymbol{X}, s)\, d\boldsymbol{x}, \tag{4.25}$$

(4.26)
$$\frac{d\langle x_i x_j\rangle}{dt} = \int \{\langle u_i\rangle x_j + \langle u_j\rangle x_i + x_j \frac{\partial}{\partial x_m}\mathsf{D}_{im} + x_i \frac{\partial}{\partial x_m}\mathsf{D}_{jm} + 2\mathsf{D}^S_{ij}\} P\, d\boldsymbol{x}$$

See Fig. 4. These approximations extend Corrsin's strong hypotheses to strongly inhomogeneous turbulence (Corrsin, 1959; Bennett, 1987), and contain more information than the identities (4.13) and (4.15), which follow (4.11). For example, (4.25) relates Lagrangian and Eulerian mean velocities, in incompressible flow. The original hypotheses are easily derived; for example

(4.27)
$$\frac{d\langle \boldsymbol{x}\rangle}{dt} = \langle \boldsymbol{u}(\boldsymbol{X}, s|t)\rangle = \langle \boldsymbol{u}(\boldsymbol{A}(\boldsymbol{X}, s|t), t)\rangle$$

(4.28)
$$= \int \langle \boldsymbol{u}(\boldsymbol{x}, t)\delta(\boldsymbol{x} - \boldsymbol{A}(\boldsymbol{X}, s|t))\rangle\, d\boldsymbol{x}$$

(4.29)
$$\sim \int \langle \boldsymbol{u}(\boldsymbol{x}, t)\rangle\langle \delta(\boldsymbol{x} - \boldsymbol{A}(\boldsymbol{X}, s|t))\rangle d\boldsymbol{x}$$

(4.30)
$$\equiv \int \langle \boldsymbol{u}(\boldsymbol{x}, t)\rangle P(\boldsymbol{x}, t|\boldsymbol{X}, s) d\boldsymbol{x}.$$

Note the Markovian assumption (4.29).

A rearrangement of (4.24) yields

(4.31)
$$\frac{\partial P}{\partial t} + \left\{\langle u_i(\boldsymbol{x}, t)\rangle + \frac{\partial}{\partial x_j}\mathsf{D}^A_{ij}(\boldsymbol{x}, t|s)\right\}\frac{\partial P}{\partial x_i} = \frac{\partial}{\partial x_i}\left\{\mathsf{D}^S_{ij}(\boldsymbol{x}, t|s)\frac{\partial P}{\partial x_j}\right\}$$

where $\mathbf{D}^S$ and $\mathbf{D}^A$ are respectively the symmetric and antisymmetric parts of $\mathbf{D}$. Notice that the latter part does not contribute to diffusion of probability. By assumption, $\partial u_i(\boldsymbol{x}, t)/\partial x_i = 0$ while $\partial^2 A_{ij}/\partial x_i \partial x_j \equiv 0$ for any smooth antisymmetric tensor $\mathbf{A}$. In contrast $\partial u_i(\boldsymbol{x}, t|s)/\partial x_i \neq 0$ in general and so $\partial^2 D^S_{ij}/\partial x_i \partial x_j \neq 0$ either.

4.3 Compressible flow: mean concentration. For compressible flow, the conventional approach is to average the concentration equation (2.10)

(4.32)
$$\langle\rho\rangle\langle C\rangle_t + \mathrm{r}\cdot(\langle\rho\boldsymbol{u}\rangle\langle C\rangle) = -\langle\rho' C'\rangle_t - \mathrm{r}\cdot\langle(\rho\boldsymbol{u})' C'\rangle.$$

To effect a closure, solve the fluctuating (2.12):

(4.33)
$$C'_t + \boldsymbol{u}\cdot \mathrm{r}C' = \langle\boldsymbol{u}'\cdot \mathrm{r}C'\rangle - \boldsymbol{u}'\cdot \mathrm{r}\langle C\rangle.$$

A Lagrangian expression for C' may be derived from (4.33). Substituting into (4.32) and approximating as before yields a diffusion equation for $\langle C\rangle$, provided the time derivative on the right hand side of (4.32) is neglected:

$$\frac{\partial}{\partial t}\langle C\rangle + \hat{u}_i\frac{\partial}{\partial x_i}\langle C\rangle = \langle\rho\rangle^{-1}\frac{\partial}{\partial x_i}\left(\langle\rho\rangle \mathsf{L}_{ij}\frac{\partial\langle C\rangle}{\partial x_j}\right) \tag{4.34}$$

where $\hat{u}_i = \langle\rho u_i\rangle/\langle\rho\rangle$ is a density-weighted mean velocity, and L_{ij} is a density-weighted diffusivity. Partitioning $\mathbf{L}$ into symmetric and antisymmetric parts leads to

$$\begin{aligned}\frac{\partial}{\partial t}\langle C\rangle+\Big\{\langle u_i\rangle + \langle\rho' u_i'\rangle/\langle\rho\rangle + \frac{\partial}{\partial x_j}\left(\langle\rho\rangle L^A_{ij}\right)\Big\}\frac{\partial\langle C\rangle}{\partial x_j}\\ = \langle\rho\rangle^{-1}\frac{\partial}{\partial x_i}\left(\langle\rho\rangle L^S_{ij}\frac{\partial\langle C\rangle}{\partial x_j}\right).\end{aligned} \tag{4.35}$$

None of the velocities multiplying the mean concentration gradient on the lhs of (4.35) is solenoidal. The interested reader may deduce that the rate of change for the center-of-concentration $[\boldsymbol{x}] \equiv \int \boldsymbol{x}\langle\rho\rangle\langle C\rangle d\boldsymbol{x} / \int\langle\rho\rangle\langle C\rangle d\boldsymbol{x}$ is an ugly formula, since the denominator is not a constant. A specific form for the second and third terms in the braces $\{\ \}$ in (4.35) has been proposed by Gent and McWilliams (1990), and tested in numerical simulation of the world ocean circulation (Danabasoglu et al., 1994). Eq. (4.34) reduces to eq. (33) of Plumb (1979), in the special case $\langle\rho\rangle = \langle\rho\rangle(z)$ only, and provided that density-weighted averages are replaced with simple averages. It may also be remarked that density-weighting of (2.10) obviates the need to neglect the time derivative on the right hand side of (4.32).

4.4 Thermodynamic coordinates. Oceanographers find thermodynamic coordinates useful. That is, $(x, y, z, t) \rightarrow (x, y, \varepsilon, t)$ where $\varepsilon = \varepsilon(x, y, z, t)$ is a thermodynamic state variable such as pressure π, density ρ or entropy η. Defining $\dot{\varepsilon}$ by

$$\dot{\varepsilon} \equiv \varepsilon_t + u\varepsilon_x + v\varepsilon_y + w\varepsilon_z, \tag{4.36}$$

the total derivative is given by

$$\partial_t + u\partial_x + v\partial_y + w\partial_z = \partial_t + u\partial_x + v\partial_y + \dot{\varepsilon}\partial_\epsilon. \tag{4.37}$$

The partial derivatives such as ∂_x on the left hand side of (4.37) refer to constant z, while ∂_x on the right refers to constant ε. Assuming that the vertical momentum balance is approximately hydrostatic: $\pi_z = -\rho g$, the mass conservation law (2.11) becomes

$$\pi_{\varepsilon t} + (u\pi_\varepsilon)_x + (v\pi_\varepsilon)_y + (\dot{\varepsilon}\pi_\varepsilon)_\varepsilon = 0, \tag{4.38}$$

where subscripts denote partial derivatives in (x, y, ε, t) coordinates. [If the Eulerian flow is incompressible in Cartesian coords, that is if $u_x + v_y + w_z = 0$, then π_ε in (4.38) should be replaced by z_ε.] One special case of interest is $\varepsilon = \pi$; then (4.38) becomes

$$u_x + v_y + \dot{\pi}_\pi = 0. \tag{4.39}$$

Thus $(u, v, \dot{\pi})$ is always solenoidal in (x, y, π) coordinates. Another special case is $\varepsilon = \eta$, the entropy. In the absence of internal sources or sinks of heat, entropy is conserved:

$$\dot{\eta} \equiv \eta_t + u\eta_x + v\eta_y + w\eta_z \equiv 0, \tag{4.40}$$

hence in entropy coordinates the mass conservation law becomes planar:

$$\pi_{\eta t} + (u\pi_\eta)_x + (v\pi_\eta)_y = 0, \tag{4.41}$$

while the scalar conservation law is

$$(\pi_\eta C)_t + (u\pi_\eta C)_x + (v\pi_\eta C)_y = 0. \tag{4.42}$$

Combining (4.41) and (4.42) yields

$$C_t + uC_x + vC_y = 0 \tag{4.43}$$

where the derivatives are taken on surfaces of constant entropy η. The law (4.43) implies that scalars can only be stirred on surfaces of constant entropy; there is no flow normal to these surfaces. Recall, however, that in general $u_x + v_y \neq 0$ in isentropic coordinates even if $u_x + v_y + w_z = 0$ in Cartesian coordinates.

Proceeding with (4.43) as in §4.3 yields

$$\frac{\partial \overline{C}}{\partial t} + \left(\tilde{u}_i + \overline{\pi}_\eta{}^{-1} \frac{\partial}{\partial x_j} \left(\overline{\pi}_\eta \mathsf{M}_{ij}^A \right) \right) \frac{\partial \overline{C}}{\partial x_i} = \overline{\pi}_\eta{}^{-1} \frac{\partial}{\partial x_i} \left\{ \overline{\pi}_\eta \mathsf{M}_{ij}^S \frac{\partial \overline{C}}{\partial x_j} \right\} \tag{4.44}$$

where i ranges from 1 to 2, $\overline{(\ )}$ is an average at constant entropy η, while $\tilde{u}_i = \overline{\pi_\eta u_i}/\overline{\pi_\eta}$ is a thickness-weighted mean velocity and M_{ij} a 2×2 thickness-weighted diffusivity tensor.

In spite of the unjustified neglect of $\overline{(\pi'_\eta C')_t}$ in the derivation of (4.44) and the nonsolenoidal nature of $\hat{u}_i$, the two-dimensional diffusion equation (4.44) is the most credible candidate for the modeling of the mixing of ocean tracers. The entropy coordinate system exploits the thermodynamic constraint: adiabatic motion is restricted to entropy surfaces. The involvement of thickness weighted velocities $\tilde{u}_i$ is not a problem since finite-difference approximations to the momentum equations may be expressed in terms of $\tilde{u}_i$.

Using the *mean* inverse transformation $(x, y, \eta, t) \to (x, y, \overline{z}, t)$, that is $z = \overline{z}(x, y, \eta, t)$, the left hand side of (4.44) becomes, in (x, y, z, t) coordinates,

$$\frac{\partial}{\partial t}\overline{C} + \mu_i \frac{\partial \overline{C}}{\partial x_i} + \omega \frac{\partial \overline{C}}{\partial z} = \dots \tag{4.45}$$

where $i = 1, 2$; $\mu_i = \tilde{u}_i + \overline{\pi}_\eta^{-1} \frac{\partial}{\partial x_j}(\overline{\pi}_\eta M^A_{ij})$ and $\omega = (\partial \overline{z}/\partial t) + \mu_i(\partial \overline{z}/\partial x_i)$. Note that $\tilde{u}_i = \overline{u_i} + \overline{(u'_i \pi'_\eta)}/\overline{\pi}_\eta$, with all averages $\overline{(\ )}$ being taken at constant η. In particular the correlation term in $\tilde{u}_i$ is distinct from the contribution made to μ_i by the antisymmetric part of the diffusivity tensor $\mathbf{M}$. Finally, none of the three contributions to μ_i is solenoidal, in general.

Thermodynamic coordinates are useful only if ε is a monotonic function of z. In fact, high-resolution vertical profiles of ocean temperature and salinity show numerous, small-scale inversions or reversals of slope. These may be smoothed away, and then ε may be defined as the "microstructure-averaged" thermodynamic variable. This induces apparent sources for entropy, in particular, in the form of "microstructure fluxes" on the right hand side of the adiabatic law (4.40). Thus, the dientropic velocity $\dot{\eta}$ no longer vanishes. Nevertheless, it is far smaller than the isentropic velocity components, and so advection in (4.41)–(4.43) remains very approximately planar (de Szoeke and Bennett, 1993).

4.5 Surface Floats. Buoys floating freely on the ocean surface can be tracked by satellites, or by GPS (Global Positioning System) and satellite altimetry. Drogues tethered to the float and set at about 100 m below the surface ensure that the float moves with the near-surface current rather than the wind. If $\boldsymbol{u}$ denotes the horizontal current, then we cannot assume $r \cdot \boldsymbol{u} = 0$ even if the ocean were of exactly constant density. Let $C(\boldsymbol{x}, t)$ denote the concentration of surface floats per unit area. Then conservation of floats implies

$$C_t + r \cdot (\boldsymbol{u}C) = 0. \tag{4.46}$$

Again $\boldsymbol{u}(\boldsymbol{x}, t)$ is a two-dimensional vector field over the surface, and $r = \frac{\partial}{\partial x_i}$ $(i = 1, 2)$. Averaging (4.46) yields

$$\langle C \rangle_t + r \cdot (\langle \boldsymbol{u} \rangle \langle C \rangle) = -r \cdot (\langle \boldsymbol{u}' C' \rangle). \tag{4.47}$$

An estimate for C' is needed, in order to close (4.47) approximately. It may be shown that

$$\left(\frac{\partial}{\partial t} + \boldsymbol{u} \cdot r \right) J^{-1} C' = J^{-1} r \cdot (\langle \boldsymbol{u}' C' \rangle - \boldsymbol{u}' \langle C \rangle) \tag{4.48}$$

where $J = |\partial(\boldsymbol{X})/\partial(\boldsymbol{x})|$ and $\boldsymbol{X} = \boldsymbol{A}(\boldsymbol{x}, t|0)$. This is a consequence of (4.46) and the identity

$$J_t + \boldsymbol{u} \cdot rJ = -Jr \cdot \boldsymbol{u}. \tag{4.49}$$

Proceeding much as previously yields

$$\langle C \rangle_t + \big(\langle \boldsymbol{u} \rangle + \boldsymbol{V} \big) \cdot r \langle C \rangle + r \cdot \boldsymbol{V} \langle C \rangle = r \cdot \big(\mathbf{D} \langle C \rangle \big) \tag{4.50}$$

where $\mathbf{D} = \mathbf{D}(\boldsymbol{x}, t|0)$ and

$$\boldsymbol{V}(\boldsymbol{x}, t) = -\int_0^t \langle \boldsymbol{u}(\boldsymbol{x}, t)(r \cdot \boldsymbol{u}')(\boldsymbol{x}, t|r) \rangle \, dr. \tag{4.51}$$

In particular the surface divergence $r \cdot \boldsymbol{u}$ leads not only to an additional advecting velocity $\boldsymbol{V}$, but to a concentration decay rate $r \cdot \boldsymbol{V}$. Floats cannot dive but are commonly seen to cluster around regions of surface *convergence.*

4.6 Corrections. Diffusion equations were derived in the preceding sections by making the simplest possible closures to formal expressions for fluxes. It is worth examining higher order corrections. Consider a conserved scalar concentration in incompressible flow:

$$\frac{\partial C}{\partial t} + \boldsymbol{u} \cdot rC = 0 \tag{4.52}$$

where

$$r \cdot \boldsymbol{u} = 0, \tag{4.53}$$

and the initial concentration is given by

$$C(\boldsymbol{x},0) = C_0(\boldsymbol{x}). \tag{4.54}$$

The mean concentration equation is

$$\frac{\partial}{\partial t}\langle C\rangle + \langle \boldsymbol{u}\rangle \cdot \mathrm{r}\langle C\rangle = -\mathrm{r}\cdot\langle \boldsymbol{u}' C'\rangle, \tag{4.55}$$

subject to

$$\langle C\rangle(\boldsymbol{x},0) = C_0(\boldsymbol{x}) \tag{4.56}$$

while the concentration fluctuations obey

$$\frac{\partial}{\partial t}C' + \boldsymbol{u}\cdot \mathrm{r}C' = \mathrm{r}\cdot\langle \boldsymbol{u}'C'\rangle - \boldsymbol{u}'\cdot \mathrm{r}\langle C\rangle, \tag{4.57}$$

subject to

$$C'(\boldsymbol{x},0) = 0. \tag{4.58}$$

The solution of (4.57)–(4.58) may be expressed in terms of a time integral, with dummy variable r, of the right hand side of (4.57) evaluated along the particle path $\boldsymbol{A}(\boldsymbol{x},t|r)$. In our simple derivation of the standard, semiempirical diffusion equation, $\boldsymbol{A}(\boldsymbol{x},t|r)$ was replaced with $\langle A\rangle(\boldsymbol{x},t|r) = \boldsymbol{x} - \int_r^t \langle \boldsymbol{u}(\boldsymbol{x},t|r')\rangle\, dr'$ in the argument of $\langle u'C'\rangle$. As a consequence, the Lagrangian solution for C' made no contribution to the mean flux $\langle u'C'\rangle$ appearing in (4.55). Also, replacing r by t in the argument of $\langle C\rangle$ in the integral caused $\boldsymbol{A}$ to be replaced by $\boldsymbol{x}$, and yielded the gradient formula for $\langle u'C'\rangle$ in (4.55). These crude approximations can be improved upon. First, the flux $\langle u'C'\rangle$ appearing in the integral may be expanded about $\langle \boldsymbol{A}\rangle$, in powers of $\boldsymbol{A} - \langle \boldsymbol{A}\rangle$, while $\langle C\rangle(\boldsymbol{A},r)$ may be replaced by $\langle C\rangle(\langle \boldsymbol{A}\rangle, r)$ instead of $\langle C\rangle(\boldsymbol{x},r)$. These better approximations lead to the following implicit equation for the flux component $F_i \equiv \langle u_i' C'\rangle$:

(4.59)

$$F_i(\boldsymbol{x},t) - \int_0^t dr\,\mathsf{D}_{ij}(\boldsymbol{x},t|r)\left(\frac{\partial^2 F_k}{\partial x_j \partial x_k}\right)(\langle \boldsymbol{A}(\boldsymbol{x},t|r)\rangle, r)$$
$$= -\int_0^t dr\,\langle u_i(\boldsymbol{x},t)' u_j(\boldsymbol{x},t|r)''\rangle \frac{\partial\langle C\rangle}{\partial x_j}(\langle \boldsymbol{A}(\boldsymbol{x},t|r)\rangle, r)$$

Davis (1987) obtained a closure equating F_i directly to a somewhat similar right hand side.

Let us examine what has been neglected in deriving (4.59). One approximation is

$$\boldsymbol{F}(\boldsymbol{A}, r) = \boldsymbol{F}(\langle\boldsymbol{A}\rangle, r) + \boldsymbol{A}' \cdot \mathrm{r}\boldsymbol{F}(\langle\boldsymbol{A}\rangle, r) + 0(|\boldsymbol{A}'|^2|\partial^2\boldsymbol{F}|) \tag{4.60}$$

where $\boldsymbol{A} = \boldsymbol{A}(\boldsymbol{x}, t|r)$ and $\boldsymbol{A}' = \boldsymbol{A} - \langle\boldsymbol{A}\rangle$. If $|\partial F_i/\partial x_j| \sim \mathcal{F}/L$, and $|A_i'| \sim \lambda$, then the relative error in (4.60) is $\lambda/\mathcal{L}$. The other approximation is, in effect,

$$|\langle u_i(\boldsymbol{x}, t)' u_j(\boldsymbol{x}, t|r)'' u_l(\boldsymbol{x}, t|w)'\rangle| \ll |\langle u_i'^2\rangle|^{\frac{3}{2}} \tag{4.61}$$

which can hold even if the Lagrangian velocity covariance does not vanish at large lag:

$$\langle u_i(\boldsymbol{x}, t|r)' u_j(\boldsymbol{x}, t|w)'\rangle = 0(\langle u_i'^2\rangle) \tag{4.62}$$

$$\text{as } |t-r|, |t-w|, |r-w| \to \infty.$$

Inequality (4.61) refers to the symmetry of the velocity distribution, not to its second moment structure.

Now let us examine the simplification of (4.59) to the gradient formula

$$F_i(\boldsymbol{x}, t) = -\mathsf{D}_{ij}(\boldsymbol{x}, t|0)\frac{\partial\langle C\rangle}{\partial x_j}(\boldsymbol{x}, t). \tag{4.63}$$

If the diffusivity $\mathsf{D}_{ij} \sim \mathcal{D}$, then the diffusion time for mean concentration fields is $\mathcal{L}^2/\mathcal{D}$, so the second term in the left hand side (lhs) of (4.59) is not negligible. If the mixed velocity covariance on the rhs of (4.59) has time scale τ, then $|\boldsymbol{x} - \langle\boldsymbol{A}\rangle| \sim \tau U$ where $|\langle\boldsymbol{u}(\boldsymbol{x}, t|r)\rangle| = 0(U)$, and the relative error in the rhs of (4.63) is $0(\tau U/\mathcal{L}) = 0\big((\lambda/\mathcal{L})(U/\mathcal{U})\big)$ where $\langle|\boldsymbol{u}'|^2\rangle^{\frac{1}{2}} = 0(\mathcal{U})$. In the ocean, $\lambda \sim 50$ km, $\mathcal{L} \sim 500$ km, $U \sim 10^{-2}$ ms^{-1}, while $\mathcal{U} \sim 10^{-1}$ ms^{-1}. Hence, lagging the mean concentration gradient, as in (4.59), yields only a 1% correction. Note that $\langle\boldsymbol{A}\rangle$ cannot be replaced with $\boldsymbol{x}$ on the lhs since $\mathsf{D}_{ij} \not\to 0$ as $t-r \to \infty$. In summary, we should equate the lhs of (4.59) to the rhs of (4.63). The resulting equation for $\langle C\rangle$ is now a higher order diffusion equation with (mean) Lagrangian arguments.

However, it may well make some sense to use only the simplest closure, rather than include a few extra levels of corrections. Indeed, if the turbulence is homogeneous and stationary, and the single-point, multi-time Lagrangian velocity field is jointly Gaussian, then the infinite series of corrections may be completely summed, yielding the gradient formula for the flux. The simplest way to see this is to return to the Lagrangian solution for the concentration:

$$C(\boldsymbol{x},t) = C_0(\boldsymbol{x} - \int_0^t \boldsymbol{u}(\boldsymbol{x},t|r)\,dr). \tag{4.64}$$

Let us assume that $\langle \boldsymbol{u}(\boldsymbol{x},t)\rangle = 0$; hence $\langle \boldsymbol{u}(\boldsymbol{x},t|r)\rangle = 0$, as a consequence of homogeneity. It follows that

$$\langle C(\boldsymbol{x},t)\rangle = \exp\,[\frac{1}{2}\langle (A_i - x_i)(A_j - x_j)\rangle \frac{\partial^2}{\partial x_i \partial x_j}]\,C_0(\boldsymbol{x}) \tag{4.65}$$

where $A_i = A_i(\boldsymbol{x},t|0)$ and the covariance is homogeneous. Now (4.65) is the formal solution of

$$\frac{\partial}{\partial t}\langle C\rangle = \mathsf{D}_{ij}\,(t|0)\frac{\partial^2}{\partial x_i \partial x_j}\langle C\rangle \tag{4.66}$$

subject to

$$\langle C(\boldsymbol{x},0)\rangle = C_0(\boldsymbol{x}). \tag{4.67}$$

In conclusion, exact analysis of homogeneous, stationary turbulence having jointly Gaussian Lagrangian velocities indicates that it is most likely misleading to add a few corrections to the simplest closure schemes in the more general setting of inhomogeneous, nonstationary, nonGaussian turbulence.

5. Stochastic Processes

The diffusion equations (4.11) and (4.24) are the Fokker-Planck equations (FPEs) for quite different stochastic processes. The process for (4.11) is defined by the linear stochastic differential equation (SDE)

$$d\boldsymbol{x} = \langle \boldsymbol{u}(\boldsymbol{X},s|t)\rangle dt + \sqrt{2\mathsf{K}^S(\boldsymbol{X},s|t)}\;d\boldsymbol{\beta}(t). \tag{5.1}$$

The square-root matrix in (5.1) is real if K^S has nonnegative eigenvalues. The components of the vector process $\boldsymbol{\beta}$ are mutually independent Wiener processes; that is, the increments $d\beta_i$ satisfy

$$\langle d\beta_i(t)\rangle = 0,\ \ \langle d\beta_i(t)d\beta_j(t')\rangle = \delta_{ij}\delta(t-t')dt\,dt'. \tag{5.2}$$

The initial condition for (5.1) is simply $\boldsymbol{x} = \boldsymbol{X}$ at $t = s$. The equation is linear since the coefficients are independent of $\boldsymbol{x}$, and so the solution for $\boldsymbol{x}$ in an unbounded domain has a Gaussian distribution. While (4.11) holds only asymptotically for $\boldsymbol{x} = \boldsymbol{A}(\boldsymbol{X}, s|t)$ as $t - s \to \infty$, it holds exactly for solutions of (5.1), for all $t \geq s$.

The process for (4.24) is the Itô SDE

$$d\boldsymbol{x} = \{\langle \boldsymbol{u}(\boldsymbol{x}, t)\rangle + r\mathbf{D}(\boldsymbol{x}, t|s)\}\, dt + \sqrt{2\mathbf{D}^S(\boldsymbol{x}, t|s)}\, dfi \tag{5.3}$$

where $(r\mathbf{D})_i = \partial D_{ij}/\partial x_j$, and where $\mathbf{D}^S$ is evaluated at $\boldsymbol{x}(t)$, rather than at $\boldsymbol{x}(t')$ where $t \leq t' \leq t + dt$ (Jazwinski, 1970; Gardner, 1985). This is a nonlinear equation, since the coefficients depend upon the solution $\boldsymbol{x}(t)$. In general, solutions are not Gaussian.

Notice that if the divergence of the diffusivity is omitted from the "drift" coefficient in the Itô SDE (5.3):

$$d\boldsymbol{x} = \langle \boldsymbol{u}(\boldsymbol{x}, t)\rangle dt + \sqrt{2\mathbf{D}^S(\boldsymbol{x}, t|s)}\, dfi \tag{5.4}$$

then the corresponding FPE is obtained by subtracting that divergence from the advecting velocity in (4.24):

$$\begin{aligned} P_t(\boldsymbol{x}, t|\boldsymbol{X}, s) + \{\langle \boldsymbol{u}(\boldsymbol{x}, t)\rangle - r\mathbf{D}(\boldsymbol{x}, t|s)\} \cdot rP(\boldsymbol{x}, t|s) = \\ r \cdot \{\mathbf{D}(\boldsymbol{x}, s|t) rP(\boldsymbol{x}, t|s)\}\,. \end{aligned} \tag{5.5}$$

An alternative process for (4.24) is the Stratonovitch SDE

$$d\boldsymbol{x} = \left(\langle \boldsymbol{u}(\boldsymbol{x}, t)\rangle + r\mathbf{D}^A\right) dt + \sqrt{2\mathbf{D}^S(\bar{\boldsymbol{x}}, t|s)}\, dfi \tag{5.6}$$

where $\bar{\boldsymbol{x}} = \frac{1}{2}\{\boldsymbol{x}(t) + \boldsymbol{x}(t + dt)\}$ and $\mathbf{D}^A$ is the antisymmetric part of $\mathbf{D}$. For future convenience, the stochastic dance card is:

SDE	FPE	Style
(5.1)	(4.11)	linear
(5.3)	(4.24)	Itô
(5.4)	(5.5)	Itô
(5.6)	(4.24)	Stratonovitch

Yamazaki and Kamykowski (1991) used the Itô SDE (5.4) in one dimension, to model vertical trajectories of motile phytoplankton in a wind-mixed water column. Holloway (1994) pointed out that the corresponding

FPE, namely (5.5), differs from the conventional empirical equation for turbulent diffusion (4.24). He advocated instead the Stratonovitch SDE (5.6) for which (4.24) is the FPE. Yamazaki and Kamykowski (1994) replied, in effect, that the derivation of (4.24) assumes incompressibility of the flow, which could only be satisfied by a one-dimensional flow if it were unrealistically uniform. Perhaps the resolution is that one must assume incompressible, three-dimensional flow, but seek a solution for P dependent only upon the vertical: $P = P(z, t|Z, s)$. This is possible only if $\langle w \rangle$ and $\mathbf{D}$ depend only upon the vertical: $\langle w \rangle = \langle w \rangle(z, t), \mathbf{D} = \mathbf{D}(z, t|s)$. Then (4.24) becomes the FPE

$$\frac{\partial P}{\partial t} + \langle w \rangle \frac{\partial P}{\partial z} = \frac{\partial}{\partial z}\left\{D_{33}\frac{\partial P}{\partial z}\right\}, \tag{5.7}$$

corresponding to the SDE

$$dz = \langle w(z,t) \rangle\, dt + \sqrt{2D_{33}(\overline{z}, t|s)}\, d\beta_3(t) \tag{5.8}$$

which is consistent with Holloway's remarks. The real issue is the validity of (4.24), that is, the conventional empirical theory of diffusion. The validity of the alternative (4.11) will be explored in subsequent sections. Of interest here is the fact that the derivation of (4.11) does not assume incompressibility. A "one-dimensional" SDE corresponding to (4.11) is

$$dz = \big\langle w(Z, s|t) \big\rangle + \sqrt{2K_{33}(Z, s|t)}\, d\beta_3(t) \tag{5.9}$$

where $z = Z$ at $t = s$. The distribution of z is Gaussian in z, but not in Z.

The stochastic processes (5.1) and (5.2) are Markovian, yet this need not be true of the Lagrangian displacement (2.3) which they model. For example the exact result (4.66) is consistent with a Markovian model, even though the Gaussian homogeneous and stationary velocity field in (4.64) may be otherwise: an *anticipating* process, perhaps (Gardner, 1985, p.86!).

6. Gaussianity

Forward Lagrangian closure (see §4.1) leads to the constant-coefficient diffusion equation (4.11) for the displacement pdf $P(\boldsymbol{x}, t|\boldsymbol{X}, s)$. The coefficients, which are the Lagrangian mean velocity and Taylor diffusivity, do depend upon the launch coordinates $(\boldsymbol{X}, s)$ but not upon the observer coordinates or field coordinates $(\boldsymbol{x}, t)$. Thus, the solution of (4.11) subject to (4.12) is a Gaussian for $\boldsymbol{x}$ at time t, with moments given by (4.13)–(4.15). Backward Lagrangian closure for incompressible flow yields (4.24), the "empirical" diffusion equation which has coefficients depending upon $(\boldsymbol{x}, t)$, and so has a nonGaussian solution for $P(\boldsymbol{x}, t|\boldsymbol{X}, s)$ in general. There is no disagreement between the backward and forward closures in homogeneous turbulence, since the Lagrange mean equals the Lagrangian mean and the Taylor diffusivity equals the mixed diffusivity. The disagreement occurs when the turbulence is inhomogeneous.

Suppose that the Lagrangian velocity $\boldsymbol{u}(\boldsymbol{X}, s|t)$ is Gaussian. That is, let $t_1, ..., t_M$ be M values of t, equally spaced for convenience. The supposition is that $\boldsymbol{u}(\boldsymbol{X}, s|t_m)$ is a multivariate Gaussian of dimension DM where D is the spatial dimension (2 or 3). It follows that $\Delta t \sum_{m=1}^{M} \boldsymbol{u}(\boldsymbol{X}, s|t_m)$ is a multivariate Gaussian of dimension D. By any reasonable definition of an integral, it is concluded that

$$\boldsymbol{A} = \boldsymbol{X} + \int_s^t \boldsymbol{u}(\boldsymbol{X}, s|r)\, dr \tag{6.1}$$

is Gaussian. That is, if $\boldsymbol{x} = \boldsymbol{A}$, then $P(\boldsymbol{x}, t|\boldsymbol{X}, s)$ is Gaussian. Score a point for (4.11). Or, pick (4.24) but accept that the Lagrangian velocity must be nonGaussian.

Suppose that the Lagrangian velocity is nonGaussian, but decorrelates rapidly in time. That is

$$\langle u_i(\boldsymbol{X}, s|r)' u_j(\boldsymbol{X}, s|w)' \rangle \longrightarrow\!\!\!\!\rightarrow 0 \tag{6.2}$$

as $| r - w | \to \infty$. The double arrow denotes a "sufficiently rapid" decorrelation rate. Suppose also that the Lagrangian variance $\langle |\boldsymbol{u}(\boldsymbol{X}, s|t)'|^2 \rangle$ does not vary "greatly", as a function of t. Then the central limit theorem "should" apply, leading to the conclusion that $\boldsymbol{A}$, as given by the integral (6.1), is asymptotically Gaussian as $|t - s| \to \infty$. Score another point for (4.11), if only asymptotically. The applicability of the central limit theorem to homogeneous and stationary turbulence has been widely accepted (see, for example, Hunt, 1985 p. 476). Should it not also apply to inhomogeneous or nonstationary turbulence?

Let us review the proof of the central limit theorem (clt) for a sample case. Consider a random walk with independent steps a_m, $(m = 1, 2, 3, ...,)$, having values of $+1$ or -1 with equal probability. The means and covariances of the a_m are assumed to be, respectively,

$$\langle a_m \rangle = 0, \quad \langle a_m a_k \rangle = \delta_{mk} = \begin{cases} 1, m = k \\ 0, m \neq k \end{cases}. \tag{6.3}$$

Let A_n be the sum of the first n steps:

$$A_n = \sum_{m=1}^{n} a_m. \tag{6.4}$$

Let P_n be the probability distribution function for $\xi_n = A_n/\sqrt{n}$. Then P_n is Gaussian, asymptotically as $n \to \infty$. The standard proof begins with the characteristic function ϕ_n, which is the Fourier transform of P_n:

$$\phi_n(k) = \int_{-\infty}^{\infty} P_n(\xi)e^{ik\xi}\, d\xi = \langle e^{ik\xi} \rangle \tag{6.5}$$

(by definition of P_n),

$$= \langle e^{ikn^{-\frac{1}{2}}(a_1+..+a_n)} \rangle \tag{6.6}$$

$$= \langle e^{ikn^{-\frac{1}{2}}a_1} \rangle .. \langle e^{ikn^{-\frac{1}{2}}a_n} \rangle \tag{6.7}$$

(since the a_m are independent),

$$= \left\{ \frac{1}{2}e^{ikn^{-\frac{1}{2}}} + \frac{1}{2}e^{-ikn^{-\frac{1}{2}}} \right\}^n = \cos^n(kn^{-\frac{1}{2}}) \tag{6.8}$$

(since $a_m = \pm 1$ with equal probability of $\frac{1}{2}$)

$$= (1 - \frac{1}{2}k^2 n^{-1} + \frac{1}{4}k^4 n^{-2} - ...)^n. \tag{6.9}$$

Hence, as $n \to \infty$,

$$\phi_n(k) \sim e^{-\frac{1}{2}k^2}. \tag{6.10}$$

The inverse Fourier transform of (6.10) is

$$P_n(\xi) \sim (2\pi)^{-\frac{1}{2}} e^{-\frac{1}{2}\xi^2}, \tag{6.11}$$

that is, P_n is asymptotically Gaussian with zero mean and unit variance. [If the values of a_m are $\pm 1 + \alpha$ with equal probability, then it is $\xi_n = (A_n - n\alpha)/\sqrt{n}$ which is distributed by (6.11)]. Notice that each step a_m makes an equal contribution of n^{-1} to the (unit) variance of the scaled sum ξ_n. Specifically, this individual contribution vanishes as $n \to \infty$.

The simple random walk described above is stationary in the sense that all steps have the same pdf, given by

$$p(a_n) = \frac{1}{2}\delta(a_n + 1) + \frac{1}{2}\delta(a_n - 1) \tag{6.12}$$

where δ is the Dirac delta function. Now suppose the walk is nonstationary, so the means $\langle a_n \rangle$ and covariances $\langle a'_n a'_m \rangle = \langle a'^2_n \rangle \delta_{nm}$ vary with n. It is still assumed that the steps are statistically independent.

Let A_n denote the sum of the first n steps, as in (6.4). Let M_n and σ_n denote, respectively, the mean and the standard deviation of A_n:

$$M_n = \sum_{j=1}^{n} \langle a_j \rangle, \quad \sigma_n^2 = \sum_{j=1}^{n} \langle a'^2_j \rangle. \tag{6.13}$$

Then the pdf of $\xi_n = (A_n - M_n)\sigma_n^{-1}$ is asymptotically Gaussian, with zero mean and unit variance provided that the Lindeberg condition holds: for any $\tau > 0$,

$$\lim_{n \to \infty} \sigma_n^{-2} \sum_{j=1}^{n} \int (a - \langle a_j \rangle)^2 p_j(a)\, da = 0 \tag{6.14a}$$

where p_j is the pdf of the jth step, and the range of integration is given by

$$|a - \langle a_j \rangle| > \tau \sigma_n. \tag{6.14b}$$

The proof of this central limit theorem may be found in, for example, Gnedenko (1976). It proceeds as in (6.5)–(6.11) but with considerably more analytical detail. The meaning of the Lindeberg condition (6.14) is of greater interest here.

It is first worth noting that if $\tau = 0$, then the integral in (6.14) is the variance of a_j, while the sum is equal to σ_n^2 and so every term in the sequence in (6.14) is equal to unity. Thus, the limit can only vanish if $\tau > 0$. An example is helpful. Suppose p_j is exponential:

$$p_j(a) = (2\lambda_j)^{-1} e^{-|a|\lambda_j^{-1}} \tag{6.15}$$

for $-\infty < a < \infty$, where λ_j is positive. Then

$$\langle a_j \rangle = 0, \qquad \langle a_j^2 \rangle = 2\lambda_j^2, \tag{6.16}$$

and the left hand side of (6.14a) becomes

$$\lim_{n \to \infty} \frac{1}{2} \sum_{j=1}^{n} e^{-\tau \lambda_j^{-1} \sigma_n} \{\tau^2 + 2\tau\lambda_j\sigma_n^{-1} + 2\lambda_j^2\sigma_n^{-2}\} \tag{6.17}$$

which vanishes for any $\tau > 0$, provided $\lambda_j \sigma_n^{-1} \to 0$ sufficiently rapidly for each $j \leq n$, as $n \to \infty$. That is, the Lindeberg condition holds if each step makes a vanishingly small contribution to the total variance of the random walk, as the number of steps becomes infinite. Gnedenko (1976) shows that (6.14) ensures, in general, that the probability

$$\text{Prob}\{\max_{1\leq j\leq n}|a_j - \langle a_j\rangle| \geq \tau\sigma_n\} \to 0 \tag{6.18}$$

as $n \to \infty$, for any $\tau \to 0$, and so ".. the Lindeberg condition is a peculiar kind of demand for the uniform smallness of the terms $(a_j - \langle a_j\rangle)\sigma_n^{-1}$ in the sum $\xi_n = (A_n - M_n)\sigma_n^{-1}$." Finally, it may be noted that, in the stationary case, the Lindeberg condition reduces to the simple requirement that the steps have finite variance.

The central limit theorem is not yet applicable to the integral in (6.1), which we shall approximate by the sum

$$\boldsymbol{A}(\boldsymbol{X}, s|s + n\Delta t) \cong \boldsymbol{X} + \sum_{m=1}^{n} \boldsymbol{u}_m \Delta t \tag{6.19}$$

where $t - s \cong n\Delta t$, $r - s \cong m\Delta t$ and $\boldsymbol{u}_m \equiv \boldsymbol{u}(\boldsymbol{X}, s|s + m\Delta t)$. The difficulty is that the steps $\boldsymbol{u}_m$ are statistically dependent, at least for a number of consecutive time steps. Fortunately there is a proof of the clt for a sum of nonstationary vector steps which are asymptotically independent for large separation (Cocke, 1972). The theorem will now be stated in terms of our notation. In particular, we shall consider the centered vector sum

$$\boldsymbol{S}_n = \sum_{m=1}^{n} (\boldsymbol{u}_m - \langle \boldsymbol{u}_m\rangle)\Delta t \tag{6.20}$$

where again, $\boldsymbol{u}_m = \boldsymbol{u}(\boldsymbol{X}, s|s + m\Delta t)$. The idea is to omit strings of terms from the sum in (6.21), so that as $n \to \infty$ the gaps become sufficiently great that the remaining clusters of steps are independent of each other, yet the gaps represent only a vanishingly small part of the variance of the complete sum. The approach follows Rosenblatt (1962). Cocke's notation has been altered slightly here, and his condition (III) has been trivially rearranged in order to reduce the considerable number of symbols involved.

Let $\boldsymbol{V}_m$ be a region in $\Re^3, m = 1, ..., n$. Denote the *probability function* by

$$P\{\boldsymbol{u}_1\Delta t \in \boldsymbol{V}_1, ..., \boldsymbol{u}_n\Delta t \in \boldsymbol{V}_n\} = F^{(n)}(\boldsymbol{V}_1, ..., \boldsymbol{V}_n). \tag{6.21}$$

The averages $\langle \boldsymbol{u}_m\rangle\Delta t$ and $\langle \boldsymbol{u}_m\boldsymbol{u}_l^*\rangle(\Delta t)^2$, where $(\ )^*$ denotes transposition, are assumed finite, but the existence of higher order moments is not assumed. Let ϕ_n be a *subset* of all the positive integers $\leq n$. Consider also that ϕ_n is composed of $K(n)$ disjoint subsets ϕ_{nk}, where $\phi_n = \bigcup_{k=1}^{K(n)} \phi_{nk}$. Let the set of "included" variables be

$$\boldsymbol{\mu}(n) = \{\boldsymbol{u}_m \Delta t : m \in \phi_n\} \tag{6.22}$$

with corresponding subsets $\boldsymbol{\mu}^{(nk)}$. Also let

$$\begin{cases} \boldsymbol{S}_{nk} &= \sum_{m \epsilon \phi_{nk}} \big(\boldsymbol{u}_m - \langle \boldsymbol{u}_m \rangle\big) \Delta t, \\ B_{nk}^2 &= \langle \boldsymbol{S}_{nk}^* \boldsymbol{S}_{nk} \rangle, \\ B_n^2 &= \sum_{k=1}^{K(n)} B_{nk}^2, \quad \boldsymbol{\xi}_{nk} = \boldsymbol{S}_{nk}/B_n, \\ \boldsymbol{\xi}_n &= \sum_{k=1}^{K(n)} \boldsymbol{\xi}_{nk}. \end{cases} \tag{6.23}$$

Thus, $\boldsymbol{S}_{nk}$ is one of $K(n)$ partial sums of the terms in $\boldsymbol{S}_n$ (see (6.20)); B_{nk}^2, B_n are scalar normalizing factors, while $\boldsymbol{\xi}_n$ is a scaled partial sum of the terms in $\boldsymbol{S}_n$. Denote the marginal probability of the set of vector variables $\boldsymbol{\mu}^{(n)}$ by

$$G^{(n)}(\boldsymbol{V}) = P\{\boldsymbol{\mu}^{(n)} \epsilon \boldsymbol{V}\} \tag{6.24}$$

and correspondingly for the subsets

$$G^{(nk)}(\boldsymbol{V}) = P\{\boldsymbol{\mu}^{(nk)} \epsilon \boldsymbol{V}\}. \tag{6.25}$$

The probability functions for the scaled vector sums $\boldsymbol{\xi}_{nk}$ are written

$$F_{nk}(\boldsymbol{V}) = P\{\boldsymbol{\xi}_{nk} \epsilon \boldsymbol{V}\}. \tag{6.26}$$

Three conditions need be imposed.

(I) For any $\tau > 0$

$$\lim_{n \to \infty} \sum_{k=1}^{K(n)} \int_{|\mu| > \tau} |\boldsymbol{\mu}|^2 F_{nk}(d\boldsymbol{\mu}) = 0; \tag{6.27}$$

(II)

$$\lim_{n \to \infty} \int |G^{(n)}(d\boldsymbol{\mu}^{(n)}) - \prod_{k=1}^{K(n)} G^{(nk)}(d\boldsymbol{\mu}^{(n)})| = 0; \tag{6.28}$$

(III)

$$\lim_{n \to \infty} \langle |(\boldsymbol{S}_n/B_n) - \boldsymbol{\xi}_n| \rangle = 0. \tag{6.29}$$

Condition (I) is a Lindeberg condition, while (II) requires that the $\boldsymbol{\mu}^{(nk)}$ become independent in the limit as the "gaps" between the $\boldsymbol{\mu}^{(nk)}$ become increasingly large. Condition (III) holds, roughly, if infinitely more vector steps are included than are excluded.

Theorem (Cocke, 1976). *If there exist an index sequence ϕ_{nk} such that conditions (I), (II), and (III) hold, and if the limiting 3×3 covariance matrix*

$$\mathsf{C} = \lim_{n\to\infty} \sum_{k=1}^{K(n)} \int \boldsymbol{\mu}\boldsymbol{\mu}^* F_{nk}(d\boldsymbol{\mu})$$

exists, then the distribution function for the scaled vector sum $\boldsymbol{S}_n/B_n$ is asymptotically Gaussian, with mean $\boldsymbol{0}$ and covariance C.

Unscrambling the notation leads to $\boldsymbol{x} = \boldsymbol{A}(\boldsymbol{X}, s|t)$ being Gaussian, and having first and second moments satisfying (4.13) and (4.15) respectively. Slam dunk for (4.11).

That Cocke's construction is ever possible, is most easily seen in the case of a stationary random walk as in (6.3), (6.4), but having correlated steps. We may define a decorrelation spacing $N \equiv \langle A_n^2\rangle n^{-1}$. Subdivide the n steps into $n^{\frac{1}{4}}$ groups each containing $n^{\frac{3}{4}}$ steps, and omit the first $n^{\frac{1}{4}}$ steps in each group. This leaves $n^{\frac{3}{4}} - n^{\frac{1}{4}}$ steps in each group. These remaining steps become independent if they are in different groups, asymptotically as $n^{\frac{1}{4}}N^{-1} \to \infty$. The total variance of the remaining steps in all groups is $n^{\frac{1}{4}}N(n^{\frac{3}{4}} - n^{\frac{1}{4}}) = N(n - n^{\frac{1}{2}}) \sim Nn = \langle A_n^2\rangle$ as $n^{\frac{1}{4}}N^{-1} \to \infty$.

For further illustration, consider a self-similar turbulent flow like a plume or jet in which the Lagrangian velocity increases exponentially with time. No dynamical basis for this hypothetical flow is offered here. Suppose the Lagrangian velocity decorrelation time has the finite value τ. The displacement at time t may be roughly approximated by an uncorrelated random walk of $n = t/\tau$ steps, where the variance of the nth step a_n is $\propto e^n$. It is easily shown that $\langle a_n^2\rangle\sigma_n^{-2} \to 1$ as $n \to \infty$, where σ_n^2 is the walk variance after n steps. The Lindeberg condition is not satisfied, and the statistics of the walk would be essentially those of just the last step. Exponential growth of variance is essential here; if the growth rate were algebraic, the Lindeberg condition would be met.

Next consider some hypothetical homogeneous turbulence, with kinetic energy decaying as $t^{-5/2}$, which behavior is integrable as $t \to \infty$. The particle displacement variance would be finite in that limit, and so again the Lindeberg condition would not be met.

It should be noted that even when the clt holds, the rate of approach to the asymptotic distribution may be slow. The rule is that the error is

$O(\nu^{-\frac{1}{2}})$, where ν is the number of degrees of freedom. In the uncorrelated walk, $\nu = n$. In the correlated case discussed previously, it would appear that $\nu \sim n^{\frac{1}{4}}$ since there are $n^{\frac{1}{4}}$ independent groups. In practice, however, the statistical dependence is often negligible if the gaps exceed about $3N$. Thus, $\nu \sim n/N$.

This section is concluded with a quotation from Monin and Yaglom (1971; § 9.3, p.541): "Recently, the central limit theorem was also proved directly, although with some mild restrictions, for integrals (of our type (6.1)); see, for example, Rozanov (1967) where an integral of a stationary random function is considered; similar theorems occur also for integrals of several nonstationary random functions. Unfortunately, direct use of these proofs is still impossible since the mild conditions imposed on the random functions which figure in them cannot be directly verified for real processes.[1] Nevertheless, these conditions are so natural that it would be extremely remarkable if the probability distribution for the displacement $\boldsymbol{A}(\boldsymbol{X}, s|t)$ when $t \gg s$ differed essentially from a normal distribution."

It may well be possible to prove that solutions of the Navier Stokes equations obey conditions sufficient for the clt. The Hopf equation for the characteristic functional of the velocity field (Monin and Yaglom, 1975) may be an appropriate starting point. This would be a major contribution to the theory of turbulence.

7. Path Integrals

All the approximate closures considered in §4, and the related stochastic differential equations in §5, assume that the Lagrangian velocities $\boldsymbol{u}(\boldsymbol{x}, t|s_1)$ and $\boldsymbol{u}(\boldsymbol{x}, t|s_2)$ decorrelate for widely separated values of s_1 and s_2. The conditions for Cocke's proof of the clt express this assumption in a rather abstract fashion. The assumption is widely accepted, as already mentioned, for stationary homogeneous turbulence. Ideally, its general validity should be resolved by an analysis of the Navier-Stokes equation. Failing that, let us address a simpler question: given that the Eulerian velocity $\boldsymbol{u}(\boldsymbol{x}, t)$

[1] "Thus, for example, several proofs use the so-called *strong mixing conditions*; as applied to the function $u(\boldsymbol{X}, s|t)$, this reduces, roughly speaking, to the requirement of the existence of a positive function $\gamma(\tau)$, such that $\gamma(\tau) \to 0$ as $\tau \to \infty$ and that the correlation coefficient between an arbitrary functional of the values $\boldsymbol{u}(\boldsymbol{X}, s|t'), -\infty < t', t_0$, and that of an arbitrary functional of the values $u(X, s|t'')$, $t_0 + \tau < t'' < \infty$, be less than $\gamma(\tau)$ in modulus." Cocke (1972) discussed other such conditions which lead to central limit theorems.

decorrelates for large spatial or temporal lags, does the Lagrangian velocity decorrelate for large temporal lags? Again, the case of special interest is nonstationary inhomogeneous turbulence. More specifically, we wish to know if particle displacement are asymptotically Gaussian, given that the Eulerian velocity decorrelates.

The probability distribution function for displacements may be related to that for the Eulerian velocity field, using a path integral. To see this, recall that

$$(7.1)\qquad \begin{aligned} P(\boldsymbol{x},t|\boldsymbol{X},s) &= \langle \delta(\boldsymbol{x}-\boldsymbol{A}(\boldsymbol{X},s|t))\rangle \\ &\cong \langle \delta(\boldsymbol{x}_N-\boldsymbol{x}_0-\sum_{n=0}^{N-1}\boldsymbol{u}(\boldsymbol{x}_0,t_0|t_n)\Delta t)\rangle \end{aligned}$$

where $\boldsymbol{x}_0=\boldsymbol{X}$, $t_0=s$, $t_n=s+n\Delta t$, $\boldsymbol{x}_N=\boldsymbol{x}$ and $t_N=t$. Or, we may write

$$(7.2)\qquad \begin{aligned} P(\boldsymbol{x},t|\boldsymbol{X},s) \cong \int \langle \delta(\boldsymbol{x}_N-\boldsymbol{x}_{N-1}-\boldsymbol{u}(\boldsymbol{x}_{N-1},t_{N-1})\Delta t) \\ \times\, \delta(\boldsymbol{x}_{N-1}-\sum_{n=0}^{N-2}\boldsymbol{u}(\boldsymbol{x}_0,t_0|t_n)\Delta t)\rangle\, d\boldsymbol{x}_{N-1}. \end{aligned}$$

Proceeding this way, and rescaling the measures, leads to

$$(7.3)\qquad P(\boldsymbol{x},t|\boldsymbol{X},s) = \prod_{n=0}^{N-1}\int \frac{d\boldsymbol{x}_n}{(\Delta t)^D}\left\langle \delta\left(\frac{\boldsymbol{x}_{n+1}-\boldsymbol{x}_n}{\Delta t}-\boldsymbol{u}(\boldsymbol{x}_n,t_n)\right)\right\rangle.$$

So the Lagrangian displacement pdf may be expressed in terms of the Eulerian velocity field $\boldsymbol{u}(\boldsymbol{x},t)$. Thus, the expectations in (7.3) invokes the Eulerian velocity pdf:

(7.4)

$$\begin{aligned} P(\boldsymbol{x},t|\boldsymbol{X},s) = \prod_{n=0}^{N-1}\int \frac{d\boldsymbol{x}_n}{(\Delta t)^D}\int d\boldsymbol{u}_n\delta\left(\frac{\boldsymbol{x}_{n+1}-\boldsymbol{x}_n}{\Delta t}-\boldsymbol{u}(\boldsymbol{x}_n,t_n)\right) \\ \times\, P_E(\boldsymbol{u}_0,\ldots,\boldsymbol{u}_{N-1};\boldsymbol{x}_0,\ldots,\boldsymbol{x}_{N-1};t_0,\ldots,t_{N-1}) \end{aligned}$$

where P_E is the joint distribution of the $\boldsymbol{u}_n$, the latter being the Eulerian velocities at the $(\boldsymbol{x}_n,t_n)$. Performing the integrations over $\boldsymbol{u}_n$ yields

$$(7.5)\qquad \begin{aligned} &P(\boldsymbol{x},t|\boldsymbol{X},s) = \\ &\prod_{n=0}^{N-1}\int \frac{d\boldsymbol{x}_n}{(\Delta t)^D}P_E(..,\frac{\boldsymbol{x}_{n+1}-\boldsymbol{x}_n}{\Delta t},..;..,\boldsymbol{x}_n,..;..,t_n,..). \end{aligned}$$

Path integrals such as (7.5) always attract obscure notation, such as

$$P(\boldsymbol{x}, t|\boldsymbol{X}, s) = \int \mathcal{D}[\boldsymbol{A}(r)] P_E(\dot{\boldsymbol{A}}, \boldsymbol{A}, r) \tag{7.6}$$

where $(\boldsymbol{A}(r), r)$ is a path from $(\boldsymbol{X}, s)$ to $(\boldsymbol{x}, t)$.

Note that the velocity field in (7.3) is evaluated at $\boldsymbol{x}_n$. No other choice of the form $\boldsymbol{x}_n + \theta(\boldsymbol{x}_{n+1} - \boldsymbol{x}_n)$, where $0 \leq \theta \leq 1$, is consistent with (4.3) unless $r.\boldsymbol{u} = 0$. In that case, the path integral is independent of the choice of θ. Equipped with (7.6), we may now choose various P_E and calculate corresponding P. Are there choices for P_E such that the Eulerian velocities decorrelate but P is not asymptotically Gaussian, as $t \to \infty$? Exact evaluation of (7.7) only seems possible when P_E is Gaussian, in which case P is also Gaussian.

It turns out in practice to be easier to approximate P if one performs the integrals over $\boldsymbol{x}_n$ in (7.4), rather than those over $\boldsymbol{u}_n$:

$$P(\boldsymbol{x}, t|\boldsymbol{X}, s) = \prod_{n=0}^{N-1} \int d\boldsymbol{u}_n P_E(\boldsymbol{u}_0, .., \boldsymbol{u}_{N-1}; \boldsymbol{A}_0, .., \boldsymbol{A}_{N-1}; t_0, .., t_{N-1}) \tag{7.7}$$

where $\boldsymbol{A}_0 = \boldsymbol{X}$, $\boldsymbol{A}_n = \boldsymbol{X} + \sum_{l=0}^{n} \boldsymbol{u}_l \Delta t$, $(0 \leq n \leq N-1)$, and $\boldsymbol{A}_N = \boldsymbol{x}$. However, subsequent discussion will be clearer if expressed in terms of (7.5), (7.6).

8. Monte Carlo Integration

Gaussianity of $\boldsymbol{x}$ may be determined by calculating various moments of $\boldsymbol{x}$. Thus, we need to evaluate

$$\langle f(\boldsymbol{x}) \rangle = \int \mathcal{D}[\boldsymbol{A}(r)] f(\boldsymbol{x}) P_E(\dot{\boldsymbol{A}}, \boldsymbol{A}, r) \tag{8.1}$$

for various functions f, where again $\boldsymbol{A}(s) = \boldsymbol{X}$ and $\boldsymbol{A}(t) = \boldsymbol{x}$. A Monte Carlo estimate of $\langle f(\boldsymbol{x}) \rangle$ is the arithmetic mean of samples of $f(\boldsymbol{x})$, where the samples of $\boldsymbol{x}$ are distributed as $P_E(\dot{\boldsymbol{A}}, \boldsymbol{A}, r)$. They may be generated using *importance sampling* (Metropolis, et al., 1953): let $\mathit{ff}(r)$ be a random perturbation to the sample $\boldsymbol{A}(r)$. Accept $\boldsymbol{A}' = \boldsymbol{A} + \mathit{ff}$ as another example, provided

$$\Re \equiv P_E(\dot{\boldsymbol{A}}', \boldsymbol{A}', r) / P_E(\dot{\boldsymbol{A}}, \boldsymbol{A}, r) > \rho \tag{8.2}$$

where ρ is a random number in the interval $[0, 1]$. If the ratio $\Re$ is less than ρ, then accept the unperturbed path $\boldsymbol{A}$ as another sample.

If the perturbation *ff* is too big, then most likely $\Re < \rho$ and the perturbed sample is rejected. If the perturbation is too small, then the sample set is unrepresentative. Numerical experimentation is essential, and massive computing resources are required as $t \to \infty$. Code, and perturbation schemes, may be verified using known results for Gaussian distributions.

In practice it is preferable to perturb the velocities: $\boldsymbol{u}'_n = \boldsymbol{u}_n + \boldsymbol{\mu}_n$ where the random perturbation $\boldsymbol{\mu}_n$ is independent of $\boldsymbol{u}_n$. Then the unperturbed and perturbed paths are $\boldsymbol{A}_n = \boldsymbol{X} + \sum_{l=0}^{n-1} \boldsymbol{u}_l \Delta t$, and $\boldsymbol{A}'_n = \boldsymbol{X} + \sum_{l=0}^{n-1} \boldsymbol{u}'_l \Delta t$ respectively. For stationary, independent isotropic vector walks the correlation coefficient for $\boldsymbol{A}_n$ and $\boldsymbol{A}'_n$ is $\gamma = \sigma_u \{\sigma_u^2 + \sigma_\mu^2\}^{-\frac{1}{2}}$, where σ_u^2 and σ_μ^2 are the variances of the components of $\boldsymbol{u}_n$ and $\boldsymbol{\mu}_n$ respectively. If the components of $\boldsymbol{\mu}$ are uniformly distributed from $-\epsilon\sigma_u$ to $\epsilon\sigma_u$, then $\sigma_\mu^2 = \epsilon^2\sigma_u^2/3$ and so $\gamma = (1 + \epsilon^2/3)^{-\frac{1}{2}}$. For example, if $\epsilon = 0.25$, then $\gamma = 0.9897$ and $\gamma^{100} = 0.36 \sim e^{-1}$. Thus, assuming every perturbed path is accepted, 10^2 perturbations would be needed in order to generate an independent path. So, out of 10^6 Monte Carlo trials, about 10^4 paths would be independent. The implied sampling error for $\langle f(\boldsymbol{x})\rangle$ would be of the order of 1%.

Some preliminary results of Monte Carlo integrations are shown in Fig. 5. The turbulence is two-dimensional. The simplest measure of the Gaussianity of the displacement $\boldsymbol{x} = (x, y)$ is the scaled, fourth-order cumulant $(\langle x'^4\rangle - 3\langle x'^2\rangle^2)/\langle x'^2\rangle^2$ where $x' \equiv x - \langle x\rangle$, and similarly for y. Both cumulants vanish if $\boldsymbol{x}$ is Gaussian. They are plotted in Fig. 5, versus the number N of vector steps. An integration of N vector steps involves $2N$ velocity components $(u_1, ..., u_N)$, $(v_1, ..., v_N)$. In these experiments, the u_N are independent of the v_N, and $\langle u'_n u'_m\rangle = \langle v'_n v'_m\rangle = \sigma_u^2\delta_{nm}$. The pdf $P_E(u'_1, ..., u'_N)=$ constant if $u_1^2 + ... + u_N^2 \le (N+2)\sigma_u^2$; otherwise P_E is zero. The v'_n have the same distribution as the u'_n. In the upper panel, $\langle \boldsymbol{u}(\boldsymbol{x}, t)\rangle = (0, 0)$ so the turbulence is nonGaussian but homogeneous and stationary. The displacements are convincingly Gaussian for $N > 64$; the expected error is $O(N^{-\frac{1}{2}})$. In the lower panel, $\langle \boldsymbol{u}(\boldsymbol{x}, t)\rangle = (e^{-y^2}, 0)$. Thus the Eulerian velocity is inhomogeneous. For $N = 511$, the y-variance $\langle y'^2\rangle$ (not shown) has almost reached the expected value of $N\sigma_u^2\Delta t^2 = 5.11$ after 7×10^8 trials (note that $\sigma_u = 1.0$, $\Delta t = 0.1$ in all experiments), so the large x-cumulant is likely correct. It may be expected that for $N > 511$, the x-cumulant will decrease as the "plume" width $N^{\frac{1}{2}}\sigma_u\Delta t$ exceeds that of the jet and so the displacements are mostly in the region of homogeneous turbulence. However, simple importance sampling is too inefficient for $N > 511$. The Hybrid Monte Carlo methods used in lattice gauge theory may be helpful.

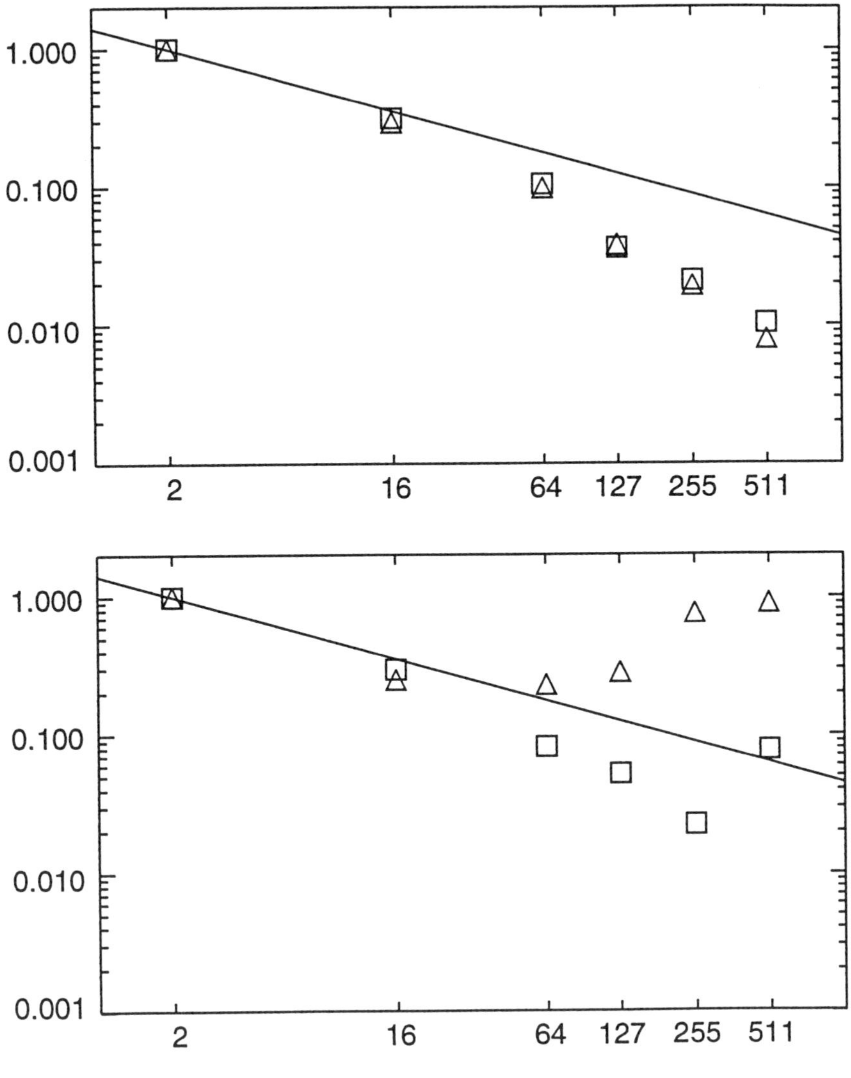

Figure 5. Fourth-order cumulants for simulated two-dimensional random walk, versus the number of vector steps N; triangles: x-coordinate; squares: y-coordinate. The straight line passes through the results for $N = 2$, and has a slope $\propto N^{-\frac{1}{2}}$. Upper panel: the random Eulerian velocity field is nonGaussian, homogeneous, stationary and uncorrelated in time or space. Lower panel: there is a mean jet in the x-direction but otherwise the random Eulerian velocity field is as in the upper panel.

9. Boundary conditions

The diffusion equation (4.11) is consistent with the clt, provided the spatial domain is entirely unbounded, and the boundary condition

$$P(\boldsymbol{x}, t|\boldsymbol{X}, s) \to 0 \quad \text{as} \quad |\boldsymbol{x}| \to \infty \tag{9.1}$$

is assumed. Then the solution of (4.11) is Gaussian. Partially and completely bounded domains are of interest, especially for inhomogeneous turbulence. Let us assume that the domain is the half-space $z = x_3 > 0$, and that the boundary plane is rigid:

$$w \equiv u_3 = 0, \qquad \text{at} \quad z = 0. \tag{9.2}$$

Assume that $\partial w/\partial z$ is finite as $z \to 0$. Then no particle initially within the fluid can reach the boundary in a finite time, and so we must have

$$P(\boldsymbol{x}, t|\boldsymbol{X}, s) \to 0, \qquad \text{as} \quad z \to 0. \tag{9.3}$$

On the other hand, the particle must always be found in the half-space, so

$$\int_{-\infty}^{\infty} dx \int_{-\infty}^{\infty} dy \int_{0}^{\infty} dz \, P(\boldsymbol{x}, t|\boldsymbol{X}, s) = 1, \tag{9.4}$$

where $x \equiv x_1, y \equiv y_2$ are coordinates in the boundary plane. Combining (4.11) and (9.4) yields the "reflecting" boundary condition

$$\langle w(\boldsymbol{X}, s|t)\rangle P(\boldsymbol{x}, t|\boldsymbol{X}, s) - K_{3j}(\boldsymbol{X}, s|t)\frac{\partial P}{\partial x_j}(\boldsymbol{x}, t|\boldsymbol{X}, s) = 0, \tag{9.5}$$

at $z = 0$. In fact, (9.5) only need hold when integrated over the plane $z = 0$. In any case, the second-order equation (4.11) is overdetermined by (9.1), (9.3) and (9.5). The normalization (9.4) is easily seen to be consistent with the exact equations (4.1), (4.2). Thus, the difficulty with (4.11) is an artifact of applying the simple second-order closure to the exact equation (4.5), in order to obtain (4.11).

The artificial contradiction may be resolved (Merilees, personal communication) by considering the fact (Lumley and Panofsky, 1964, p.131), that the variance of vertical velocity is quite significant, even very close to the ground. In other words, $\partial w/\partial z$ is very large near the ground. Thus (9.3) should be set aside, in favor of (9.1) and (9.5). It seems likely that the mean vertical advection by the Lagrangian velocity is much smaller than diffusion, and so (9.5) may be simplified to a symmetric reflection condition:

$$\frac{\partial P}{\partial z}(x, y, 0, t|\boldsymbol{X}, s) = 0. \tag{9.6}$$

The corresponding solution of (4.11) is simply the sum of two Gaussian profiles. One arises from the real source at (X, Y, Z, s); the other arises from a virtual source at $(X, Y - Z, s)$. As to the validity of (4.11) in a half-space, we can only rely on the closure in §4.1 as the clt is not available.

A similar discussion leads to the following boundary condition at $z = 0$ for (4.24):

$$\langle w(\boldsymbol{x}, |t)\rangle P(\boldsymbol{x}, t|\boldsymbol{X}, s) - D_{3j}(\boldsymbol{x}, s|t)\frac{\partial P}{\partial x_j}(\boldsymbol{x}, t|\boldsymbol{X}, s) = 0, \tag{9.7}$$

assuming incompressible flow. The coefficients in (9.5) are Lagrangian, bearing the source coordinates $(\boldsymbol{X}, s)$. Those in (9.7) are Eulerian or mixed, and bear the Eulerian coordinates $(x, y, 0, t)$. If the boundary plane at $z = 0$ is rigid, then the mean vertical velocity $\langle w(x, y, 0, t)\rangle$ vanishes. The fluctuating vertical velocity $w'(x, y, 0, t)$ also vanishes, hence $D_{3j}(x, y, 0, t|s)$ vanishes and (9.7) is no constraint. However, the practical situation is that $\langle w'^2\rangle$ is significant even close to real, and therefore rough, ground so while it is appropriate to assume $\langle w\rangle = 0$ at $z = 0$, a nonvanishing value should be chosen for D_{3j} at $z = 0$. Thus the boundary condition is again a reflection symmetry (9.6), but the corresponding solution of (4.24) is not, in general, a reflected Gaussian profile.

The validity of (4.24) rests solely on the closure in §4.2, which does not depend explicitly on the shape of the domain.

Observations of mean concentration downwind of steady point sources will be discussed in §11. A needed result analogous to (3.11) is that

$$\langle C(\boldsymbol{x}, t)\rangle = S_0 \int_s^t P(\boldsymbol{x}, t|\boldsymbol{X}, r)\, dr, \tag{9.8}$$

where S_0 is the strength of a steady source which was turned on at $\boldsymbol{X}$, at time s. If the turbulent flow is statistically stationary and horizontally homogenous, it may be deduced from (4.11) and (9.6) that $\langle C\rangle$ has an asymptotically steady state $\langle C(\boldsymbol{x}, \boldsymbol{X})\rangle_\infty$ as $(t - s) \to \infty$, satisfying

$$\begin{aligned}\langle u_j(Z, 0|\infty)\rangle \frac{\partial}{\partial x_j}\langle C\rangle_\infty =& K_{ij}(Z, 0|\infty)\frac{\partial^2}{\partial x_i \partial x_j}\langle C\rangle_\infty \\ &+ S_0 \delta(\boldsymbol{x} - \boldsymbol{X}),\end{aligned} \tag{9.9}$$

subject to

$$\frac{\partial}{\partial z}\langle C\rangle_\infty = 0, \qquad \text{at} \quad z = 0, \tag{9.10}$$

Standard boundary-layer approximations reduce (9.9) to

$$\begin{aligned}\langle u(Z,0|\infty)\rangle\frac{\partial\langle C\rangle_\infty}{\partial x} &= K_{22}(Z,0|\infty)\frac{\partial^2\langle C\rangle_\infty}{\partial y^2}\\ &\quad + K_{33}(Z,0|\infty)\frac{\partial^2\langle C\rangle_\infty}{\partial z^2} + S_0\delta(\boldsymbol{x}-\boldsymbol{X}),\end{aligned} \tag{9.11}$$

which is parabolic with $x \to \infty$ being the time-like direction, provided all coefficients are positive. An analogous equation may be derived from (4.24); its coefficients are functions of z.

10. Observations

A classic analysis of paths subsurface freely drifting buoys near the Gulf Stream was made by Freeland, Rhines and Rossby (1975). The floats were tracked acoustically for at least 201 days, and as long as 328 days. The time series of Lagrangian velocities had decorrelation times of about 20 days, and the displacement variance was closely $\propto t$ for large t, strongly indicative of linear random walks (5.1) with temporally constant diffusivities. This is particularly striking as the circulation near the Gulf Stream is significantly inhomogeneous and nonstationary.

Surface drifters were tracked over the California continental shelf by Davis (1985) using radio range-finding; 164 tracks were followed for 5–10 days. The Lagrangian decorrelation time was about 1.5 days, much less than the Eulerian (moored current meter) decorrelation of about 5 days. Well-defined, temporally constant diffusivities were derived. These exhibited strong anisotropy ($K_{ij} \neq K_0\delta_{ij}$) and strong cross-shelf inhomogeneity. However, linear random walks were observed (the displacement variances were $\propto t$), so the inhomogeneities can only have been a consequence of launch position. Davis (1985) also derived histograms of drifter-pair scalar separations $R = \left|\boldsymbol{A}(\boldsymbol{X}_1, s|t) - \boldsymbol{A}(\boldsymbol{X}_2, s|t)\right|$, with $(t-s) = 4$ days, for initial separations in two ranges (Fig. 6). The lower curve is closely consistent with Gaussian distribution for vector separations $\boldsymbol{R} = \boldsymbol{A}_1 - \boldsymbol{A}_2$. The simplest interpretation of this result is that the two drifters were moving independently, their individual positions $\boldsymbol{A}_1$ and $\boldsymbol{A}_2$ having Gaussian distributions in spite of the strong cross-shelf inhomogeneity.

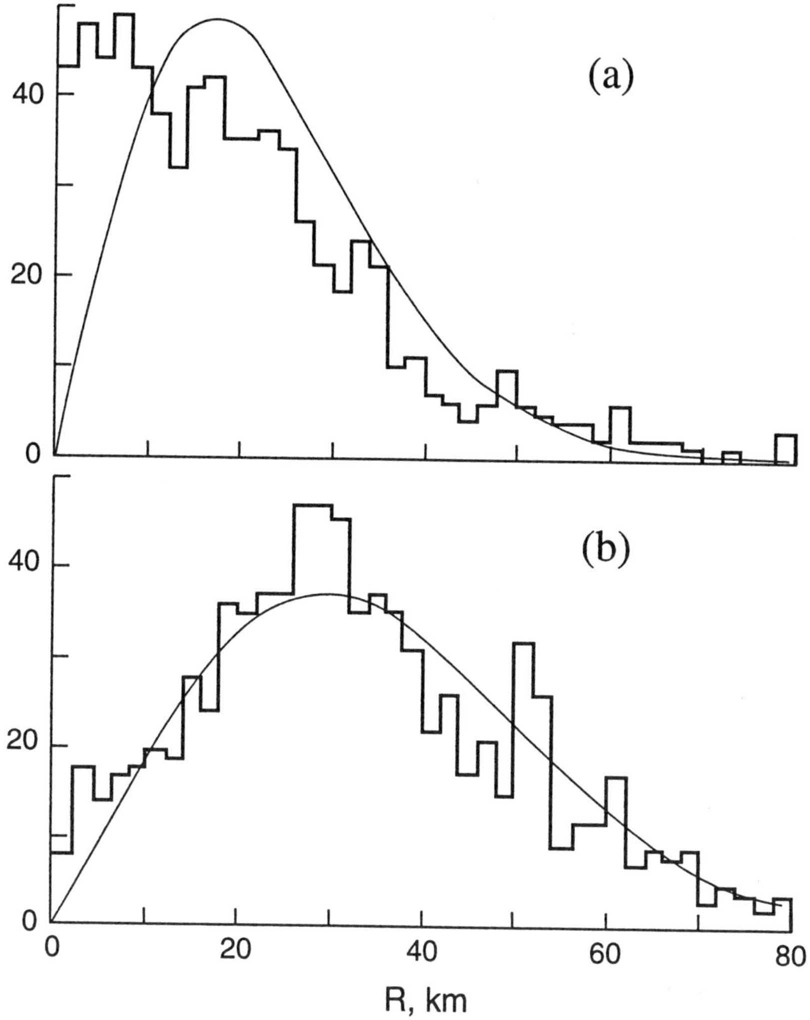

Figure 6. Histograms of scalar separations R of pairs of ocean surface drifters, 4 days after release off the California coast: (a) initial separations in the range 4–16 km; (b) initial separations in the range 16–30 km. The histograms are based on bins 2 km wide. The smooth curves correspond to a Gaussian distribution for vector separations $\boldsymbol{R}$ (after Davis, 1985).

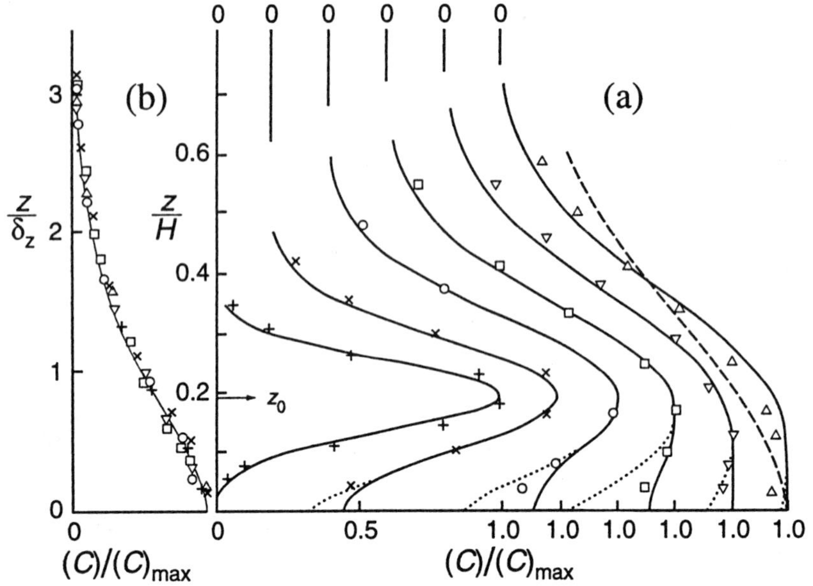

Figure 7. (a) Vertical profiles of $\langle C\rangle$ at six stations downwind ($y/H = 0$) of an elevated source ($Z/H = 0.19$) : $(x - X)/H =$ 9.96, $+$; 1.92, x; 2.88, 0; 3.83, ; 4.79, ∇; 6.52, Δ. Solid curves: reflected Gaussian (10.1); dotted curve: single Gaussian; dashed curve: empirical formula (10.2).

(b) Vertical profiles of $\langle C\rangle$ at seven stations downwind ($y/H = 0$) of a ground level source $z_0 H \simeq 0$) : $(x - x_0)/H = 0.83$; $\diamond$; 1.67, ; 2.50, Δ; 3.33, 0; 4.17, ∇; 5.00, $\times$; 5.92, $+$. Solid curve: empirical formula (10.2) after Fackrell and Robins (1982).

Vertical profiles of mean concentration were measured in a wind-tunnel by Fackrell and Robins (1982). A ground-level source (GLS: Fig. 7a), and an elevated source (ELS: Fig. 7b) were employed in separate experiments. The ELS height was $Z = 0.19H$, where H was the boundary layer height. Measured profiles at six locations downwind of the ELS were compared with the reflected Gaussian

$$(10.1) \qquad \langle C(z)\rangle \propto \exp[-0.693(z - Z)^2/\delta_z^2] + \exp[-0.693(z + Z)^2/\delta_z^2].$$

Note that $e^{-0.693} = \frac{1}{2}$. The half-width δ_z depended on the downwind distance roughly as $(x - X)^{\frac{1}{2}}$. Thus (10.1) is the centerline solution ($y = 0$) of (9.11) subject to (9.6). The agreement between measurements and (10.1) is good, except very close to the ground. Fackrell and Robins (1982) also measured vertical profiles of turbulent fluxes of concentration, that is, $\langle w'(\boldsymbol{x},t)C'(\boldsymbol{x},t)\rangle$. These varied significantly with height, as is consistent with down-gradient transfer. However, the vertically-varying eddy diffusivities so inferred would cause the solutions of semiempirical theory to be nonGaussian. Thus, the observations of mean concentration are inconsistent with the semiempirical theory.

Far downward, the ELS profiles resemble the GLS profiles. The latter were well-fitted by

$$\langle C(z)\rangle \propto \exp[-0.693(z/\delta_z)^\theta] \tag{10.2}$$

where $\theta \cong 1.5$ and $\delta_z \propto (x - X)^{0.75}$. It can only be inferred that the clt does not apply to particles which spend a significant amount of time very close to the surface.

Releases of a chemical tracer (SF_6) into the Eastern North Atlantic at a depth of about 310 m (Ledwell, Watson and Law, 1993) yielded Gaussian-like vertical distributions of mean concentration. The mean profile was based on 20 to 30 vertical samples, collected in an area of about 2° of longitude by 2° of latitude. More recent releases and measurements (Watson et al., 1993) yielded mean concentration profiles that are close to Gaussian: see Fig. 8. The turbulent vertical velocity should be significantly inhomogeneous over the vertical scale of the mean concentration profile, yet the central limit theorem seems to have prevailed.

11. Summary

The basics of fluid particle kinematics and conserved scalars were outlined in §2. Various Lagrangian and mixed first and second order statistics for particle displacement were defined in §3 for compressible and incompressible flow. Forward and backward closure approximations were described in §4; the resulting diffusion equations differ significantly. Forward closure for nonstationary, inhomogeneous, nonGaussian turbulent flow yielded a linear equation for which the fundamental solution is the Gaussian distribution with Lagrangian parameters. Backward closure yielded a nonlinear diffusion equation, that is, the coefficients depend upon the field or Eulerian coordinates rather than the launch or Lagrangian coordinates. The fundamental solution is not Gaussian. Implications for ocean modeling were reviewed, and higher corrections examined. In particular, little justification could be found for including a few such corrections. The stochastic processes consistent with the two diffusion equations were given in §5. Gaussianity (or otherwise) of displacement statistics being a distinguishing feature of the forward and backward closures, central limit theory was

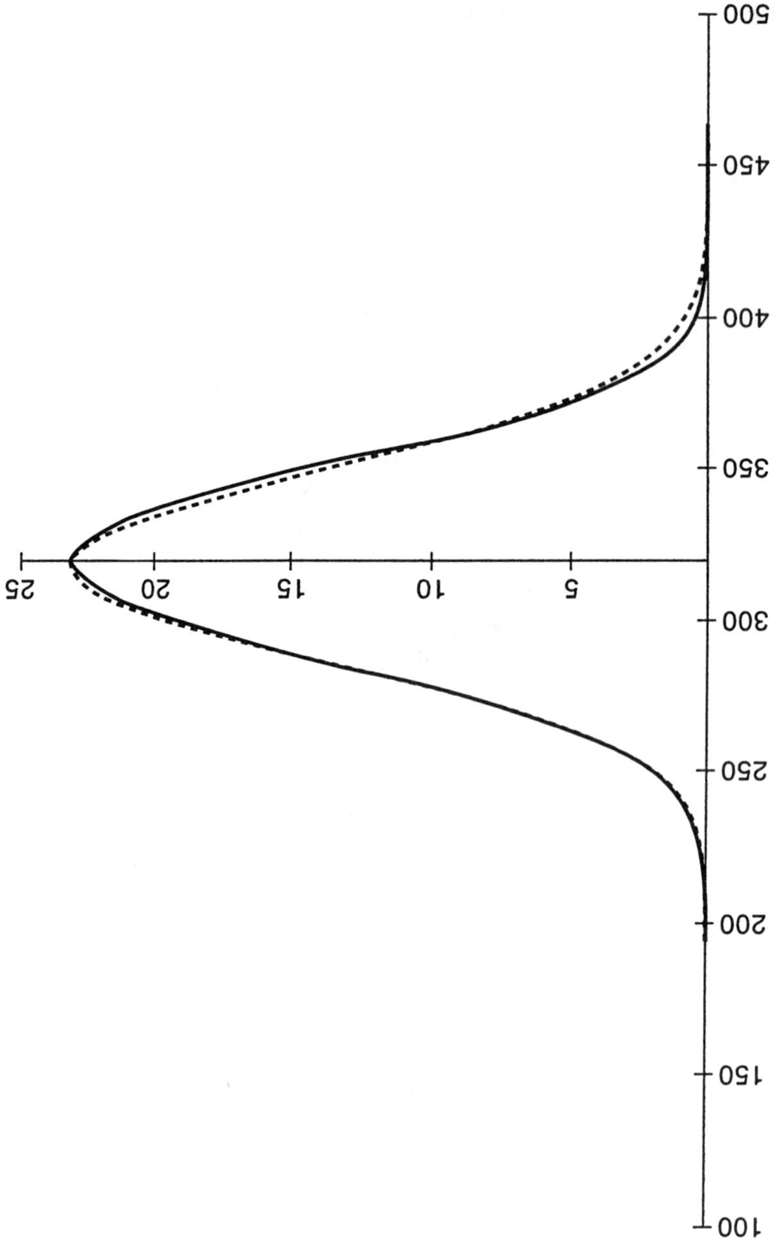

Figure 8. Mean SF_6 profile (solid line), compared to a Gaussian with the same standard deviation (broken line). Units: SF_6 concentration in femtoMoles, depth in meters. After Watson et al. (1993).

reviewed in §6. Cocke's (1972) theorem for nonstationary vector coupled sums is consistent with forward closure. Preliminary Monte Carlo calculations reported in §7 and §8 were inconclusive but seemed to support backward closure (nonGaussian displacement statistics). Boundary conditions were discussed in §9. The atmospheric and oceanic data described in §10 supported Gaussian statistics. The emerging theoretical challenge is deriving the assumptions of Cocke's central limit theorem from the Navier-Stokes equations.

Acknowledgements . I am grateful to Roland de Szoeke for many thoughtful comments. Florence Beyer typed the manuscript. Comments from an anonymous referee were helpful. The Monte Carlo calculations were performed in part using a grant of HPC time from the DoD Shared Resource Center, Maui, HI, IBM SP-2 computer. Rodney James and Boon Chua assisted with the calculations. This work was supported in part by ONR Grants N00014-90-J-1181 and N00014-93-1-0422.

References

[1] Bennett, A. F., 1987: A Lagrangian analysis of turbulent diffusion. *Rev. Geophys.*, **25**, 799–822.

[2] Bennett, A. F., and K. L. Denman, 1985: Phytoplankton patchiness: Inferences from particle statistics. *J. Marine Res.*, **43**, 307–355.

[3] Cocke, W. J., 1972; Central limit theorems for sums of dependent vector variables. *Ann. Math. Statist.*, **43**, 968–976.

[4] Corrsin, S., 1959: Progress report on some turbulent diffusion research. *Adv. Geophys.*, **6**, 161–164.

[5] Danabasoglu, A., J. C. McWilliams, and P. R. Gent, 1994: The role of mesoscale tracer transports in the global ocean circulation. *Science*, **264**, 1123-1126.

[6] Davis, R. E., 1985: Drifter observations of coastal surface dynamics during CODE: The statistical and dynamical views. *J. Geophys. Res.*, **90**, 4756–4772.

[7] Davis, R. E., 1987: Modeling eddy transports of passive tracers. *J. Marine Res.*, **45**, 635-666.

[8] Davis, R. E., 1991: Lagrangian ocean studies. *Ann. Rev. Fluid Mech.*, **23**, 43–64.

[9] de Szoeke, R. A., and A. F. Bennett, 1993: Microstructure fluxes across density interfaces. *J. Phys. Oceanogr.*, **23**, 2254–2264.

[10] Drummond, I. T., 1982: Path-integral methods for turbulent diffusion. *J. Fluid Mech.*, **123**, 59-68.

[11] Fackrell, J. E., and A. G. Robins, 1982: Concentration fluctuations

and fluxes in plumes from point sources in a turbulent boundary layer. *J. Fluid Mech.*, **117**, 1–26.

[12] Freeland, H. J., P. B. Rhines, and T. Rossby, 1975: Statistical observations of the trajectories of neutrally buoyant floats in the North Atlantic. *J. Marine Res.*, **33**, 383–404.

[13] Gardner, C. W., 1985: *Handbook of Stochastic Methods*, 2nd ed., Springer-Verlag, Berlin, 442 pp.

[14] Gent, P. R., and J. C. McWilliams, 1990: Isopycnal mixing in ocean circulation models. *J. Phys. Oceanogr.*, **20**, 150–155.

[15] Gnedenko, B., 1976: *The Theory of Probability*, Mir, Moscow, 376 pp.

[16] Haidvogel, D. B., and P. B. Rhines, 1983: Waves and circulation driven by oscillatory winds in an idealized ocean basin. *Geophys. Astrophys. Fluid Dyn.,* **25**, 1–63.

[17] Holloway, G., 1993: On modeling vertical trajectories of phytoplankton in a mixed layer. *Deep-Sea Res.*,

[18] Hunt, J. C. R., 1985: Turbulent diffusion from sources in complex flow. *Ann. Rev. Fluid. Mech.*, **17**, 447–486.

[19] Jazwinski, A. H., 1970: *Stochastic Processes and Filtering Theory.* Academic, New York, 392 pp.

[20] Kraichnan, R. H., 1964: Lagrangian history closure for turbulence. *Phys. Fluids*, **8**, 575–598.

[21] Ledwell, J. R., A. J. Watson, and C. S. Law, 1993: Evidence for slow mixing across the pycnocline from an open-ocean tracer-release experiment. *Nature*, **364**, 701–703.

[22] Lumley, J. L., 1962: The mathematical nature of the problem of relating Lagrangian and Eulerian statistical functions in turbulence. Mècanique de la turbulence (Coll. Intern. du CNRS à Marseille), Paris, Ed. CNRS.

[23] Lumley, J. L., and H. A. Panofsky, 1964: *The Structure of Atmospheric Turbulence*, Interscience, New York, 239 pp.

[24] Lundgren, T., 1981: Turbulent pair dispersion and scalar diffusion. *J. Fluid Mech.*, **111**, 25–57.

[25] Metropolis, N., A. W. Rosenbluth, M. N. Rosenbluth, A. W. Teller, and E. Teller, 1953: Equation of state calculations by fast computing machine. *J. Chem. Phys.*, **21**, 1087-1092.

[26] Monin, A. S., and A. M. Yaglom, 1971: *Statistical Fluid Dynamics*, vol. **1**, MIT, Cambridge, 769 pp.

[27] Plumb, R. A., 1979: Eddy fluxes of conserved quantities by small-amplitude waves. *J. Atmos. Sci.*, **36**, 1699–1704.

[28] Port, S. C., and C. J. Stone, 1976: Random measures and their application to motion in an incompressible fluid. *J. Appl. Prob.*, **13**,

498–506.

[29] Rhines, P. B., 1977: The dynamics of unsteady currents. In *The Sea*, Vol **6**: Marine Modelling, E. D. Goldberg, I. N. McCave, J. J. O'Brien and J. H. Steele, editors, Wiley, New York, pp

[30] Rosenblatt, M., 1962: *Random Processes*, Oxford, Oxford, 198 pp.

[31] Rozanov, Y. A., 1967: *Stationary Random Processes*, Holden-Day, San Francisco, pp.

[32] Taylor, G. I., 1921: Diffusion by continuous movements. *Proc. London, Math. Soc.,* **20**, 196-211.

[33] Watson, A. J., K. Van Scoy, J. R. Ledwell, C. S. Law, D. Jones, C. Marquette, T. Donoghue, S. Watts, M. J. Liddicoat, C. Fernandez, J. Bouthilette, D. Ciochetto, J. Donoghue, S. T. Bolmer, K. Tedesco, K. Smith, and J. Scott, 1993: *R.S.S. Charles Darwin Cruise* CD 78 (22/4/93 – 24/5/93). *The North Atlantic Tracer Release Experiment: Final Sampling Leg.* Plymouth Marine Laboratory, Plymouth, 71 pp.

[34] Yamazaki, H., and D. Kamykowski, 1991: The vertical trajectories of mobile phytoplankton in a wind-mixed column. *Deep-Sea Res.*, **32**, 219–241.

[35] Yee, E., D. J. Wilson, and B. W. Zelt, 1993: Probability distributions of concentration fluctuations of a weak diffusive passive plume in a turbulent boundary layer. *Boundary-Layer Meteorology*, **64**, 321–354.

College of Oceanic and Atmospheric Sciences
Oregon State University
Oceanography Administration Bldg 104
Corvallis, OR 97331-5503

bennett@oce.orst.edu

Massively Parallel Simulations of Motions in a Gaussian Velocity Field *

René A. Carmona, Stanislav A. Grishin,
Stanislav A. Molchanov

Abstract

The purpose of the present note is to describe the details of a set of simulation tools which we developed in order to study the statistical properties of the solutions of the equation:

$$dX_t = \vec{v}(t, X_t)dt$$

when $\{\vec{\mathbf{v}}(t, \mathbf{x});\ \mathbf{t} \geq \mathbf{0}, \mathbf{x} \in \mathbb{R}^2\}$ is a stationary and homogeneous Gaussian field with a spectrum of Kolmogorov type. The study is motivated by problems of transport of passive tracer particles at the surface of a two dimensional medium. We are mostly concerned with mathematical modeling of problems from oceanography and we think of the surface of the ocean as a physical medium to which our modeling efforts could apply. For this reason we shall sometimes use the terminology *drifters* for the passive tracers. The programs have been written for the MASPAR II. We describe the different forms of the simulations and we give numerical results which illustrate the transport properties of such a random medium. These results lead to the formulation of several precise mathematical conjectures which we discuss in the last section.

1. Introduction

The purpose of the present note is to describe the details of numerical simulations intended for the study of the statistical properties of the solutions of the equation:

$$dX_t = \vec{v}(t, X_t)dt$$

where $\{\vec{\mathbf{v}}(t, \mathbf{x});\ \mathbf{t} \geq \mathbf{0}, \mathbf{x} \in \mathbb{R}^2\}$ is a stationary and homogeneous mean zero Gaussian field with a prescribed spectrum. This analysis is motivated

*Partially supported by ONR N00014-91-1010

by the problems of transport of passive tracer particles at the surface of a two dimensional medium. Since we are mostly concerned with problems of oceanography, we think of the surface of the ocean as a physical medium to which our modeling efforts could be applied. For this reason we shall sometimes use the terminology *drifters* for the passive tracers. In this respect one should mention the work [12] where simulations were carried out in the same spirit to study eddy diffusivity.

Computer intensive simulations have been reported in the special case of time independent velocity fields. See for example the recent works [16] or [17]. The results of these simulations cannot be compared to ours because the time independence of the velocity field changes drastically the nature of the simulation algorithms and the typical properties of the tracers. Simulations (and theoretical results) similar to ours have been obtained in the case of velocity fields constructed from a Brownian flow. See [15], [5], [21] and [11]. The main difference is the independence of the time increments which does not hold in the case of time dependent velocity fields with a Kolmogorov spectrum. In fact, the Brownian model can be considered as a diffusion approximation limit of the more realistic models which we consider here. See [3] and/or [8].

The paper is organized as follows. Section 2 contains the details of the theoretical models which we consider for the velocity field. We restrict ourselves to stationary in time and homogeneous in space Gaussian velocity fields with a spectrum of the Kolmogorov type. Section 3 gives the explanations concerning the program we develop to analyze the transport properties of such a velocity field. The simulation algorithms rely on the Markov property in time of the field. We describe in detail the discretization procedures (both in the time and the wave number domains) and the implementation on a massively parallel architecture. The advantages of the parallelism are discussed at the end of this section. The simulation programs which we describe in this note have been written for the MASPAR II. This is a Single Instruction Multiple Data (SIMD) machine with 4096 processors each having 64 K of memory. Numerical results are given in Sections 4 and 5 and the final Section 7 is devoted to the formulation and the discussion of a couple of mathematical conjectures suggested by the results of the simulations.

2. Gaussian Velocity Fields of Kolmogorov Type

We present the assumptions of a commonly used mathematical model for the velocity field at the surface of the ocean. We assume that $\{\vec{\mathbf{v}}(t,x);\ t \geq 0, x \in \mathbb{R}^2\}$ is a mean zero Gaussian vector field. The hypothesis on the mean is chosen because we are merely interested in the random fluctuation around a mean deterministic motion.

We assume that the velocity field is homogeneous in space (i.e. its distribution is invariant under the shifts of the space variable x) and stationary (i.e. its distribution is invariant under the shifts of the time variable t). These three assumptions (Gaussian character, homogeneity and stationarity) are quite common in the random mathematical models of fluid mechanics. See for example [20].

The distribution of the random field is completely characterized by the covariance:

$$\langle \vec{\mathbf{v}}(s,\mathbf{x})\vec{\mathbf{v}}(\mathbf{t},\mathbf{y})^{\mathbf{t}}\rangle = \mathbf{\Gamma}(\mathbf{t}-\mathbf{s},\mathbf{y}-\mathbf{x}), \qquad \mathbf{s},\mathbf{t}\geq \mathbf{0}, \quad \mathbf{x},\mathbf{y}\in \mathbb{R}^{\mathbf{2}}, \tag{1}$$

where we use the notation ^{t} for the transpose of a column vector and where we use the notation $\langle\Phi\rangle$ for the expectation of a random variable (or more generally a random vector) Φ on the probability space of the velocity field,

2.1 Spectral Representation Because of its positive definiteness, the matrix $\Gamma(t,x)$ is the Fourier transform of a nonnegative definite matrix:

$$\nu(d\omega,d\mathbf{k}) = \begin{bmatrix} \nu_{1,1}(d\omega,d\mathbf{k}) & \nu_{1,2}(d\omega,d\mathbf{k}) \\ \nu_{2,1}(d\omega,d\mathbf{k}) & \nu_{2,2}(d\omega,d\mathbf{k}) \end{bmatrix}$$

of measures $\nu_{i,j}(d\omega,d\mathbf{k})$ on $\mathbb{R}\times\mathbb{R}^2$ in such a way that:

$$\Gamma(t,\mathbf{x}) = \int\int_{\mathbb{R}\times\mathbb{R}^2} \mathbf{e}^{\mathbf{i}(\omega\mathbf{t}+\mathbf{k}\cdot\mathbf{x})}\nu(\mathbf{d}\omega,\mathbf{dk}).$$

We use ω to denote the frequency variable associated to the time variable t and we use the variable $\mathbf{k}$ for the wave number variable associated to the space variable $\mathbf{x}$. We assume that the (mild) integrability condition:

$$|\Gamma(t,\mathbf{x})| \leq \frac{\mathbf{c}}{|\mathbf{t}|^{\mathbf{a_1}}|\mathbf{x}|^{\mathbf{a_2}}} \tag{2}$$

is satisfied for some constants $c>0$, $a_1>1$ and $a_2>2$. This condition gives the integrability of the covariance $\Gamma(t,\mathbf{x})$ and this implies the existence of densities for the measures $\nu_{i,j}(d\omega,d\mathbf{k})$. So we can assume the existence of measurable functions $E_{i,j}(\omega,\mathbf{k})$ such that:

$$\nu_{i,j}(d\omega,d\mathbf{k}) = \mathbf{E_{i,j}}(\omega,\mathbf{k})\mathbf{d}\omega\mathbf{dk}.$$

It is possible to choose the functions $E_{i,j}(\omega,\mathbf{k})$ so that the matrix:

$$E(\omega,\mathbf{k}) = \begin{bmatrix} E_{1,1}(\omega,\mathbf{k}) & E_{1,2}(\omega,\mathbf{k}) \\ E_{2,1}(\omega,\mathbf{k}) & E_{2,2}(\omega,\mathbf{k}) \end{bmatrix}$$

is positive definite for each couple $(\omega,\mathbf{k})$. We shall use the notation $E(\omega,\mathbf{k})^{\mathbf{1/2}}$ for its self-adjoint square root.

$$E(\omega,\mathbf{k})^{\mathbf{1/2}} = \begin{bmatrix} (E^{1/2})_{1,1}(\omega,\mathbf{k}) & (E^{1/2})_{1,2}(\omega,\mathbf{k}) \\ (E^{1/2})_{2,1}(\omega,\mathbf{k}) & (E^{1/2})_{2,2}(\omega,\mathbf{k}) \end{bmatrix}.$$

In particular, $E = E^{1/2}(E^{1/2})^* = E^{1/2}E^{1/2}$ if we use the notation * for the adjoint matrx, i.e. the complex conjugate of the transpose. We shall assume below that the supports of the functions $E_{i,j}(\omega, \cdot)$ and consequently of the functions $(E^{1/2})_{i,j}(\omega, \mathbf{k})$ are bounded and bounded away from the origin.

The standard Fourier analysis of homogeneous (Gaussian) fields gives the existence of two complex Gaussian white noise orthogonal measures $W_1(d\omega, d\mathbf{k})$ and $W_2(d\omega, d\mathbf{k})$ satisfying:

$$< W_i(d\omega, d\mathbf{k})\overline{\mathbf{W_j}(\mathbf{d}\omega', \mathbf{dk'})} >= \delta_{\mathbf{i},\mathbf{j}}\delta(\omega - \omega')\delta(\mathbf{k} - \mathbf{k'})\mathbf{d}\omega\mathbf{dk}.$$

and such that:

$$\vec{\mathbf{v}}(t, \mathbf{x}) = \int\int_{\mathbb{R}\times\mathbb{R}^2} \mathbf{e}^{\mathbf{i}(\omega \mathbf{t} + \mathbf{k}\cdot\mathbf{x})}\mathbf{E}(\omega, \mathbf{k})^{1/2}\mathbf{W}(\mathbf{d}\omega, \mathbf{dk}) \tag{3}$$

where we use the notation $W(d\omega, d\mathbf{k})$ for the vector of white noise measures defined by:

$$W(d\omega, d\mathbf{k}) = \left[\begin{array}{c} W_1(d\omega, d\mathbf{k}) \\ W_2(d\omega, d\mathbf{k}) \end{array} \right].$$

See for example [1]. Further assumptions on the distribution of the velocity field, such as *incompressibility* and *isotropy* can be used to specify more precisely the form of the covariance $\Gamma(t, x)$ and of its Fourier transform. See for example [20] and/or [3]. Indeed, if we assume further that the velocity field is *Markovian* in time, the form of the spectral density matrix is essentially determined. We get:

$$E_{j,j'}(\omega, \mathbf{k}) = \frac{\beta(|\mathbf{k}|)}{\omega^2 + \beta(|\mathbf{k}|)^2}\mathcal{E}(|\mathbf{k}|)\frac{\mathbf{1}}{|\mathbf{k}|}\left(\delta_{\mathbf{j},\mathbf{j}'} - \frac{\mathbf{k_j}\mathbf{k_{j'}}}{|\mathbf{k}|^2}\right). \tag{4}$$

The function $\mathcal{E}(\nabla)$ is usually chosen to be zero near the origin, near infinity and to decay like a power inside an interval $[r_0, r_1]$ for some $0 < r_0 < r_1 < \infty$. To bridge our notations with those of [3] and [4] we shall sometimes use the notation $\mathcal{E}_\epsilon(\nabla)$ for a function $\mathcal{E}(\nabla)$ which behaves like $r^{1-\epsilon}$ inside the interval $[r_0, r_1]$. The rate of decay of the function $\mathcal{E}(\nabla)$ controls the spatial correlation in the velocity field. The numbers r_0 and r_1 are related to the so-called integral and dissipation scales respectively. The function $\beta(r)$ is usually chosen of the form:

$$\beta(r) = ar^z$$

for some $z > 0$. It controls the time correlation in the velocity field. Throughout the paper we shall call ϵ and z the spectral parameters of the velocity field.

2.2 Incompressibility and the Stream Function

The spectral representation formula (3) is very often the basic starting point for the computer simulations of a stationary homogeneous Gaussian field. We shall instead use a simpler form due to the fact that we are only dealing with the two-dimensional case and the fact that we restrict ourselves to velocity fields which are incompressible in the sense that the divergence vanishes identically. In other words we assume that:

$$\mathrm{div}\{\vec{\mathbf{v}}(t,x)\} = \partial_1 v_1(t,\mathbf{x}) + \boldsymbol{\partial_2 \mathbf{v}_2(t,\mathbf{x}) \equiv 0}.$$

Using the spectral representation we find that:

$$\mathrm{div}\{\vec{\mathbf{v}}(t,x)\} = i\int\int_{\mathbb{R}\times\mathbb{R}^2} e^{i(\omega t+\mathbf{k}\cdot\mathbf{x})}\Big[[k_1(E^{1/2})_{1,1}(\omega,\mathbf{k}) + \mathbf{k_2}(\mathbf{E}^{1/2})_{\mathbf{2},\mathbf{1}}(\omega,\mathbf{k})],$$
$$[k_1(E^{1/2})_{1,2}(\omega,\mathbf{k}) + \mathbf{k_2}(\mathbf{E}^{1/2})_{\mathbf{2},\mathbf{2}}(\omega,\mathbf{k})]\Big]\, W(d\omega,d\mathbf{k})$$

so that, because of the independence of the white noise measures W_1 and W_2, the divergence free assumption is equivalent to the system:

$$\begin{cases} k_1(E^{1/2})_{1,1}(\omega,\mathbf{k}) + \mathbf{k_2}(\mathbf{E}^{1/2})_{\mathbf{2},\mathbf{1}}(\omega,\mathbf{k}) &= 0 \\ k_1(E^{1/2})_{1,2}(\omega,\mathbf{k}) + \mathbf{k_2}(\mathbf{E}^{1/2})_{\mathbf{2},\mathbf{2}}(\omega,\mathbf{k}) &= 0. \end{cases} \tag{5}$$

In the two-dimensional case it is a simple fact from calculus that the divergence free condition is equivalent to the existence of a real valued function $\phi(t,\mathbf{x})$, called the *stream function,* which determines completely the velocity vector $\vec{\mathbf{v}}(t,\mathbf{x})$ by means of the formula:

$$v_1(t,\mathbf{x}) = -\boldsymbol{\partial_2\phi(t,\mathbf{x})} \qquad \textbf{and} \qquad \mathbf{v_2(t,x)} = \boldsymbol{\partial_1\phi(t,\mathbf{x})}. \tag{6}$$

We used the notation:

$$\partial_i = \frac{\partial}{\partial x_i} \qquad \text{for} \quad \mathbf{x} = (\mathbf{x_1},\mathbf{x_2}).$$

Formula (6) is extremely convenient because it reduces the computations on vector valued functions to simpler computations with real valued functions. This is especially useful for simulation purposes.

Because of the fact that the spectral densities $E_{i,j}(\omega,\ \cdot\ )$ vanish near the origin, the stream function $\phi(t,\mathbf{x})$ can be realized as a stationary and homogeneous mean zero Gaussian field. Indeed one can set:

$$\phi(t,x) = \int\int_{\mathbb{R}\times\mathbb{R}^2} e^{i(\omega t+\mathbf{k}\cdot\mathbf{x})}[\mathcal{E}_\infty(\omega,\mathbf{k}),\mathcal{E}_\in(\omega,\mathbf{k})]\mathbf{W}(\mathbf{d}\omega,\mathbf{dk}) \tag{7}$$

with (recall the conditions (5)):

$$\mathcal{E}_{\infty}(\omega, \mathbf{k}) = -\frac{\mathbf{1}}{\mathbf{ik_2}}(\mathbf{E}^{\mathbf{1/2}})_{\mathbf{1,1}}(\omega, \mathbf{k}) = \frac{\mathbf{1}}{\mathbf{ik_1}}(\mathbf{E}^{\mathbf{1/2}})_{\mathbf{2,1}}(\omega, \mathbf{k})$$

and:

$$\mathcal{E}_{\in}(\omega, \mathbf{k}) = -\frac{\mathbf{1}}{\mathbf{ik_2}}(\mathbf{E}^{\mathbf{1/2}})_{\mathbf{1,2}}(\omega, \mathbf{k}) = \frac{\mathbf{1}}{\mathbf{ik_1}}(\mathbf{E}^{\mathbf{1/2}})_{\mathbf{2,2}}(\omega, \mathbf{k}).$$

Since the stream function is a stationary and homogeneous mean zero Gaussian scalar field one can start the analysis and the simulations from its own spectral representation:

$$\phi(t, x) = \int_{I\!R}\int_{I\!R^2} e^{i(\omega t + \mathbf{k}\cdot\mathbf{x})} E_{\phi}(\omega, \mathbf{k})^{\mathbf{1/2}} \mathbf{W}'(\mathbf{d}\omega, \mathbf{dk}) \tag{8}$$

for some mean zero Gaussian white noise measure complex $W'(d\omega, dk)$. We used the notation $E_{\phi}(\omega, \mathbf{k})$ for the Fourier transform of the covariance function of the field $\phi(t, \mathbf{x})$. As explained earlier, we shall assume that the velocity field is incompressible, isotropic and Markovian in time. This gives a specific form to the spectral density of $\phi(t, \mathbf{x})$ and the spectral representation gets the form:

$$\phi(t, \mathbf{x}) = \int_{I\!R}\int_{I\!R^2} \mathbf{e}^{\mathbf{i}(\omega \mathbf{t} + \mathbf{k}\cdot\mathbf{x})} \left[\frac{\beta(|\mathbf{k}|)}{\omega^2 + \beta(|\mathbf{k}|)^2}\mathcal{E}_{\phi}(|\mathbf{k}|)\right]^{\mathbf{1/2}} \mathbf{W}'(\mathbf{d}\omega, \mathbf{dk}) \tag{9}$$

for some nonnegative function $\mathcal{E}_{\phi}(\nabla)$ which has properties similar to those of $\mathcal{E}(\nabla)$. In fact, one can compute the derivatives $\partial_1\phi(t, x)$ and $\partial_2\phi(t, x)$ from formula (9) and recover the form (4) of the spectral representation of the velocity field. This simple calculation shows that:

$$\mathcal{E}(\nabla) = \nabla^{\lceil + \infty}\mathcal{E}_{\phi}(\nabla). \tag{10}$$

As the discussion of the simulation algorithm will show, it is somehow possible to perform the integration in $d\omega$ first and get a representation of the form:

$$\phi(t, x) = \int_{I\!R^2} e^{i\mathbf{k}\cdot\mathbf{x}} \sqrt{\mathcal{E}_{\phi}(|\mathbf{k}|)}\, \xi_t(d\mathbf{k}) \tag{11}$$

where the $\{\xi_t(d\mathbf{k})\}$ are random measures with orthogonal increments such that $\{\xi_t(A); t \geq 0\}$ is a stationary scalar Ornstein Uhlenbeck process for each bounded Borel set A.

3. The Simulation Program

The general idea of the random simulation is based of the Markov property in time of the process $\{\vec{\mathbf{v}}(t);\ t \geq 0\}$. We use the notation $\vec{\mathbf{v}}(t)$ to denote the entire configuration of the values $\vec{\mathbf{v}}(t, x)$ of the velocity field at all the locations simultaneously.

The idea is to discretize the time into a sequence $(t_n)_{n=0,1,\cdots}$ and to use the Markov property to construct the next value $\vec{\mathbf{v}}(t_{n+1})$ from the current value $\vec{\mathbf{v}}(t_n)$ ignoring the way the field "got there" i.e. ignoring the values of $\vec{\mathbf{v}}(t_0)$, $\vec{\mathbf{v}}(t_1)$, $\cdots$, $\vec{\mathbf{v}}(t_{n-1})$. Moreover, since each $\vec{\mathbf{v}}(t)$ is itself an infinite dimensional object, we shall need to use other finite dimensional approximations to construct the configurations $\vec{\mathbf{v}}(t)$.

3.1 Discrete Approximations There are many possible way to address the discretization problem. ¿From the point of view of the spectral theory of stationary and/or homogeneous random fields, the simplest approach is to approximate the spectral measure $\nu(d\omega, dk)$ by simpler measures more amenable to random simulations on a computer. In particular, it is natural to approximate ν by a finite sum of point masses $\delta_{(\omega_i, k_i)}$ in the frequency-wave number domain. This approach is especially useful because it makes it possible to simulate directly the white noise without any concern with the possible dependence of the random variates to generate. Nevertheless, we refrain from using this approach directly, especially in the time variable. Indeed, in order to save memory (and computing time) we want to use an adaptive approach generating new values of the velocity field as we need them. More precisely, we want an approximation scheme which uses the fact that $\vec{\mathbf{v}}(t)$ is a time homogeneous Markov process. For this reason we shall separate the problems of the discretization of the time and the space variables.

The **discretization of time** is straightforward. We choose a time interval Δt and we sample the continuous time by restricting its values to:

$$t_0 = 0,\ t_1 = \Delta t,\ t_2 = 2\Delta t,\ \cdots, t_n = n\Delta t,\ \cdots\cdots$$

Because of the Markov property and because of our choice for the discretization of the time, at each instant t_j, the program will use the knowledge of the characteristics of the field $\vec{\mathbf{v}}(t_{j-1})$ at the preceding time to generate the samples of the velocity field $\vec{\mathbf{v}}(t_j, \mathbf{x})$ at the locations $\mathbf{x}$ where we need them. Notice that the choice of the time interval Δt has to depend upon the spectral characteristics, and especially of the radial lower bound r_0 on the support of the spectral density. This remark will be especially useful when the program simulates velocity fields for different values of the spectral parameters ϵ and z. See below for details.

The **discretization of the space variable** is much more involved. In the same way we replaced the continuous time t by a grid of discrete values, it is possible to replace the space variable $\mathbf{x} \in \mathbb{R}^2$ by a discrete variable taking finitely many values in a lattice $h\mathbb{Z}^2$ for some mesh $h > 0$. This strategy has several drawbacks, especially from the point of view of Monte-Carlo simulation. Indeed, for each fixed time t, there is a strong dependence between the values of $\vec{\mathbf{v}}(t, \mathbf{x})$ and $\vec{\mathbf{v}}(t, \mathbf{y})$ even when the points $\mathbf{x}$ and y are far apart, making the generation of the configuration $\vec{\mathbf{v}}(t_j)$ from $\vec{\mathbf{v}}(t_{j-1})$ extremely costly in memory storage and in computing time: this approach is not reasonable for our application.

Instead we shall use a strategy based on formulas (6) and (11) from the definition ot the stream function and a partial Fourier transform in the t variable in the spectral representation of the stream function. We choose a finite set $\mathcal{K}$ of wave vectors in $\mathbb{R}^2$ (and more precisely in the support of the function $\mathcal{E}(|\mathbf{k}|)$) to approximate the integral in (11) by a finite sum. We get:

$$\phi(t, \mathbf{x}) = \sum_{\mathbf{k} \in \mathcal{K}} \mathrm{e}^{\mathbf{i}\mathbf{k} \cdot \mathbf{x}} \mathrm{e}(\mathbf{k}) \xi_{\mathbf{t}}(\mathbf{k}) \tag{12}$$

where the $e(\mathbf{k})$'s are complex coefficients and the $\{\xi_t(\mathbf{k}) \ : \ \mathbf{t} \geq \mathbf{0}\}$ are independent Ornstein Uhlenbeck processes. The choice of the set $\mathcal{K}$ of wave vectors and the choice of the corresponding coefficients $e(\mathbf{k})$ will be explained at the end of our discussion of the discretization error. Formula (12) is still written with the continuous variable t, but as we explained earlier the time variable will be sampled on a regular grid. The effect of the sampling will just be to turn the continuous time Ornstein Uhlenbeck processes into discrete time autoregressive processes of order 1. See below for details.

3.2 The Simulation Algorithm We first discuss the simulation of the trajectories of passive tracers.

Once the choice of a sampling rate Δt has been made, the positions of the tracer particles are computed at the times $t_j = t_0 + j\Delta t$. In order to update the position of a particle one uses the simplest Euler scheme:

$$\mathbf{x}(\mathbf{t_j}) = \mathbf{x}(\mathbf{t_{j-1}}) + \vec{\mathbf{v}}(\mathbf{t_j}, \mathbf{x}(\mathbf{t_{j-1}}))\mathbf{\Delta t}. \tag{13}$$

The simulation of the time evolution trajectories reduces to the problem of the generation of the velocity field, which in turn reduces to the generation of the stream function because of formula (6). Simple arithmetic on formula (12) leads to the real representation:

$$\phi(t, \mathbf{x}) = \sum_{\mathbf{k} \in \mathcal{K}} [\mathbf{a_k}(\mathbf{t}) \cos(\mathbf{k} \cdot \mathbf{x}) + \mathbf{b_k}(\mathbf{t}) \sin(\mathbf{k} \cdot \mathbf{x})] \tag{14}$$

where the $\{a_{\mathbf{k}}(t) \ : \ t \geq 0\}$ and the $\{b_{\mathbf{k}}(t) \ : \ t \geq 0\}$ are independent scalar Ornstein Uhlenbeck processes. The latter become auto regressive processes after discretization of the time. The velocity vector $\vec{\mathbf{v}}(t, \mathbf{x})$ can then be computed according to the formulas:

$$v_1(t,\mathbf{x}) = \sum_{\mathbf{k}\in\mathcal{K}} \mathbf{k_2}[\mathbf{a_k(t)}\sin(\mathbf{k}\cdot\mathbf{x}) - \mathbf{b_k(t)}\cos(\mathbf{k}\cdot\mathbf{x})] \tag{15}$$

and:

$$v_2(t,\mathbf{x}) = \sum_{\mathbf{k}\in\mathcal{K}} \mathbf{k_1}[-\mathbf{a_k(t)}\sin(\mathbf{k}\cdot\mathbf{x}) + \mathbf{b_k(t)}\cos(\mathbf{k}\cdot\mathbf{x})]. \tag{16}$$

As formula (13) shows, we shall need to compute the velocity field at the instants t_j at the positions $\mathbf{x_{j-1}}$ of the tracer particle. But notice that, once the program has knowledge of the auto-regressive processes $a_\mathbf{k}$ and $b_\mathbf{k}$ labeled by the set $\mathcal{K}$ of wave vectors, the velocity field can be computed at the point $\mathbf{x} = \mathbf{x(t_{j-1})}$. The updates of the auto regressive coefficients are done via the formulas:

$$a_\mathbf{k}(t_j) = \alpha_\mathbf{k} a_\mathbf{k}(t_{j-1}) + \sigma_\mathbf{k}\epsilon_\mathbf{k}^{(a)}(j) \tag{17}$$

and:

$$b_\mathbf{k}(t_j) = \beta_\mathbf{k} b_\mathbf{k}(t_{j-1}) + \sigma_\mathbf{k}\epsilon_\mathbf{k}^{(b)}(j) \tag{18}$$

where the $\epsilon_k^{(a)}(j)$ and the $\epsilon_k^{(b)}(j)$ are independent standard normal random variables and where the constants $\alpha_\mathbf{k}$, $\beta_\mathbf{k}$ and $\sigma_\mathbf{k}$ can be computed from the spectral density of the stream function:

$$\sigma_\mathbf{k} = |\mathbf{k}|^{-2-\epsilon}\mathbf{\Delta t}, \qquad \alpha_\mathbf{k} = \beta_\mathbf{k} = \mathbf{1} - |\mathbf{k}|^\mathbf{z}\mathbf{\Delta t} \approx \mathbf{e}^{-|\mathbf{k}|^\mathbf{z}\mathbf{\Delta t}}.$$

3.3 Implementation The details of the first form of the simulation program are given in the following paragraphs.

3.3.1 Inputs The spectral characteristics of the Eulerian spectrum, the number and the initial positions of the drifters are the inputs of the program.

• The (Eulerian) spectral characteristics of the velocity field have to be given in the form of:

- the list $\mathcal{K}$ of the wave vectors k and of the corresponding values of $\mathcal{E}(|\mathbf{k}|)$.
- the list of the parameters α_k, β_k and σ_k introduced earlier.

The initial values of the a_k's and b_k's are generated by the program as samples of $N(0,\sigma_k^2/(1-\alpha_k^2))$ and $N(0,\sigma_k^2/(1-\beta_k^2))$ random variables. Notice that this choice is consistent with the stationarity requirement of autoregressive processes. A function is also provided to generate a set of initial parameters when the Eulerian spectrum is intended to be of the form (4) with a function $\mathcal{E}(\nabla)$ with a logarithmic slope $-\epsilon$. In this case, the points k are uniformly placed on concentric circles and the only inputs needed are the radii of these circles, the numbers of k per circle and the parameters ϵ and z.

• The initial positions of the drifters can be supplied as an input file but they can also by generated randomly as a sample from a homogeneous Poisson cloud with an intensity specified by the user.

3.3.2 The Main Loop At the beginning of each cycle j, the program generates, independently of all the preceding random generations, the $N(0,1)$ independent random noise terms $\epsilon_{\mathbf{k}}^{(a)}(t_j)$ and $\epsilon_{\mathbf{k}}^{(b)}(t_j)$ for all the values of the wave vectors $\mathbf{k}$ in the selected set $\mathcal{K}$. Then, it computes the new values of the parameters $a_{\mathbf{k}}(t_j)$ and $b_{\mathbf{k}}(t_j)$ in terms of their previous values according to the formulas (17) and (18).

The second step in the main loop is to update the positions of the drifters. This is done as follows. If a drifter is at the location $\mathbf{x(t_{j-1})}$ at time t_{j-1}, then the new position $\mathbf{x(t_j}$ of this drifter will be computed from formula (13) the value of the velocity field being computed from the formulas (15) and (16).

3.3.3 Options The program can operate in two modes.

• **Eulerian mode:** the limits of the surface of the ocean which is to be considered in the simulation are defined, boundary conditions are specified (they include killing, elastic reflection and periodic extension) and the region of interest is partitioned into $N = 4096$ subsets (typically rectangles). Each of the $N = 4096$ PE's of the MasPar is assigned to one of these regions.

• **Lagrangian mode:** in this mode, the ocean has no limits! Each of the $N = 4096$ PE's of the MasPar is assigned a certain number of drifters, typically $[n/N]$ if n is the total number of drifters, and it follows the drifters wherever they go. In other words, at each cycle, each PE updates the positions of the drifters he has the control of.

The Lagrangian mode has a certain number of advantages over the Eulerian mode.

◇ Each PE has the same computing load and the same memory constraints.

◇ The PE's do not risk to run out of memory.

The numerical results (as well as the mathematical conjectures presented in the last section) indicate that points are drifting apart in the sense that the distance separating two given points is typically growing. If the time evolution of a finite number of points is followed via the plots of their positions at each cycle, when the points form a smooth curve at the initial time, gaps appear rapidly in between separate segments of the curve and after a relatively small number of cycles it is very difficult to visualize the set of points as a curve. We wanted to provide a remedy to this shortcoming of having the same number of drifters at each cycle.

• **Insertion Option:** The program has an option which allows to insert drifters in between two drifters whose separation reaches a preassigned threshold. The main drawback of the use of this option is that it gives output files with a nonrectangular format. But the main advantage is that the visual impressions left by the plots are very nice. For example, if one follows the time evolution of a set of drifters which form a (smooth) continuous curve at time $t = 0$, then at each subsequent time, the plots of the positions of the points form a (smooth) continuous curve without artificial gaps.

More options are available which influence the output of the program. In particular it is possible to choose the times t_j at which the positions of the drifters are output to a file.

3.3.4 Outputs The output of the program is a file in a standard ASCII format. It is a rectangular array when the insertion option is not used: two columns per drifter, one for each coordinate. The number of row is the number of cycles specifed in the input and the index of the row is related to the time t_j by the number of cycles before dumping the position of the drifter to the output file. The output file is not rectangular when the insertion option is used and a modicum of care is required for the interpretation of its contents.

3.4 Mean Square Error and Choice of the Parameters The estimation of the mean square error:

$$\mathbb{E}\{||\vec{\mathbf{v}}(t,\mathbf{x}) - \vec{\mathbf{v}}_{\mathcal{K}}(\mathbf{t},\mathbf{x})||^{2}\} \tag{19}$$

is one of the remaining important theoretical problems. Indeed, a precise upper bound for (19) in terms of the values chosen for the wave vectors in $\mathcal{K}$ and the spectral parameters ϵ and z would be a great help in the difficult problem of the choice of $\mathcal{K}$. From the rough upper bounds which we derived for (19) it is clear that the set $\mathcal{K}$ can be chosen as a finite subset of a small set of concentric circles with radii in the inertial range $[r_0, r_1]$. Moreovoer $\mathcal{K}$ should be chosen appropriately if the isotropy condition is desired. The optimal choices of the number of circles, of the values of the radii and of the numbers of wave vectors $\mathbf{k}$ per circle depend upon the values of the spectral parameters. Our rough upper bounds were not sharp enough to define a systematic procedure for choosing these values. Nevertheless, we shall fix these values once for all when we study the dependence of the Lyapunov exponents with respect to the spectral parameters ϵ and z.

4. Numerical Results

The first figure shows the time evolution of the boundary of a closed smooth curve under the flow generated by an incompressible isotropic ve-

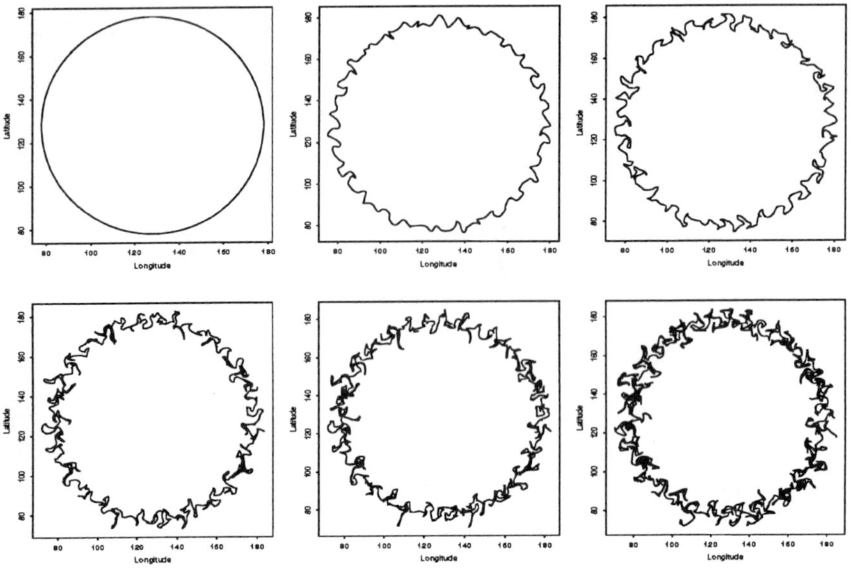

Figure 1: Time evolution of the boundary of a circle under the flow generated by an incompressible isotropic velocity field with a spectrum of the Kolmogorov's type.

locity field with a spectrum of the Kolmogorov's type. It was produced by output of the positions of the drifters at the times $T_j = (1 + 30j)\Delta t$ with $\Delta t = .25$. The velocity field was generated by the approximation described above. We chose $z = 0.125$ and we chose circles of radii 0.01, 0.125, 0.5, 0.75, 1.0 and 1.25. For each value of the radius r we chose 0.001, 0.975, 0.1, 0.45, 0.001 and 0.025 respectively for the values of the function $\mathcal{E}(\nabla)$. Notice that we did not try to mimick a function $\mathcal{E}(\nabla)$ of the type described earlier to characterize what we call spectrum of the Kolmogorov's type. Indeed, the values of $\mathcal{E}(\nabla)$ were chosen explicitly, not through the choice of the value of the spectral parameter ϵ. On each of these circles we chose 4, 4, 8, 8, 8 and 8 wave vectors in a symmetric way.

The first plot shows 1000 points equidistributed on the circle. The insertion option was used to make sure that the set of drifters would form a continuous curve at each time, even though they do tend to drift apart because of the randomness of the velocity field. More precisely, a drifter was added in between two existing ones each time the distance between the latter would reach twice its initial value. The last of the six curves shown in Figure 1 is made of approximately 20,000 drifters. This shows how useful this insertion option is.

At any time T_j the curve remains smooth. Because the divergence of the velocity field vanishes identically, Lebesgue's measure is left invariant by the flow. As a consequence, the area enclosed in the circle remains constant over time while the boundary evolves according to the flow. Notice that the parameters of the run (i.e. the choice of ϵ, z and the set $\mathcal{K}$ of wave numbers) are such that the general shape of the enclosed region remains mostly circular. While this general shape is preserved, the boundary is changing drastically. As it could be seen from the six snapshots, the length of the boundary grows, and the rate of growing is presumably exponential. Figure 2 is a blow up of the last of the six snapshots.

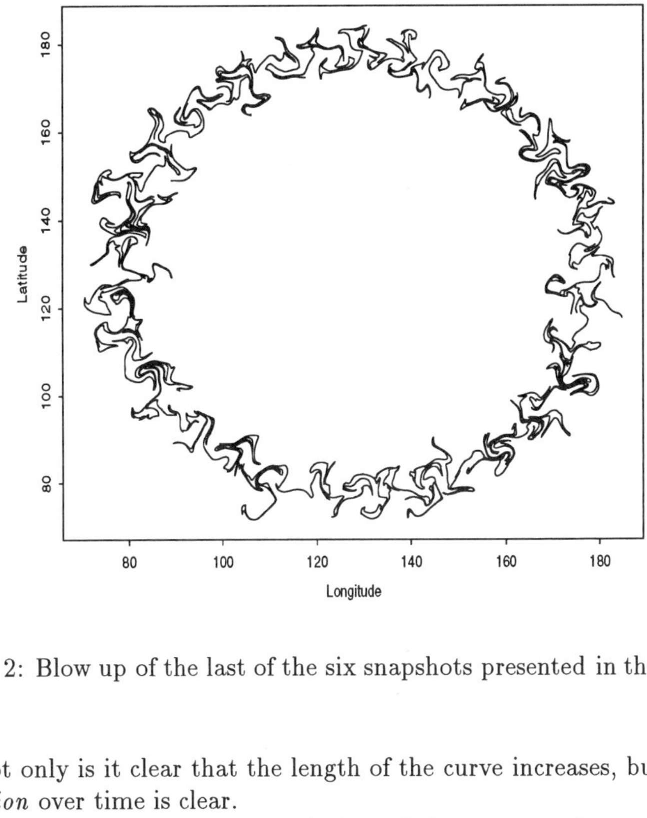

Figure 2: Blow up of the last of the six snapshots presented in the previous figure.

Not only is it clear that the length of the curve increases, but its *fractalization* over time is clear.

Figure 3 shows the time evolution of the same circle considered in Figure 1, but this time the isotropy condition has been altered. The velocity field generates a shear flow. Its vertical component vanishes identically and its horizontal component is an Ornstein Uhlenbeck process with a spectrum of the Kolmogorov type parametrized by the horizontal coordinate (i.e. longitude in the plot). As before, the incompressibility condition forces the enclosed area not to change value but the length of the boundary does not grow as fast as before. In fact it is a simple consequence of standard

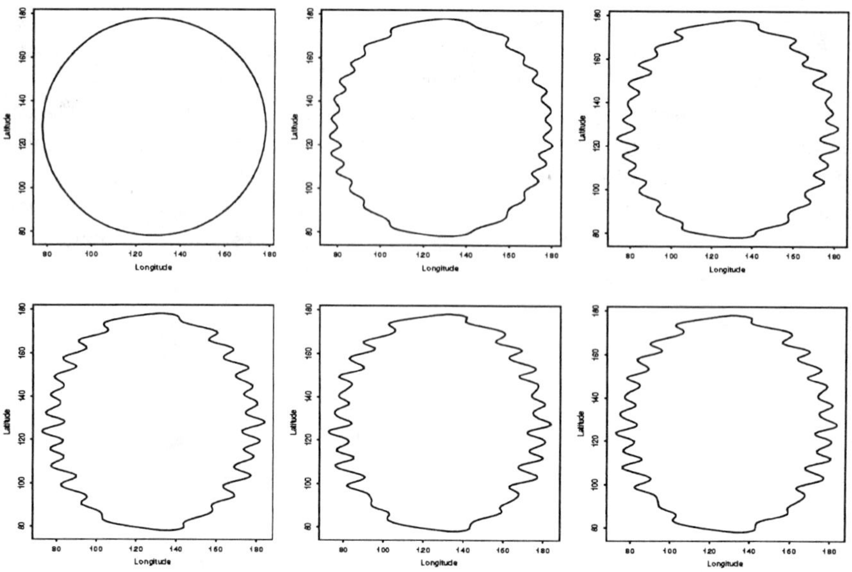

Figure 3: Time evolution of a circle under a shear flow.

properties of stationary Gaussian processes that the length goes to infinity linearly. See [7] for details. The simulation of a shear flow is done by merely setting to zero one of the components, say the first one, of all the wave vectors in $\mathcal{K}$! Modulo this last remark, the values of the parameters of this simulation run were the same as in the case of the time evolution of the circle under the Ornstein Uhlenbeck flow described above.

5. Jacobian Flow and Lyapunov Exponents

The form of the program described above was chosen to illustrate the transport properties of the flow generated by the velocity field $\vec{\mathbf{v}}(t, \mathbf{x})$. These properties are strongly dependent upon the (asymptotic) characteristics of the Jacobian matrices $J_{\mathbf{x}}(t)$ defined as the derivatives of the solution $\mathbf{x}(\mathbf{t})$ when the latter is regarded as a function of the initial condition $\mathbf{x}$. These matrices are solutions of a random linear equation of the form:

$$J'_{\mathbf{x}}(t) = A_{\mathbf{x}}(t) J_{\mathbf{x}}(t), \qquad J_{\mathbf{x}}(0) = I, \tag{20}$$

where the coefficient matrix $A_{\mathbf{x}}(t)$ is given by the formula:

$$A_{\mathbf{x}}(t) = \nabla \vec{\mathbf{v}}(t, \mathbf{x}(\mathbf{t}))$$

where $\mathbf{x}(\mathbf{t})$ denotes the position at time t of a drifter which starts from the position $\mathbf{x}$ at time $t = 0$. This coefficient matrix is given by the Lagrangian observation of the gradient of the velocity field along the solution trajectory. A result of [21] implies that the distribution of this coefficient matrix is stationary in time. Consequently, the subadditive ergodic theorem says that:

$$||J_{\mathbf{x}}(t)|| \approx e^{\lambda_1 t} \qquad \text{as} \quad t \to \infty$$

for some nonnegative deterministic constant λ_1 which is called the (upper) Lyapunov exponent of the system. The spatial homogeneity of the velocity field implies that the Lyapunov exponent does not depend upon the initial position $\mathbf{x}$. Consequently, one Lagrangian trajectory suffices (provided it is long enough) to determine the value of this exponent.

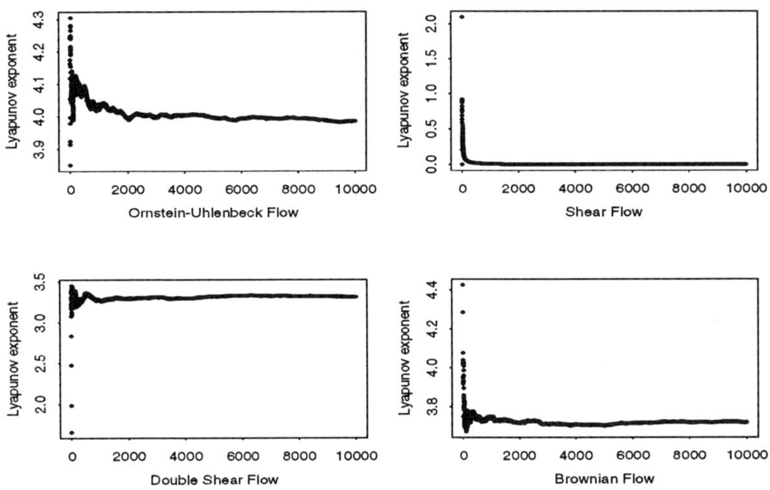

Figure 4: Approximations of the upper Lyapunov exponents in four different models.

Figure 4 shows the values of:

$$\frac{1}{n\Delta t} \log ||J_{\mathbf{x}}(n\Delta t)|| \tag{21}$$

for $n = 1, \cdots, 10000$ in the case of four types of velocity fields. The convergence seems to be confirmed in all cases and the rate of convergence seems to be very fast. The limit seems to be equal to 4.0 in the case of the flow generated by a velocity field of Ornstein Uhlenbeck type, to 3.7 in the case of the Brownian flow we simulated, to 3.3 in the case of a velocity field made up as the sum of two independent fields which would have generated shear flows in the horizontal and vertical direction respectively. This flow does not have the same isotropy property the first two ones have but it still has enough built in symmetry to give a positive Lyapunov exponent. On the other hand the results reported for the shear flow indicate that the limit should be zero!

The following remark is in order at this stage. The numerical computations of the Lyapunov exponents of stochastic systems is a very touchy business. This fact is especially well documented in [19] and [13]. The results reported above are very much in accordance with our intuition, and the theoretical results which are known in the case of white velocity fields (see for example [15] and [5]). Nevertheless, we want to warn the reader that our numerical investigations are at a very early stage and that some of the numerical results which we get today may not be confirmed by subsequent theoretical analysis and/or further numerical work.

The computations of the approximations of the Lyapunov exponents which we mentionned so far did not take advantage of the parallel architecture but massively parallel computations in the spirit of the first form of the program can be used to study the dependence of λ_1 upon the spectral parameters ϵ and z. See discussion below.

6. Advantages of the Massive Parallelism

For the purpose of the present discussion we only consider the advantages of the massively parallel simulations in the two applications presented above.

6.0.1 Transport of Passive Tracers The first obvious advantage of the use of the parallel architecture of the MasPar is the fact that one can get a very large number of simulated drifter trajectories at essentially the same cost (in storage space and in computing time) as if one were to generate one single drifter trajectory. This feature is extremely attractive for the statistical analysis of these time series for which the Lagrangian mode was chosen. See for example the analysis of spectral comparison tests presented in [10]. But, as we shall illustrate below, the most natural application is undoubtedly the possibility to follow the time evolution of curves whether or not they are the boundaries of regions of interest. This feature of the massively parallel simulations of the transport makes it possible to simulate the time evolution of *concentrations*. The latter are solutions of stochastic partial differential equations and we might have here a new tool to investigate the fine structure of the solutions.

Remark 0.1 *The advantages of the Eulerian mode are more of a computer science nature. Mapping the topology of the space into the array of processing units may become a necessity if one jumps from a 2 dimensional ocean surface to a 3 dimensional ocean and if one has to distribute the processing task over interconnected units. In particular, the 3D grid of processors pro-*

posed by Cray Computers for the next generation of super computers could be a good test for the benefits of the Eulerian mode.

6.0.2 Lyapunov Exponents The computation of the value of (or to be more precise of an approximation of) the Lyapunov exponent for one choice of the spectral characteristics of the velocity field is not something special the massively parallel simulations can do that could not be done on a standard workstation. But the simultaneous computation of the Lyapunov exponents over a large grid of values of the spectral parameters is something that only parallel computations can provide so easily. This lead us to the second form of the program which we describe now.

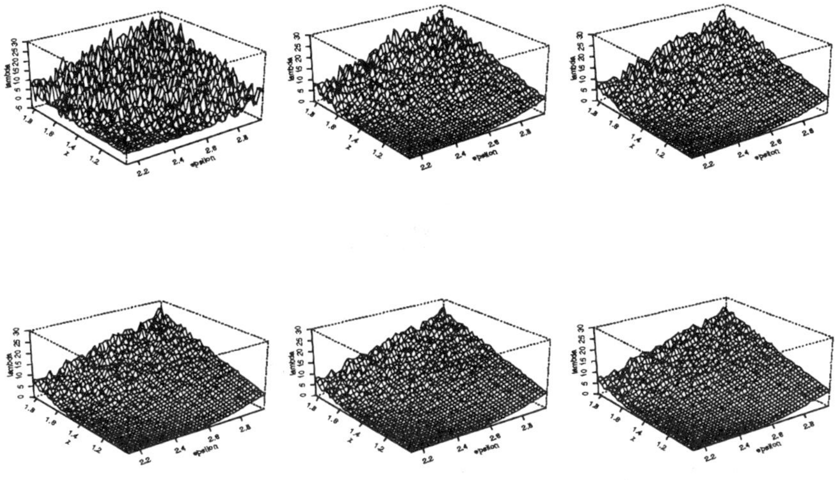

Figure 5: Approximations of the upper Lyapunov exponents for 42×42 models corresponding to a grid of 42×42 values of the couple (ϵ, z) of spectral parameters.

We define a finite grid of couple (ϵ_j, z_k) of spectral parameters. Let us assume that, for the purpose of the present discussion we have 4096 such couples. Each of them will be assigned to a processor and each of these processor is going to keep track of the values of the parameters $a_{\mathbf{k}}$ and $b_{\mathbf{k}}$ of the velocity field. These coefficients were computed once only and all the processors were reading their values to compute the value of the velocity at the positions they had a drifter to update. Now each processor does update the values of the auto regressive coefficients using the values of the spectral parameters it was assigned. But except for this difference, the computations are the same and, in the same time it took to compute the Lyapunov exponent for one model one gets the values of the Lyapunov exponents for 4096 models.

Figure 5 shows the evolution of the approximation (21) for $n = 1000$, $n = 100,000$, $n = 200,000$, $n = 300,000$, $n = 400,000$ and $n = 500,000$ iterations over a square grid of 42×42 values of the spectral parameters (ϵ, z). The values of ϵ were chosen to increase by the amount $\Delta\epsilon = 0.02$ from $\epsilon_{min} = 2.136667$. The 42 values of z were chosen to increase by the amount $\Delta z = 0.02$ from $z_{min} = 1.00667$. The approximation of the velocity field was based on the use of $|\mathcal{K}| = \nabla$ wave vectors. 4 of these $\mathbf{k}$ were chosen symmetrically on the circle of radius .01, 4 more on the circle of radius .10 and then, on each of the circles of radii 0.30, 0.50, 0.70, 0.90, 1.30 and 2.10 we chose 8 wave vectors. The time increment Δt was chosen to be 0.1. A warning is in order: the decay of the surface near the small values of ϵ and z could misleadingly give the false impression that the Lyapunov exponent vanishes. A closer look at the numerical values show that this is not the case.

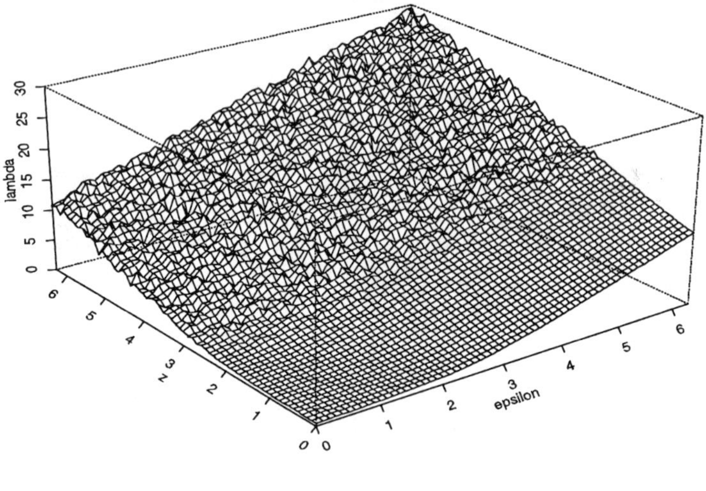

Figure 6: Another example of the approximations of the upper Lyapunov exponents over a grid of spectral parameters.

In order to show that the convergence need not be so slow and that one can get faster the impression of a smooth surface $\lambda_1 = \lambda_1(\epsilon, z)$, we made several attemps. Figure 6 shows the surface obtained for $n = 82,000$ iterations over a square grid of 64×64 values of the spectral parameters (ϵ, z). The values of ϵ and z were chosen to increase by the amount $\Delta\epsilon = \Delta z = 0.1$ from $\epsilon_{min} = z_{min} = 0.0$. The approximation of the velocity field was based on the use of $|\mathcal{K}| = \nabla\Delta$ wave vectors. We chose 4, 6, 8, 10, 12 and 14 of them on the circles of radii 0.2, 0.35, 0.5, 0.65, 0.8 and 1 respectively. The time increment Δt was chosen to be .8. Moreover, we would like to reiterate the fact that we are at a very early stage of the numerical investigation of the Lyapunov exponents. Consequently, we do not have a complete understanding of the dependence of our numerical results upon the various choices in the approximations, whether they involve the choices of the number and/or the locations of the wave vectors $\mathbf{k}$'s or even the value of the time increment Δt.

7. Mathematical Conjectures

We discuss two of the mathematical conjectures suggested by the numerical results reported in the previous section.

7.1 Fractal Dimensions Let us consider a smooth curve $\{\gamma(s)\}_{0\leq s\leq 1}$ in the plane and let us analyze the time evolution of this curve under the flow. We shall use the notation $\{\gamma_t(s)\}_{0\leq s\leq 1}$ for the curve formed by the positions of the points at time t. Let us assume, as in the case of Figure 1, that the curve is enclosing a nice (simply connected) domain. Since the divergence of the velocity field is identically 0, the domain evolves in time into domains with possibly different shapes, but with the same area. As illustrated above the curve (which remains at all times the boundary of the domain) remains smooth but its length grows (presumably exponentially) without apparent bounds. This forces a phenomenon of *fractalization* of the boundary. Figure 2 confirms this speculation. This plot suggests that the dimension of the subset of the plane formed by the points $\gamma_t(s)$ may, at least in an appropriate asymptotic regime, become strictly larger than 1. Similar problems have been considered in [4] where a solution was obtained in a renormalization regime for time independent shear flows. We are now beginning theoretical and numerical investigations of the singularity spectrum of γ_t in the multifractal formalism in the spirit of [6].

7.2 Lyapunov Exponents As before we consider a smooth curve $\{\gamma(s)\}_{0\leq s\leq 1}$ in the plane and we analyze its time evolution under the flow. As seen from FigureOB 1 and Figure 3 for example, the length of the curve at time t, say $\ell(t)$, seems to grow to ∞. As explained in [7], this length can be expressed as an integral (over the initial positions $\mathbf{x} = \gamma(\mathrm{s})$) containing the modulus of a random 2×2 matrix $Z_x(t)$ which satisfies the same random linear equation (20). As shown in [7], a simple argument shows that:

$$\lambda_1 \leq \liminf_{t\to\infty} \frac{1}{t} \log \ell(t). \tag{22}$$

The numerical simulations reported above, and especially the results shown in Figure 1, show that the hypothesis that the length $\ell(t)$ grows exponentially is very reasonable in all the models considered here, except for the shear flows. Inequality (22) suggests that the reason could be that the Lyapunov exponent λ_1 is strictly positive. So the mathematical conjecture is that:

> along each Lagrangian trajectory $\mathbf{x(t)}$ the stationary flow $\nabla\vec{\mathbf{v}}(t, \mathbf{x(t)})$ of 2×2 random matrices has a strictly positive Lyapunov exponent.

The numerical results reported above (recall Figure 4) show that this conjecture should be true under a mild condition (necessary to exclude pathological cases such as the shear flows). More compelling arguments are given in [7] but no rigorous proof of the strict positivity of λ_1 is known in general, let alone of its behavior as a function of the spectral parameters ϵ and z. See nevertheless [15] and [5] for the case of incompressible and isotropic white velocity fields. The numerical evidence provided by the computer simulations is limited to velocity fields with a finite frequency content. The latter are investigated in [9].

Beyond the positivity, the results reported in Figure 5 and Figure 6, show that the dependence of the upper Lyapunov exponent upon the spectral parameters ϵ and z could be of a specific type. Moreover they seem to confirm that the upper Lyapunov exponent cannot see the various types of renormalization regimes exhibited by Avellaneda and Majda in [3] and [4] for shear flows with Kolmogorov spectra.

Acknowledgments: The code was written in C for the MASPAR II by Dave MacDonald and Steve Seiden from the ICS Department at U.C. Irvine. The support of ONR N00014-91-1010 is acknowledged.

References

[1] R.J. Adler (1980): Geometry of Random Fields. Wiley, New York, N.Y.

[2] A. Antoniadis and R. Carmona (1985): Infinite Dimensional Ornstein Uhlenbeck Processes *Probab. Th. Rel. Fields* **74**, 31-54.

[3] M. Avellaneda and A. Majda (1990): Mathematical models with exact renormalization for turbulent transport. *Commun. Math. Phys.,* **131**, 381-429.

[4] M. Avellaneda and A. Majda (1991): An integral representation and bounds on the effective diffusivity in passive advection by laminar and turbulent flows. *Commun. Math. Phys.,* **138**, 339-391.

[5] P.H. Baxendale (1986): Asymptotic Behavior of Stochastic Flows of Diffeomorphisms: Two Case Studies. *Proba. Th. Rel. Fields* **73**, 51-85.

[6] J.F. Buzy, E. Bacry and A. Arneodo (1992): Multifractal formalism for fractal signals: the structure function approach versus the wavelet transform modulus maxima method. (preprint)

[7] R. Carmona (1995): Transport Properties of Gaussian Velocity Fields. First S.M.F. Winter School in Random Media. Rennes 1994. (to appear)

[8] R. Carmona and J. P. Fouque (1994): Diffusion-Approximation for the Advection-Diffusion of a Passive Scalar by a Space-Time Gaussian Velocity Field. Proc. Intern. Conf. on SPDE's, Ascona, June 1993. Birkhäuser, Basel.

[9] R. Carmona, S. Grishin and S.A. Molchanov (1995): Analysis of Ornstein Uhlenbeck Stochastic Flows with Finite Frequency Content. (in preparation)

[10] R. Carmona and A. Wang (1994): Comparison Tests for the Spectra of Dependent Multivariate Time Series. (this volume)

[11] E. Cinlar and C.L. Zirbel (1994): Dispersion of Particle Systems in Brownian Flows. (preprint)

[12] R. Davis (1992): Observing the general circulation with floats. *Deep Sea Res.* **38**, S531-S571.

[13] A. Grorud and D. Talay (1995): Lyapunov Exponents of Nonlinear Stochastic Differential Equations. *SIAM J. Appl. Mat.* (to appear).

[14] H. Kunita (1990): Stochastic Flows and Stochastic Differential Equations. Cambridge Univ. Press. Boston, MA.

[15] Y. Le Jan (1984): Equilibre et exposants de Lyapunov de certains flots browniens. *C. R. Acad. Sci. Paris Ser. A* **298**, 361-364.

[16] F. Elliott and A. Majda (1994): A wavelet Monte Carlo method for turbulent diffusion with many spatial scales. (preprint)

[17] F. Elliott, A. Majda, D. Horntrop and R. McLaughlin (1994): Hierarchical Monte Carlo methods for fractal random fields. (preprint)

[18] S.A. Molchanov (1994): Lectures on Random Media. in St Flour Summer School in Probability, Lect. Notes in Math. Springer Verlag (to appear)

[19] D. Talay (1991): Apprixination of Upper Lyapunov Exponents of Bilinear Stochastic Differential Systems. *SIAM J. Numer. Anal.* **28**(4), 1141-1164.

[20] A.M. Yaglom (1987): Correlation Theory of Stationary and Related Random Functions. vol. I: Basic Results. Springer Verlag, New York, N.Y.

[21] C.L. Zirbel (1993): Stochastic Flows: Dispersion of a Mass Distribution and Lagrangian Observations of a Random Field. Ph.D. Princeton.

RENE A. CARMONA
STANISLAV A. GRISHIN
Department of Mathematics
University of California at Irvine

rcarmona@phoenix.princeton.edu

STANSILAV A. MOLCHANOV
Department of Mathematics
UNC at Charlotte
Charlotte, NC 28223

smolchan@unccsun.uncc.edu

Comparison tests for the spectra of dependent multivariate time series*

René A. Carmona and Andrea Wang

Abstract

Statistical test procedures are proposed to compare the spectra of multivariate time series. We are mostly concerned with the general case of dependent series. The study was motivated by physical oceanography problems for which daily satellite measurements of drifter positions are readily available. Within the mathematical models used for the velocity field at the surface of the ocean, the Lagrangian velocities (time series of the velocities of the drifters along their trajectories) are realizations of stationary stochastic processes having the same one-dimensional marginal distributions. The theory leads to the conjecture that they also should have the same spectra. The tools presented in this paper were intended to test this hypothesis on real data and on data obtained from numerical simulations of transport properties in a two dimensional incompressible Gaussian velocity field.

1. Introduction and Notations

We first describe the notations and the mathematical assumptions which are in force throughout the present study. We consider two d-dimensional finite time series $\{x^{(1)}(t);\ t = 1, \cdots, n\}$ and $\{x^{(2)}(t);\ t = 1, \cdots, n\}$. We shall use the notations $x^{(i)}(t) = [x_1^{(i)}(t), \cdots, x_d^{(i)}(t)]^t$ to emphasize the components in a standard vector notation, the superscript ^{t} being used for the vector transpose. We want to compare the statistics of the two series, and especially their spectra. It is mathematically convenient to assume that the time series are measurements from two $\mathbb{R}^d$-valued stochastic processes $\{X^{(1)}(t);\ t \in \mathbb{Z}\}$ and $\{X^{(2)}(t);\ t \in \mathbb{Z}\}$ which in turn are viewed as the marginals of a $2d$-dimensional $\mathbb{R}^{2d}$-valued stochastic process $\{X(t);\ t \in \mathbb{Z}\}$. We assume that this process is of order 2, that it is

*Partially supported by ONR N00014-91-1010

stationary (at least in the weak sense) and that its covariance is integrable. This guarantees the existence of a spectral density which we denote by $f(\omega)$. In fact:

$$f(\omega) = \begin{bmatrix} f_{1,1}(\omega) & f_{1,2}(\omega) \\ f_{2,1}(\omega) & f_{2,2}(\omega) \end{bmatrix}$$

where $f_{1,1}(\omega)$ is the $d \times d$ spectral density matrix of the process $\{X^{(1)}(t)\}$, $f_{2,2}(\omega)$ is the $d \times d$ spectral density matrix of $\{X^{(2)}(t)\}$ and $f_{1,2}(\omega) = f_{2,1}(\omega)^*$ is the $d \times d$ spectral coherence matrix between the processes $\{X^{(1)}(t)\}$ and $\{X^{(2)}(t)\}$. We use the notation * to denote the complex adjoint, i.e. transpose of the matrix whose entries are the complex conjugates of the entries of the original matrix. We consider the problem of the test of the hypothesis:

$$H_0 : \qquad f_{1,1}(\omega) \equiv f_{2,2}(\omega)$$

against a general alternative. This problem has been considered in the univariate case $d = 1$ when the time series are independent (in which case $f_{1,2}(\omega) \equiv f_{2,1}(\omega) \equiv 0$). See for example [4]. We are interested in the general case of (possibly) dependent multivariate time series. Both generalizations create mathematical difficulties which prevent us from extending ideas from the univariate case. In the case of independent time series, we were able to give an approximate test for which one can determine rigorously the asymptotic distribution of the test statistic. We describe this test procedure in Section 4. Since we are mostly interested in real and simulated data for which the independence assumption is not realistic, we consider the general case of dependent time series in Section 5. We derive another approximate test but unfortunately, we cannot determine rigorously the asymptotic distribution of the test statistic. We rely on Monte Carlo simulations (parametric bootstrap to be specific), to determine the approximate p-values of the tests. Numerical results are given in Section 6.

There are many practical situations in which the comparison of spectra is of importance. Some examples are mentioned in [4] and [5]. Our motivation comes from the analysis of the Lagrangian velocities of drifters at the surface of the ocean. For the sake of simplicity we regard these drifters as passive drifters floating at the surface of a 2-dimensional incompressible fluid. The positions of these drifters are recorded at regular time intervals (say every day), and their velocities are estimated from these measurements. We shall illustrate the properties of the test procedures on real data collected during 1991 and 1992 in the South Atlantic and Indian Oceans. Figure 1 gives the trajectories of the drifters in question.

There is a theoretical model for which the answer to the equality of spectra is known. We review the basics of this model in Section 2 where

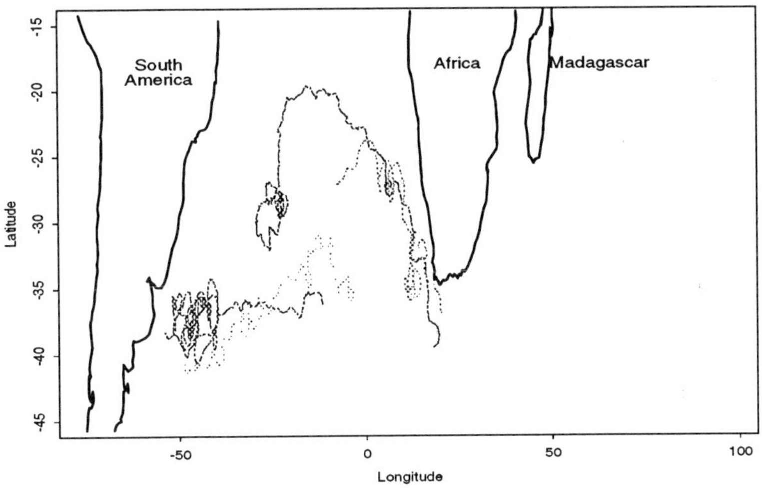

Figure 1: Plot of the daily positions of 4 of the drifters we use in this paper for the purpose of illustration.

we also explain how numerical simulations of this models can be used to evaluate the power of our statistical procedures.

2. Motion in a Gaussian Velocity Field

This section is devoted to a short discussion of a standard mathematical model in fluid mechanics. Our interest in such a model is that, according to the theoretical results recalled below, the individual Lagrangian velocities have the same spectra. Consequently, numerical simulations should provide a testbed for our statistical procedures.

We consider a Gaussian random field $\{\vec{v}(t,x);\ t \geq 0, x \in \mathbb{R}^d\}$. $\vec{v}(t,x)$ represents the velocity field at time $t \geq 0$ at the location $x \in \mathbb{R}^d$. For this reason, $\vec{v}(t,x)$ is a d-dimensional vector ($d = 2$ in the application we have in mind). The trajectory $\{x(t)\}_t$ of a particle dropped at x_0 at time $t = 0$ is given by the solution of the ordinary differential equation:

$$dx(t) = \vec{v}(t, x(t))dt, \qquad x(0) = x_0.$$

The d-dimensional multivariate time series $\{\vec{v}(t, x(t));\ t \geq 0\}$ gives the velocity along the trajectory: it is called the Lagrangian velocity (as opposed to the Eulerian velocity $\{\vec{v}(t,x);\ t \geq 0, x \in \mathbb{R}^d\}$). An old result of Lumley says that, when the velocity field is homogeneous (in the sense that its distribution is invariant under the shifts of the space variable x) and incompressible (i.e. divergence free), then for each time t and initial position x_0, the random vector $\vec{v}(t, x(t))$ has the same distribution as $\vec{v}(t, 0)$. This result was extended in [6] by Port and Stone (see also [?]) who showed that if the velocity field is also stationary (i.e. its distribution is invariant under the time shifts) then all the time series $\{\vec{v}(t, x(t));\ t \geq 0\}$ are stationary.

So, in the framework of this model, two different initial positions x_0 and x'_0 give two trajectories $\{x(t) : t \geq 0\}$ and $\{x'(t) : t \geq 0\}$ which lead to two stationary d-dimensional multivariate time series $\{\vec{v}(t, x(t));\ t \geq 0\}$ and $\{\vec{v}(t, x'(t));\ t \geq 0\}$ which have the same one-dimensional marginal distributions in the sense that $\vec{v}(t, x(t))$ and $\vec{v}(t, x'(t))$ have the same distribution for each fixed time t.

As explained earlier, the following question is the subject of the present study: do the whole time series have the same distributions and in particular, do they have the same spectra? The theoretical results which we just recalled say that, if the divergence free Gaussian model applies for the velocity field, then the answer should be yes. Notice that, as in the case of real life drifter Lagrangian velocity time series, the time series from the mathematical model are multivariate and (strongly) dependent since they are integral curves of the same realization of the random velocity field.

Massively parallel simulations of motions in a Gaussian velocity field are reported in [3]. These simulations produce large numbers of bivariate trajectories $\{x(t)\}$ computed from the simulations of a Eulerian velocity field $\{\vec{v}(t, x);\ t \geq 0, x \in \mathbb{R}^d\}$ with a Kolmogorov spectrum. Figure 2 shows the trajectories of 4 of the simulated data produced by this simulation program. Since the purpose of this note is to derive tests for equality of spectra and since the above theoretical results show that any two Lagrangian velocities simulated from this model should indeed have the same spectra, we did not try to make our the results of our simulations look like real drifter trajectories. Some of these plots may not look very realistic because the input parameters of the program were chosen at random.

This feature is irrelevant to our use of these trajectories (see Section 6 below) to illustrate the statistical test procedures which we designed.

3. Set Up from the Spectral Theory of Time Series

We now return to the notations defined in the introduction. We use the standard periodogram to estimate the spectral density matrices. For each real number λ we define the quantity $d(\lambda)$ by:

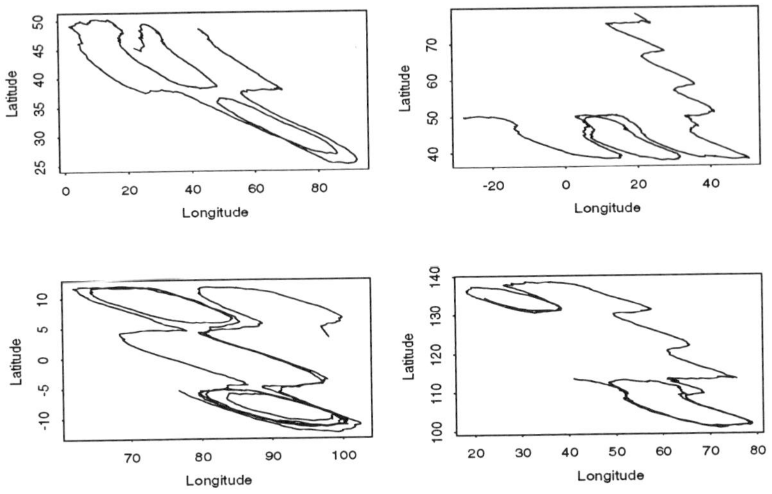

Figure 2: Plot of the trajectories of 4 of the simulation drifter data we use in this paper for the purpose of illustration.

$$d(\lambda) = \sum_{t=1}^{n} e^{i\lambda t} X(t)$$

and we denote by $d_1(\lambda)$ (resp. $d_2(\lambda)$) the d-dimensional vector formed by the first (resp. last) d components of $d(\lambda)$.

Without loss of generality we assume that the number of observations is a multiple of K, say $n = NK$ (we choose $K = 10$ in the applications described below), and we group them K by K to get estimates at N fundamental frequencies ω_j. Let us denote by $\{\lambda_{j,k}; k = 1, \cdots, K\}$ the set of K frequencies which we choose to converge to ω_j. In most applications the frequencies ω_j at which we want to estimate the spectra are regularly spaced and so are the $\lambda_{j,k}$. We consider the quantity $\hat{f}(\omega_j)$ defined by:

$$\hat{f}(\omega_j) = \frac{1}{K} \sum_{k=1}^{K} d(\lambda_{j,k}) d(\lambda_{j,k})^*$$

as an estimate of $f(\omega_j)$. $\hat{f}(\omega_j)$ is a $2d \times 2d$ complex matrix. In fact, it is a standard result from the spectral theory of multivariate time series that, asymptotically as $n \to \infty$ (while K remains fixed):

$$\hat{f}(\omega_1), \cdots\cdots, \hat{f}(\omega_N) \tag{1}$$

become independent with the complex Wishart distributions $W_d^C(K, f(\omega_1))$, $\cdots$, $W_d^C(K, f(\omega_N))$ with K degrees of freedom and the respective covariance matrices $f(\omega_1)$, $\cdots$, $f(\omega_N)$. See for example [1]. Notice that the same construction starting from $d_1(\lambda)$ (resp. $d_2(\lambda)$) instead of $d(\lambda)$ shows that (in the same asymptotic regime):

$$\hat{f}_{1,1}(\omega_1), \cdots\cdots, \hat{f}_{1,1}(\omega_N) \tag{2}$$

and:

$$\hat{f}_{2,2}(\omega_1), \cdots\cdots, \hat{f}_{2,2}(\omega_N) \tag{3}$$

are realizations of independent complex Wishart distributions:

$$W_d^C(K, f_{1,1}(\omega_1)), \cdots\cdots, W_d^C(K, f_{1,1}(\omega_N))$$

and:

$$W_d^C(K, f_{2,2}(\omega_1)), \cdots\cdots, W_d^C(K, f_{2,2}(\omega_N))$$

respectively. The $d \times d$ lower left corner of the matrix $\hat{f}(\omega_j)$ will be denoted by $\hat{f}_{2,1}(\omega_j)$. It should be viewed as an estimate of the coherence $f_{2,1}(\omega_j)$ of the two time series at the frequency ω_j.

4. A Comparison Test in the Independent Case

We now consider the particular case of independent time series. Under this assumption $f_{2,1}(\omega) \equiv f_{1,2}(\omega) \equiv 0$ and the spectral density matrix $f(\omega)$ is block diagonal. This independence assumption is often realistic and the test procedures which we derive in this section are of independent interest. Under this assumption, the estimates $\hat{f}_{1,1}$ and $\hat{f}_{2,2}$ are constructed from independent samples. In particular the $d \times d$ matrices:

$$M_1(\omega_j) = f_{1,1}(\omega_j)^{-1/2}\hat{f}_{1,1}(\omega_j)f_{1,1}(\omega_j)^{-1/2}, \qquad j = 1, \cdots, N$$

and:

$$M_2(\omega_j) = f_{2,2}(\omega_j)^{-1/2}\hat{f}_{2,2}(\omega_j)f_{1,1}(\omega_j)^{-1/2}, \qquad j = 1, \cdots, N$$

form independent samples of independent matrices having the standard complex Wishart distribution $W_d^C(K, I)$ with K degrees of freedom. Notice that, under the null hypothesis $f_{1,1}(\omega_j) = f_{2,2}(\omega_j)$ and the distribution of the random number:

$$T(\omega_j) = \text{trace}[M_1(\omega_j)M_2(\omega_j)^{-1}] \tag{4}$$

is free (i.e. independent of the particular spectra of the individual time series). This distribution, say $F_{d,K}$, is also independent of the frequency ω_j. In fact, $F_{d,K}$ is the distribution of the trace of a random matrix $M_1 M_2^{-1}$ provided that M_1 and M_2 are independent $d \times d$ matrices with the standard Wishart distribution $W_d^C(K, I)$. It is not easy to compute analytically the density of the distribution $F_{d,K}$, but it is very easy to determine its percentiles by Monte Carlo techniques. The value of T can be rewritten in the form:

$$\begin{aligned} T(\omega_j) &= \text{trace}[M_1(\omega_j)M_2(\omega_j)^{-1}] \\ &= \text{trace}[f_0(\omega_j)^{-1/2}\hat{f}_{1,1}(\omega_j)\hat{f}_{2,2}(\omega_j)^{-1}f_0(\omega_j)^{1/2}] \\ &= \text{trace}[\hat{f}_{1,1}(\omega_j)\hat{f}_{2,2}(\omega_j)^{-1}] \end{aligned}$$

provided we set $f_0(\omega_j)$ for the common value of the spectral matrices $f_{1,1}(\omega_j)$ and $f_{2,2}(\omega_j)$. This computation shows that the value of T can be computed without any knowledge of the unknown spectral matrices $f_{1,1}(\omega_j)$ and $f_{2,2}(\omega_j)$. Consequently, starting from a sample of size n from the stochastic process $\{X(t)\}$, we can compute

$$T(\omega_1), \cdots\cdots, T(\omega_N)$$

and the latter is an independent sample of size $N = n/K$ from the distribution $F_{d,K}$. Consequently, one can compute the one sample Kolmogorov-Smirnov distance statistic, say T_{KS}, from the empirical distribution of this sample to the theoretical distribution $F_{d,K}$ and determine the p-value of the test from the tables.

Remark 4.1 *It is possible to base the test on the one-sample Kolmogorov-Smirnov distance statistic D_{KS} for the sample formed by the random numbers:*

$$D(\omega_j) = det[M_1(\omega_j)M_2(\omega_j)^{-1}] \tag{5}$$

computed from the determinant instead of the trace. The same considerations apply and a test can be based on the values of this statistic. We do not report numerical results using this test procedure because they were not superior to the ones obtained from the T_{KS} statistic and most importantly, because the computations are much longer: computing the trace of a matrix is faster than computing its determinant.

5. A Test in the General Case

The tests derived in the preceding section are not appropriate for the analysis of Lagrangian spectra, whether they come from simulations of time evolutions in a random velocity field or real life drifter data. Indeed the independence assumption is not satisfied in these application and the results are poor. In order to build the dependence structure into the test procedures, we propose some modifications of the main ideas used so far.

Even though we do not assume any longer that $f_{2,1}(\omega) \equiv 0$, we nevertheless restrict ourselves to coherence structures which are constant with respect to the frequency variable ω. In other words our model includes the assumption of the existence of a complex $d \times d$ matrix F such that:

$$f_{2,2}^{-1/2}(\omega) f_{2,1}(\omega) f_{1,1}(\omega)^{-1/2} \equiv F \tag{6}$$

for all the frequencies ω. We suspect (but we were not able to prove) that this property is characteristic of linear models. Figure 3 below shows that this assumption is reasonable in the case of the Lagrangian velocities obtained from the drifter data considered in this study.

We propose to use the same statistic T_{KS} as before. But we need to be more careful in the computation of the p-value of the test. The first task is to estimate the spectral coherence F. We choose to use:

$$\hat{F} = \frac{1}{N} \sum_{j=1}^{N} f_{2,2}^{-1/2}(\omega_j) f_{2,1}(\omega_j) f_{1,1}(\omega_j)^{-1/2}.$$

Then we generate B (we used $B = 200$ in the numerical experiments which we report in the next section) independent samples $\{X^{(b)}(t);\ 1 \leq t \leq n\}$ for $b = 1, \cdots, B$ of a mean zero stationary Gaussian time series $\{X(t);\ 1 \leq t \leq n\}$ with spectral density matrix:

$$f(\omega) = \begin{bmatrix} I_d & F^* \\ F & I_d \end{bmatrix}$$

where we used the standard notation I_d for the $d \times d$ identity matrix. We compute the T_{KS} statistic for each of the B time series and form a histogram. This parametric bootstrap can then be used to compute the approximate p-value of the test based on the observed value of T_{KS}.

Remark 5.1 *The p-value is computed from Monte Carlo simulations of Gaussian time series. The Gaussian assumption can be regarded as restrictive at first, but is fully justified in the applications we have in mind. Indeed, it is well known that, at least for turbulent flows, the Lagrangian velocities series can be reasonably assumed to be normal. This assumption was first proposed by Kolmogorov in the case of high Reynolds numbers and it has been checked experimentally in many situations. Moreover, in the* *mathematical analysis of the transport properties of the Gaussian velocity fields, the result of Lumley which we recalled earlier states that the one time marginal distributions of the Lagrangian velocities are normal. Of course, this does not implies that the whole time series are normal, but it makes it reasonable to make such an assumption for a parametric bootstrap.*

Remark 5.2 *As before one can choose to use the determinant instead of the trace in the definition of the test statistic and the same conclusions apply as well.*

6. Numerical Results

Figure 3 is a matrix of plots. The (j, j')-th plot gives the values of the modulus of the estimate of the coherence of the j-th and j'-th components of $X(t) = [X_1(t), X_2(t)]^t$ for a specific choice of Lagrangian velocities $X_1(t)$ and $X_2(t)$ computed from a typical couple of drifters from our data set. In particular, the 4×4 upper left corner gives the plots of the entries of the estimate $\hat{f}_{1,1}(\omega_j)$ of the spectral matrix of the first Lagrangian velocity, the 4×4 lower right corner shows the entries of the estimate $\hat{f}_{2,2}(\omega_j)$ of the spectral matrix of the second Lagrangian velocity while the lower left 4×4 corner gives the plots of the modulus of the entries of the estimate $\hat{f}_{2,1}(\omega_j)$ of the spectral coherence between the two Lagrangian velocities. All the plots are on the same scale. One can see that the coherence is significantly different from zero. At the same time one can appreciate the validity of the assumption that this coherence is constant over the frequencies.

6.1 Simulated Power The power of the test based on the statistic T_{KS} was computed by Monte Carlo simulation for a certain number of alternatives. Since we suspect that our model is exact for linear processes, we computed the power of the tests (based on both statistics T_{KS} and D_{KS}) when the distribution of the processes were given by bivariate moving average processes ($MA_2(2)$ for short)of order 1 and 2.

Here is a typical example of the results one gets: We consider a 4 dimensional $MA_4(2)$ processes of the form:

$$X(t) = \epsilon(t) + A(1)\epsilon(t-1) + A(2)\epsilon(t-2)$$

for a $N(0, I_2)$ bivariate white noise sequence $\{\epsilon(t)\}_t$ and for the coefficient matrices:

$$A(1) = \begin{bmatrix} 1 & 0 & a & 0 \\ 0 & 1 & 0 & a \\ 1 & 0 & 1 & 0 \\ 0 & 1 & 0 & 1 \end{bmatrix} \qquad \text{and} \qquad A(2) = \begin{bmatrix} 1 & 0 & b & 0 \\ 0 & 1 & 0 & b \\ 1 & 0 & 1 & 0 \\ 0 & 1 & 0 & 1 \end{bmatrix}.$$

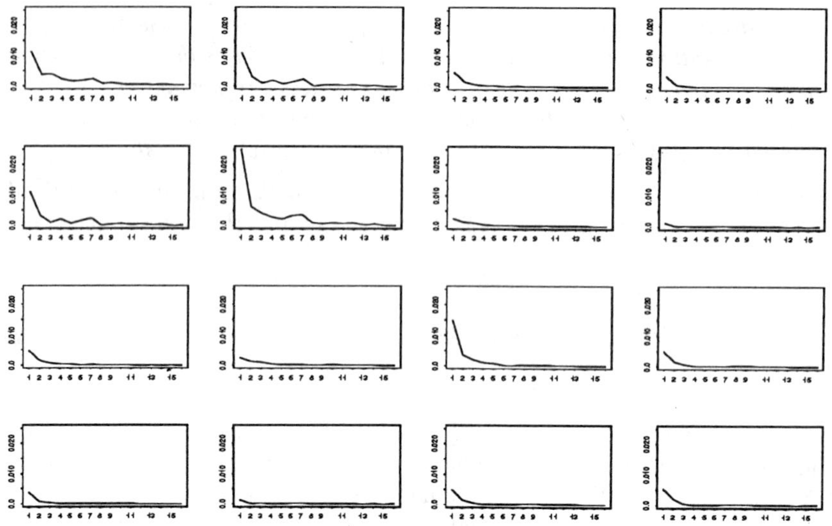

Figure 3: Plot of the modulus of the entries of the 4×4 matrix $\hat{f}(\omega_j)$ versus the frequency ω_j, for a typical couple of drifters. This plot was produced by the commands "`f <- spec.matrix(v80,v91)`" and "`plot.spec(f)`" described in the appendix.

The numbers a and b are parameters which we chave the freedom to choose. The following tables give, for each of the values of the parameters a and b, the probabilities (computed empirically from 50 runs) that the null hypothesis is accepted. In the present situation, the null hypothesis states that the bivariate time series formed by the first 2 components of the 4 dimensional time series $\{X(t)\}_t$ has the same spectrum as the bivariate time series formed by the last 2 components. Notice that these two bivariate time series are dependent.

	b=0.0	b=.2	b=,4	b=.6	b=.8	b=1.0
a=0.0	0	0	0	0	.1	.9
a=.2	0	0	0	0	0	.8
a=.4	0	0	0	0	0	0
a=.6	0	0	0	0	0	0
a=.8	0	0	0	0	0	0
a=1.0	1.0	.7	.1	0	0	0

Results for the test based on the statistic T_{KS} an the independence assumption.

	b=0.0	b=.2	b=.4	b=.6	b=.8	b=1.0
a=0.0	0	0	0	.08	.26	.6
a=.2	0	0	0	.02	.14	.6
a=.4	.02	0	0	0	0	.28
a=.6	.04	0	0	0	0	.1
a=.8	.18	.08	0	0	0	.01
a=1.0	.62	.6	.34	.18	.1	1.0

Results for the test based on the statistic T_{KS} computed without the independence assumption.

Notice that two bivariate times series have the same spectra in the case $a = b = 1$ but this was not picked up by the test based on the independence of the series.

Here is another example which shows that the general test should be preferred to the test based on the independence assumption. We use the same notation as above and the coefficient matrices:

$$A(1) = \begin{bmatrix} 1 & 0 & a & 0 \\ 0 & 1 & 0 & a \\ a & 0 & 1 & 0 \\ 0 & a & 0 & 1 \end{bmatrix} \quad \text{and} \quad A(2) = \begin{bmatrix} 1 & 0 & b & 0 \\ 0 & 1 & 0 & b \\ b & 0 & 1 & 0 \\ 0 & b & 0 & 1 \end{bmatrix}$$

In this case, the two bivariate time series always have the same spectra. They are independent when the numbers a and b are zero and the dependence increases with the values of these parameters. The results of runs made in the same conditions as above are contained in the following tables.

	b=0.0	b=.2	b=,4	b=.6	b=.8	b=1.0
a=0.0	.94	.98	.92	.96	.94	.92
a=.2	.92	.94	.96	.94	.98	..96
a=.4	.98	.96	.94	.98	.96	.98
a=.6	.76	.86	.76	.8	.84	.84
a=.8	.06	.08	.1	.18	.08	.08
a=1.0	0	0	0	0	0	0

Results for the test based on the statistic T_{KS} an the independence assumption.

	b=0.0	b=.2	b=.4	b=.6	b=.8	b=1.0
a=0.0	.98	1.0	.96	1	1	.98
a=.2	.94	.98	1	.96	.96	1
a=.4	.98	.98	.98	.96	.96	.96
a=.6	.98	.98	.98	.94	.96	1
a=.8	1	.9	.94	.96	.98	.94
a=1.0	1	1	1	1	1	1

Results for the test based on the statistic T_{KS} computed without the independence assumption.

One immediately sees that the dependent test performs better, especially when there is a strong dependence between the two time series. Indeed, this test continues to accept the null hypothesis while the test based on the independence of the time series see its level deteriorate in a drastic manner.

Remark 6.1 *We also investigated the influence of the window parameter K on the power of the tests. Our numerical simulations show that small values of K (by which we mean $3 \leq K \leq 6$ give a smaller power of detection than larger values (typically $8 \leq K \leq 10$).*

Acknowledgments We would like to thank Don Olson from the Marine Station in Miami for providing us with the drifter data used in the analysis.

Appendix

A. Drifter Data Documentation

The drifter dataset consists of 19 files, and each file consists of following fields: days, latitude (degree), longitude (degree), surface temperature (degree Celsius), ease-west velocity (m/s), and north-south velocity (m/s). All data were collected from the South Atlantic and Indian Oceans. The file name begins with a two-digit number followed by ".vel", e.g., 80.vel. The two-digit numbers range from 80 to 99, with 94 skipped.

B. S Function Documentation

This appendix contains documentation for the S functions developed during the study. It is organized into three categories: data structure/input, data display, and test procedures.

B.1 Data Structure/Input When the data file is in table format like ours, it can be read in as a data frame via the S command `read.table`. The columns of the resulting data frame are the corresponding fields in the file. Appropriate names should be assigned to different columns for easy access. If the dataset contains more than one file, all the data frames can be combined into a list.

Although the following two functions were written specifically for the dataset we have in hand, users can use them as an example to read in and organize their own data files. Our functions require that each data frame at least consists of two fields named "latitude" and "longitude".

make.vel.frame

USAGE

```
make.vel.frame()
```

DESCRIPTION

Combines 19 data frames, which are the output of read.vel, into one data frame. This function should be called after the drifter data files have been read into individual data frames.

VALUE

The output is a data frame with seven fields (columns). They are called tag (the two-digit number associated with each data file), days, longitude, latitude, temperature, v1 and v2. 19 data frames are combined in increasing order of the numbers associated with the file names.

read.vel

USAGE

```
read.vel()
```

DESCRIPTION

Reads drifter data files into data frames and combines all 19 data frames into a list called "drifter."

VALUE

The output consists of 19 data frames with names starting with "v" followed by the two-digit number corresponding to the file name and a list named "drifter". Each data frame corresponds to one drifter file, and there are six fields (columns) in each each frame. They are called days, longitude, latitude, temperature, v1 and v2.

B.2 Data Display

drifter.world

USAGE

```
drifter.world()
```

DESCRIPTION

Plots part of the world that all the drifters travel, which include the South Atlantic and Indian Oceans.

plot.drifter

USAGE

```
plot.drifter(drifter.list, scale=F)
```

DESCRIPTION

Plots a list of drifters. If the option "scale" is TRUE, all the drifters are plotted with the same scale. The graphic window should be formatted according to the number of drifters before plotting. For example, use the S command "par(mfrow=c(4,5))" to divide the graphic window into a 4 by 5 rectangular array of plots, so that all 19 drifters can be viewed at one time.

ARGUMENTS

drifter.list: a list of data frames where each data frame corresponds to one drifter data and has at least two fields, longitude and latitude. It is like the output of read.vel.

scale: logical flag: if TRUE, all the drifters are plotted with the same scale.

plot.traj

USAGE

```
plot.traj(..., inter=F)
```

DESCRIPTION: Plots the trajectories of the given drifters and superimposes the parts of the drifters that intersect in time if the option "inter" is TRUE.

ARGUMENTS

...: any number of drifter data frames to be plotted. Each data frame consists of at least two fields named "latitude" and "longitude".

inter: logical flag: if TRUE, the parts of the drifters that intersect in time are superimposed on the trajectories. This option works properly only if each data frame has another field named "days" that record the time of the drifter data.

B.3 Test Procedures

dep.spec.test

USAGE

```
dep.spec.test(f, g, method=''determinant'')
```

DESCRIPTION: Tests if either the trace or the determinant sample computed from the spectral density matrix, f, has the same distribution as the corresponding sample computed from g. It returns the p-value of the Kolmogorov-Smirnov two-sample test.

ARGUMENTS

f: spectral density matrix estimate of the drifters.
g: spectral density matrix estimate of the simulated data with constant coherence.
method: a character string indicating the type of statistic used in the test, "determinant" or "trace".

det.stat

USAGE

```
det.stat(f)
```

DESCRIPTION: Computes the determinant statistic of the spectral density matrix f.

ARGUMENTS

f: spectral density matrix estimate of the drifters.

est.spectra.matrix

USAGE

```
est.spectral.matrix(x1, y1, x2, y2, K=10)
```

DESCRIPTION: Computes the 4 by 4 spectral density matrix estimate for the given vector-valued series.

ARGUMENTS

x1: a vector of numerical numbers.
y1: a vector of numerical numbers.
x2: a vector of numerical numbers.
y2: a vector of numerical numbers.
K: a positive number indicating the number of points to be grouped for the estimate at one frequency.

indep.spec.test

USAGE

```
indep.spec.test(f, K=10, method=''determinant'')
```

DESCRIPTION: Tests if either the trace or the determinant sample computed from f has the same distribution as the corresponding sample generated from Monte Carlo simulation. It returns the p-value of the Kolmogorov-Smirnov two-sample test.

ARGUMENTS

f: spectral density matrix estimate of the drifters.
K: the same positive number used when f is computed.
method: a character string indicating the type of statistic used in the test, "determinant" or "trace".

kstwo

USAGE

```
kstwo(sample1, sample2)
```

DESCRIPTION: Performs Kolmogorov-Smirnov two-sample test and returns the p-value.

ARGUMENTS

sample1: a numerical vector.
sample2: a numerical vector with the same length of sample1.

ma1.sim

USAGE

```
ma1.sim(N, a1=diag(4))
```

DESCRIPTION: Generates an N by 4 MA(1) series with coefficient matrix a_1.

ARGUMENTS

N: a positive number indicating the length of MA(1) series to be simulated.
a1: a 4 by 4 coefficient matrix for the MA(1) series.

ma2.sim

USAGE
```
ma2.sim(N, a1=diag(4), a2=diag(4))
```

DESCRIPTION: Generates an N by 4 MA(2) series with coefficient matrices a_1 and a_2.

ARGUMENTS

N: a positive number indicating the length of the MA(2) series to be simulated.
a1: a 4 by 4 coefficient matrix for the MA(2) series.
a2: a 4 by 4 coefficient matrix for the MA(2) series.

myqq

USAGE
```
myqq(sample1, sample2, xlab, ylab)
```

DESCRIPTION: Compares the distributions of two samples by `qqplot`. "xlab" and "ylab" are optional arguments.

ARGUMENTS

sample1: a numerical vector.
sample2: a numerical vector.
xlab: a character string.
ylab: a character string.

plot.coh

USAGE
```
plot.coh(f)
```

DESCRIPTION: Plots the coherency, both the real part and the imaginary part, computed from the spectral density matrix estimate, f, at all calculated frequencies.

ARGUMENTS

f: spectral density matrix estimate of the drifters.

plot.spec

USAGE

```
plot.spec(f)
```

DESCRIPTION: Plots the modulus of the 4 by 4 spectral density matrix estimate, f, at all calculated frequencies with the same scale in each panel.

ARGUMENTS

f: spectral density matrix estimate of the drifters.

sim.dep

USAGE

```
sim.dep(f, N=500)
```

DESCRIPTION: Simulates N by 4 data with the given spectral density matrix f.

ARGUMENTS

f: spectral density matrix estimate of the drifters.
N: a numerical number indicating the length of the simulated series.

spec.matrix

USAGE

```
spec.matrix(drifter1, drifter2, K=10)
```

DESCRIPTION: Computes the 4 by 4 spectral density matrix estimate for the given drifters.

ARGUMENTS

drifter1: a data frame containing at least two fields, longitude and latitude.
drifter2: a data frame containing at least two fields, longitude and latitude.
K: a positive number indicating the number of points to be grouped for the estimate at one frequency.

trace.stat

USAGE

```
trace.stat(f)
```

DESCRIPTION: Computes the trace statistic of the spectral density matrix f.

ARGUMENTS

f: spectral density matrix estimate of the drifters.

wishart.ratio

USAGE

```
wishart.ratio(N=100, r=2, K=10, method=''determinant'')
```

DESCRIPTION: Computes N trace statistics or determinant statistics of $W_1 W_2^{-1}$, where W_1 and W_2 are independent and have a Wishart distribution of dimension r and degrees of freedom K.

ARGUMENTS

N: a positive number indicating the size of the sample.
r: a positive number indicating the dimension of the Wishart distribution.
K: a positive number indicating the number of points to be grouped for the estimate at one frequency.
method: a character string indicating the type of statistic used in the test, "determinant" or "trace".

wishart.sample

USAGE

```
wishart.sample(N, method=''determinant'')
```

DESCRIPTION: Computes N trace statistics or determinant statistics of $W_1 W_2^{-1}$, where W_1 and W_2 are independent and have a Wishart distribution of dimension 4 and degrees of freedom K, ranging from 2 to 10.

ARGUMENTS

N: a positive number indicating the size of the sample.
method: a character string indicating the type of statistic used in the test, "determinant" or "trace".

References

[1] D.R. Brillinger (1973): The Frequency Analysis of Vector-valued Time Series. *Holt* New York, N.Y.

[2] R. Carmona (1994): Ornstein-Ulhenbeck Stochastic Flows. (in preparation)

[3] R. Carmona and S. Grishin (1994): Massively Parallel Simulations of Motions in a Gaussian Velocity Field. (preprint)

[4] D.S. Coates and P.J. Diggle (1986): Test for Comparing Two Estimated Spectral Densities. *J. of Time Series Analysis,* **1**, 7-20.

[5] P.J.Diggle and N.I. Fisher (1991): Nonparametric Comparison of Cumulative Periodograms. *Appl. Statist.* **40**, 423-434.

[6] S.C. Port and C.J. Stone (1976): Random Measures and their Applications to Motion in an Incompressible Fluid. *J. Appl. Prob.* **13**, 498-506.

[7] M. Rosenblatt (1988): Stationary Sequences and Random Fields. *Birkhaüser*, Boston, MA.

[8] y A.M. Yaglom (1987): Correlation Theory of Stationary and Related Random Functions. vol. I: Basic Results. Springer Verlag. New York, N.Y.

[9] z C.L. Zirbel (1993): Stochastic Flows: Dispersion of a Mass Distribution and Lagrangian Observations of a Random Field. Ph. D. Princeton Univ.

RENE A. CARMONA
Department of Mathematics
University of California at Irvine

ANDREA WANG
Department of Statistics
University of California at Berkeley

rcarmona@phoenix.princeton.edu

A statistical approach to ocean model testing and tuning

Claude Frankignoul

1. Introduction

For many purposes such as the forecasting of oceanic conditions or the prediction of climate changes, it is important to have *realistic* models of the ocean, that is models which are able to reproduce sufficiently well the main features of interest, and their time variability. This is particularly critical in the design of coupled ocean-atmosphere models where model errors are generally exacerbated by the coupling, so that artificial "flux-corrections" (Sausen et al, 1988) are often used to prevent a drift toward unreasonable climate conditions, even though they may distort the dynamics. Although more complex models should represent reality more correctly than simpler ones, they do not necessarily do so, as illustrated by the El Niño-Southern Oscillation (ENSO) phenomenon where simple ocean-atmosphere models have so far provided forecasts as good as general circulation models (GCMs), in part because of the drift of the latter (Latif et al, 1993). Provided models are consistent with known physics within the tolerance allowed by the approximations made, model adequacy should thus be judged by the ability at reproducing the relevant observations.

The question of whether an oceanic model is able to reproduce the mean oceanic state (climate), or its variability, is often referred to as model validation, but it should rather be called model testing, as one can only verify whether a model is *consistent* with reality, as described by the observations. One may also ask whether a change in a parameterization improves the model, or whether a model is *more realistic* than another one. Although visual comparisons between simulated and observed climatologies can reveal obvious differences, they become less effective as the fidelity of the models increases and can lead to an incorrect assessment of their agreement. Furthermore, they are inadequate for separating the effect of model flaws from that of the observational data uncertainties. Hence, a quantitative measure of agreement based on statistical techniques is needed to test models against observations.

A difficulty is that the oceanic circulation is primarily forced by the atmosphere, but the atmospheric forcing is poorly known, in particular in the tropics and at high latitudes where the synoptic global analysis remains to be improved and marine observations are sparse. The bulk formulae used to estimate the air-sea fluxes of momentum and heat are also major sources of uncertainty, and precipitation is barely known above the ocean. As the oceanic initial conditions are little documented at best, even the predictions of a perfect ocean model would not be consistent with the observations.

The forcing uncertainties induce uncertainties in oceanic model response that have large correlation scales. Furthermore, ocean models, like reality, often have a substantial natural variability in the form of mesoscale eddies. In addition, the oceanic fields used for validation are mostly analyzed fields with highly correlated errors. Hence, distinguishing between model inadequacies and data uncertainties requires a *multivariate* statistical viewpoint, rather than the usual univariate (point-by-point) comparisons with the observations. Information on model validity can also be derived from data assimilation procedures, but they require substantial programming and computer ressources, and the distinction between model errors and forcing uncertainties is difficult.

The problem with multivariate model testing is that the ocean models and most data fields have a very large dimension, whereas the number of simulations and observations is limited, whence the sample size is much smaller than the dimension. Standard multivariate statistical methods are generally adapted to the large sample case, so new strategies had to be devised for the oceanic context. Frankignoul et al. (1989) have developed a multivariate model testing procedure which compares the main space and time structure of the model-observation differences to that of the data uncertainties. This is done by simulating the effects of the atmospheric forcing uncertainties using a Monte-Carlo approach and then calculating a misfit in a subspace of much reduced dimension. If forcing and observational errors are well-represented, the misfit provides a quantitative measure of model performance, and it can also be used for model intercomparison.

All oceanic models contain parameterizations of such processes as mixing or convection which are based on physical ideas, but yield forms that contain parameters whose value are not known precisely from theory. Surface forcing also depends on poorly known parameters. In view of their inherent imprecision, the uncertain model parameters need to be tuned against observed data. This is usually done by comparing visually a few key features of the simulations to the observations, and exploring their sensitivity to small parameter changes. However, only a limited number of parameters can be changed simultaneously, and it is difficult to consider the effect of the forcing uncertainties. For a finer tuning that takes into account the data uncertainties, the model testing method can be used,using the same trial and error approach. However, this also requires that the number of adjustable parameters be small. A more efficient tuning approach for linear parameter dependence is that of Blumenthal and Cane (1989), who used inverse modeling procedures to simultaneously determine the values of all the adjustable parameters that were required to optimally fit sea surface temperature (SST) in a simplified SST model. A priori knowledge constraining the parameter range was included in the calculation, but only highly idealized models were used for the data and modeling errors. Yet, the error models enter the measure of the misfit between observed and predicted data which is minimized in the best-fit calculation. When a more realistic model is used for the data errors and the atmospheric forcing uncertainties, however, the inverse problem becomes nonlinear.

To better represent the data errors, Sennéchael et al. (1994) have developed an adaptive tuning procedure, where the SST model that is being tuned is also used to construct a statistical model of the observational errors for the best-fit calculation. This is done by performing the optimization on the averaged seasonal SST cycle, and using the dispersion of the model response for different years and plausible forcing fields as independent information to construct a sample estimate of the observational error covariance matrix. Since the degree of freedom of the latter is limited, the optimization is done in reduced space and the number of parameters that can be tuned is somewhat limited. The procedure yielded refined estimates for the model parameters, but there were still some inconsistencies in the inverse calculation, primarily because the modeling errors were strongly idealized.

In this chapter, the statistical approach to model testing and tuning is reviewed. In section 2, the testing for the mean is introduced; the difficulties of applying univariate statistics to ocean model are stressed and the mutivariate approach introduced. In section 3, ocean model testing and intercomparison are discussed and illustrated in the context of the tropical oceans. The adaptive model tuning procedure is described in section 4, and some concluding remarks given in section 5.

2. Testing for the mean

2.1 Univariate approach. Oceanic simulations are often compared to observations by using a univariate statistical approach and, since most oceanic variables are very nearly gaussian, parametric tests. Let us consider the case where we have n independent simulations of a normal variable, denoted by $m_1, ...m_n$. Their dispersion is assumed to be representative of the model natural variability, if any, and the effects of the forcing uncertainties, which can be derived from a Monte-Carlo approach where several independent and equally plausible atmospheric fields are used as forcing (Braconnot and Frankignoul, 1993). Suppose that we also have p independent observations of the same variable, denoted by $d_1, ...d_p$. If simulations and observations have the same variance, one can test the null hypothesis H_0 that the true means are equal, $\mu_m = \mu_d$, by considering the test statistic

$$t = \frac{\overline{m} - \overline{d}}{(n^{-1} + p^{-1})^{1/2}\, s_p} \tag{2.1}$$

where an overbar denotes the sample mean and ${s_p}^2$ is an unbiased pooled estimate of the variance given by

$${s_p}^2 = \frac{1}{n+p-2}\left(\sum_{i=1}^{n}(m_i - \overline{m})^2 + \sum_{i=1}^{p}\left(d_i - \overline{d}\right)^2\right). \tag{2.2}$$

If H_0 is true, t is distributed as a Student t variable with $n+p-2$ degrees of freedom, hence H_0 is rejected in favor of the alternative hypothesis H_A (say $\mu_m \neq \mu_d$) at a given level of significance if $|t|$ is larger than the corresponding critical value; otherwise H_0 is accepted. Stuart and Ord (1991) discuss the case where simulations and observations have different variance.

The t-test is powerful, and it has been widely used to analyze oceanic and atmospheric simulations. However, its interpretation normally requires evaluating the significance of differences in fields composed of many grid points and, possibly, variables, which raises the question of the collective or field significance of an ensemble of univariate tests. Even in the simple case where all the univariate tests are independent, the overall rate of rejection of the null hypothesis at the α% level should be larger than α% for global rejection of the null hypothesis, if the number of local tests is finite. The critical rejection rate can be inferred from the binomial distribution and, for a small number of tests, the threshold for field significance can be large (von Storch, 1982; Livezey and Chen, 1983). In the more realistic case where the data are not independent but spatially correlated, H_0 tends to be rejected in "pools" of grid points, and one expects the critical rejection rate to be larger since the *effective* number of independent tests is smaller. This number is difficult to estimate, however, because the data correlations are usually poorly known. As reviewed by Livezey (1995), field significance can be established by using permutation techniques. An alternative is to use a multivariate approach.

2.2 Multivariate approach. Let us denote by $\boldsymbol{m}_i$, $i = 1, n$, the n independent simulations of a vector variable of dimension N, which represent the oceanic variables in the space-time domain where the model is supposed to be realistic and the p independent observations, denoted by $\boldsymbol{d}_j$, $j = 1, p$, are reliable. As before, the dispersion of the simulations represents both their natural variability and the effects of the forcing uncertainties, defining a probability region for the model response to the true forcing. If simulations and observations are multinormal variables with the same covariance matrix $\sum$, the null hypothesis $\boldsymbol{\mu}_m = \boldsymbol{\mu}_d$ can be tested if $n + p - 2 \geq N$ against the alternative hypothesis H_A, say $\boldsymbol{\mu}_m \neq \boldsymbol{\mu}_d$, by considering the two-sample Hotelling T^2 statistic

$$T^2 = \frac{np}{n+p} \left(\overline{\boldsymbol{m}} - \overline{\boldsymbol{d}}\right)^T \boldsymbol{S}_p{}^{-1} \left(\overline{\boldsymbol{m}} - \overline{\boldsymbol{d}}\right) \tag{2.3}$$

where $\boldsymbol{S}_p$ is a pooled estimate of $\sum$ with $n+p-2$ degrees of freedom, defined as in (2.2), and the upper index T indicates vector transpose. When the null hypothesis is true, the quantity $F = \frac{n+p-1-N}{N(n+p-2)} T^2$ has the F distribution with N and $N - n - p - 1$ degrees of freedom (e.g., Morrison, 1976), and H_0 is rejected at the $\alpha\%$ level if

$$T^2 > \frac{N\,(n+p-1)}{n-N} F_{\alpha; N, n+p-1-N} \tag{2.4}$$

where $F_{\alpha;N,n+p-1-N}$ denotes the 100 α upper percentage point of the F distribution. T^2 is the multivariate analogue of the univariate $t - ratio$, reducing to t^2 when $N = 1$. Except for skewed distributions, the T^2 test is rather robust against departures from normality, but it is ineffective unless the sample size $n+p-2$ is much larger than N, which is seldom the case for ocean models. When the true covariance matrix $\sum$ is known, it replaces $\boldsymbol{S}_p$ in (3); the test statistic is then distributed as a χ^2_m variable when H_0 is true.

In the oceanic context, the error covariance matrix of simulations and observations are rarely comparable. As discussed e.g. in Seber (1984), the effects of differences in the covariance matrices on the significance level and the power of the T^2 test (2.4) are serious when n and p are not comparable, and a more appropriate test statistic is then

$$T^2 = (\bar{\boldsymbol{m}} - \bar{\boldsymbol{d}})^T \left(\frac{\boldsymbol{S}_m}{n} + \frac{\boldsymbol{S}_d}{p}\right)^{-1} (\bar{\boldsymbol{m}} - \bar{\boldsymbol{d}}) \tag{2.5}$$

where $\boldsymbol{S}_m$ and $\boldsymbol{S}_d$ are sample estimates of the covariance matrix of the simulations and the observations, respectively. When n and p are very large, (2.5) becomes asymptotically distributed as χ^2_N if H_0 is true. Otherwise, (2.5) is approximately distributed as Hotelling's T^2, and the degrees of freedom can be coarsely estimated. Note that if the observations come from an analysis, $\bar{\boldsymbol{d}}$ in (2.5) should be replaced by $\boldsymbol{d}$, since only one observed field, namely the analyzed field, enters the comparison. Correspondingly, the sampling uncertainty $\boldsymbol{S}_d/p$ may be replaced by the analysis error covariance matrix $\boldsymbol{D}$.

The test statistics (2.3) and (2.5) define the misfit between observations and simulations, measuring the differences between the two fields relative to their accuracy. The misfit does not solely characterize model adequacy, as it depends on the data accuracy: a good model will fare poorly if the observations are sufficiently accurate, and a bad one well, if the data are inaccurate.

In practice, the dimension of the intercomparison space must first be strongly reduced. Indeed, $\boldsymbol{S}_m$ and $\boldsymbol{S}_d$ are generally estimated from a small sample and are of much reduced rank, or else $\boldsymbol{D}$ is provided by a limited error analysis and its details cannot be trusted. Thus, the inverse in (2.3) or (2.5), even if it exists, is dominated by unreliable information, and a strong data compression is needed for the problem to be well-posed. This severely limits the amount of model details that can be investigated in practice.

The need for data compression was first stressed by Hasselmann (1979) in the context of response studies with atmospheric GCMs. He pointed out that, because of the very large dimension of the GCM fields, a model response would normally fail a multivariate significance test, even if $\sum$ were known, since too many (noisy) grid-points are needed to describe circulation patterns. A signal will thus be very difficult to recognize from noise, unless filtering technics are used to reduce the dimension and increase the signal-to-noise ratio. Because of the subjective choice of a highly truncated representation, however, the data reduction must be done *a-priori*. As reviewed by Frankignoul (1995) this is a stringent limitation for atmospheric response studies, as it requires a priori insights on the expected model response. The multivariate test of significance can tell whether the GCM response is consistent with our assumptions, but it does not indicate whether there is any significant response at all. In the model testing context, however, this arbitrariness disappears since the reduced base should be that which allows to best represent the main features of both modeled and observed fields.

3. Application to tropical ocean models

3.1 Model testing. Frankignoul et al. (1989) have developed a multivariate model testing procedure which compares the main space and time structures of the model-observation differences to those of the data uncertainties. It is illustrated here by an investigation of the ability of the LODYC GCM (Blanke and Delecluse, 1993) at simulating the 1982-1984 evolution of the tropical Atlantic thermocline, as determined by the depth of the $20°C$ isotherm. Details can be found in Braconnot and Frankignoul (1994) and Frankignoul et al. (1995).

The observed thermocline depth $d(\boldsymbol{x}, t)$ was derived from temperature measurements by Reverdin et al. (1991). As the data coverage coverage was very sparse, the field was mapped monthly using a function fitting algorithm and a cut-off period of 3 months. The error covariance matrix $\boldsymbol{D}$ of the analyzed fields was calculated, so that the non-locality of the interpolation procedure is accounted for in the error field. The intercomparison domain was limited to the $12°N - 12°S$ region, so $\boldsymbol{d}$ is a space-time vector with 224 (number of grid points) $\times 36$ (number of months) $= 8064$ components.

The tropical oceans are primarily wind-driven, and the thermocline variations could be simulated deterministically for the most part if the wind stress were accurately known. However, wind observations are inaccurate and gappy, and the bulk formulae used to estimate the wind stress are uncertain. To represent the forcing uncertainties, three independent, equally plausible monthly wind stress fields were constructed by a Monte Carlo method and used to force the ocean model. The fields are consistent with the original wind measurement statistics and are based on three different drag laws. Although their dispersion only represents part of the forcing uncertainties, the corresponding thermocline depth uncertainty has a root mean square (rms) value of several meters, nearly as large as its interannual variability. The averaged model response $\overline{\boldsymbol{m}}$ was calculated in the space-time intercomparison domain, as well as its sample error covariance matrix $\boldsymbol{M}$, defined by

$$\text{(3.1)} \qquad \mathrm{M} = \frac{1}{6} \sum_{t=1}^{3} (\boldsymbol{m}_t - \overline{\boldsymbol{m}}) (\boldsymbol{m}_t - \overline{\boldsymbol{m}})^T .$$

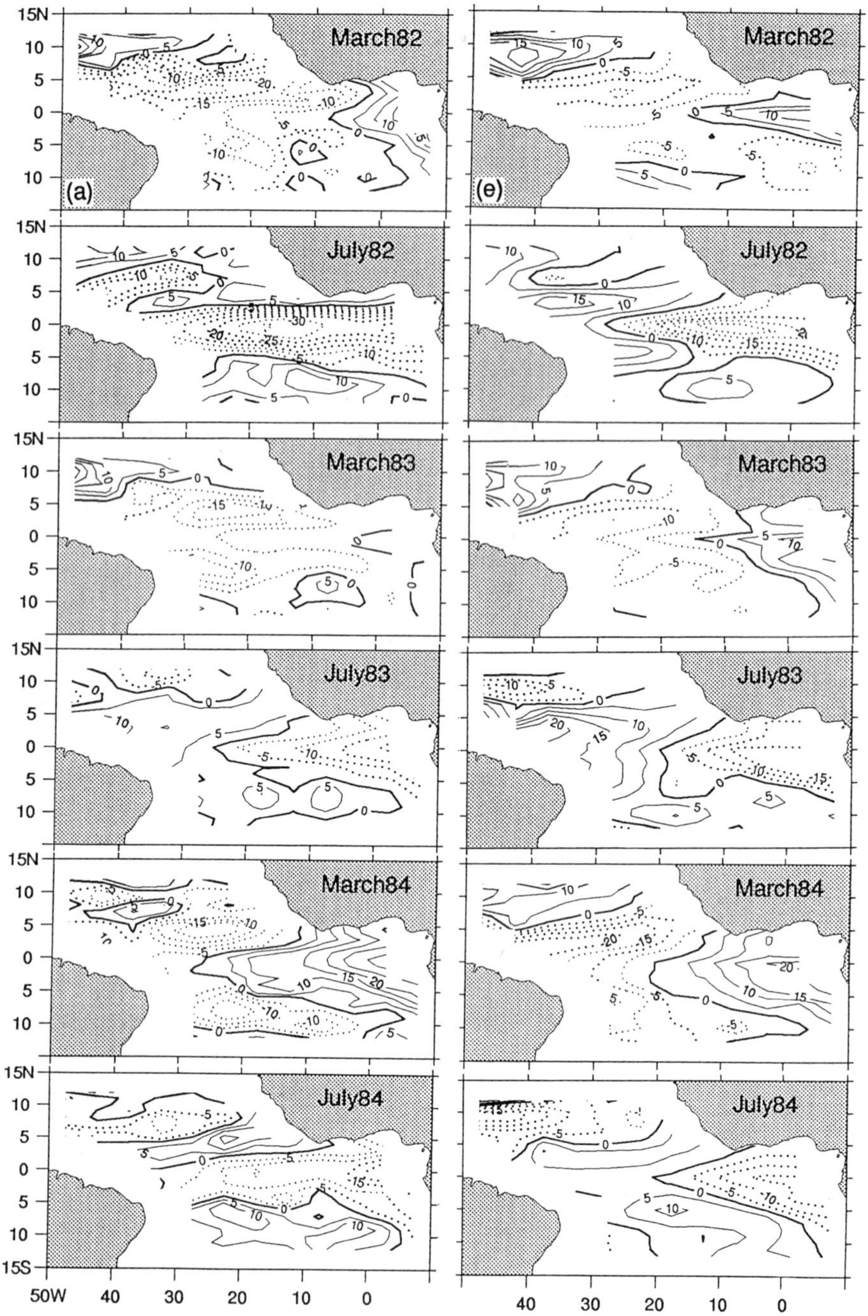

Figure 1. Left: Observed departures (in m) of the 20°C isotherm from the 1982-1984 mean for March and July, from 1982 to 1984. Right: Corresponding simulations with the LODYC GCM. (From Frankingnoul et al., 1995).

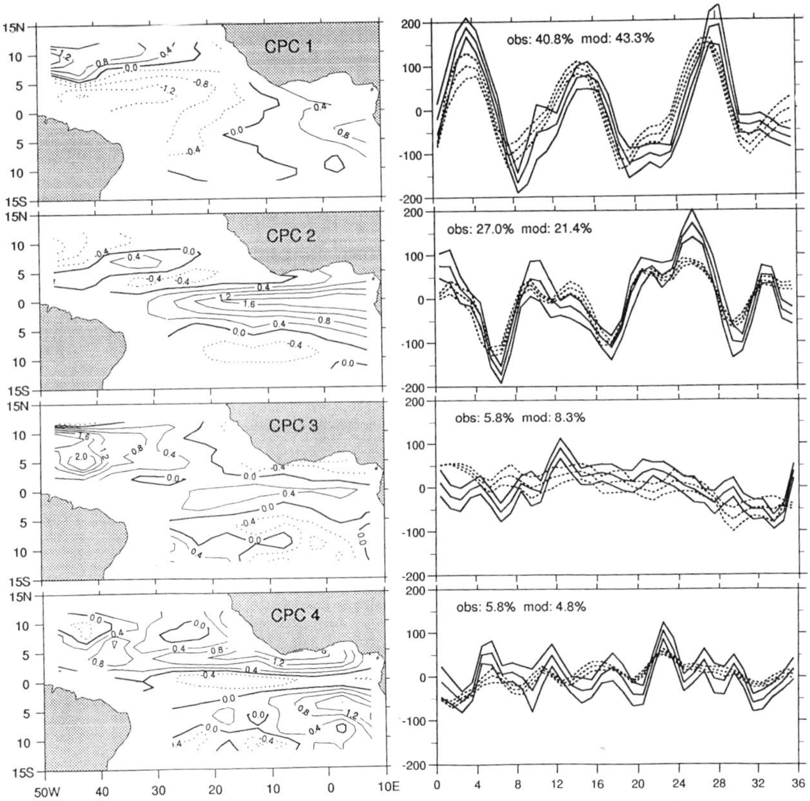

Figure 2. Left: basis vectors for the spatial reduction. Right: projections of observed (continuous middle line) and modeled (dotted middle line) fields onto the spatial base, with univariate 95% confidence intervals.

Here we only consider the observed and simulated departures of the thermocline depth from the 1982-1984 annual mean, which are compared in Fig.1. Since the degrees of freedom are limited, a strong data compression is needed to meaningfully compute the misfit (2.5). Using common principal component analysis, which is a generalization of principal component analysis that applies simultaneously to two or more fields (Flury, 1989), an orthonormal base is first defined where the main spatial features of simulations and observations are well represented. The first four common empirical orthogonal functions (EOFs) have large spatial scales (Fig.2, left) and account for more than 80% of the observed and simulated fields. Observations and simulations are represented in this subspace by the time series in Fig.2 (right), where the 95% confidence intervals, estimated from the diagonal terms of the error covariance matrices in the reduced space, give an (univariate) indication of the data errors. The first common EOF primarily describes a zonal redistribution of mass across the Atlantic basin, which occurs at the annual period, while the second describes a latitudinal mass redistribution and the occurrence of upwelling, with mostly annual and semi-annual variations. Both features are well simulated. The third common EOF describes systematic differences between model and observations, which is expected since common principal component analysis tries to best project two fields onto a small number of spatial patterns, thereby favoring either in-phase or out-of-phase behaviors. The fourth common EOF is somewhat more noisy, while the higher structures that were filtered out showed an increasing amount of noise in both space and time.

A data compression is then done in the time domain. The most efficient way is to work with the *differences* between the observed and model time series in Fig.2 (right), thus defining for each of the four common EOFs a unique 36-month difference time series. These four time series are then orthonormalized using EOF analysis, yielding four new temporal patterns that suffice to completely represent the four difference time series. The time dimension is thus reduced from 36 to 4, with no further loss of signal variance. The dimension of the reduced intercomparison space is 4 (number of spatial patterns) $\times 4$ (number of temporal behaviors) $= 16$, which is substantially less than the initial dimension (8064), at little loss (less than 20%) of variance. The signal $(\overline{\boldsymbol{m}} - \boldsymbol{d})$ and the error covariance matrices $\boldsymbol{M}$ and $\boldsymbol{D}$ are calculated in, or projected onto, respectively, this subspace to estimate the misfit. Note that such double data reduction is only valid if the errors are reasonably homogenous in space and time, as local variations in the errors can only affect the overall uncertainty of the global space-time patterns. A slightly more efficient way to reduce the spatial dimension would be to work directly with the difference fields and use standard principal component analysis, but we prefer the present approach as it emphasizes the signal rather than the errors, thereby giving more physical insights on the model-reality discrepancies.

Although the details of the error covariance matrix $\boldsymbol{D}$ provided by Reverdin et al. (1991) cannot be trusted as they depend on a number of simplifying assumptions, they affect little its projection in the reduced space. Hence, $\boldsymbol{D}$ is treated for simplicity as a true error covariance matrix. The estimated number of degrees of freedom of the inverse in (2.5) turns out to be very large when using Yao's (1965) formula, hence T^2 behaves approximately as a χ^2 variable with 16 degrees of freedom. As shown by the solid square on the left of Fig. 3, the misfit is much larger than the critical value at the 5% level (dashed line), and H_0 is rejected. Although the misfit would be smaller if all of the forcing uncertainties had been represented, it is clear that the observational errors cannot explain all the model-reality discrepancies. This is a very stringent test, however, as the LODYC model is one of the better oceanic GCMs.

3.2 Model intercomparison. Since only part of the forcing uncertainties can generally be represented by the Monte-Carlo approach and no oceanic model is expected to be perfect, the test statistic (2.5) will normally be much larger than the critical value at a reasonable level of significance and the null hypothesis rejected, at least if sufficiently accurate observations are available. Nonetheless, the misfit indicates how closely the model is able to reproduce the observations, and it is thus handy for model intercomparisons.

This is briefly illustrated by results from Frankignoul et al. (1995), who compared four tropical Atlantic ocean models of increasing complexity: the linear 3-model of Cane (1984), the LODYC nonlinear 2-layer model (Février et al, 1995), the KNMI GCM (Allaart and Kattenberg, 1990), and the LODYC GCM, in the same context of the 1982-1984 thermocline depth evolution. All models were forced by the same three independent wind stress fields and, for each model, the misfit between simulations and observations was calculated as above in four different regions: the $12°N - 12°S$ domain of Fig. 1, the equatorial waveguide between $3°N$ and $3°S$, and two well-sampled meridional "sections". The reduced spaces differed slightly from model to model, but they were found to represent comparable amount of variance. The results for the fluctuations around the 3-year mean are summarized in Fig. 3, where the approximate error bars on the misfits are based on the non-central T^2. Since the representation of the observational and forcing errors are the same in each comparison, the error bars should only be taken into account once in the intercomparison. Although H_0 is rejected in all the cases, the differences in model performance are statistically significant, and consistent with our expectation that more complex models (the GCMs) perform generally better than simpler ones.

4. Model tuning

4.1 Inverse modeling approach. Blumenthal and Cane (1989) have used inverse modeling procedures to determine the parameter values required to optimally fit sea surface temperature in a tropical ocean model. They used a highly idealized model for the data errors, hence Sennéchael et al. (1994) extended their approach, developing an adaptive tuning procedure, where the observational errors and the forcing uncertainties could be taken into account more realistically. The method is based on a sampling approach, and is general as long as the parameter dependence is linear.

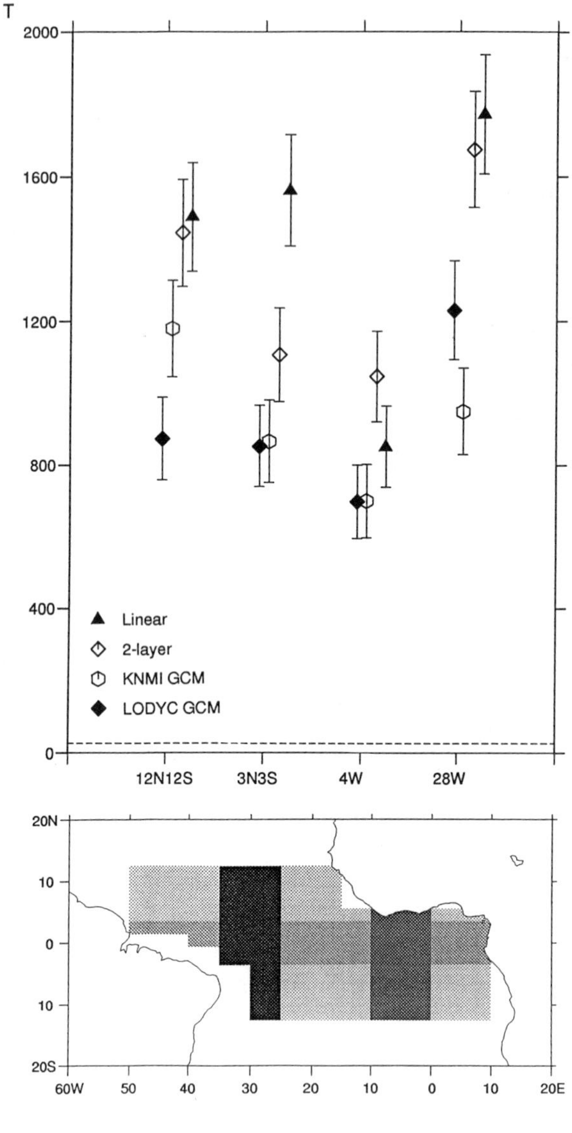

Figure 3. Misfit between four tropical Atlantic models and observations in the four testing regions indicated on the map below, for the 1982-1984 thermocline depth variations around the 3-year mean. The error bars represent the 95% confidence intervals, and the dotted line the critical value for rejecting the null hypothesis of no model error at the 5% level. (From Frankignoul et al., 1995).

In the simplified SST model of Blumenthal and Cane (1989), the mixed-layer temperature is determined from the net balance of horizontal advection, upwelling, horizontal diffusion and surface heat exchanges:

$$(4.1)\quad \partial_t T + u\partial_x T + v\partial_y T + \gamma w\,(T - T_d)\,/h = k\,(\partial_{xx} + \partial_{yy})\,T + Q/pC_p h$$

where w is the vertical velocity at the mixed layer base in the case of entrainment, and zero otherwise; γ is the entrainment efficiency, T_d the temperature below the mixed layer, which is parameterized as a function of the thermocline depth, κ the horizontal diffusion coefficient, and the surface heat flux Q is written as (Seager et al. 1988)

$$(4.2)\quad Q = 0.94Q_0\,(1 - a_c C + a_\alpha \alpha) - \rho C_E L v^a a_{rh} q_s\,(T) - a_T\,(T - T_r)\,.$$

Here Q_0 is the clear sky solar flux, C the cloud amount and α the solar angle. The latent heat flux is computed from a bulk formula using a fixed percentage a_{rh} of the saturation humidity $q_s(T)$ as the evaporation potential, which assumes that the moisture content of the air has equilibrated with the SST. Sensible heat flux and back radiation are represented by the last term, where T_r is a constant.

In Sennéchael et al. (1994), seven parameters were tuned. As they enter the SST calculation linearly, it is convenient to write (4.1) - (4.2) in matrix form

$$(4.3)\qquad \boldsymbol{L}(\boldsymbol{T}) + \boldsymbol{M}(\boldsymbol{T})\,\boldsymbol{a}_p = \boldsymbol{0}$$

where the tunable parameters are represented by the vector $\boldsymbol{a} = (\gamma, \kappa, a_c, a_\alpha, a_{rh}, a_T, a_T T_r)$, which was given the a priori value $\boldsymbol{a}_p$. The vector $\boldsymbol{T}$ represents temperature at all the p points in space and time where the SST model is considered, $\boldsymbol{M}(\boldsymbol{T})$ and $\boldsymbol{L}(\boldsymbol{T})$ are linear operators determined at these points by retaining the terms of the model equations (4.1) and (4.2) that would and would not be affected by parameter changes, respectively. The model was considered between $10°S$ and $20°N$ in January, April, July and October, so the dimension p of $\boldsymbol{L}$ is 322 (number of grid points) $\times 4$ (number of months) $= 1288$; $\boldsymbol{M}$ has dimension 1288×7.

The model was forced with a monthly wind stress for the 22-year period 1965-1986, using a cloud cover climatology. To represent the forcing uncertainties, five different, equally plausible drag coefficients were used to estimate the stress and a normal noise of 0.1 standard deviation was added to the cloud cover, yielding 5×22 simulations of the SST annual cycle, whose dispersion is (crudely) representative of the interannual variability and the forcing uncertainties.

The averaged simulated SST, written as $<\overline{\boldsymbol{T}}>$, where the overbar denotes the 22-year mean and angle brackets the average over the five 22-year samples, is warmer than the observations of Servain et al. (1985) (Fig.4). The differences are due to errors in the observational data, model shortcomings, and poor choice of the model parameters.

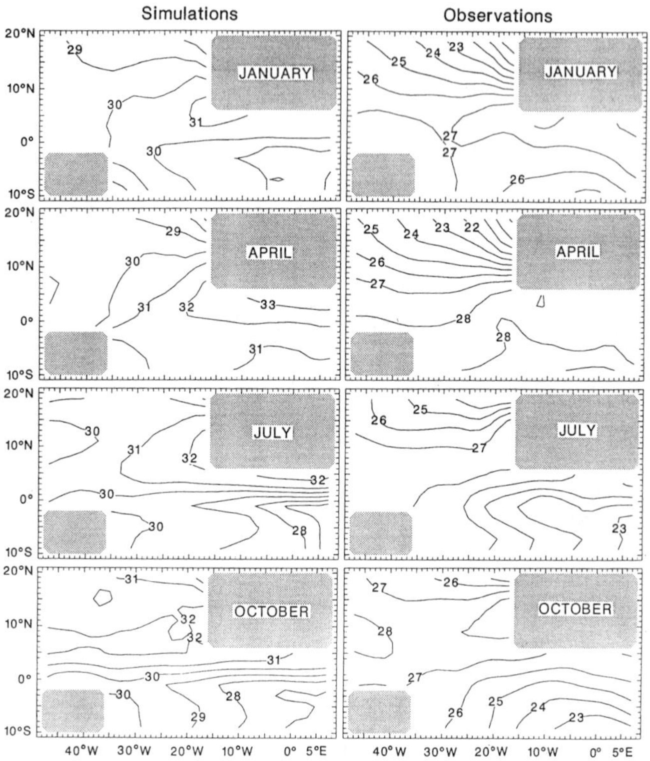

Figure 4. Left: Mean SST in $°C$ during January, April, July and October as predicted by the SST model using the a priori values of the parameters for the period 1965-1986. Right: Corresponding SST as derived from the observations by Servain et al. (1985). (From Sennéchael et al., 1994).

Let us define the "corrective heat flux" $\delta\boldsymbol{q}$ that, for the a-priori values of the uncertain model parameters, would be needed to make the modeled SST match the observed SST exactly. This is obtained by runing the model again, using the observed SST, denoted by $\boldsymbol{T}_0$, instead of the calculated one. Equation (4.3) is then only satisfied for each year $t\,(t=1,22)$ and forcing $i\,(i=1,5)$ by adding a heat flux correction $\delta\boldsymbol{q}^{t,i}$ defined by

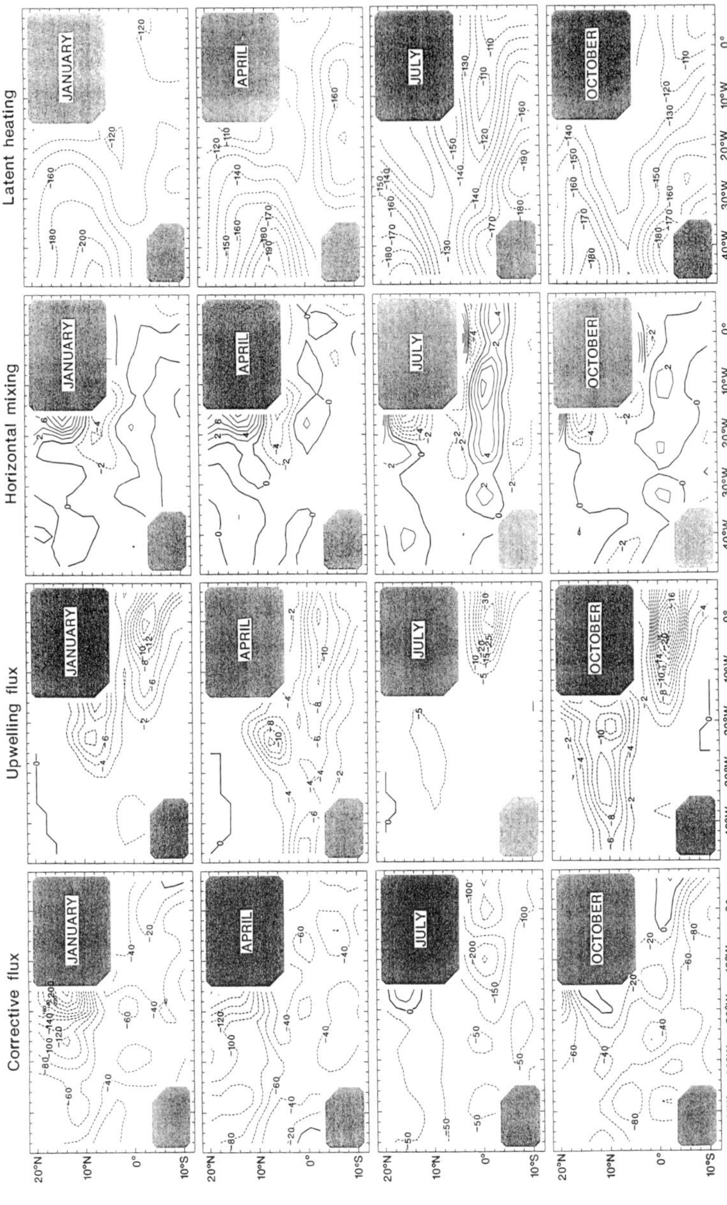

Figure 5. (a) Mean heat flux correction in WM^{-2} during January, April, July and October for the period 1965-1986, when using the a priori values of the model parameters. Corresponding values of (b) the upwelling flux, (c) horizontal diffusion, (d) latent heat flux. (From Sennéchael et al., 1994).

(4.4) $$\boldsymbol{L}^{t,i}(\boldsymbol{T}_0) + \boldsymbol{M}^{t,i}(\boldsymbol{T}_0)\boldsymbol{a}_p + \delta\boldsymbol{q}^{t,i} = 0.$$

Averaging yields

(4.5) $$< \overline{\boldsymbol{L}(\boldsymbol{T}_0)} > + < \overline{\boldsymbol{M}(\boldsymbol{T}_0)} > \boldsymbol{a}_p + < \overline{\delta\boldsymbol{q}} > = \boldsymbol{0}.$$

The averaged heat flux correction $< \overline{\delta\boldsymbol{q}} >$ is rather large, and negative values in excess of $-100\ Wm^{-}2$ are found off Africa and in the Gulf of Guinea, mostly where the largest SST differences are observed (Fig. 5a).

Since $< \delta\boldsymbol{q} >$ depends linearly on the tunable model parameters, the estimation of their optimal value can be formulated as an inverse problem:

(4.6) $$< \overline{\delta\boldsymbol{q}} > = < \overline{\boldsymbol{M}(\boldsymbol{T}_0)} > \delta\boldsymbol{a} + < \bar{\boldsymbol{e}} >,$$

where $\delta\boldsymbol{a} = (\delta\gamma, \delta\kappa, ..., \delta a_T T_r)$ represents the parameter changes that minimize the heat flux correction and $\boldsymbol{e}$ is a random error. The tuning can be viewed as determining the best fit in some least squares sense of the averaged heat flux correction vector in Fig.5a by the seven column vectors of $< \overline{\boldsymbol{M}(\boldsymbol{T}_0)} >$ (remember that these vectors are defined in space and time). Three of the latter are represented in Fig.5b-d: the "upwelling pattern", which has a large signal in the Gulf of Guinea, with maximum amplitude in July, and a smaller signal in the intertropical convergence zone off Africa, the "diffusion pattern", and the "evaporation pattern".

A good estimator of $\delta\boldsymbol{a}$ must take errors into account, as well as our knowledge of the expected parameter range. The wind stress, SST and cloud data induce significant errors in (4.6), mostly with large correlation scales. The averaged seasonal cycle has also sampling errors, which reflect the interannual variability. Finally, there are *irreducible* modeling errors inherent in the ocean model formulation, due to the oversimplification of the dynamics and the air-sea fluxes, which cannot be reduced by model tuning. The modeling errors (called system errors in the Kalman filter) represent the errors that would exist if there was no observational errors and the uncertain parameters were at their true value.

To solve a simplified form of (4.6) (the model response to the mean seasonal forcing was considered), Blumenthal and Cane (1989) assumed that there were only modeling errors in $< \overline{\boldsymbol{M}} >$, described by the covariance $\boldsymbol{C}_m$, and statistically independent observational errors in $< \overline{\delta\boldsymbol{q}} >$, described by the covariance $\boldsymbol{C}_0$. For zero a priori value of $\delta\boldsymbol{a}$, the optimal solution is given in the normal case by the minimum of the misfit function (Tarantola, 1987)

(4.7)
$$S(\delta\boldsymbol{a}) = [(< \overline{\boldsymbol{M}} > \delta\boldsymbol{a} - < \overline{\delta\boldsymbol{q}} >)^T \boldsymbol{C}^{-1} (< \overline{\boldsymbol{M}} > \delta\boldsymbol{a} - < \overline{\delta\boldsymbol{q}} >) + \delta\boldsymbol{a}^T \boldsymbol{C}_{\boldsymbol{a}}^{-1} \delta\boldsymbol{a}]/2$$

with $\boldsymbol{C} = \boldsymbol{C}_m + \boldsymbol{C}_0$, where the covariance matrix $\boldsymbol{C}_a$ describes the a priori uncertainty of $\delta \boldsymbol{a}$. This minimization yields

$$\delta \boldsymbol{a} = \left(< \overline{\boldsymbol{M}} >^T \boldsymbol{C}^{-1} < \overline{\boldsymbol{M}} > + \boldsymbol{C}_a^{-1} \right)^{-1} < \overline{\boldsymbol{M}} >^T \boldsymbol{C}^{-1} < \overline{\delta \boldsymbol{q}} > \tag{4.8}$$

However, the assumed error models were highly idealized. In particular, it is easy to see that $< \overline{\delta \boldsymbol{q}} >$ and $< \overline{\boldsymbol{M}} >$ are both affected by the data uncertainties and the modeling errors. Their errors are thus *not statistically independent*, and the model matrix really is a stochastic regression matrix.

4.2 The adaptive procedure. Since the tuning is sensitive to the assumed error models, Sennéchael et al. (1994) adopted a more elaborate strategy, using the dispersion of the 5×22 seasonal responses about their mean as independent information to construct a model for the observational errors. However, because of the statistical dependence between $< \overline{\delta \boldsymbol{q}} >$ and $< \overline{\boldsymbol{M}} >$ an estimate of $\delta \boldsymbol{a}$ is required to estimate the random errors from the sample. Thus, an adaptive approach was used, where the estimates of the observational error covariance and the model parameters were updated as part of an iterative procedure.

Using an a-priori estimate of $\delta \boldsymbol{a}$, say $\delta \boldsymbol{a}_0$, the mean error over the different forcings can be estimated for each year t by

$$< \boldsymbol{e}_1{}^t > = < \delta \boldsymbol{q}^t > - < \boldsymbol{M}^t > \delta \boldsymbol{a}_0 \tag{4.9}$$

A first sample estimate of the error covariance matrix associated with the interannual variability and the random observational errors can be derived from the dispersion of the seasonal cycles of each of the 22 years around their long-term average. This dispersion may be estimated by the covariance matrix

$$\boldsymbol{S}_{i1} = \frac{1}{21} \sum_{t=1}^{22} \left(< \boldsymbol{e}_1{}^t > - \overline{< \boldsymbol{e}_1 >} \right) \left(< \boldsymbol{e}_1{}^t > - \overline{< \boldsymbol{e}_1 >} \right)^T \tag{4.10}$$

so the error covariance matrix, say $\boldsymbol{S}_{r1}$, is given by $\boldsymbol{S}_{r1} = \boldsymbol{S}_{r1}/22$, if for simplicity the observations are assumed to be independent at yearly intervals. The error covariance matrix $\boldsymbol{S}_{f1}$ associated with the drag coefficient uncertainties can be estimated similarly from the 22-year mean seasonal cycle in each of the five runs with different forcings, if one considers for simplicity that the five drag coefficients are random samples from some normal distribution. Hence, the first estimate of the error covariance $\boldsymbol{S}_{01}$ associated with the observational and forcing uncertainties, given by

$$\boldsymbol{S}_{01} = \boldsymbol{S}_{r1} + \boldsymbol{S}_{f1}, \tag{4.11}$$

can be used to compute a generalized least squares estimate of $\delta\boldsymbol{a}$, say $\delta\boldsymbol{a}_1$. As in (4.8), the modeling errors and our a priori knowledge on the model parameters are incorporated, and normality assumed:

$$(4.12) \qquad \delta\boldsymbol{a}_1 = \left(< \overline{\boldsymbol{M}} >^T \boldsymbol{S}_1{}^{-1} < \overline{\boldsymbol{M}} > + \boldsymbol{C}_a^{-1}\right) < \overline{\boldsymbol{M}} >^T \boldsymbol{S}_1{}^{-1} < \overline{\delta\boldsymbol{q}} >,$$

with $\boldsymbol{S}_1 = \boldsymbol{S}_{01} + \boldsymbol{C}_T$. As in Blumenthal and Cane (1989), a simplified model was used for the modeling errors (normal noise with short correlation scales), lacking precise information.

The procedure is repeated by using $\delta\boldsymbol{a}_1$ to get an improved estimate $\boldsymbol{S}_{02}$, leading to the parameter correction $\delta\boldsymbol{a}_2$, and so on. If $\delta\boldsymbol{a}_0$ represents a reasonable first guess and the inverses in (4.12) are well conditioned, the procedure converges rapidly. The end result is based on an error model that represents most of the non-systematic data errors. On the other hand, the weighting is not affected by the systematic errors that recur every year: model deficiencies and systematic data biases must be dealt with explicitly.

By construction, the weighting in the least squares fit takes into account, at least approximately, the lack of statistical independence between $< \overline{\boldsymbol{M}} >$ and $< \overline{\delta\boldsymbol{q}} >$. However, ordinary and generalized least-squares estimators may not be consistent in the case of nonlinear coupling between model and data errors, and they can have a large variance in the case of multicollinearity. Alternative estimating procedures that give more consistent estimators have been suggested for nonlinear problems (Seber and Wild, 1989; Van Huffels and Vanderwalle, 1991), but they require that the data be transformed to a standard form where the errors are independent and have a similar error covariance matrix. Hence, they are hardly applicable to a case where the (dependent) covariance matrices have to be estimated iteratively.

Because of the limited sample, the error covariance matrices $\boldsymbol{S}_i$ at each iteration i are of strongly reduced rank and the inverse dominated by unreliable information. To circumvent the ill-conditioning, the dimension of the fields is drasticly reduced by working in the space defined by orthonormalizing the eight space-time vectors consisting of $< \overline{\delta\boldsymbol{q}} >$ and the seven column vectors of $< \overline{\boldsymbol{M}} >$. The iterative method is directly implemented in the 8-dimensional reduced space (see details in Sennéchael et al, 1994).

Convergence is reached in two or three iterations, with the largest changes occurring after the first iteration. Fig.6 shows the a priori and a posteriori values of the adjustable parameters with twice their standard deviation (an approximation to the 95% confidence interval). Two parameters undergo large changes and reach values which are largely outside their expected range: the upwelling efficiency γ, which strongly decreases, and the horizontal diffusion that remains positive, but not significantly so at

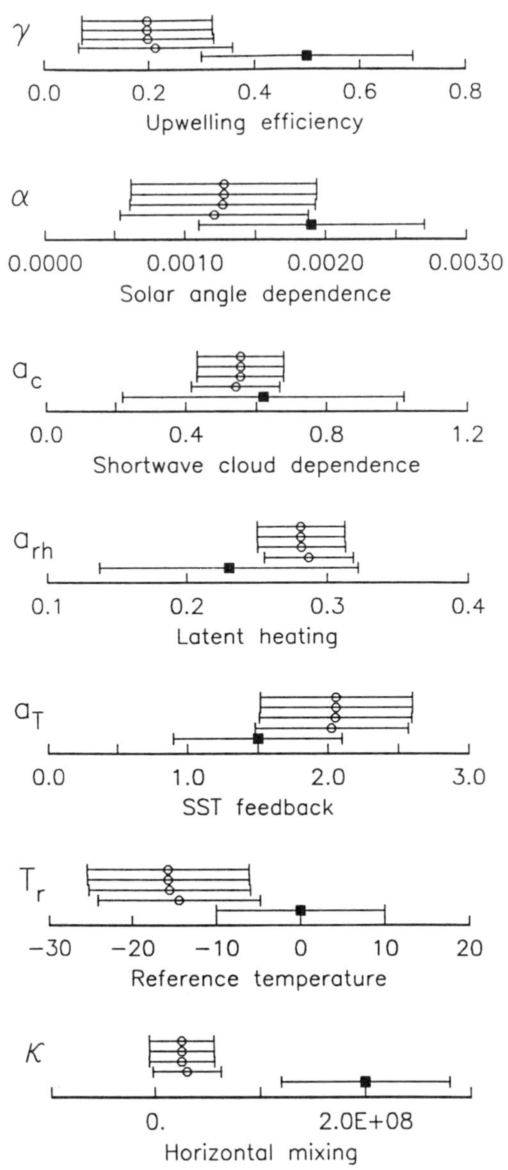

Figure 6. Changes in the parameters as a function of the number of iterations. The solid square is the initial choice and the open circles proceeding upward are the iterations. The error bars represent the 95% confidence intervals. (From Sennéchael et al., 1994).

the 5% level. Both parameters are well resolved by the data set and independently resolved, and their changes point to model limitations. A reason for the large changes is apparent in the flux patterns in Fig.5. In the upwelling zone off Africa, the model is too warm and cooling is needed in the first part of the year, but horizontal mixing heats the off-shore waters and it thus tends to be decreased by the fit. At the same time, cooling by upwelling takes place too far south and off-shore to provide the required cooling. In the Gulf of Guinea, horizontal mixing is again out of phase with the heat flux correction, while the upwelling flux is alternatively in (January, April, July) and out (October) of phase, and too far east at its maximum in July, so it also tends to be reduced. The other parameters are not independently resolved, and their changes are small or not statistically significant.

Figure 7a shows the heat flux correction after tuning. The space-time average has dropped from -61 to -8 Wm^{-2}, suggesting that the warm SST bias has been mostly corrected. However, the tuning is unable to do more, and heat flux corrections larger than 100 Wm^{-2} can still be seen. These are too large to be explainable by the assumed modeling errrors and the data uncertainties. In particular, the model does not work well near the African coast, because of the simplification of the coastal geometry and the assumption that the air temperature has equilibrated with the SST, and it poorly represents the equatorial upwelling.

The consistency of the inverse calculation can be tested from the minimum of the misfit function (4.7), given by

$$S\left(\delta \boldsymbol{a}\right) = < \overline{\delta \boldsymbol{q}} >^{T} \left(< \overline{\boldsymbol{M}} > \boldsymbol{C}_{a} < \overline{\boldsymbol{M}} >^{T} + \boldsymbol{S}_{n} \right)^{-1} < \overline{\delta \boldsymbol{q}} > \tag{4.13}$$

If all the errors have been correctly represented, (4.13) should be approximately distributed as Hotelling's T^2 with degrees of freedom given by the reduced dimension and the equivalent degrees of freedom of $\boldsymbol{S}_n$. Since the observational uncertainties are represented by an error model which is, by construction, consistent with the observations, the null hypothesis is that the modeling errors have been correctly assessed. As expected, the critical value of the test statistic is largely exceeded, and it can be concluded that the modeling errors have been underestimated. This testing procedure is analogous to that of section 3, but more general since it includes model tuning. A SST hindcast with the optimized parameters confirmed that the SST simulations had been improved by the parameter changes (Fig.7b). The model improvements are mostly limited to a strong decrease of the warm SST bias, although it still averages to $1.1°C$, possibly reflecting a small bias of the least squares estimation. A multivariate test as in section 3 confirmed that the SST model remained largely inconsistent with the

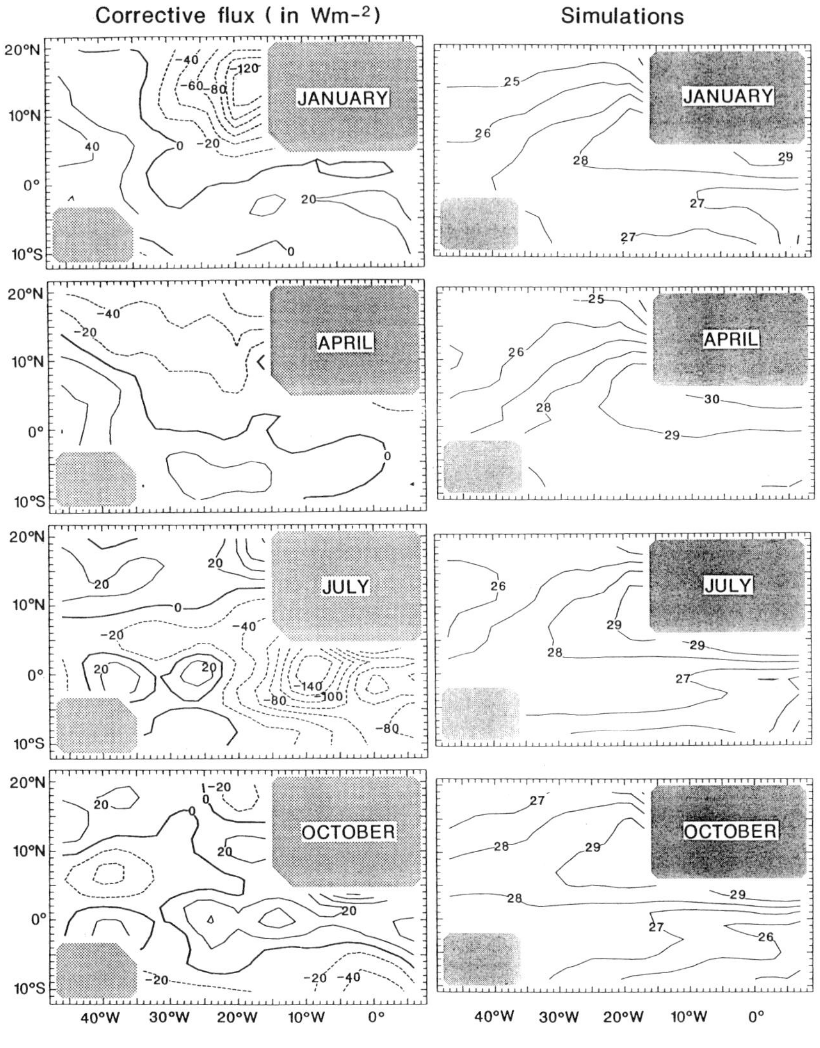

Figure 7. Left: Mean heat flux correction as in Fig. 5a, after optimization. Right: Mean SST predicted by the optimized multimode model, as in Fig. 4. (From Sennéchael et al., 1994).

5. Discussion

In this paper, we have reviewed two applications of multivariate statistical analysis to ocean modeling which both required a strong data compression to compensate for the limited observational sample and the large dimension of the oceanic fields.

The model testing procedure allows to compare simulations and observations, while taking into account the uncertainties in the atmospheric forcing and the large correlation scales of error fields. The procedure is straightforward, but requires a simulation of the effects of the forcing errors on the model response. Hence, several model runs with different, equally plausible forcing fields are needed. The test is very stringent, but the model-reality misfit estimated in the procedure can be easily used for model intercomparison. It can also be used for model tuning, although this may be cumbersome if there are many tunable parameters.

The adaptive tuning procedure is a more efficient way to tune and test a model, and it is general as long as the parameter dependence is linear and sufficient observations are available. However, because of the nonlinear nature of the inverse problem, the least squares solutions may be asymptotically biased, and alternative optimization procedures should be searched. Note that the method can be used iteratively to treat nonlinear parameter dependence, but it may not always converge in this case (Sennéchael, 1994); thus, tuning with nonlinear parameter dependence may still require the trial and error approach.

Model testing and adaptive tuning yield a consistent picture: modeling errors and biases are generally larger than assumed in tropical ocean models, and the model-reality discrepancies cannot be solely attributed to the effect of the observational and forcing uncertainties. To achieve an optimal model tuning, however, a good representation of the modeling errors would be needed, as a best fit should properly weight observations and simulations. A similar weighting is also needed for data assimilation, hence more effort should be devoted to better representing the modeling errors.

Acknowledgement. The work reviewed here is the result of fruitfull and enjoyable collaborations with Pascale Braconnot, Mark Cane, Christine Duchêne, and Nathalie Sennéchael, who are gratefully acknowledged. It has been carried out at the University Pierre et Marie Curie, the Lamont-Doherty Earth Observatory and the Massachusetts Institute of Technology, and mainly supported in France by grants from the P.N.E.D.C. and the Environment Program of the Commission of the European Communities, and in the USA by grants from National Oceanic and Atmospheric Administration and the Office of Naval Research. Thanks are due to Peter Müller for inviting the author to discuss the work at two Aha Huliko'a workshops, and to Hans von Storch for his careful reading of the manuscript.

References

[1] Allaart, M.A.F. and A. Kattenberg, 1990: A primitive equation model for the equatorial Pacific. Technical Report **TR-124**, KNMI.

[2] Blanke, B., and P. Delecluse, 1993: Low frequency variability of the tropical Atlantic ocean simulated by a general circulation model with mixed layer physics. *J. Phys. Oceanogr.*, **23**, 1363-1388.

[3] Blumenthal, M.B. and M.A. Cane, 1989: Accounting for parameter uncertainties in model verification: an illustration with tropical sea surface temperature. *J. Phys. Oceanogr.*, **19**, 815-830.

[4] Braconnot, P., and C. Frankignoul, 1993: Testing model simulations of the thermocline depth variability in the tropical Atlantic from 1982 through 1984. *J. Phys. Oceanogr.*, **23**, 626-647.

[5] Braconnot, P. and C. Frankignoul, 1994: On the ability of the LODYC GCM at simulating the thermocline depth variability in the equatorial Atlantic. *Climate Dynamics*, **9**, 221-234.

[6] Cane, M.A., 1984: Modeling sea level during El Niño. *J. Phys. Oceanogr.*, **14**, 1864-1874.

[7] Février, S., J. Sirven and C. Frankignoul, 1995: An intermediate resolution model of the seasonal variability of the tropical Atlantic. *The Global Atmosphere-Ocean System, to appear.*

[8] Flury, B.N., 1989: *Common Principal Components and Related Multivariate Models.* Wiley and Sons.

[9] Frankignoul, C., 1995: Statistical analysis of GCM output. In *Analysis of Climate Variability. Applications of Statistical Techniques.* H. von Storch and A. Navarra, Ed., Springer Verlag, 139-152.

[10] Frankignoul, C., C. Duchêne and M. Cane, 1989: A statistical approach to testing equatorial ocean models with observed data. *J. Phys. Oceanogr.*, **19**, 1191-1208.

[11] Frankignoul, C., S. Février, N. Sennéchael, J. Verbeek and P. Braconnot, 1995: An intercomparison between four tropical ocean models. Part 1: Thermocline variability. *Tellus*, **47A**, 351-364.

[12] Hasselmann, K., 1979: On the signal-to-noise problem in atmospheric response studies. In *Meteorology of the Tropical Oceans*, D.B. Shaw, Ed., *Roy. Meteor. Soc.*, 251-259.

[13] Latif, M., A. Sterl, E. Maier-Reimer, and M.M. Junge, 1993: Structure and predictability of the El Niño/ Southern Oscillation phenomenon in a coupled ocean-atmosphere general circulation model. *J. Clim.*, **6**, 700-708.

[14] Livezey, R.E., 1995: Field intercomparison. *In Analysis of Climate Variability. Applications of Statistical Techniques.* H. von Storch and A. Navarra, Ed., Springer Verlag, 159-175.

[15] Livezey, R.E. and W.Y. Chen, 1983: Statistical field significance and its determination by Monte Carlo techniques. *Mon. Wea. Rev.*, **11**, 46-59.

[16] Morrison, D.F., 1976: *Multivariate Statistical Methods.* McGraw-Hill.

[17] Stuart, A. and Ord, J.K., 1991: *Kendall's Advanced Theory of Statistics.* 5th edition. Edward Arnold.

[18] Reverdin, G., P. Delecluse, C. Levi, A. Morlière, and J. M. Verstraete, 1991: The near surface tropical Atlantic in 1982-1984. Results from a numerical simulation and data analysis. *Progress in Oceanogr.*, **27**, 273-340.

[19] Sausen, R., K. Barthels and K. Hasselmann, 1988: Coupled ocean-atmosphere models with flux correction. *Climate Dynamics*,**2**, 154-163.

[20] Seager, R., S.E. Zebiak and M.A. Cane, 1988: A model of the tropical Pacific sea surface temperature climatology. *J. Geophys. Res.*, **93**, 11,587-11,601.

[21] Seber, G.A.F., 1984: *Multivariate Observations.* Wiley and Sons.

[22] Seber, G.A.F. and C.J. Wild, 1989: *Nonlinear Regression.* Wiley and Sons.

[23] Sennéchael, N., C. Frankignoul and M.A. Cane, 1994: An adaptive procedure for tuning a sea surface temperature model. *J. Phys. Oceanogr.*, **24**, 2288-2305.

[24] Sennéchael, N., 1994: Optimisation d'un modèle de l'ocan atlantique tropical par méthode inverse adaptative. Thèse de doctorat, Université Pierre et Marie Curie, Paris.

[25] Servain, J., J. Picaut and A.J. Busalacchi, 1985: Interannual and seasonal variability of the tropical Atlantic ocean depicted by 16 years of sea surface temperature and wind stress. In *Coupled Ocean-atmosphere Models*, J.C.J. Nihoul, Ed., Elsevier, 211-237.

[26] Tarantola, A. 1987: *Inverse problem theory.* Elsevier.

[27] von Storch, H., 1982: A remark on Chervin-Schneider's algorithm to test significance. *J. Atmos. Sci.*, **39**, 187-189.

[28] Van Huffel, S. and J. Vandewalle, 1991: The Total Least Squares Problem. *Frontiers in Applied Mathematics*, **9**, SIAM, Philadelphia.

[29] Yao, Y., 1965: An approximate degrees of freedom solution to the multivariate Behrens-Fisher problem. *Biometrika*, **52**, 139-147.

Laboratoire d'Océanographie Dynamique et de Climatologie
Unité mixte de recherche CNRS-ORSTOM-UPMC
Université Pierre et Marie Curie
4 place Jussieu, 75005 Paris
FRANCE

cf@lodyc.jussieu.fr

Applications of stochastic particle models to oceanographic problems

Annalisa Griffa

Abstract

Three Markovian particle models are reviewed, providing a hierarchy of increasingly detailed descriptions of particle motion and dispersion. Model 1 assumes that the scales of turbulent motion are infinitesimal, and it is equivalent to the advection-diffusion equation. Model 2 introduces a finite scale T for the turbulent velocity,, and model 3 introduces an additional scale for the acceleration, $T_a < T$. The models are compared with oceanographic data from drifting buoys, which satisfactorily approximate the motion of ideal particles in mesoscale turbulent fields. Model 2 appears to provide a satisfactory description of the second order particle statistics in the upper ocean. Model 3 appears to be applicable to deep ocean data with some questions still remaining open. Some examples of analytical calculations of dispersion using the models are shown for some simple oceanographic flows. The results indicate that the introduction of finite scales of turbulence plays an important role not only at initial times, $t < T$, but also for dispersion at longer times if the mean flow is strongly dependent on space and time, so that the scales of the mean flow and of the turbulence are of the same order. In these situations, which are characteristics of important current systems in the ocean, the advection-diffusion equation is not accurate, and the use of stochastic models such as 2 and 3 is especially indicated. Two different classes of applications for the models are reviewed: "direct" applications, where the models are directly integrated to compute dispersion, and "inverse" applications where the models are used to extract information about the velocity field from the Lagrangian data. A discussion is also provided on future applications of the models to study more general classes of oceanic flows including coherent structures.

1. Introduction

The use of stochastic particle models has a long history in dispersion studies (e.g. Chandrasekhar, 1943). Here we focus on some examples of models which are useful in the study of the ocean. The specific nature of the oceanic processes is discussed, a review of recent progress is provided, and a number of open questions is pointed out, which can lead to new areas of research and applications.

The main difficulty in the study of the ocean is the infinite number of scales participating in the motion, ranging from the planetary scales of the general circulation (thousands of km) to the scales of molecular motion. Analytical and numerical treatments involving all these scales are not feasible, so that the oceanographic problems are usually approached isolating first the scales of interest (characterized by the mean flow $\mathbf{U}$) and deriving approximated equations for them. In order to obtain a closed formulation for the scales of the mean field, the smaller scales, (usually indicated as turbulence $\mathbf{u}$), have to be parameterized in terms of the mean field.

In the study of tracer dispersion, the simplest parameterization is obtained using the same "scale separation" arguments that are used for molecular diffusion, i.e. by assuming that the scales of the turbulence are infinitesimal compared with the scales of the mean field (Taylor, 1921). Under this assumption, the equation which describes the evolution of the mean tracer concentration C is the same as for molecular diffusion, i.e. it is the advection-diffusion equation (5) with the molecular diffusivity coefficient replaced by an "eddy-diffusivity" coefficient K. The eddy-diffusion coefficient is several orders of magnitude larger than the molecular diffusivity and it parameterizes the action of the turbulent vortices (eddies) on the mean field. In the following, we focus on applications valid for mesoscale dispersion, i.e. for dispersion due to eddies with scales of 50-100 km acting on the large scales of the general circulation.

As noticed by a number of authors (e.g. Davis, 1987; Holloway, 1989; Zambianchi and Griffa, 1994), the applicability of the advection-diffusion equation is often questionable for oceanographic problems, because in most flows there is no clear gap betweeen scales, so that the scale separation hypothesis does not strictly hold. Despite this problem, the advection-diffusion equation is widely used in oceanography, mainly because it is simple and straightforward to implement. Generalizations of the advection-diffusion equation (generalized "K-models"), where the scale separation hypothesis is relaxed, are available in the literature (e.g. Davis, 1987) , but they are not frequently used in practical applications because of the difficulties in their implementation.

In this paper an alternative class of models for dispersion is discussed, based on the use of stochastic equations for particle motion. These models combine the advantage of being generalizable to less stringent assumptions than the advection-diffusion equation, while maintaining a simple implementation. The models describe the motion of "single" particles, i.e. of ensembles of particles independently launched in different realizations of the turbulent flow. Since the particles can be thought of as belonging to a tracer, their concentration corresponds to the (normalized) ensemble average concentration C of the tracer itself (e.g. Csanady, 1980).

We remark that the particle models can also be used as a diagnostic tool for the interpretation of data from Lagrangian (i.e. current followers) instruments. These instruments, which have become increasingly popular in the last two decades, play a major role in providing oceanographic data at global scales. They consist of drifing buoys designed to follow the currents at the ocean surface or in the interior, reporting their position at discrete times either acoustically or via satellite (for a review on Lagrangian measurements the reader is referred to Davis, (1991a)). They provide, at least in the mesoscale range in which we are interested, a satisfactory approximation of the motion of ideal particles, so that they are perfectly suitable to be interpreted and explained in terms of particle models.

In this paper, three Markovian models are discussed in some depth, providing a hierarchy of increasingly detailed descriptions of particle motion. The first model (model 1) corresponds to the classic "random walk" model, and it assumes that the particle position $\mathbf{x}$ is a Markov variable. This model assumes that the scales of the turbulence velocity are infinitesimal and it is equivalent to the advection-diffusion description (5). Model 2, which is sometimes referred to as a "random flight" model (van Dop et al., 1985), assumes that the particle position $\mathbf{x}$ and the turbulent velocity $\mathbf{u}$ are jointly Markovian. Finally, model 3 (Sawford, 1991) assumes that $\mathbf{x}$, $\mathbf{u}$ and the turbulent acceleration $\mathbf{a}$ are jointly Markovian. Physically, models 2 and 3 correspond to the introduction of a finite scale for the velocity and the acceleration, respectively. Extensive literature is available for these models, including a number of meterological applications focused on dispersion at relatively small scales (discharges in local regions and valleys, e.g. Thomson, 1986). Here we focus on those aspects of the models that are relevant for oceanographic applications and in particular for applications at the scales of the general circulation and of the mesoscale. We stress the comparison of the models with the data, and illustrate typical oceanographic scales and situations where the use of models 2 and 3 is especially indicated.

In Section 2, a review of the general characteristics of the three models is given, while the details for Model 1, 2, 3 are given in Section 3, 4, 5 respectively. A comparison between the performance of the advection-diffusion equation and model 2 is shown in Section 6, whereas in Section 7 a discussion is presented of the practical applications of the models. A summary and a general discussion are provided in Section 8.

2. General characteristics of the models 1, 2, 3.

The three models 1, 2 and 3 describe the motion of single particles. Each particle is independently launched in a different realization of **u**, and the statistics are computed by averaging over ensembles of particles launched from the same position in space. In these models, since the particles are independent, **u** can be considered as a purely time-dependent process, which represents the turbulent velocity encountered by the particle during its motion. This allows for a substantial simplification compared with the space-dependent representation of **u** necessary to study two-particles or higher order statistics.

The three models are Markovian, i.e. they describe processes whose conditional probability density at time t_n depends only on the process value at the earlier time t_{n-1}. The Markov property is valid for a wide class of physical processes. It represents the generalization to probabilistic systems of the deterministic property of unique dependence on the initial conditions, which is characteristic of all the systems obeying first-order differential equations. Considered in this light, the assumption of Markovian turbulent velocity appears motivated and natural. The real velocity field, in fact, obeys the Navier Stokes equations, which are of first order in time. As a consequence, the time evolution of the velocity is uniquely determined by the initial conditions, and in the presence of a stochastic forcing or pertubation, it is a Markovian process.

The models 1, 2 and 3 are simple examples of a general class of stochastic models called generalized Langevin equations (e.g. Risken, 1989), which can be nonlinear and have arbitrary dimensions N:

$$ds_i = h_i(\mathrm{s}, t)dt + g_{i,j}(\mathrm{s}, t)d\mu_j \tag{1}$$

where $i = 1, N$, $d\mu_i$ is a random increment, and h_i and $g_{i,j}$ are continuous functions. The behaviour of stochastic processes such as s is characterized by their probability density function P. The evolution equation for the probability density P of a Markovian process as (1) is called the Fokker-Planck equation (or forward Kolmogorov equation). The Fokker-Plank equation, which has the form of a partial differential equation, can be derived from the stochastic equations (1) using a procedure which is conceptually quite difficult, but by now well known and standadized (e.g. Risken, 1989). The Fokker-Plank equations for our specific models will be introduced and discussed in the following.

We terminate this Section remarking that the main discussion of the three models in Sections 3, 4 and 5 will be done using some simplifyng assumptions. More general forms of the models valid for more general conditions are available (a discussion is included in Section 4), but the simplyfing assumptions allow for a clear presentation of the fundamental properties of the models and of their relationships, while maintainig a general physical validity. The assumptions can be summarized as follows:

1) the velocity field is 2-dimensional and on an infinite domain. The assumption of 2-dimensional motion is appropriate for mesoscale phenomena (e.g. Davis, 1991b), since the particles move mainly on isodensity surfaces (e.g. on the ocean surface or on interior isopycnals). The simplification of an infinite domain is acceptable for the description of the ocean interior, away from the boundaries,

2) the turbulent velocity field is homogeneous in space and stationary in time. The condition of homogeneity can be verified, at least partially, in selected regions of the ocean, typically in the subtropical interior regions, away from the boundaries and from the equatorial regions. The stationarity assumption is more difficult to satisfy, because the presence of low frequency variability makes the ocean circulation an essentially red spectrum process. This is a problem for any statistical description or analysis concerned with the description of average quantities or "typical" aspects of the ocean. This difficulty is usually overcomed acknowledging that the mean quantities are in reality only "averages over some finite time, representing a particular ocean climate" (Davis, 1991b). Notice that we allow inhomogeneity and nonstationarity to occur in the mean flow $\mathbf{U}(\mathbf{x}, \mathbf{t})$. This a suitable choice, since the main sources of inhomogeneity and nonstationarity in the ocean are indeed given by the mean flow.

3) the two components of the velocity are independent. This condition is the least essential of the three, and it is introduced solely in order to simplify the notation in the following. Since each component is independent, in fact, the models can be written in 1-dimensional form, for each component separetely.

3. Model 1. Markovian x

Under the assumptions stated in Section 2, the equations that describe the particle motion for Model 1 can be written, for a single component, in incremental form as

$$dx = U\,dt + \hat{dx} \tag{2}$$
$$d\hat{x} = (K)^{1/2} dw \tag{3}$$

with

$$K = \sigma T, \tag{4}$$

and $x(0) = 0$. In the above expressions, for the selected component, dx is the total displacement of the particle during the time dt; $d\hat{x}$ is the displacement due only to the turbulent velocity; $U(\mathbf{x}, t)$ is the mean flow; K is the diffusion coefficient; σ is the turbulent velocity variance; T is the turbulent time scale; $dw(t)$ is a random increment from a normal distribution with zero mean and second order moment $< dw \cdot dw >= 2dt$. Notice that the turbulent time scale T in (4) must be $T = dt/2$ in order to have the correct turbulent velocity variance from (3), $< (d\hat{x}/dt)^2 >= \sigma$.

Physically, the equations (2)-(3) describe the displacement of a particle as resulting from two contributions. The first one is due to the mean flow and it is represented deterministically by the increment Udt, the second one, $d\hat{x}$, is due to the turbulence and it is represented as a stochastic process uncorrelated from one time step to the next. This means that the particle moving through the fluid receives at each time step a random impulse due to the action of the incoerent turbulent motions, and it "loses memory" of its previous turbulent momentum.

Another way of saying this is that the model describes a turbulent motion whose typical scale T is infinitesimal (of order dt) with respect to the other time scales of the problem. This means that T is small with respect to the scales of U, and it is also small with respect to the actual time t. In other words, the initial transient, $t < T$ is not described by the model (2)-(3) which applies only for asymptotic times $t >> T$.

The Fokker-Planck equation associated with model 1, in 2-dimensions and for an incompressible flow ($\nabla \cdot \mathbf{U} = 0$) is

$$\partial P/\partial t = -\mathbf{U} \cdot \nabla P + \nabla(\mathbf{K} \nabla P) \tag{5}$$

with $P(x_1, x_2, 0) = \delta(x_1)\delta(x_2)$. In (5), $P(x_1, x_2, t)$ is the probability density function that a particle launched at $(0,0)$ at time 0 is found at (x_1, x_2) at time t, and $\mathbf{K}$ is a diagonal tensor with non zero elements given by (4).

Notice that in the case of a tracer released at time $t = 0$ from a point source located at $\mathbf{x} = (0,0)$, the average tracer concentration C at $(\mathbf{x}, t)$ is connected to P in the following way:

$$C(\mathbf{x}, t) = QP(\mathbf{x}, t)$$

where Q is the total tracer mass released (see, e.g. Csanady, 1980). As a consequence, (5) is also the advection-diffusion equation for the average concentration C, with the eddy-diffusion coefficient K defined by (4). To have an idea of the order of magnitude of K for mesoscale motion, consider that T is of the order of 2-10 days, and σ is of the order of $10 - 10^2\ cm^2/sec^2$, so that K is of the order of $10^6, 10^7 cm^2/sec$.

4. Model 2. Joint Markovian x and u

The incremental equations for particle motion for Model 2 under the same assumptions as before are

$$dx = (U + u)dt \tag{6}$$
$$du = -(1/T)udt + (\hat{K})^{1/2} d\hat{w} \tag{7}$$

with

$$\hat{K} = \sigma/T = K/T^2 \tag{8}$$

and $x(0) = 0, u(0) = \hat{u}$, where $\hat{u}$ are drawn from a Gaussian distribution with mean zero and variance σ. In (7) and (8), $\hat{K}$ is the diffusion coefficient, and $d\hat{w}$ is a random increment with the same statistical characteristics as dw in (3).

Model 2 differs from Model 1 in the treatment of the turbulent velocity, which is not assumed to be uncorrelated from one time step to another. Rather, the turbulent velocity obeys the classical Langevin equation (7), stating that at each time step the particle loses only a fraction of its momentum, $U(dt/T)$, and in turn receives a random impulse $d\hat{w}$. As a consequence, the particle "conserves the memory" of its initial turbulent velocity during a finite time of order T. The autocorrelation function of u decays exponentially (e.g. Risken, 1989),

$$R(\tau) = \frac{< u(t)u(t+\tau) >}{\sigma} = e^{-\tau/T} \tag{9}$$

From (9) the time scale of the memory of the turbulent velocity T turns out to be also the integral time scale. T is arbitrary in the model (6)-(7), and it can be of the same order or larger than the other time scales in the problem, i.e. of the scales of U and of the actual time t.

Notice that, even though model 2 introduces the scale for the velocity which is absent in model 1, the acceleration is still assumed to have an infinitesimally small scale, as it is shown by (7) where the acceleration has a discontinuity at each time step dt. This limitation is physically more acceptable than the limitation of small velocity scales, since in real flows the time scales over which the acceleration is correlated are usually much shorter than the velocity time scales. In the following, a comparison with data will be performed to verify for which flows in the ocean the assumptions of infinitesimal acceleration scales are indeed acceptable.

The Fokker-Planck equation associated with model 2, in 2-dimensions is (van Dop et al., 1985):

$$\partial P/\partial t = -(\mathbf{U}+\mathbf{u}) \cdot \nabla P + \nabla_u(\mathbf{u}P/T) + \nabla_u(\hat{\mathbf{K}}\nabla_u P) \tag{10}$$

with $P(x_1, x_2, u_1, u_2, 0) = M\delta(x_1)\,\delta(x_2)\,e^{-\frac{u_1{}^2}{\sigma_1}}e^{-\frac{u_2{}^2}{\sigma_2}}$ where M is a normalization factor. In (10) ∇_u represents the gradient with respect to $\mathbf{u}$ and $\hat{\mathbf{K}}$ is a diagonal tensor with nonzero elements given by (8).

Notice that the probability density function $P(x_1, x_2, u_1, u_2, t)$ is now a function not only of the space variables $\mathbf{x}$ but also of the turbulent velocities $\mathbf{u}$. This is because the particle "remembers" the value of $\mathbf{u}$ for a finite time, so that the motion at each instant depends not only on the previous position but also on the turbulent velocity. Equation (10) can effectively describe also the transient processes occurring at time $t < T$, unlike equation (5) which is valid only for $t > T$.

4.1 Comparison of model 2 with numerical and oceanographic data. Model 2 describes a process with a turbulent velocity characterized by the exponential autocorrelation (9) and by an associated spectrum with normalized density

$$S(\omega) = (\frac{1}{\pi})\frac{1/T}{1/T^2 + \omega^2}. \tag{11}$$

The spectrum (11) is approximately white for frequencies lower than the cutoff frequency, $1/T$, and it decreases as ω^{-2} for higher frequencies.

A simple way to test the applicability of model 2 (at least for second order statistics) for real flows consists in verifying whether or not the autocorrelation (9) and the spectrum (11) are present in simulated and experimental data of turbulent motion.

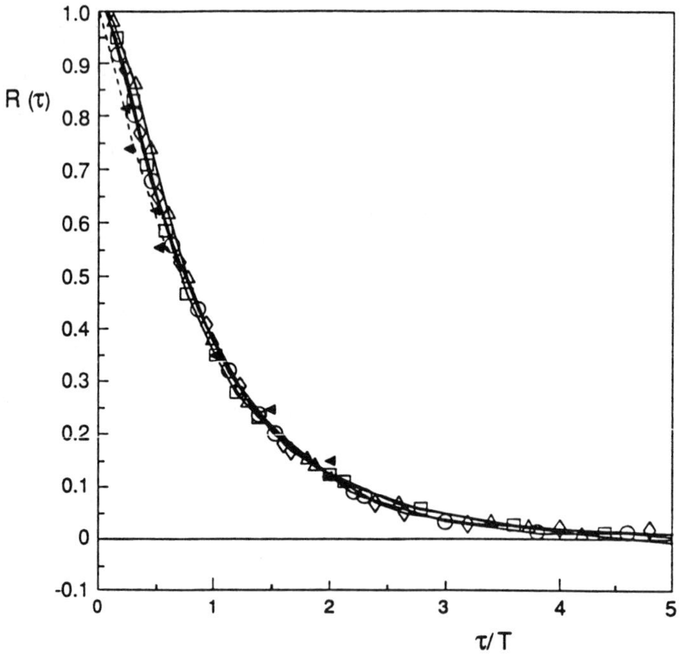

Figure 1. Velocity autocorrelations from particle simulations in isotropic turbulence (from Yeung and Pope, 1989). The dashed line is the exponential autocorrelation of model 2; the open symbols and lines are the Direct Numerical Simulations of Yeung and Pope (1989); the full symbols are from the experiments of Sato and Yamamoto (1987).

Tests in this direction have been done in the framework of 3-dimensional isotropic turbulence, considering the statistics of ensembles of particle trajectories in simulated velocity fields (Sato and Yamamoto, 1987; Yeung and Pope, 1989). For high Reynolds number, the results show that the autocorrelation is well represented by the exponential shape (9), except at very short time lags where the scale of the acceleration cannot be neglected (Fig. 1). For these scales, close to the origin $\tau \approx 0$, the autocorrelation computed from the simulated trajectories is smoother than the exponential curve predicted by model 2. For all the other lags, the exponential approximation appears to be very satisfactory. Positive results have also been found in numerical simulation (Verron and Nguyen, 1989; Davis, 1991b) reproducing mesoscale turbulent flows in the upper ocean. The exponential shape is found to represent very accurately a wide class of flows of this type, characterized by various wavenumber spectra.

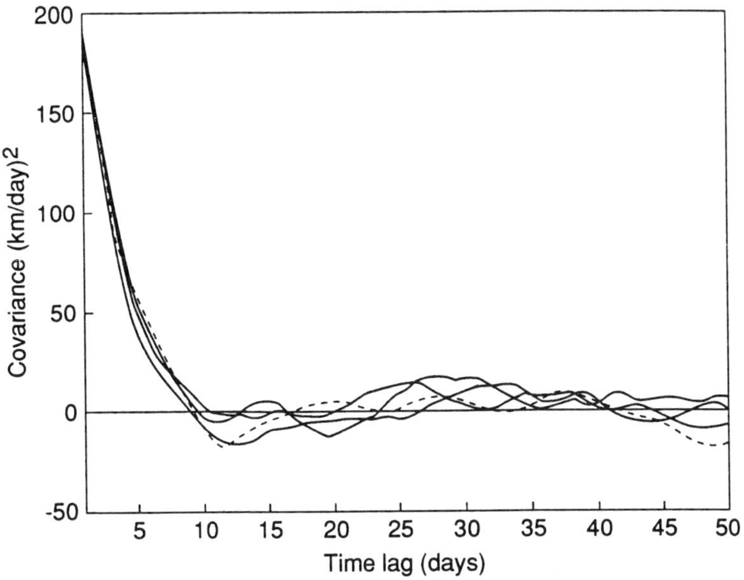

Figure 2. Velocity covariance from drifting buoys in the Brazil Malvinas Current (dashed line) and from particle simulations using model 2 (continuous lines) (from Griffa et al., 1995).

Concerning the experimental oceanographic data, a good agreement is found with model 2 for upper ocean mesoscale flows in approximately homogeneous regions (e.g. Krauss and Boning, 1987; Colin de Verdiere, 1983). Most data of this type, in fact, are characterized by spectra with a ω^{-2} slope at frequencies higher than a cutoff value, and the corresponding autocorrelations have a shape qualitatively similar to the exponential curve. A closer inspection of the autocorrelations shows some additional interesting features. First of all, the autocorrelations computed from the oceanographic data usually do not deviate significantly from the exponential shape at small time lags (unlike the simulated isotropic turbulence data) (Fig. 2). This can be understood by considering that for these flows the scale of the velocity T is of the order of 2-10 days, whereas the scale of the acceleration is of the order of one day or less. Motions at scales less than one day are usually filtered out and neglected in mesoscale studies, because they are characterized by different dynamics, such as inertial and tidal oscillations. In the mesoscale range, then, the scales of the acceleration are not considered and the exponential approximation for the autocorrelation holds, even close to the origin. At longer time lags ($\tau > T$), on the other hand, the experimental autocorrelations usually show some deviations from exponential, characterized by zero crossings, negative lobs and oscillations (Fig. 2). An important question is whether these deviations are significant (indicating that the dynamics are indeed different from the description of model 2), or whether they are simply due to the relatively small number of data points available that do not allow the proper resolution of the longer time lags.

A first step in the direction of addressing this question in a quantitative way has been taken by Griffa et al. (1995, GOPR), and applied to the analysis of a set of surface drifting buoys in the Brazil Malvinas current. GOPR have tested whether or not the measured data can be considered as a specific realization of model 2, characterized by a given length and a sampling interval. Their procedure can be summarized in the following way. First they use the data to estimate the model parameters σ and T (with a method that is reviewed in Section 7.2), and then they integrate the model forward, using the estimated values of σ and T in (6)-(8). A set of realizations of simulated velocities are generated, with the same number of measurements and the same sampling interval as for the data. The data and the simulations are then compared by plotting together the autocorrelation functions (examples for three realizations are shown in Fig. 2) The results strongly suggest that the data are indeed consistent with the model, since data and simulations appear to be essentially indistinguishable, with the data correlation falling right into the envelope of the simulated ones. Even though this test is still qualitative, it certainly indicates that the model is a reasonable and valid starting point. Similar results have been obtained also with data in the California Current (Zambianchi, private communication) and in the Equatorial Pacific. An interesting future development can be envisioned involving more accurate and quantitative statistical tests of the model along this line.

All the results summarized so far indicate that model 2 is well suited to describe mesoscale particle motion in the upper ocean. An important remark, though, is that model 2 does not appear equally suited for deep

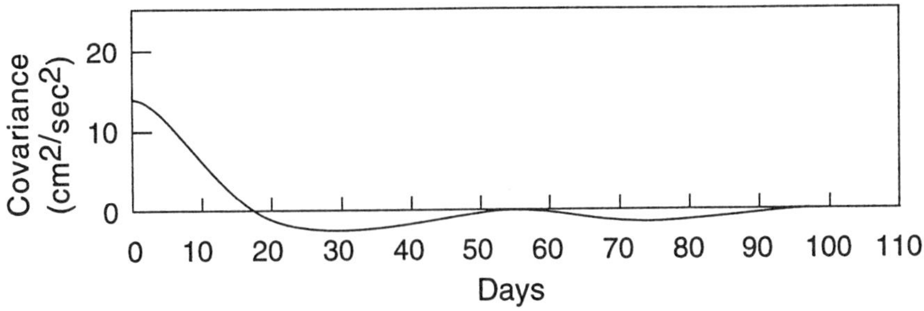

Figure 3. Velocity covariance from particle simulations in the lower layer of a 2 layer quasi-geostrophic model of the ocean circulation (from Verron and Nguyen, 1989).

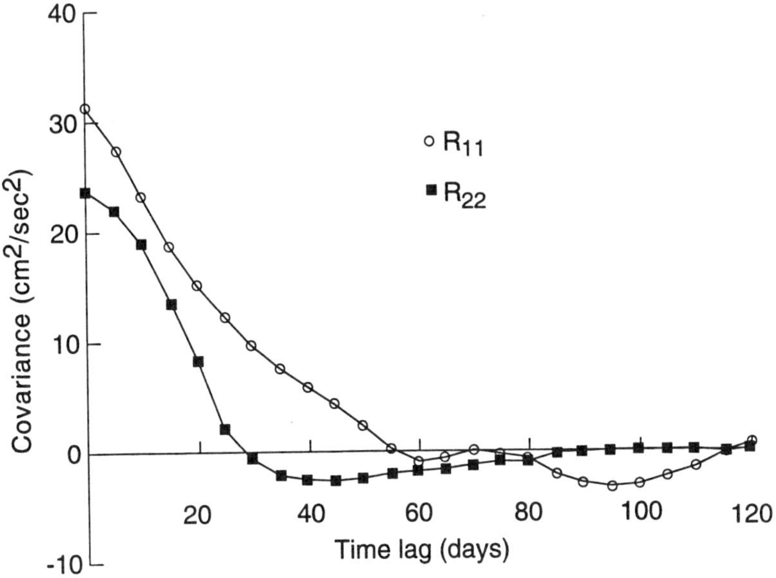

Figure 4. Velocity autocorrelation from subsurface floats in the Western Atlantic at 700 mt. R11 indicate the east-west component and R22 the north-south component (from Riser and Rosby, 1983).

ocean flows (see Fig. 3, Fig. 4). Deep flows (e.g. Riser and Rossby, 1983), in fact, are characterized by much longer time scales of evolution (T is of the order of 10 days or more), so that the scale of the acceleration cannot generally be neglected. We will come back on this point in Section 5. In addition to this, we also notice that model 2 is not appropriate when the motion is dominated by strong coherent structures or waves. As an examples, if the particles move in a ring with a very long life, the model is not suited to describe them. This can be easily understood considering that the autocorrelation in coherent structures is characterized by more complex and persistent patterns than the exponential decay. Also this point will be discussed in more detail in the following.

4.2 Generalization of model 2 to inhomogeneous and nonstationary turbulence. The results presented so far have been obtained under the assumption of homogeneous and stationary turbulence. As discussed in Section 2, this assumption is acceptable in midlatitude regions and in the interior of the ocean. When considering regions close to the western boundary currents and their extensions or close to the equatorial currents, this assumption breaks down. The intense shears and the associated instabilities, in fact, generate intensifications of the turbulent activity in well defined geographical regions. Also, when considering coastal regions, the dependence of the turbulent flows on the distance from the coast plays an important role and it cannot be neglected. For practical applications which include these type of flows, then, it appears very important to relax the assumption and to study the stochastic models in the presence of inhomogeneous turbulence.

The generalization of stochastic particle models such as model 1 and 2 to the case of inhomogenous and nonstationary flows is not completely straightforward, because the basic Langevin equation is incorrect in presence of inhomogeneity and nonstationarity. When the turbulent parameters are space and time dependent, $\sigma(\mathbf{x}, \mathbf{t})$ and $T(\mathbf{x}, \mathbf{t})$, eq. (2)-(3) and (6)-(7) predict an incorrect particle distribution, characterized by unphysically high concentrations in regions of lower variance (e.g. van Dop et al., 1985). The problem of adequately correcting the stochastic models has been addressed by several authors in the past decade (e.g. Legg and Raupach, 1982; Thomson, 1984). The main motivation for these type of studies has been provided by atmospherical applications, such as studies of dispersion from point sources close to the ground, where the spatial dependence of the turbulent parameters in the boundary layer is very important. The studies show that model 1 is fundamentaly unable to produce the correct disributions in presence of inhomogeneity, whereas model 2 can be satisfactorily modified. Several modified forms of model 2 have been proposed in the literature, and a certain number of criteria have been indicated in order to identify the "good" models (e.g. Janicke, 1983; Durbin, 1984). The relationships between the various models and criteria have been finalized by Thomson (1987), with the introduction of a generalized "well mixed" criterium, which unify the previous results and guarantees that the particles have the correct distribution.

The simplest form of model 2 for inhomogeneous and nonstationary turbulence obeying the "well mixed" condition, and having a Gaussian distribution for the turbulent velocity, is a direct generalization of (6)-(7), and can be written as

$$dx = (U + u)dt \tag{12}$$

$$du = -(1/T(x,t))udt + (\frac{\partial \sigma(x,t)}{\partial x})dt + (\hat{K})^{1/2} d\hat{w} \tag{13}$$

with $\hat{K}$ and $d\hat{w}$ defined as in (6-8).

Equation (13) differs from the homogeneous model (7) by the term $\frac{\partial \sigma}{\partial x}dt$ on the r.h.s.. This term describes a mean acceleration acting on the particles and directed toward the regions of larger variance. This acceleration counterbalances the tendency of the particles to spend more time in the lower variance regions. As a result, the particle distribution remains well mixed, with no formation of unphysical concentrations in the lower variance regions.

We conclude remarking that, in order to correctly use (12)-(13) in practical applications, the time and space dependence of σ and T has to be known with accuracy. In mesoscale oceanographic applications, the geographic distribution of σ and T is approximately known in some regions of the ocean, but more information is needed in order to have a reliable and detailed description. Concerning smaller scale, e.g. coastal applications, not much is known on the characteristics of the coastal boundary layer. Specific experiments focused on these aspects are presently underway (Olson, private communication) and hopefully they will provide sufficient insights for a correct application of the model.

5. Model 3. Joint Markovian x, u and a

The equations for model 3, written in incremental form and under the assumptions in Section 2, are

$$dx = (U + u)dt \tag{14}$$

$$du = adt \tag{15}$$

$$da = -(1 + \frac{T_a}{T}a)\frac{dt}{T_a} - \frac{u}{T}\frac{dt}{T_a} + (K^*)^{1/2} dw^* \tag{16}$$

with

$$K^* = \frac{1}{T_a}(\sigma_a(1 + \frac{T_a}{T})), \tag{17}$$

$$\sigma_a = \frac{\sigma}{T_a T}$$

and $x(0) = 0, u(0) = \hat{u}$, $a(0) = \hat{a}$ where $\hat{u}$ and $\hat{a}$ are drawn from a Gaussian distribution with mean zero and variance σ and σ_a respectively. In (14-17), K^* is the diffusion coefficient, σ_a is the acceleration variance, and dw^* is a random increment with the same statistical characteristics as dw in (3) and $d\hat{w}$ in (7).

Model 3 assumes that the turbulent acceleration a obeys the first order autoregressive equation (16), and it is characterized by the finite scale T_a. The model has then two distinct time scales: T which characterizes the velocity, and T_a which characterizes the acceleration. The quantity da/dt, i.e. the time rate of change of the acceleration, is instead considered infinitesimal in the model. Since the changes in the acceleration are characterized by shorter scales than the acceleration itself, model 3 is expected to be more accurate than model 2.

In order to understand the basic characteristics of model 3, it is useful to consider the form of the autocorrelation for the velocity (e.g. Pope, 1994)

$$R(\tau) = (e^{-\tau/T} - \frac{T_a}{T}e^{-\tau/T_a})/(1 - \frac{T_a}{T}) \tag{18}$$

For simplicity, we discuss the case where there is a complete separation between the scale of the velocity T and the scale of the acceleration T_a, $T_a << T$. For time lags $\tau >> T_a$, R reduces to $R(\tau) \approx e^{-\tau/T}$, as (9) for model 2. This corresponds to a spectrum having an ω^{-2} slope for $\omega T_a << 1$. For time lags of the order of T_a, instead, R has a different shape than in model 2 and is characterized by a quadratic behaviour at $\tau \approx 0$. This is the correct behaviour, expected in a real flow. The autocorrelation for the acceleration, R_a, has an approximately exponential shape for small lags, $R_a(\tau) \approx e^{-\tau/T_a}$, and correspondinly, the velocity spectrum has a ω^{-4} slope at high frequencies, $\omega T_a >> 1$

Concerning the physical interpretation of model 3 and in particular of the acceleration time scale T_a, we note that Model 3 has been first introduced (Sawford, 1991) in the framework of isotropic turbulence to correct the small scale behaviour of model 2 in the description of flows at high but finite Reynolds number. In the isotropic turbulence context, the velocity time scale T is the scale of the energy containing eddies, and the acceleration scale T_a is determined by the dissipation. In the mesoscale oceanographic context we are interested in, the molecular dissipation range is too removed to be able to play a direct role, but eddy dissipation by internal waves and other "subgrid-scales" processes might be a posible candidate. If model 3 has to be applied to oceanographic problems, then, the interpretation of the acceleration time scale T_a has to be rethought and redefined. It might help to consider that the system of equations (15)-(16) in model 3 is equivalent to a Langevin equation for u with an exponentially correlated noise instead of white noise (Krasnoff and Peskin, 1971; Sawford, 1991). The acceleration time scale, then, could be physically dependent on the nature of the forcing represented by the correlated noise.

As a last remark on model 3, we notice that even though the model appears to be in principle generalizable also to inhomogeneous turbulence problems, applications of this type have not been done yet, at least to the author's knowledge.

5.1 Comparison of model 3 with numerical and oceanographic data Model 3 has been tested with results of direct numerical simulations of 3-dimensional isotropic turbulence at high but finite Reynolds number (Sawford, 1991). The model appears to represent very well the second-order statistics of the simulated particles. There are of course some differences between the model and the simulations, such as for instance the form of the acceleration autocorrelation at the origin, which is not analytical in the model, but altogether the ability of the model to describe the second order Lagrangian statistics is definitely impressive.

Regarding oceanographic simulations and data, a close comparison with model 3 has not yet been performed, so that only a preliminary and qualitative discussion can be done. In the following, we focus on oceanographic flows in the deep ocean which, as noted in Section 4.1, have long acceleration time scales and which are not correctly represented by model 2 (see Fig. 3, Fig. 4). Model 3 is a possible candidate to describe them. At first inspection, Model 3 seems suitable to represent the simulated and measured velocity autocorrelations at small time lags, $\tau \approx 0$, since it is characterized by the correct quadratic behaviour. The interesting (and hard) question to address, though, is whether or not the model can adequately describe the oceanographic autocorrelations at longer times, $\tau > T$. The data and the simulations of the deep flows, in fact, show pronounced oscillations and negative lobes at $\tau > T$, which are not present in the model autocorrelation (18) characterized by the exponential decay. As already stated in Section 4.1 for the upper ocean flows, the question is whether or not the oscillations are significant, given the finite number of data available. With respect to the upper ocean flows, the lobes and the oscillations of the deep flows seem more pronounced, but on the other hand the time scales T are much longer so that the resolution is harder to acheive from the data. Only a direct statistical comparison of the model and the data will enable one to answer this question.

The question of significance of the oscillations in the autocorrelation is conceptually very important. If they are found to be significant, in fact, this is a clear indication that the deep velocity field is dominated by coherent structures or waves. If instead the oscillations are not significant, it is justified to hypothize that the deep ocean has a similar, random structure as the upper ocean, except that the dominant time scales are much longer. In both cases, the results will open further questions which are of fundamental interest in oceanography, such as what is the link between time scales and generation mechanisms, and what is the persistence and the relevance of coherent structures in the ocean.

6. Dispersion estimates from the models. Accuracy of the eddy-diffusion parameterization.

The advection-diffusion equation (5) with the eddy-diffusion parameterization (4) is commonly used in oceanography, even in situations where the basic assumption of infinitesimal scales of turbulence is questionable, and the exact limits of its applicability have been the object of many debates in the literature (e.g. Holloway, 1989). A simple way to quantitatively define the accuracy of the advection-diffusion parameterization is to compare estimates of dispersion obtained with the advection-diffusion equation (5) (or equivalently with model 1) with estimates obtained using the stochastic models 2, 3 which are not based on the assumption of infinitesimal turbulent scales. Model 2 and 3 are not completely realistic, but, as shown in Sections 4 and 5, they describe satisfactorily the second order statistics of a vast class of flows. For these flows, the results of the models can be used as a valid approximation of the real turbulent dispersion.

In the following we focus mainly on the comparison between the results of model 1 and 2 for some simple flows at oceanographic scales. This comparison has been analyzed in a recent paper by Zambianchi and Griffa (1994, ZG in the following), and it will be summarized in the following. The results apply to mesoscale turbulent flows in the upper ocean, where model 2 has been shown to provide an accurate description. When considering flows with longer acceleration scales, the dispersion estimates should be corrected, for instance using model 3, and this could cause modifications to the presented results (Sawford, 1991).

ZG compute analytically the first two moments of particle displacements, i.e. the mean displacement $< x >$ and the "dispersion" $S =< (x- < x >)^2 >$, for the two models 1 and 2, using the method of moments applied on the Fokker-Planck equations (5) (the advection-diffusion equation) and (10). In all the cases considered by ZG, the mean displacement $< x >$ is the same in both models, whereas the dispersion S is different. In order to quantify this difference, ZG compute the normalized difference

$$\Delta = \frac{| S_1 - S_2 |}{| S_1 |} \tag{19}$$

where the subscripts 1 and 2 indicate the model which is used to compute the dispersion. Δ is used as a measure of the action of finite scales of turbulence, and therefore of the accuracy of the advection-diffusion equation.

Three simple and idealized mean flows $\mathbf{U}$ are considered by ZG: a constant mean flow ($\mathbf{U} = \mathbf{C}$), a linear shear ($\mathbf{U} = \mathbf{C} + \mathbf{Bx}$), and a linear shear oscillating in time ($\mathbf{U} = (\mathbf{C} + \mathbf{Bx})\mathbf{sin}\omega_\mathbf{s}\mathbf{t}$). In all the cases, the turbulent field $\mathbf{u}$ superimposed to the mean shear is considered homogeneous and stationary. The first case, i.e. the flow with constant mean, corresponds to the classic and well known case of purely homogeneous turbulence (e.g. Taylor, 1921). We briefly review it first, because it provides a good basis for the understanding of more complex situations.

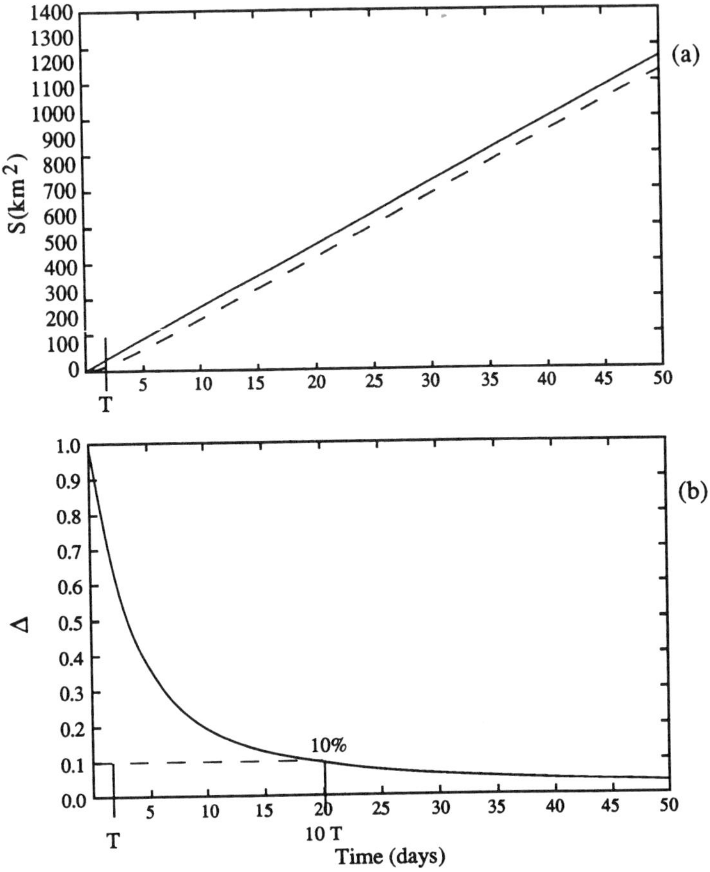

Figure 5. a) Dispersion versus time for homogeneous turbulence with constant mean flow. The solid line represents the result from model 1, S_1; the dashed line represent the result from model 2, S_2. b) Relative difference between the results from the two models, $\Delta = \frac{|S_1 - S_2|}{|S_1|}$ (from Zambianchi and Griffa, 1994).

The dispersion S computed for $\mathbf{U} = \mathbf{C}$ is, (in one direction),

$$S_1 \quad = \quad 2Kt \tag{20}$$

for model 1, and

$$S_2 \quad = \quad 2Kt - 2KT(1 - e^{-\gamma t}) \tag{21}$$

for model 2, where $\gamma = 1/T$ is the inverse of the turbulent velocity time scale. As expected, the two models differ in a transient term which is not present in model 1, and it has an exponential form in model 2. Expanding the exponential in (21) for $t << T$ we obtain

$$S_2 \sim \sigma t^2 \tag{22}$$

which is the classical result of Taylor (1921) for initial time dispersion, with the quadratic behaviour due to the initially high correlation of the turbulent velocity u. Comparing the linear expression (20), rewritten as $2\sigma Tt$, with the quadratic expression (22), it is clear that model 1 overestimates the dispersion for initial times $t < T$. This behaviour is shown in Fig. 5a. The initial overestimate of model 1 influences the dispersion also for $t > T$ producing an offset $2KT$ between the two curves S_1 and S_2 (Fig. 5a). The relative importance of this offset is illustrated by the quantity Δ (19) at the limit $\lim_{t\to\infty}$

$$\Delta_\infty = \frac{2KT}{2Kt} = \frac{T}{t}, \tag{23}$$

which shows how the results converge as t goes to infinity, as expected (see Fig. 5b). Quantitatively, the overestimate of model 1 is less than 10% for $t > 10T$, which for oceanographic values of T, $T \simeq 2-10$ days, corresponds to $t > 20 - 100$ days.

The previous results suggest that the use of model 1, and equivalently the use of the advection-diffusion equation (5), are appropriate in homogeneous situations in the ocean for dispersion studies at scales of one month or more (i.e. at the scales of the mesoscale and of the general circulation), while caution should be used in studies over shorter times.

When a linear shear mean flow $\mathbf{U}$ is considered, the behaviour of the dispersion S changes (Taylor, 1953), but the relative difference Δ still behaves asymptotically as T/t. This suggests that for mean flows with weak spatial dependence, the accuracy of the advection-diffusion equation is the same as for purely homogeneous flows.

Very different results are found, instead, when the time-dependence is introduced in the mean flow $\mathbf{U}$ (Okubo, 1987). For the oscillating linear shear with frequency ω_s, the relative difference Δ between the two models does not converge to zero in the asymptotic limit $\lim_{t\to\infty}$, but rather it converges to the oscillating function

$$\Delta_\infty = \frac{{\omega_s}^2}{(\gamma^2 + {\omega_s}^2)}\frac{1}{(2 + cos\ 2\omega_s t)}, \tag{24}$$

with constant average and constant extrema. The graphs of the dispersion estimates from the two models and the corresponding Δ are shown in Figs (6a) and (6b).

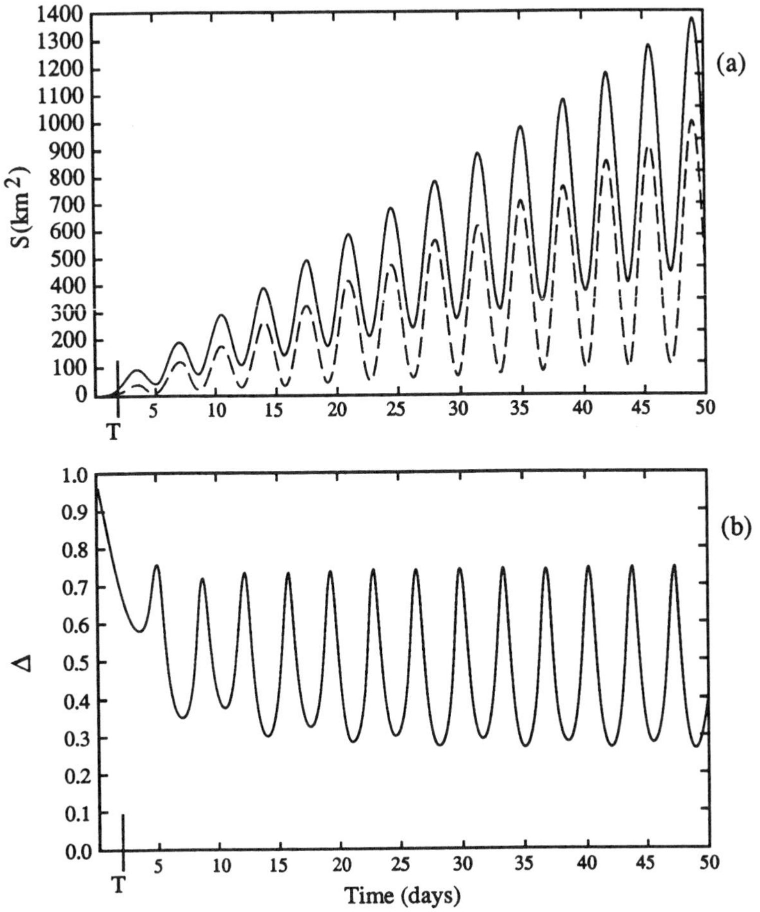

Figure 6. As in Figure 5, but for dispersion with oscillating linear shear. The time scales of the turbulence and of the shear are comparable, $T \sim \frac{T_s}{2\pi}$.

As it is shown from (24), the value of Δ depends on the ratio between the turbulent time scale $T = \frac{1}{\gamma}$ and the time scale of the shear $T_s = \frac{2\pi}{\omega_s}$.

When

$$\gamma^2 >> {\omega_s}^2, \qquad \Delta_\infty \sim 0,$$

which indicates that the two estimates essentially coincide when the current oscillation is slow compared with the turbulent correlation time. In the other limit

$$\gamma^2 \leq {\omega_s}^2, \qquad \Delta_\infty \sim O(1),$$

which indicates that when the two time scales are comparable, $T \geq T_s/2\pi$, the estimates are significantly different at all times. In this case the use of the advection-diffusion equation (5) can lead to an order one error in the estimate of the dispersion even for asymptotically large times. Intuitively, this arises because each reversal of the mean velocity results in a change of regime in the dispersal flow. During each regime, for time $t < T$, the motion of the particles is described differently by the two models, with model 2 being characterized by a correlation between particle position and turbulent velocity which is absent in model 1. If the reversals occur at time intervals of the same order of T, this initial time difference in particle behaviour becomes permanent, and the dispersion between the two models is different at all times.

Examples of flows which can be idealized as oscillating mean shears and for which $T \sim T_s$ are along-shore coastal currents characterized by fluctuations with a period of few days and cross-stream turbulent motion with similar time scales (Kundu and Allen, 1976), and tidal currents with highly variable cross-stream motion.

The results obtained analytically by ZG for oscillating linear shear are likely to be generalizable to a much wider class of flows, including shear flows with complex spatial structure. Unfortunately analytical calculations using the method of moments are not feasible for such flows in the framework of model 2, so that more detailed results will have to rely either on future numerical studies or on analytical studies using different techniques. By now, only a qualitative discussion can be given. It appears reasonable to assume that in the presence of strong inhomogeneity in the shear flows, the interplay of the space scales of turbulence and shear (L and L_s respectively) can introduce an additional mechanism for permanent differences between the two models. Classical examples of flows of this type are the intense western boundary currents and their open ocean extensions (e.g. the Gulf Stream, the Kuroshio Current, the Brazil Current), characterized by strong meandering currents. For flows of this type, even though the impact of finite scales of turbulence cannot be clearly assessed at this point, the previous arguments suggest that caution should be used in applying the advection-diffusion equation with the eddy-diffusion parameterization.

7. Practical applications.

As discussed in Section 6, the introduction of finite scales for the turbulence plays an important role in the study of dispersion for initial times, $t < T$, and also for longer times, provided that the mean flow is strongly space or time dependent so that the scales of the mean flow and of the turbulence are of the same order. In these situations, that are characteristics of important current systems in the ocean, the eddy-diffusion parameterization is likely to be inaccurate, while its generalizations (generalized K-models) are usually not simple to implement. The stochastic particle models 2 and 3, instead, provide a convenient and valid tool of investigation. Two main classes of applications of the models can be identified in this general framework.

The first class includes the "direct" applications, where the stochastic particle models are used to study and simulate dispersion problems in mean velocity fields that can have arbitrarily complex space and time dependence. The equations for single particle motion are directly integrated in time to simulate ensemble of tracer particles in the prescribed mean velocity fields, and the concentration of the particles corresponds to the ensemble average concentration of the tracer. The models 2 and 3 are especially suitable to study dispersion from localized sources, since the statistics have to be computed over a large number of particles (of the order of thousands) for each release point. An example of an application of model 2 to a dispersion study in an idealized model of the Gulf Stream is reviewed in Section 7.1 (Dutkiewicz et al., 1993).

An other class of promising applications of the stochastic particle models is related to the "inverse" problem of extracting information about the velocity field from Lagrangian data from drifting buoys. The Lagrangian data are usually available only for relatively short intervals of time, so that their statistics are often satisfactory only for times of the order of T. As a consequence, as noticed by Davis (1991b), the Lagrangian data should be studied in the framework of a model that describes satisfactorily the behaviour of turbulent particles at scales of the order of T. Davis (1987) suggests to use an elaborated form of the advection-diffusion equation with time-dependent diffusivity. We, instead, advocate the use of the stochastic particle models as a more natural and simple tool. An example of an application of model 2 to surface drifter data to estimate turbulence parameters is summarized in Section 7.2 (GOPR).

7.1 Diffusion in a meandering jet. A direct problem. Dutkiewicz et al. (1993) have performed a numerical study on the turbulent mixing across an ideal model of a meandering Gulf Stream extension, using model 2 to simulate the turbulent motion. This is an important problem from both the physical and biological point of view, because the Gulf Stream divides two bodies of water (the Sargasso Sea and the Slope Water) which differ in physical characteristics such as temperature and salt, and in biological properties such as zooplankton and nekton species.

In the study of Dutkiewicz et al., (as in other previous studies such as Bower 1991), the Gulf Stream extension is crudely represented by a meandering jet (see Fig. 7), steadily propagating eastward with velocity c. Dutkiewicz et al. consider the distribution of ensemble of particles, simulated using model 2 and launched in several regions of the jet, over typical mesoscale times of the order of one month. The particle distribution patterns turn out to depend crucially on the initial conditions. Particles launched in the meander bends, where the flow is recirculating, tend to be trapped in the bends and the distribution tends to homegenize. Particles launched in the jet core (defined as the region where the local velocity U is higher than the meander phase speed c, $U > c$) tend to be lost from the jet in plumes at the extrema of the meanders and to be entrained in successive recirculating regions (Fig. 7).

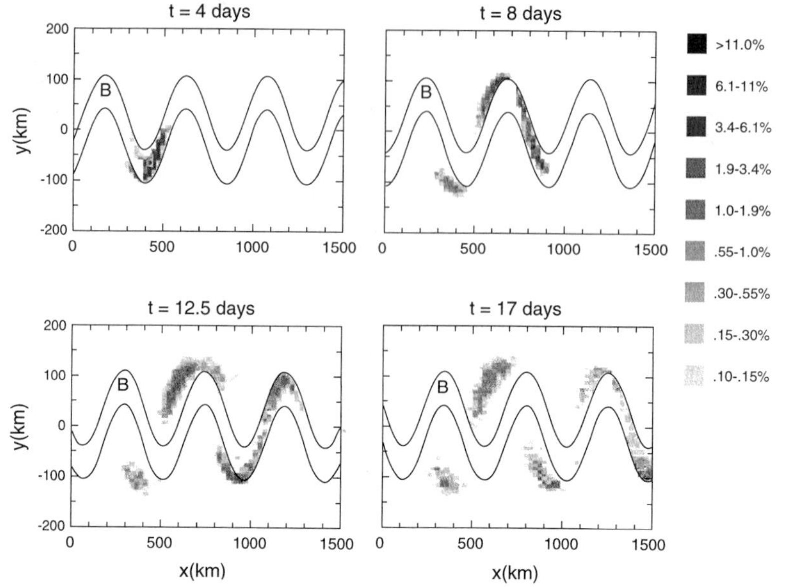

Figure 7. Evolution of the concentration of particles in a meandering jet propagating eastward (toward the right) at constant speed c. The solid lines indicate the core of the jet (defined as the region where the particle velocity U is greater than c). The particles are launched at point B, inside the jet core, and their concentration is indicated by the shading as in the right panel (from Dutkiewicz et al., 1993)

The net result is thus a buildup of homogenized particle distributions in the meander regions. This has important biological consequences, as it has been tested by Dutkiewicz et al. incorporating a simple biological model in the particle model. The biological model, which represent the interaction of two species, one dominant in the Sargasso Sea and the other in the Slope Water, shows the observed tendency to produce patches of species in the recirculating meander regions, on the side of the jet each species favors.

7.2 Estimates of turbulence parameters from Lagrangian data. An inverse problem. GOPR have proposed a method based on model 2 to study Lagrangian data and to estimate turbulence parameters from the data. The method, which is based on a "parametrical" approach, consists of assuming that the data obeys the model, so that the velocity autocorrelation is assumed to have the known exponential shape which depends on the two turbulent parameters σ and T. The classical method of moments is then used to estimate from the data the parameters σ and T, together with the mean value of the velocity U. Also an estimate of the diffusivity parameter K is obtained using (4)

The main advantage of this approach is that it provides more accurate parameter estimates than other existing methods (e.g. Figueroa and Olson, 1989), by using the a-priori knowledge of the model. In particular, the estimates of T and K depend on the slope of the autocorrelation and are computed using only the first two time lags. The other estimates in the literature, instead, are based on the integration of the autocorrelation and turn out to depend on subjective procedures of truncation of the long time lags of the correlation. By contrast the GOPR estimates are "objective" (in the sense that, given the model assumption, the method provides unique values for the estimates) and have a well defined associated error. The method, in fact, provides a complete error analysis of the estimates and is valid in the presence of observational errors. On the other hand, the validity of the a-priori assumptions has to be checked before the application of the method in order to obtain reliable estimates. As discussed in Section 4, model 2 appears to be generally appropriate for mesoscale upper ocean flows, so that the method is expected to have a large applicability for these type of flows.

Note that, even though the results obtained by GOPR are valid for the specific model 2, and in the additional assumption of homogeneous turbulent flows with constant mean U, the methodology is very general and can be applied to more complete and less restrictive models. Work is presently underway to generalize the model in the case of space and time dependent U. Also, the identification of a suitable model for deep ocean flows is under consideration. Once the model is identified, the same method of moments used by GOPR can be used to estimate the parameters.

8. Summary and discussion.

In this paper we discuss a hierarchy of three Markovian particle models. Model 1 assumes that the scales of turbulent motion are infinitesimal, and it is equivalent to the advection-diffusion equation with the eddy-diffusion parameterization. Model 2 introduces a finite scale T for the turbulent velocity, still neglecting the acceleration scales. The particle velocity obeys the classical Langevin equation, and its autocorrelation is characterized by an exponential behaviour. Model 3 introduces in addition a scale for the acceleration, $T_a < T$, while assuming infinitesimal scales for the acceleration derivatives. The velocity autocorrelation is exponential as in model 2 at time lags $> T_a$, and it has a quadratic behaviour near the origin.

A comparison of model 2 with oceanographic Lagrangian data shows that the model gives a satisfactory description of mesoscale turbulent flows in the upper ocean. For these flows, the velocity autocorrelation following particles is typically exponential as in model 2. The exponential behaviour is found to hold even at small time lags, suggesting that the acceleration scales are small and can be neglected. The typical velocity time scale T is of the order of 2-10 days.

Model 2, on the other hand, is not appropriate for deep ocean flows, where the time scales are much longer (T is of the order of 10 days or more). The acceleration scales are not negligible for these flows, and the autocorrelation is characterized by a smoother, quadratic behaviour at the origin. Model 3 reproduces correctly the autocorrelation behaviour for small lags, but, on the other hand, the data autocorrelations show strong oscillations at longer lags, which are not present in the model. Specific statistical tests on the data have not been performed yet, so that it is not clear at this stage whether the oscillations signal a significant deviation from model 3, or whether they are only due to the relatively short record lengths available.

A comparison between dispersion estimates performed with the three stochastic models shows that the introduction of finite scales of turbulence plays an important role not only for dispersion at initial times, $t < T$, but also for dispersion at longer times, $t > T$, if the mean flow is characterized by scales of the same order than T. This suggests that model 1, and therefore the advection-diffusion equation with the eddy-diffusion parameterization, is not accurate when used in higly variable mean flows where the scales of the mean flow and of the turbulence are comparable. Flows of this type are quite common in oceanography. They range from tidal or shelf oscillating currents, to large scale meandering jets such as the Gulf Stream or other western boundary currents, which play a major role in the distribution of properties in the ocean. For these type of flows, the use of stochastic models such as 2 or 3 appears more indicated and accurate than the advection-diffusion equation.

Two main classes of applications of the stochastic particle models are indicated. The first one consists of "direct" problems, where the models are directly integrated in time to study dispersion problems in flows with high space and time complexity. An example of a study on an idealized Gulf Stream model is discussed in some detail. The other class of applications involve "inverse" problems, where the models are used to infer information on the statistics of the velocity field from Lagrangian oceanographic data. An example of an application of model 2 to the estimate of turbulence parameters from surface drifting buoys is discussed. New applications and problems are suggested, that can be addressed with a combination of physical and statistical modeling.

The summarized results show that the particle models provide a valuable and flexible tool which can be used in a wide range of problems in oceanography. It is important to recognize, though, that Markovian models such as 2 or 3 are limited to ergodic processes where the autocorrelation decays exponentially to zero, and thus they cannot explain phenomena related to the presence of strong coherent structure with more complex autocorrelations. The presence and the importance of coherent structures in the ocean have been the object of works and discussions in the literature (e.g. Mc Williams, 1984). Oceanographic data, and in particular Lagrangian data in mid and deep ocean, show a remarkable incidence of eddies persisting for a long time (Richardson, 1993). On the other hand, a clear assessment of their impact on the statistical properties of the oceanic flows, and in particular on the dispersion, is not yet avaiable, and work is in progress on this subject.

We believe that the particle models discussed here can provide, at least indirectly, some contributions to this investigation. The comparison between the oceanographic data and model 2, for instance, suggests that the coherent structures do not play an important role in the upper ocean, whereas in the deep ocean the question is still open. As previously noted, an accurate statistical analysis is necessary to establish whether or not the data significantly deviate from model 3 at long times, $t > T$. If they do, this strongly suggests that coherent structures or waves dominate the flow. If they do not, instead, it is possible that the deep ocean has a random structure similar to the upper ocean, but with longer time scales. The question of how the time scales are connected to different forcing and generation mechanisms in the upper and deep ocean is still open.

We conclude remarking that, even though we have confined our discussion to the three models 1,2 and 3, an extremely interesting and challenging question is how to formulate more general particle models that can include the presence of coherent strucures (e.g. Bennett, 1986; Carnevale et al., 1990). Contributions to this problem, which is still unsolved in the literature, are expected to help improve our understanding of turbulent mechanisms and transport.

9. Acknowledgements The author wish to thank G.Buffoni, N.Pinardi and E.Zambianchi for interesting discussions. This work was supoorted by the Office of Naval Research (Grant N00014-91-JI346) and by the Italian Consiglio Nazionale delle Ricerche, Istituto Metodologie Geofisiche Ambientali (CNR-IMGA, Grant CE MAS2 CT92-0055).

References

[1] Bennett, A.F., 1986. Random walks with manholes: simple models of dispersion in turbulence with coherent vortices. *J. Geophys. Res.*, 91(C9), 10,769-10,770.

[2] Bower, A.S., 1991. A simple kinematic mechanism for mixing fluid parcels across a meandering jet. *J. Phys. Oceanogr.*, 21, 173-180.

[3] Carnevale, G.F., Y. Pomeau and W.R. Young, 1990. Statistics of ballistic agglomeration. *Phys. Rev. Lett.*, 64(24), 2913-2937.

[4] Chandrasekhar, S., 1943. Stochastic problems in physics and astronomy. *Rev. Modern Physics*, 15, 1-89.

[5] Colin de Verdiere, A., 1983. Lagrangian eddy statistics from surface drifters in the eastern North Atlantic. *J. Mar. Res.*, 41, 375-398.

[6] Csanady, G.T., 1980. *Turbulent diffusion in the environment.* Dordrecht, D. Reidel Publ. Co., 248 pp.

[7] Davis, R.E., 1987. Modelling eddy transport of passive tracers. *J. Mar. Res.*, 45, 635-666.

[8] Davis, R.E., 1991a. Lagrangian ocean studies. *Ann. Rev. Fluid Mech.*, 23, 43-64.

[9] Davis, R.E., 1991b. Observing the general circulation with floats. *Deep-Sea Res.*, 38, S531-S571.

[10] Durbin, P.A., 1984. Comment on papers by Wilson et al. (1981) and Legg and Raupach(1982). *Boundary-Layer Met.*, 29, 409-411.

[11] Dutkiewicz, S., A. Griffa and D.B. Olson, 1993. Particle diffusion in a meandering jet. *J. Geophys. Res.*, 93, 16,487-16,500.

[12] Figueroa, H.A. and D.B. Olson, 1989. Lagrangian statistics in the South Atlantic as derived from SOS and FGGE drifters. *J. Mar. Res.*, 47, 525-546.

[13] Griffa, A., K. Owens, L. Piterbarg and B. Rozovskii, 1995. Estimates of turbulence parameters from Lagrangian data using a Stochastic Particle Model. *J. Mar. Res.*, 53, 371-401.

[14] Holloway, G., 1989. Subgridscale representation. In:*Oceanic Circulation Models: Combining Data and Dynamics.* L.T. Anderson and J. Willebrand, eds., NATO ASI Series, 513-587.

[15] Janicke, L., 1983. Particle simulation of inhomogeneous turbulent diffusion. In:*Air Pollution Modelling and its Application II*, Plenum, 527-535.

[16] Krasnoff, E. and R.L. Peskin, 1971. *Geophys. Fluid Dyn.*, 2, p. 123.

[17] Krauss, W. and C.W. Boning, 1987. Lagrangian properties of eddy fields in the northern Atlantic as deduced from satellite-tracked buoys. *J. Mar. Res.*, 45, 259-291.

[18] Kundu, P.K. and J.S. Allen, 1976. Some three-dimensional characteristic of low-frequency current fluctuations near the Oregon coast. *J. Phys. Oceanogr.*, 6, 181-199.

[19] Legg, B.J. and M.R. Raupach, 1982. Markov-Chain simulation of particle dispersion in inhomogeneous flows: the mean drift velocity induced by a gradient in Eulerian velocity variance. *Boundary-Layer Meteor.*, 24, 3-13.

[20] McWilliams, J.C., 1984. The emergence of isolated coherent vortices in turbulent flow. *J. Fluid Mech*, 146, 21-43.

[21] Okubo, A., 1987. The effect of shear in an oscillatory current on horizontal diffusion from an instantaneous source. *Int. J. Oceanogr. Limnol*, 1, 194-204.

[22] Pope, S.B., 1994. Lagrangian PDF methods for turbulent flows. *Ann. Rev. Fluid Mech*, 26, 24-63.

[23] Richardson, P.L., 1993. A census of eddies observed in North Atlantic SOFAR float data. *Prog. Oceanogr*, 31, 1-50.

[24] Riser, S.C. and H.T. Rossby, 1983. Quasi-Lagrangian structure and variability of the subtropical western North Atlantic circulation. *J. Mar. Res*, 41, 127-162.

[25] Risken, H., 1989. *The Fokker-Planck Equation* Springer-Verlag.

[26] Sato, Y. and K. Yamamoto, 1987.*J. Fluid Mech*, 175, p. 183.

[27] Sawford, B.L., 1991. Reynolds number effects in Lagrangian stochastic models of turbulent dispersion. *Phys. Fluids A* 3(6), 1577-1586.

[28] Taylor, G.I., 1921. Diffusion by continuous movements. *Proc. Lond. Math. Soc.*, 20, 196-212.

[29] Taylor, G.I., 1953. Dispersion of soluble matter in solvent flowing slowly through a tube. *Proc. R. Soc. Lond. A*, 219, 186-203.

[30] Thomson, D.J., 1984. Random walk modelling of diffusion in inhomogeneous turbulence. *Quart. J.R. Met. Soc.*, 110, 1107-1120.

[31] Thomson, D.J., 1986. A random walk model of dispersion in turbulent flows and its application to dispersion in a valley. *Quart. J.R. Met. Soc.*, 112, 511-530.

[32] Thomson, D.J., 1987. Criteria for the selection of stochastic models of particle trajectories in turbulent flows. *J. Fluid Mech.*, 180, 529-556.

[33] van Dop, H., F.T.M. Niewstand and J.C.R. Hunt, 1985. Random walk models for particle displacements in inhomogeneous unsteady turbulent flows. *Phys. Fluids*, 28, 1639-1653.

[34] Verron, J. and K.-D. Nguyen, 1989. Lagrangian diffusivity estimates from a gyre-scale numerical experiment on float tracking *Oceanologica Acta*, 11(2), 167-176.

[35] Yeung, P.K. and S.B. Pope, 1989. Lagrangian statistics from direct numerical simulations of isotropic turbulence. *J. Fluid Mech.*, 207, 531-586.

[36] Zambianchi, E. and A. Griffa, 1994. Effects of finite scales of turbulence on dispersion estimates. *J. Mar. Res.*, 52, 129-148.

Rosenstein School of Marine and Atmospheric Science
University of Miami
4600 Richenbacker Causeway
Miami, Florida, 33149

anna@mamey.miami.edu

Sound through the internal wave field

Frank S. Henyey and Charles Macaskill

Abstract

We review the subject of sound propagation through the random ocean internal wave field. Among the assumptions that are usually made, two are identified as not always valid: the Markov approximation and the expansion around a deterministic ray. When these approximations are valid, and the internal wave field known, acoustic phase fluctuations are very accurately predicted and intensity fluctuations reasonably well calculated. For the calculations of statistical moments of the acoustic field, the two techniques of moment equations and path integrals have been developed to a high degree, and give similar results. Among all other statistical quantities, the probability density function for intensity has received the most attention, but no satisfactory theory has been developed for this or other quantities.

1. The direct and inverse problems

The subject of the propagation of sound through the ocean internal wave field, as well as closely related topics of wave propagation in random media, has been presented in a number of reviews and books (Flatté et al, 1979, Flatté, 1983, Ishimaru, 1978, Prokhorov et al, 1975, Rytov et al, 1988, Tatarski, 1971, Tatarski et al, 1992, Uscinski, 1977). There is no good reason to present another review of the accomplishments in this field. We instead concentrate on what is not known, from our personal points of view, with the hope of stimulating fresh thought along unexplored lines.

This subject involves a stochastic partial differential equation. A statistical description of the coefficients of the acoustic wave equation is assumed given from oceanographic studies, and one wishes to obtain the statistics of the solution.

The equation for the pressure, P, is

$$D_t \frac{1}{c^2} D_t P = \rho \nabla \cdot \frac{1}{\rho} \nabla P \tag{1.1}$$

where the total time derivative operator is

$$D_t = \frac{\partial}{\partial t} + \boldsymbol{v} \cdot \nabla \quad . \tag{1.2}$$

In these equations, the sound speed c, the density ρ, and the water flow velocity $\boldsymbol{v}$ are functions of space and time. Equation (1.1) already contains approximations. The $\frac{1}{\rho}\nabla P$ on the right side should be accompanied by the acceleration of gravity $\boldsymbol{g}$, which cancels the vertical component of the hydrostatic pressure gradient, but since $\nabla \cdot \boldsymbol{g} = \frac{3\rho g}{R_E \rho_E}$, where R_E and ρ_E are the radius and average density of the Earth, this term is justifiably neglected. More significantly, source, nonlinear acoustics, and loss terms have been neglected. Source terms provide some of the ambient noise, which the experimentalist attempts to exclude from measurements; thus the theoretician should also exclude sources. Sound pressures used in random propagation studies are generally low enough to justify the neglect of nonlinear terms. Loss terms are important in ocean acoustics for propagation over very many wavelengths. It is assumed that the effect of the loss can be adequately described by Fourier transforming from time to the frequency domain, multiplying by $e^{-\alpha R}$, and transforming back, where α is the frequency dependent attenuation coefficient which includes a number of loss effects, and R is the distance propagated. The coefficient α varies from the essentially negligible $3 \times 10^{-6}\text{km}^{-1}$ at a frequency of 10 Hz to 80 km^{-1} at 1 MHz. This simplified approach works well in practice, although the correct method would be to solve differential equations for the concentrations of the chemicals responsible for the attenuation.

At this point, further approximations are made. In equation (1.1), P includes not only the acoustic pressure but the hydrostatic pressure and the dynamic pressure associated with internal waves and vortical motion. There are linear acoustic terms from the fluctuations in c, ρ, and $\boldsymbol{v}$ produced by the acoustic wave itself. These are small effects in the deterministic problem, and their effects in the random problem are small compared to the dominant random effects. Thus, equation (1.1) with P interpreted as the acoustic pressure, and the coefficients only depending on other dynamics is taken as the fundamental equation.

We assume given statistical descriptions of $c(\boldsymbol{r}, t)$, $\rho(\boldsymbol{r}, t)$, and $\boldsymbol{v}(\boldsymbol{r}, t)$. Both c and ρ have a large mean value and small fluctuations around that value. The flow velocity $\boldsymbol{v}$ is likely to be specified as a zero-mean fluctuating quantity. For fluctuations caused by deep-water internal waves, fairly reasonable statistical models are available.

A number of rather different subjects fall under the heading of wave propagation in random media. The ocean acoustics problem is similar to the problem of the twinkling of starlight; the waves generally propagate as they would without the randomness, but acquire fluctuations due mostly to rather large-scale (relative to a Fresnel zone) temperature variations. What to ocean acoustics and astronomical observation is very strong scattering, is to other subjects extremely weak scattering. With strengths well beyond our concerns are the subjects of wave diffusion (as in light propagation through fog) and Anderson localization, when long-distance propagation no longer occurs. The acoustic case is discussed in Sornette (1989). With scatterers smaller (in a dimensionless sense) than those of our concern are the fog or haze problem and the problem of the blue sky. There are also stochastic problems in ocean acoustics that are not the subject of this review. Those include scattering from random surface waves (Thorsos, 1990) and scattering from the ocean bottom and fluctuations under the bottom (Jackson et al, 1986) and scattering from random bubble clouds (Henyey, 1991). Often, different random effects may combine, and one can study how the combined effect differs in very nontrivial ways from the individual effects considered separately, e.g. the combined effects of surface and volume scattering at the ocean bottom (Jackson and Briggs, 1992, Lyons et al, 1994).

One would like an efficient approximate statistical description of the pressure field P resulting from the input ensemble of c, ρ, and $\boldsymbol{v}$ values for a variety of initial and boundary values. This description should reflect the physics of the problem, in preference to being merely phenomenological. This is the direct problem.

Also of considerable interest are inverse problems, in which one attempts to obtain information about the ocean medium from acoustics measurements (Uscinski, 1986, Ewart and Reynolds, 1991). The internal wave strength, its spectrum, its large-scale distribution in space, and its intermittency on small scales are all interesting information that one would like to extract from fluctuations in acoustic transmission measurements.

Up to now, the problem has been described in a rather general sense, as we wish to point out possible research directions in the future. In practice, a more narrow definition has been used in theoretical studies. We now turn to a detailed discussion of the existing work.

2. What is actually done

A major simplification is to neglect any time dependence of the medium (i.e. one neglects the t in $c(\boldsymbol{r},t)$, $\rho(\boldsymbol{r},t)$, and $\boldsymbol{v}(\boldsymbol{r},t)$) during the passage of a single acoustic pulse. With this assumption, the acoustic field can be Fourier transformed from time to frequency, and each frequency propagates independently; there is no frequency shift. That is not to say that frequency shift statistics are lost. One can consider propagation of a single frequency at different times with different configurations of the medium, and calculate the temporal autocorrelation function. The Fourier transform of that autocorrelation function gives the frequency shift spectrum. The approximation consists of neglecting the effects on propagation from that frequency shift. Thus, for example, if there is an increase in frequency during propagation, the first Fresnel zone is a little smaller than if the frequency hadn't changed, and the wave responds slightly differently to fluctuations on the scale of the Fresnel zone. As discussed below, the frequency shift is very small for ocean acoustics, and the slight change in propagation is almost certainly of no consequence. Thus, the operator in equation (1.2) becomes

$$D_t = -i\omega + \boldsymbol{v} \cdot \nabla \tag{2.1}$$

where $\omega = 2\pi f$ is the angular frequency and f is the frequency of the Fourier component.

This approximation cannot be avoided by ray tracing. As discussed in the previous paragraph, it is only the diffractive effects which are approximated, and ray tracing has no diffractive effects; it assumes the first Fresnel zone has zero width.

An approximation that is central to all techniques for random propagation, with the exception of ray tracing, is the "parabolic" approximation. This name is somewhat misleading, as the classification of equations into hyperbolic, elliptic, and parabolic applies to real equations, and the "parabolic" approximation makes the equation complex. Accordingly, some authors use the term "paraxial" approximation to describe this approximation. The purpose of the parabolic approximation is to cause the propagation to proceed through space in only one direction along some coordinate (and arbitrarily along the others). Normally this coordinate is chosen to be a horizontal coordinate. In most experiments to explore internal wave effects on acoustic propagation, the depth difference between sources and receivers is small compared to their horizontal range, and the transverse horizontal extent of either source or receiver arrays is also small compared to the range, allowing a single Cartesian coordinate to be selected as the dominant direction of propagation. Some authors prefer a cylindrical coordinate system centered on the source, because energy "really" propagates radially out from the source. In our opinion, there is no particular advantage to using a cylindrical coordinate system, but for most purposes there is no particular harm either. As practiced, results in the two coordinate systems tend to be identical.

Clearly, the justification of the parabolic approximation must involve considerations of the direction of energy propagation. One generally finds vertical propagation angles within about 15° of the horizontal, with horizontal transverse angles considerably smaller.

The original parabolic approximation leads to the Schrödinger equation (Leontovich and Fock, 1946). In ocean acoustics, this is called the narrow-angle parabolic approximation. Let x be the selected horizontal coordinate. Then the $\left(\frac{\partial}{\partial x}\right)^2$ operator in the frequency space version of equation (1.1) is

(2.2)

$$\left(\frac{\partial}{\partial x}\right)^2 = e^{ik_0x}\left(\frac{\partial}{\partial x} + ik_0\right)^2 e^{-ik_0x} = e^{ik_0x}\left(-k_0^2 + 2ik_0\frac{\partial}{\partial x} + \frac{\partial^2}{\partial x^2}\right)e^{-ik_0x}$$

where all the factors are taken to be operators and the multiplications as operator products. The parameter k_0 is chosen to be some average or nominal value of the horizontal wavenumber, and the narrow-angle parabolic approximation is achieved by dropping $\left(\frac{\partial}{\partial x}\right)^2$ from the right side of equation (4). By algebraically reorganizing equation (1.1) we get the parabolic (Schrödinger) equation

$$i\frac{\partial}{\partial x}\Psi = -\frac{1}{2k_0}\left[\left(\frac{\partial}{\partial y}\right)^2 + \left(\frac{\partial}{\partial z}\right)^2\right]\Psi + k_0\frac{1-n^2}{2}\Psi \qquad (2.3)$$

in the case $\boldsymbol{v} = 0$, $\rho =$ constant. The notation in equation (2.3) is as follows:

$\Psi =$ reduced acoustic field $= e^{-ik_0x}P$

$n =$ index of refraction $= \frac{c_0}{c}$ where $c_0 =$ reference sound speed $= \frac{\omega}{k_0}$

Usually, the transverse horizontal derivative $\left(\frac{\partial}{\partial y}\right)^2$ is neglected, except for contributing a factor $\frac{1}{x}$ in the intensity from a point source.

The narrow angle parabolic approximation is not very good. As an example, consider an infinite medium with a sound speed of 1500 m/s, which we also take as the reference sound speed, so a horizontally propagating plane wave is exactly described by the parabolic approximation. Now consider a plane wave traveling at an angle above the horizontal of arcsin(.1)≈ 6° with a low frequency of 150 Hz. Its wavelength is 10 m, its vertical wavelength is 100 m, and its horizontal wavelength is

$\frac{\lambda}{\sqrt{1-\frac{\lambda^2}{\lambda_v^2}}} = 10.05037815\ldots$m. In the parabolic approximation with this vertical wavelength, its horizontal wavelength is $\frac{\lambda}{1-\frac{\lambda^2}{2\lambda_v^2}} = 10.05025126\ldots$m. This might seem like a small difference, but after traveling 400 km, a short distance in ocean acoustics at this frequency, it is half a wavelength wrong. With the wrong sign, it interferes with other components incorrectly, and the sound is calculated to be in the wrong part of the ocean. Of course, by choosing the reference sound speed optimally, we could get the 6° phase exactly right, at the expense of getting the horizontal propagation wrong.

Figure 1 shows a more realistic example actually worked out. The sound speed is taken to be a function of depth similar to that found at mid latitudes. The frequency is 400 Hz, and the range is 300 km. The intensity as a function of depth is compared for the narrow angle parabolic approximation at two different choices of reference sound speed, one at the minimum value which is a common choice in the literature, and one optimally chosen for the particular narrow beam of the calculation. The large difference between them is consistent in order of magnitude with the simple example in the last paragraph; the parabolic approximation breaks down at large ranges.

There are a number of wide-angle parabolic approximations that avoid the inaccuracies of the narrow-angle parabolic equation. These approximations lead to equations with operators going beyond the second derivative in the vertical. This leads to issues with the boundary conditions at the top and bottom of the ocean. These wide-angle equations are used in simulations, in which random realizations of the ocean fluctuations are constructed, and the equation is numerically solved. They are not, however, used in analytic work. There also is the possibility of embedding the Helmholtz equation in a parabolic eqation in a space of one extra dimension. This extra dimension can be thought of as a pseudo-time. The resulting equation can be truly parabolic in the mathematical sense, in which case the Helmholtz solution is recovered as the field as pseudo-time $\rightarrow \infty$. Or, the equation can be a Schrödinger equation, in which case the Helmholtz solution is obtained by Fourier transform (Palmer, 1978).

How can the narrow-angle parabolic equation be used if it is not an accurate approximation? It will give an acoustic field that resembles the actual one at some range not too different from the actual range. The modeled fluctuations in the acoustics at the modified range are used to predict the actual fluctuations. Generally, one is interested in energy that propagates close to some ray of a range-independent sound speed profile, and one can choose the reference wavenumber to be equal to the horizontal wavenumber of that ray.

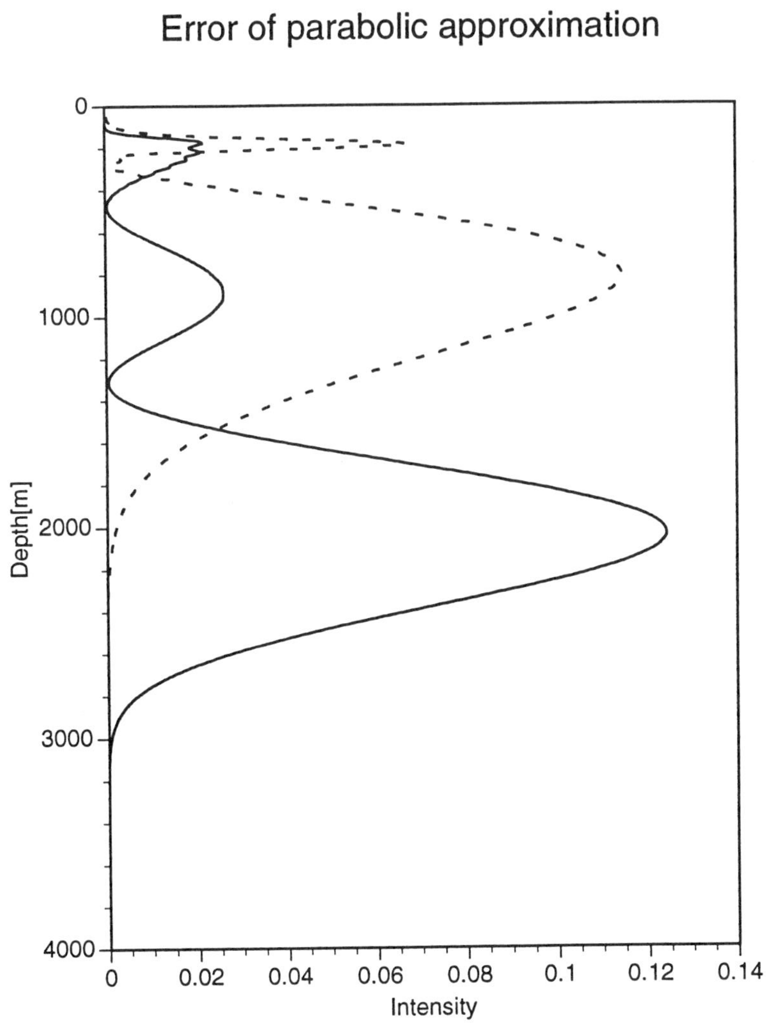

Figure 1. A comparison between two choices of the reference sound speed in the parabolic approximation for the same problem. 400 Hz propagation from a 200 m deep source was calculated in a Munk sound speed profile, which resembles the real ocean. The range is 300 km. The difference between these curves is indicative of the error made by the parabolic approximation.

The next simplification in actual practice is to neglect the fluctuations in the density $\rho(\boldsymbol{r}, t)$. The density contribution is indeed small. By including a factor of $\frac{1}{\sqrt{\rho}}$ in the reduced pressure Ψ, there is a term $\frac{1}{2}\frac{\nabla^2 \delta\rho}{\rho}\Psi$

in equation (1) while the sound speed term is $2k_0^2\frac{\delta c}{c}$. For a temperature fluctuation, $2\frac{\delta c}{c} \approx 100 \times \frac{1}{2}\frac{\delta\rho}{\rho}$, so the medium wavelength must be about $\frac{1}{10}$ of the acoustic wavelength to make the density term important. However, a $\frac{1}{10}$ wavelength perturbation scatters extremely weakly; double scattering is required to even satisfy a Bragg condition.

The velocity $\boldsymbol{v}(\boldsymbol{r}, t)$ is also rather small and hence is neglected. Typical internal-wave caused sound speed fluctuations in the upper ocean are about 1m/s, whereas velocities are a few cm/s. However, there is a way of separating out the velocity effect (Worcester et al, 1985). Without velocity or time dependence during the travel time of a pulse, there is reciprocity, which means that the transmission amplitude from a point A to a point B is identical to that for transmission from B to A. The difference between the two is due to a combination of time decorrelation and velocity. The time decorrelation effect can be estimated by delayed transmission in the same direction, leaving the velocity effect to be inferred from a reciprocal transmission experiment. The velocity effect is handled by replacing the sound speed with $c + \hat{\boldsymbol{k}} \cdot \boldsymbol{v}$, where $\hat{\boldsymbol{k}}$ is the unit vector in the propagation direction, which, in the spirit of the parabolic approximation can be taken to be the horizontal direction from source to receiver. For the two parts of the reciprocal experiment, the values of $\hat{\boldsymbol{k}}$ are negatives of each other. In this approximation, δc of the one-way experiment would be replaced by $2\hat{\boldsymbol{k}} \cdot \boldsymbol{v}$ for the difference between reciprocal transmissions.

Almost all published work has modeled the sound speed fluctuations as a Gaussian process. This does not imply that the acoustic fluctuations are Gaussian, except for scattering so weak that the first Born approximation is valid. The first Born approximation is linear in the sound speed fluctuations, but any stronger scattering is nonlinear, leading to non-Gaussian acoustics. Gaussian statistics for a natural process (in our case, the ocean internal wave field) are associated with dispersion of a linear wave field. The dispersion can be either different wavelength waves having different group speeds, or angular dispersion, waves traveling in different directions. If there is a set of many independent sources of the waves, and the waves have undergone dispersion, then the waves in one region will consist of the interference of contributions from the different sources. The central limit theorem implies that the wave field viewed locally in space and time has Gaussian statistics. Nearby sources and nonlinearities can cause deviations from Gaussian behavior. In particular, nonlinear effects on small-scale internal waves are thought to cause a certain type of non-Gaussian behavior. This non-Gaussian behavior can be described as a result of a coordinate transformation; in a vertically Lagrangian coordinate system, the waves seem to be much more nearly Gaussian. A vertically Lagrangian coordinate system uses the equilibrium level $z_0 = z - \zeta$ as the vertical coordinate rather than z, where ζ is the vertical displacement from static equilibrium. The displacement $\zeta_0(x, y, z_0, t)$ in the vertically Lagrangian coordinate system is assumed to be a Gaussian process in four dimensions, and its nonlinear transformation to Eulerian coordinates, $\zeta(x, y, z, t) = \zeta_0(x, y, z_0, t)$, induces more complicated statistics. Pinkel and coworkers (e.g. Pinkel and Anderson, 1992) have gathered considerable experimental evidence for this view. Unpublished work by Ewart and Henyey has examined, by means of simulations, the effect of this non-Gaussian nature of the internal wave field. They compare Gaussian and non-Gaussian models with the same spectrum, and find focusing effects occurring at smaller ranges for the non-Gaussian model. This happens because the transformation skews the distribution of strain $\frac{d\zeta}{dz}$, causing converging ($\frac{d\zeta}{dz} < 0$) strains to have smaller vertical scales than diverging strains.

In the previous section we described the ideal of an efficient statistical description of the acoustic fluctuations. The search for better descriptions is not a prominent part of the literature, except for the result that phase ϕ (and according to some, log intensity) has a simpler behavior than $e^{i\phi}$. Rather, the bulk of the work in this field has been to calculate various moments, i.e. low order polynomials in the acoustic field. Two of the major theoretical tools, moment equations and path integrals, are formulated in terms of moments, and are not useful for other purposes. Signal processing practice also is based on moments. The standard signal processing algorithms are of the form

$$I_j = \left| \int P dW_j \right|^2$$

Here the subscript j represents an index for a pixel in an image, and the integral is over time as well as a sum over different receivers. The choice of the "filter" dW_j reflects the expected time dependence of the signal, the noise present, and the time differences expected at the different receivers. The expectation $\langle I_j \rangle$ is a second moment. The performance of a signal processing system is often measured by the variance of the pixel intensity; that variance is a fourth moment. This type of signal processing method can be optimum when the signal and noise are both Gaussian processes. However, as discussed below, the signal is very definitely not a Gaussian process. We are aware of only one sonar system that exploits some non-Gaussian aspect of the signal (Uscinski, private communication).

Ray tracing and parabolic equation (narrow or wide angle) simulation, on the other hand, can be used for quantities other than moments. Travel time statistics unweighted with intensity or higher moments are calculated from ray tracing. Intensity probability distributions are calculated from simulations.

The probability distributions of acoustic intensity, or joint intensities at several points are long-tailed distributions; very large intensities happen much more frequently than given by an exponential distribution. Mathematical idealizations of long-tailed distributions (i. e., where the tail continues indefinitely) are not uniquely described by their moments, even for one-variable distributions. For example, Rytov theory for weak scattering predicts a log-normal distribution of intensity, $P_{LN}(I)$ (see e.g. Flatté et al, 1979, p 131). A simple generalization of this model is the family of distributions

$$P(I) = P_{LN}(I)\,[1 - \alpha \sin(\ln(I) - \langle \ln(I) \rangle)] \qquad -1 \leq \alpha \leq 1$$

for which all of the moments are independent of α. The case of non-zero alpha does not necessarily have direct application to wave propagation through random media, but it does illustrate the fact that even if one could calculate all the moments, one could not determine the distribution uniquely. (See Feller, 1971 p 227, for further discussion.)

Other approximations are made in dealing with moment equations and path integrals. One is the Markov approximation which assumes that the correlation function of sound speed fluctuations (assumed to be a Gaussian process) can be taken to be a delta function of the separation in the direction of propagation The correlation function is assumed to have a factor $L_p \delta(x - x')$ relative to the correlation function at $x = x'$. The correlation length L_p is assumed to depend only on the depth and the average direction of energy propagation (that direction being calculated by ray tracing without the fluctuations being present). This approximation allows the derivation of moment equations, since fluctuations are independent at each range, and a range derivative of expectation values can be expressed in terms of the fluctuations and deterministic refraction at the current value of the range. The Markov approximation is not fundamental to the path integral formulation; useful expressions can be obtained without it. Nonetheless, in practice, it is made in order to slightly simplify the expressions.

The Markov approximation requires that the relative change of any relevant variable over a distance L_p be small. Applied to the variable L_p itself, this implies $\left|\frac{dL_p}{dx}\right| < 1$. For sound that gets close to the ocean surface, this condition might not hold. L_p depends sensitively on angle due to the anisotropy of internal waves, and is much larger near the horizontal; a typical value is 10 km. Rays that become horizontal near the surface turn to steeper angles in distances on the order of 1 km, due to the strong sound speed gradient. Figure 2 shows L_p for two paths at the MATE experiment (Ewart and Reynolds, 1984, 1992). The lower path is consistent with the Markov approximation, but the upper path is not. The medium fluctuations in the region where $\left|\frac{dL_p}{dx}\right| > 1$ on the upper path are considerably larger than elsewhere along the path, making the problem more severe.

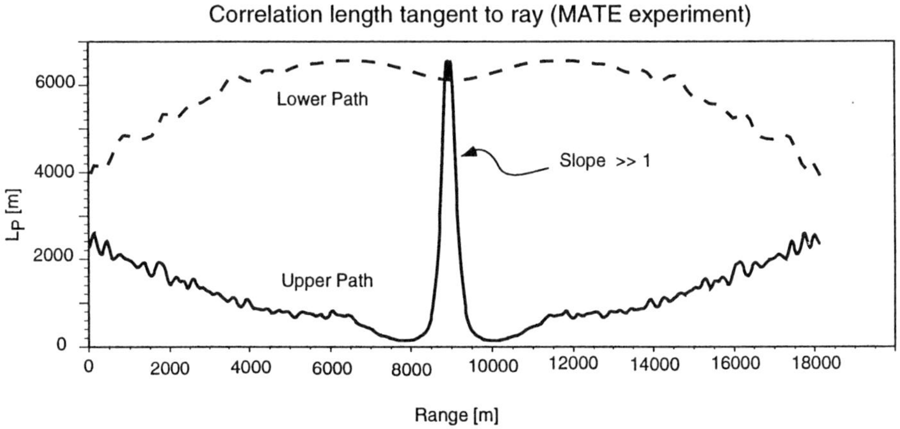

Figure 2. The value of the correlation length L_p calculated for the two paths in the MATE experiment. The Markov approximation requires $dL_p/dx \ll 1$, which is badly violated by the upper path.

Another approximation is made in the path integral formulation to convert an infinite-dimensional integral into a finite-dimensional one. This is a saddle-point approximation to the path integral, which is a high-frequency approximation. The expansion is generally done around the ray in the absence of fluctuations (which is the saddle point in that limit), rather than around the actual saddle point including the fluctuations. The actual saddle point for a moment is a frequency-dependent complex ray, identical to that described by Macaskill (1986) from another point of view. We expect situations such as the upper path in the MATE experiment to have the largest deviations from the deterministic ray. An approximation which leads to the same results is made in finding approximate solutions of the moment equation. In Macaskill and Uscinski (1981), an eikonal approach is used for the second moment equation, where the eikonal equation is just that corresponding to the deterministic case, in the presence of a sound channel. Using this approach, a perturbation method is developed that gives rise to solutions for the second moment equivalent to those obtained using path integrals by Dashen (1979). However, because the rays are those for the deterministic case, the approximation involved, which is fundamentally the slow variation of the intensity with depth, breaks down at caustics, and gives an inadequate description of the penetration into shadow zones due to scattering. In Macaskill (1986) a modified eikonal approach to the second moment equation is used and the ray paths then became complex-valued because of the medium fluctuations. It is assumed there that the effect of scattering is strong enough so that the average ray paths should be significantly altered by the effects of random inhomogeneities.

It is conventional to report dimensionless values for the moments, by normalizing by an appropriate power of an intensity I_0. For the experimentalist, I_0 is the average intensity $\langle I \rangle$, while the theoretical practice has been to use the deterministic intensity (i.e. that in the absence of fluctuations), and more particularly, that estimated from ray tracing. For the same example used in figure 1, these three values are shown in figure 3 (calculated using the narrow-angle parabolic approximation). For a receiver at a depth of 2500 m (within a few hundred meters) the values are zero for the ray trace approximation, .0005m^{-1} for the deterministic calculation, and .0012m^{-1} for the fluctuating average. These are normalized so that $\int I \, dz = 1$. Fortunately for the agreement of theory and experiment, no experiment approximating these conditions has been carried out. This example also shows a pitfall of expanding about the deterministic ray for any moment; most of the energy in the fluctuating case is in a region where there are no real rays.

3. General Results

Consider equation (2.3), with the dependence on y suppressed. We can write

$$(3.1) \quad (1 - n^2) = (1 - [\frac{c_0}{c_0 + \hat{c}(x,z) + \delta c(x,z)}]^2) \approx 2\frac{\hat{c}(x,z)}{c_0} + 2\frac{\delta c(x,z)}{c_0}$$

where $\hat{c}(x,z)$ represents the deterministic refractive index or sound speed fluctuations, while $\delta c(x,z)$ gives the small scale fluctuations which are modeled by a Gaussian random process. Here the average sound speed c_0 is around 1500 m/s, the deterministic variations are of order 1-10 m/s, and the random fluctuations are smaller again by a factor of 10 or more. The background sound channel is often modeled as being independent of the range x, but of course this breaks down over the large distances that sound can travel in the ocean. However, averaging the sound-speed profile corresponding to $\hat{c}(x,z)$ over range gives a simple model that gives a good

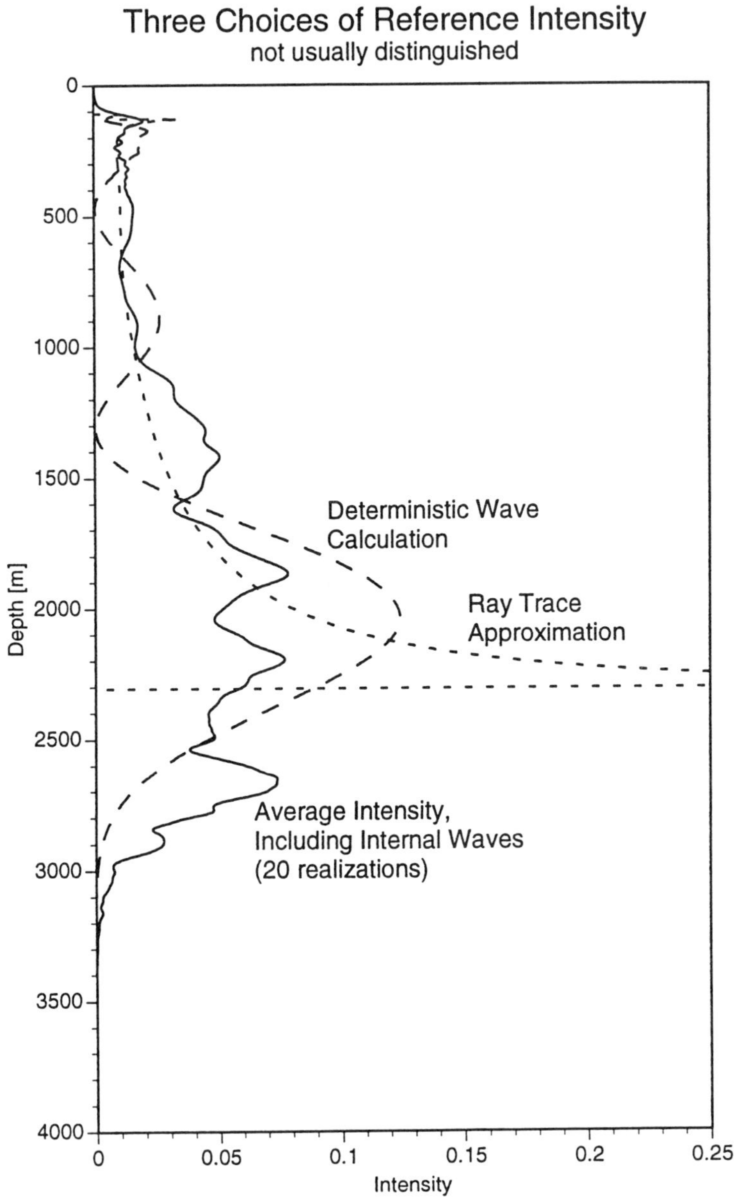

Figure 3. The effect of diffraction and fluctuations on the average intensity. The problem of figure 1 is evaluated by ray tracing, by deterministic wave calculation, and by averaging a number of simulations containing realistic fluctuations. All calculations were based on the same parabolic equation. equation.

understanding of the basic phenomena involved, and we can then picture the sound propagating along deterministic (and for the most part uncorrelated) ray-paths given by $\hat{c}(z)$, with random fluctuations superposed due to the $\delta c(x, z)$ terms. These refractive index variations directly give rise to phase fluctuations, or travel time fluctuations, and then as the wavefield propagates further, these phase fluctuations give rise to amplitude variations, or intensity fluctuations, as will be discussed later.

The largest fluctuations cause most of the phase variance, but do not do much focusing. Somewhat smaller fluctuations dominate the intensity fluctuations. The transfer function in the Rytov regime has been worked out by Desaubies (1978). As a consequence, for the MATE lower path, travel times, which are phases divided by frequencies, are identical to a few microseconds at frequencies from 2 to 13 kHz, while intensities are nearly uncorrelated. Simulations show the higher frequency intensity to have finer structure. When the intensity is averaged over a vertical extent, the correlations between different frequencies get much larger (Ewart, private communication).

Travel time statistics are of particular interest in ocean acoustics. The "tomography" experiments use travel time to attempt to determine changes in the temperature on the part of the ocean through which the sound has traveled. A surprise occurs at the very lowest moment, the average travel time for weak (Rytov) scattering in a uniform medium, as described in Codona et al (1985). For mean-zero fluctuations in the index of refraction, there is no first order travel time shift, but there is a second order bias. The mean "travel time" along a straight line connecting source and receiver is unchanged. Fermat's principle says the actual ray has the shortest time, so the bias must be toward earlier times. Yet when the calculation is done using moment equation or path integral methods, the predicted bias is toward later times. The resolution of this puzzle turns out to be in the meaning of "average". Moment-based techniques calculate the intensity-weighted average, unlike the unweighted average for which the Fermat argument applies. There is a positive correlation between the first-order travel-time fluctuations and the first-order intensity fluctuations, giving a second-order bias in the intensity weighted average travel time which turns out to reverse the sign from the unweighted average. Remarkably, except for sign, the expressions are identical in the weak scattering approximation, and the magnitude is the same as the bias in the extra Euclidean path length. As the scattering strength increases, Fermat's principle becomes a stationary-time, rather than least-time, principle, and the Euclidean path length increase becomes the dominant effect.

We now turn to a general treatment of equation (2.3) using regular perturbation techniques, under the assumption that the scattering is sufficiently weak, which in practice means that the acoustic amplitude variations must be small. Under this approximation, it is convenient to treat the deterministic and random refractive index variations separately. Define Ψ_0 to be the solution of the deterministic form of (2.3):

$$2ik_0\frac{\partial\Psi_0}{\partial x}+\frac{\partial^2\Psi_0}{\partial z^2}=2k_0^2\frac{\hat{c}(x,z)}{c_0}\Psi_0, \tag{3.2}$$

with $\Psi_0 \to \Psi_i = \frac{1}{\sqrt{x}}e^{ik_0z^2/2x}$ as $(x,z)\to(0,0)$. This is the initial condition corresponding to the parabolic expansion of a point source at the origin, with a factor of e^{ik_0x} absorbed in our definition of Ψ. For a uniform linear profile, say $\hat{c}=az$, with a a constant, (3.2) can be solved exactly, by Fourier transform for example. However, for more general profiles we need an approximate solution of (3.2), and it is common (see e.g. Flatté et al, 1979) to use a ray or eikonal model, where the substitution $\Psi_0 = A_0(x,z)\exp[ik_0\phi_0(x,z)]$ is made, which leads to

$$\frac{\partial\phi_0}{\partial x}+\frac{1}{2}\Big(\frac{\partial\phi_0}{\partial z}\Big)^2=-\frac{\hat{c}(x,z)}{c_0}. \tag{3.3}$$

at leading order in $1/k_0$, with a similar (transport) equation for A_0 at the next order. This is a high-frequency, non-uniform approximation, and equation (3.3) is the parabolic form of the *eikonal* equation for the phase. This first order nonlinear PDE for ϕ_0 can be solved explicitly, along with the next order transport equation, and we find

$$\phi_0(x,z)=\int_0^x\Big(-\frac{\hat{c}(x',z(x'))}{c_0}+\frac{1}{2}\Big[\frac{dz(x')}{dx'}\Big]^2\Big)\,dx' \tag{3.4}$$

along the rays $z(x)$ given by the solution of

$$\frac{d^2z}{dx^2}=-\frac{1}{c_0}\frac{\partial\hat{c}(x,z)}{\partial z}, \tag{3.5}$$

and the intensity is proportional to the reciprocal of the spacing of infinitesimally close rays (Landau and Lifshitz, 1975, Chapter 7, and Whitham, 1974). Thus in the parabolic approximation a linear sound-speed profile gives rise to rays that are parabolas, while a quadratic profile gives sinusoidal rays. We see from equation (3.4) that the phase acquired by the field as it propagates through the sound channel is just $k_0\phi_0(x,z)$. For sound-speed profiles that give rise to caustics, such as cubic or higher degree polynomials, diffractive corrections are required at the caustics that are formed (Landau and Lifshitz, 1975, Chapter 7). Alternatively, one can return to a direct numerical solution of (3.2) using a number of standard techniques, often based around the use of the FFT, which we discuss in section IV.

We can now proceed to the perturbative treatment of the stochastic equation (2.3), treating the deterministic sound-speed variation by using $\Psi_0 = A_0 \exp(ik_0\phi_0)$ as our reference solution. If the expansion $\Psi = \Psi_0 + \Psi_1 + \ldots$ is made, where Ψ_1 is assumed small when compared with Ψ_0, then we find

$$2ik_0\frac{\partial\Psi_1}{\partial x} + \frac{\partial^2\Psi_1}{\partial z^2} = 2k_0^2\frac{\delta c(x,z)}{c_0}\Psi_0 + 2k_0^2\frac{\hat{c}(x,z)}{c_0}\Psi_1, \tag{3.6}$$

where the right hand side has been approximated on the basis that the scattering is very weak. Here we are expanding around the deterministic solution for a single ray; in general a sum over rays is required, and this clearly breaks down if two deterministic rays are close enough so that the random sound-speed fluctuations along them are correlated with one another, as is the case for rays passing through a caustic. Neglecting the deterministic term on the right hand side of (3.6) (this can be included with a little extra effort) we get the Born approximation

$$\Psi_1(x,z) = -k_0\sqrt{\frac{ik_0}{2\pi x}}\int_0^x\int_{-\infty}^{\infty}\exp\frac{ik_0(z-z')^2}{2x}\frac{\delta c(x',z')}{c_0}\Psi_0(x',z')dz'dx' \tag{3.7}$$

which could also be obtained as the first term in a Neumann series expansion of the integral equation corresponding to (2.3), with use of the Green's function for the deterministic problem. Further terms of such an iteration give multiple integrals. An approximate summation of these integrals as described in Flatté et al (1979, p166, Chapter 10) gives the *Rytov* approximation. Alternatively, put $\Psi = \exp(\Phi)$ in (2.3), and use (3.1) to get

$$2ik_0\frac{\partial\Phi}{\partial x} + \frac{\partial^2\Phi}{\partial z^2} + \left(\frac{\partial\Phi}{\partial z}\right)^2 = 2k_0^2\frac{\hat{c}(x,z)}{c_0} + 2k_0^2\frac{\delta c(x,z)}{c_0}. \tag{3.8}$$

We have now converted the linear stochastic PDE of (2.3), where the 'noise' corresponding to the refractive index variations is multiplicative, to a nonlinear stochastic PDE with additive noise, which is of course no easier to solve! (see the discussion in N.G. Van Kampen, 1981, Chapter 14). If we expand $\Phi = (\Phi_0 + \Phi_1 + \ldots)$, and identify Ψ_0 with e^{Φ_0}, then we obtain a transformed version of the deterministic background equation (3.2) to leading order and to the next order we find that

$$2ik_0\frac{\partial\Phi_1}{\partial x}+\frac{\partial^2\Phi_1}{\partial z^2}+2\frac{\partial\Phi_0}{\partial z}\frac{\partial\Phi_1}{\partial z}=2k_0^2\frac{\delta c(x,z)}{c_0}, \tag{3.9}$$

where it is implicitly assumed that random sound-speed fluctuations are small compared to deterministic variations. We then use the transformation $\Phi_1 = u\exp(-\Phi_0)$ in (3.9), which gives equation (3.6), so that u is precisely Ψ_1 where as previously we ignore the second order correction to the deterministic solution. Therefore the Rytov solution is given by $\Psi_0\exp(\frac{\Psi_1}{\Psi_0})$, and we see that the Born approximation is regained if we expand the exponential.

All these approximations implicitly make the assumption that the fluctuations of the acoustic field are small in some sense. So which approximate solution is better? Consider the balance

$$2ik_0\frac{\partial\Psi}{\partial x}=2k_0^2\frac{\delta c(x,z)}{c_0}\Psi, \tag{3.10}$$

where we ignore diffraction and the deterministic sound-speed variations in (2.3). If we integrate (3.10) from x_0 to x_1 we find

$$\Psi(x_1,z)=\Psi(x_0,z)\exp\Big(-ik_0\int_{x_0}^{x_1}\frac{\delta c(x',z)}{c_0}dx'\Big)=\Psi(x_0,z)e^{-ik_0\phi(z)}, \tag{3.11}$$

say. Therefore, in the absence of any other phenomena, the sound-speed fluctuations appear immediately as a phase modulation in the acoustic field. This feature is correctly modeled by the Rytov approximation (or by geometric optics) but not by the Born approximation, and so long as the phase modulation is additive in the exponent, the Rytov approximation will remain correct. Thus, for moderate ranges, it is found that the phase spectrum of the acoustic field mirrors the sound-speed spectrum precisely, in the sense that the sound-speed fluctuations integrated over range appear as phase fluctuations, as in (3.11). This has been explicitly demonstrated in the MATE experiment (Ewart and Reynolds, 1984, 1992), over a range of 18km, at a depth of around 1000m, for acoustic frequencies between 2 and 13kHz.

However, the amplitude fluctuations are not so straightforward. For amplitude fluctuations due to the randomness of the medium, we need to study the evolution of Ψ_1 with range in the Born approximation. Here, for the perturbation to make sense, we require $|\Psi_1| \ll |\Psi_0|$, so the method is limited to small fluctuations in both phase and amplitude. By contrast, using the Rytov approximation, the total phase modulation due to random scattering may be large, because Φ_1 may be large, and yet the ordering $|\Phi_1| \ll |\Phi_0|$ may be preserved. Consider the simple case where there is no sound channel and the initial condition is that corresponding to a plane wave. Then $\Phi_0 = ik_0z^2/2x$. Thus $|\Psi_0| = |\exp \Phi_0| = 1$, and amplitude fluctuations, as determined by $\exp \Phi_1$ must give a multiplicative factor close to unity for any perturbative approximation to be appropriate i.e. for the amplitude fluctuations we require $|\Phi_1|$ to be small, not just small compared with $|\Phi_0|$. Thus, both the Born and Rytov techniques are limited to small fluctuations of amplitude, with the intensity variance significantly less than unity, as was discussed extensively by Tatarski (1971). For propagation over greater distances, or through larger fluctuations, when the scattering is strong and the intensity variance may easily reach a value of two or more in the ocean, deterministic equations describing the propagation of moments of the complex field are required, as described in the next section. In general, these are far more difficult to solve, but allow treatment of multiple scattering and strong intensity fluctuations.

Let us now extend the model of (3.10) by calculating the effect of diffraction for $x > x_1$, in the absence of any further phase modulation, i.e. an approximate model for the effect of a thin slab of inhomogeneous medium, of width $x_1 - x_0$. (The limiting form of this approximation, when the slab thickness goes to zero, but the phase modulation remains finite, is the so-called phase screen model, where the effects of diffraction on a wavefield traveling through free space, with some initial phase modulation are studied as in e.g. Mercier (1962), Jakeman and McWhirter (1977), Roberts (1985). The patterns cast on a wall by light that has passed through an irregular window pane are qualitatively similar to the patterns of intensity variation caused by sound-speed fluctuations in the ocean.) The parabolic wave operator of (2.3) applied to such an initial phase modulation gives:

$$(3.12)\quad \Psi(x,z) = -\frac{1}{2k_0}\sqrt{\frac{ik_0}{2\pi x}}\int_{-\infty}^{\infty} \exp\frac{ik_0(z-z')^2}{2x}\Psi(x_0,z')e^{-ik_0\phi(z')}dz'.$$

Clearly, as the propagation distance x increases, we find amplitude fluctuations arising. A stationary phase approximation to (3.12) shows that the major contribution to the amplitude fluctuations is given by $z \approx z'$, i.e. small vertical scales, and only as the acoustic wave propagates further do the larger scales become significant (i.e. as k_0/x becomes smaller). This is true globally for phase-screen modulation problems. However, the basic ideas follow through for the extended medium, if we imagine the medium divided into slabs, with the lumped phase modulation for each slab applied as in (3.11), and with diffraction for each slab following. In fact, this idea has been developed further in numerical simulations (Macaskill and Ewart, 1984a, Martin and Flatté, 1988, 1990), and in approximate analytical solutions for the extended medium problem (Uscinski, 1985).

4. Methodology

4.1 Moment equations and path integrals. In this section we discuss the deterministic equations describing moments of the complex field that can be derived from the stochastic PDE (2.3). First we give the form of the moment equations for a medium with both large-scale deterministic and small-scale random sound speed variations. This is used in practice for ocean acoustic propagation studies where scattering is significant. In the 2-D case, where there is no y-dependence, the n-th moment of the field is defined to be $m_n = \langle \Psi_1 \Psi_2^* \Psi_3 \Psi_4^* \cdots \Psi_{n-1} \Psi_n^* \rangle$, where the subscripts correspond to different transverse locations z_j and also possibly wavenumbers k_j. However, here we only consider the special case $k_j = k_0$ for all j. If the mean squared phase fluctuation $\langle |\Phi|^2 \rangle \gg 1$, then only moments with equal numbers of unconjugated and conjugated fields are measurably large, so we restrict ourselves to that case. Then the equation describing the evolution with range of the n-th moment is given in the Markov approximation by (see, e.g. Codona et al, 1986a)

$$\frac{\partial m_n(x)}{\partial x} = -iL_0 m_n(x) - \frac{1}{2}\Big[\int_{-\infty}^{\infty} dx' \langle M(x) M_{shift}(x') \rangle\Big] m_n(x). \tag{4.1}$$

Here the deterministic part of the evolution is described by the operator

$$L_0 = \sum_{j=1}^{n} \pm \frac{1}{k_0} \Big(-\frac{1}{2} \frac{\partial^2}{\partial z_j^2} + k_0^2 V_0(z_j) \Big) \tag{4.2}$$

with the plus sign corresponding to the Ψ_j's and the minus sign corresponding to the Ψ_j^*'s, and the deterministic background $V_0(z) = \hat{c}(z)/c_0$ assumed range-independent. The randomly fluctuating part of the medium is given by

$$M(x) = \sum_{j=1}^{n} \pm k_0 \mu(x, z_j) \tag{4.3}$$

with the same convention on the signs as for L_0 and with $\mu(x, z) = \frac{\delta c(x,z)}{c_0}$. $M_{shift}(x)$ corresponds to $M(x)$ shifted to the transverse location that a deterministic ray would move to in going from x to x' (see below for a more detailed discussion).

Derivations of the moment equations for an *isotropic* medium, specifically the turbulent atmosphere, were given by a number of authors in the late sixties (Beran and Ho, 1969, Chernov, 1969, Shishov, 1968, Tatarski, 1969). Extensive discussion is also given in the standard reference books of Tatarski (1971) and Ishimaru (1977) where the medium is assumed delta-correlated in the range-direction so that the Markov approximation may be used and results from functional analysis applied. A complementary and physically revealing derivation is given in the book by Uscinski (1977), where free-space propagation and the scattering effects are treated independently over the scale of a single medium irregularity.

Uscinski identifies two physical requirements that allow the moment equations to be derived. The first is that angles of scatter should be small. The characteristic angle of scatter for irregularities with vertical scales much smaller than horizontal scales is determined by the vertical scale and is roughly $1/kL_v$. Therefore, this requirement has already been imposed by the use of the parabolic approximation. His second requirement is that the phase perturbation imposed in a single irregularity should be small, i.e. that $k\langle\mu^2\rangle^{1/2}L_p$ should be small (as is also required for the neglect of higher order terms in the development given by Van Kampen, 1981, described below); this requirement allows the use of the Born or Rytov approximations *locally*, i.e. over a single irregularity, as in Chernov's method, described in Prokhorov et al (1975) as the 'local method of small perturbations'. Uscinski then expands $\psi(x + \Delta x)$ in a Taylor series, and then assumes that the field at x is uncorrelated with the fluctuations in the slab of medium $(x, x + \Delta x)$ to determine a differential relation for the effect of weak scattering. Adding in the deterministic propagation operator then gives the isotropic form of the moment equation under consideration. Note that although the effect of scattering over a single irregularity is weak, the integrated effect of the scattering through many such irregularities may still be strong.

For anistropic scattering, where (typically) the horizontal scale size greatly exceeds the vertical scale size, Uscinski's treatment may be extended (Uscinski, private communication) provided that his two basic assumptions are still satisfied. This simplified treatment gives rise to an integro-differential equation similar to those given by Besieris and Tappert (1973) and Beran and McCoy (1976). This formulation was explored for the second moment problem in McCoy and Beran (1979). Here, by considering the plane-wave expansion of the propagating moment, they showed that a quasi-isotropic formulation was adequate so long as $L_p/kL_v^2 \ll 1$, i.e. the characteristic scattering angle $1/kL_v$ multiplied by the anistropy ratio L_p/L_v should be small, which is a much stronger restriction than just requiring the characteristic scattering angle to be small (for convenience, we use here the inverse of the parameter α discussed in McCoy and Beran). For example, in the ocean case, the anistropy ratio may vary between 10 and 100 or so.

They also discussed what they called a 'highly anistropic approximation' in the case $L_p/kL_v^2 \gg 1$, but even in this case, as they note, their results still require weak scattering in a single irregularity. The extension to problems where strong scattering takes place while the medium remains correlated in the range direction has been explored using direct numerical simulation for an idealized problem in Spivack and Uscinski (1989) and for a realistic ocean case in unpublished work of Ewart and Kaczkowski.

An alternative way to describe the moment equation is to start with the lowest order transport equation (Van Kampen, 1981). Starting with the product of field values whose expectation is the moment of interest, we can take its range derivative, and use the parabolic equation to express that in terms of an operator acting on that product. The operator is a sum of a deterministic operator and a mean-zero fluctuation, assumed to be a Gaussian process. In the limit that the interaction is weak, we can derive the moment equation (Codona et al, 1986a)

$$\frac{\partial}{\partial x} m = -iL_0 m - Gm \tag{4.4}$$

where m is the moment under consideration, L_0 is the deterministic part of its evolution as in equation (4.2) and in terms of the fluctuation M,

$$G = \int_0^x \langle M(x) U_0 M(x') U_0^{-1} \rangle dx' \tag{4.5}$$

where U_0 is the deterministic evolution operator

$$U_0 = e^{-iL_0(x-x')} \quad . \tag{4.6}$$

If U_0 were replaced by the full evolution operator U, with all correlations taken between the M's in U, U^{-1}, and G included, the equation would be exact. The approximation is thus that the propagation from x' to x and back again after application of M can be replaced by the same free propagation. The correction involves the commutator of $U_0 M(x') U_0^{-1}$ with the same expression at a different value of x'. By estimating the commutator as a product, we obtain the condition

$$k_0^2 \langle \mu^2 \rangle L_p^2 \ll 1 \tag{4.7}$$

as guaranteeing the correctness of the approximation. It is possible that the commutator is considerably smaller than the product, in which case the approximation is even better. Besieris and Tappert (1973) derived a similar moment equation, which is an integro-differential equation rather than a differential equation. Theirs is related to the one above by replacing U_0^{-1} by U^{-1} but leaving U_0 alone. Their corrections are products rather than commutators. Contrary to an assertion in Codona et al (1986a), a further approximation by Besieris and Tappert which they called the "long-time Markov approximation" is not accurate for ocean acoustics with anisotropic fluctuations, and will be ignored.

The moment equation in actual use is related to equation (4.4) by a simple approximation. The idea is that the propagation is dominated by the deterministic ray propagation, and that the ray is approximately a straight line over a distance L_p.

$$U_0 M(x') U_0^{-1} = M_{shift}(x') \tag{4.8}$$

where $M_{shift}(x')$ is obtained from $M(x')$ by replacing any depth z by $z - \frac{dz_{ray}}{dx}(x - x')$. Finally, the integration over $x' < x$ is replaced by half the integral over all x', as a backwards transport equation with $x' > x$ could have been derived, and the average of the two taken. At this point, we have the Markov approximation described above. The trouble comes because that replacement is not always very good. The scattered wave from x' may spread over a vertical extent at x large enough that its average correlation is considerably different from the straight-line correlation, due to the strong anisotropy of the correlation function. Secondly, the ray might not be straight enough over a distance L_p. That was the cause of the large $\frac{dL_p}{dx}$ for the upper path at the MATE experiment, where strong anisotropy of the fluctuations and strong curvature of the ray worked together. Finally, there can be a bias of the fluctuations, favoring deviations toward the stronger fluctuations than given by the ray; $\langle \mu^2 \rangle$ is a rapidly varying function of depth near the apex of the upper MATE ray. None of these difficulties cause concern with equation (4.4), which has not been used in practice. The calculation of the operator G is likely to be rather difficult. The condition $k^2 \langle \mu^2 \rangle L_p^2 \ll 1$ is often quoted in the literature as the condition for the Markov approximation. As we have discussed, the deterministic sound speed profile and the anisotropy of the fluctuations make this condition insufficient.

One might have a much better approximation by shifting M by $z_{ray}(x) - z_{ray}(x')$ instead of its tangent straight line; the worst problem with the MATE upper path would thereby be avoided.

Since the parabolic equation is identical to the Schrödinger equation, techniques that have been developed for nonrelativistic quantum mechanics can be used for ocean acoustics. One such technique is Feynman's path integral (Feynman and Hibbs, 1965, Schulman, 1981). The path integral is a formal expression which represents a phase screen method for solving the Schrödinger equation. The phase screen method imagines concentrating the $\frac{1-n^2}{2}$ at a set of specific ranges. At those ranges, the wave undergoes a phase change of $\frac{1-n^2}{2}k_0\delta x$, where δx is the spacing between screens. Between the phase screens, free propagation occurs. In the limit that the spacing between screens, δx, goes to zero, the phase screen approximation approaches the solution of the Schrödinger equation. (Mathematically, this is a time-ordered extension of the Trotter product formula.) The product of integrals over the depth (and transverse horizontal, if it is to be included) at each phase screen, multiplied by a constant factor from the free propagation between the screens, is written as $\int \mathcal{D}\text{paths}$. The quantity $\frac{z(x+\delta x)-z(x)}{\delta x}$ is written as $\frac{dz}{dx}$, indicating the limit $\delta x \to 0$. The summation over phase screens including a factor δx is written as $\int dx$. Then the phase screen solution is

$$\Psi = \int \mathcal{D}\text{paths}\exp\left(ik_0\int\left[\left(\frac{1}{2}\frac{dz}{dx}\right)^2 - \frac{1-n^2}{2}\right]dx\right) \tag{4.9}$$

for a unit point source. Other initial states can be handled by integrating over the depth at the initial x. The path integral for an n-th moment is obtained by multiplying n such path integrals together, forming the complex conjugate where necessary, and taking a statistical average. Here the assumption that the fluctuations in $\frac{1-n^2}{2}$ comprise a Gaussian process enters; the expectation of the exponential is the exponential of half the variance. For example, the first moment is

$$\langle\Psi\rangle = \int \mathcal{D}\text{paths}\exp\left(ik_0\int\left[\left(\frac{1}{2}\frac{dz}{dx}\right)^2 - V_0\right]dx\right.$$
$$\left. - \frac{k_0^2}{2}\int\int\langle\mu(x,z(x))\mu(x',z(x'))\rangle dx dx'\right) \tag{4.10}$$

where V_0 is the ensemble average of $\frac{1-n^2}{2}$ and μ is its deviation from that average.

In the Markov approximation

$$\int\langle\mu(x,z(x))\mu(x',z(x'))\rangle dx' = \langle\mu^2\rangle L_p \quad , \tag{4.11}$$

and the exponent of equation(4.10) becomes a single integral over the range.

This form allows an alternate derivation of the moment equation. The path integral for $\langle\Psi\rangle$ is just like the equation for Ψ with $V = V_0 + \mu$ replaced by $V_0 + i\frac{k}{2}\langle\mu^2\rangle L_p$, so as long as the extra i doesn't bother the reader, the relationship of the path integral and the equation must be the same in both cases. Since the original path integral solves the Schrödinger equation, the averaged path integral solves the Schrödinger equation with the same replacement; see equation (4.1). Higher moment equations are derived from the path integral in a similar way in Codona et al (1986a).

4.2 Moment equation solutions. The deterministic equation for the mean field, and the equation describing the propagation of the second moment, derived as above, can be solved exactly in a number of cases. For example, in the quasi-isotropic case the second moment equation can be solved exactly with a background refractive index of quadratic form (Beran and Whitman, 1975). However, the equation describing the propagation of the fourth moment of the complex field, which must be solved if information on the intensity correlation or the intensity variance is required, has proved far more intractible.

In the absence of refractive effects, the equation for the fourth moment $m_4 = \langle\Psi_1\Psi_2^*\Psi_3\Psi_4^*\rangle = \langle\Psi(x,z_1)\Psi^*(x,z_2)\Psi(x,z_3)\Psi^*(x,z_4)\rangle$ can be reduced to the form (e.g. Uscinski,1977):

$$\frac{\partial m_4}{\partial X} = i\frac{\partial^2 m_4}{\partial Z\partial u} + i\frac{\partial^2 m_4}{\partial\zeta_1\partial\zeta_2} + \Gamma h(\zeta_1,\zeta_2,u)m_4 \tag{4.12}$$

where

$$\begin{aligned} h(\zeta_1,\zeta_2) = & f(\zeta_1 + u/2) + f(\zeta_1 - u/2) + f(\zeta_2 + u/2) + f(\zeta_2 - u/2) \\ & - f(\zeta_1+\zeta_2) - f(\zeta_1-\zeta_2) - 2 \end{aligned}$$

and $f(\zeta)$ is the transverse correlation of intensity fluctuations, with

$$\begin{aligned} \zeta_1 &= (z_1 - z_2 - z_3 + z_4)/2L_v \\ \zeta_2 &= (z_1 + z_2 - z_3 - z_4)/2L_v \\ u &= (z_1 - z_2 + z_3 - z_4)/L_v \\ Z &= (z_1 + z_2 + z_3 + z_4)/4L_v \end{aligned}$$

for the transverse locations z_1, z_2, z_3, z_4. (A word of caution: note that many books and papers interchange the rôle of locations 3 and 4, following Tatarski, 1971). Here $X = x/kL_v^2$ is the scaled range and $\Gamma = k^3\langle\mu^2\rangle L_p L_v^2$ is the parameter describing the strength of the scattering. This simplified form of the fourth moment equation holds only for statistically homogeneous media, where there is no dependence of the medium correlation on the average location Z; this is not a good approximation near the surface but is reasonable for the deep ocean. In this case it is also reasonable to ignore any dependence on y. This is justified on the basis that the sound speed fluctuations are highly anistropic, with the horizontal correlation far greater than the vertical correlation. Therefore one expects angles of scatter in the horizontal to be very small compared to those in the vertical, so that the scattering problem can be treated as two-dimensional. If in addition, the initial condition for m_4 is independent of Z, as is the case for a point source or plane wave initial condition, then the first term on the right hand side can be ignored. Once this has been done, the independent variable u only appears parametrically. The choice $u = 0$, corresponding to the special choice of the four observation points such that $\zeta_1 = (z_1 - z_2)/L_v = (z_4 - z_3)/L_v$ and $\zeta_2 = (z_1 - z_4)/L_v = (z_2 - z_3)/L_v$, contains the interesting case of intensity correlations as follows: on the axis $\zeta_2 = 0$, we have $z_1 = z_4$ and $z_2 = z_3$, so that the fourth moment reduces to the correlation of intensity $m = \langle \Psi_1 \Psi_2^* \Psi_2 \Psi_1^* \rangle = \langle I_1 I_2 \rangle$; similarly on the axis $\zeta_1 = 0$, $m = \langle \Psi_1 \Psi_1^* \Psi_3 \Psi_3^* \rangle = \langle I_1 I_3 \rangle$. Finally at the origin, $\zeta_1 = \zeta_2 = 0$, so that the fourth moment reduces to the second moment of intensity, $m_4 = \langle I^2 \rangle$.

Approximate solutions for the fourth moment equation may be determined by straightforward perturbation analysis, using the scattering parameter Γ as a small parameter; the first-order solution is then equivalent to the Born approximation. A more general approximate result, was obtained by Uscinski (1982) by writing each term of the series expansion as a multiple convolution, then approximately evaluating each of these multiple convolutions, and then finally forming an approximate sum over all orders. This gives the approximate solution for the fourth moment $m_4(X, \zeta_1, \zeta_2)$, e.g. for a plane wave initial condition, where the corresponding initial condition is $m_4(X = 0) = 1$:

(4.13)
$$m_4(X, \zeta_1, \zeta_2) = \frac{1}{2\pi} \iint\limits_{-\infty}^{\infty} \exp\left[\Gamma \int_0^X h(\zeta_2', \zeta_1 + \nu s)\, ds\right] \exp\left[i\nu(\zeta_2 - \zeta_2')\right] d\nu\, d\zeta_2'$$

This approximate form has also been found by a number of different techniques, in Macaskill (1983), Frankenthal et al (1984), Uscinski (1985), Codona et al (1986b), Gozani (1987), Furutsu (1988), and in these papers and others various corrections to this basic form have been proposed. Without exception, these corrections are very complicated, involving multiple integrals, and are very difficult to evaluate accurately. The basic approximation, by contrast, is relatively easy to evaluate, and it has been shown to reproduce weak fluctuation results successfully. It also approaches the saturation limit, as the range approaches infinity, but not with the correct functional behaviour. At intermediate ranges, for strong scattering, where the scintillation index, or variance of intensity reaches its maximum value, it gives qualitatively correct forms for the spectra of intensity fluctuations, but under-predicts the total scintillation index, particularly for point source initial conditions. Moreover, as the scattering strength increases, this under-prediction becomes more rather than less serious, so early claims that the method is asymptotic in the scattering strength are not correct. The various correction terms, while generally giving improved comparisons with simulations and experimental results, also show the greatest discrepancies near the peak of the scintillation index, with this discrepancy increasing with the scattering strength Γ. Moreover, the contribution from the correction terms appears to be important at all spatial frequencies, i.e. uniformly across the spectrum of intensity fluctuations, which explains why good qualitative predictions are possible using the simple approximation. Because of these difficulties, numerical techniques for the solution of (4.12) are important.

4.3 Numerical techniques and simulation. In the analytic treatments, the medium is assumed to be unbounded. In a numerical approach, one needs to consider the boundary conditions in the finite domain carefully. For unbounded media, the appropriate boundary conditions for large ζ_1, ζ_2 are that m or its normal derivative should go to zero. In the deep ocean, this may still be appropriate, for the following reasons. At some fixed range x, the intensity correlation $\langle I_1 I_2 \rangle$ between the two points z_1, z_2, say, which can be recognized as $m_4(X, \zeta_1, \zeta_2 = 0)$, has a characteristic scale that at short range is the same as that of the medium, say 150m in the vertical at the lower path of the MATE experiment. The correlation length of the intensity decreases markedly as the wavefield propagates further into the medium and the integrated effect of the scattering becomes more significant, with this effect becoming more marked if the acoustic frequency and hence the scattering strength is increased. Therefore, for a ray path that does not interact with the surface or the bottom, it is reasonable to treat the numerical solution of the fourth moment using the boundary conditions appropriate for an unbounded medium.

For an unbounded medium, the fourth moment equation was first solved numerically by Dagkesamanskaya and Shishov (1970), Brown (1972) and then by Tur (1982, 1985) using finite difference techniques on a fixed grid. Encouraging agreement with approximate analytical solutions was reported for atmospheric propagation in Gozani (1985). Further progress was made when Spivack and Uscinski (1988) provided numerical solutions using a split-step Fourier technique on a fixed grid. However, all these results were limited to moderate values of the scattering strength Γ, because for large scattering strengths, as the scaled range X increases, extremely high resolution is needed near the axes $\zeta_1 = 0$ and $\zeta_2 = 0$, and yet at the same time a large calculational domain is required to ensure that all features of the solution are resolved. If this last requirement is not satisfied, inaccuracy and instability result.

In order to overcome these problems adaptive grid methods (see e.g Dorfi and Drury, 1987 for a discussion of the application of similar methods in gas dynamics) were used for a medium with a Gaussian correlation function for three cases. These were the plane wave initial condition, Leonard et al (1990), the point source initial condition, Reeve et al (1990) and the two frequency problem with a plane wave initial condition, Leonard and Uscinski (1993). In these papers the grid was allowed to adjust automatically with range, so that points were clustered where the fourth moment was changing most quickly in the transverse directions. The numerical approach has the advantage that it can be used to deal with the decorrelation between two signals of different acoustic frequencies; as discussed in Miller and Uscinski (1986), existing approximate analytic solutions of the fourth moment problem are very inaccurate for such problems, although they show some of the correct qualitative behaviour.

In the above papers the method of lines was used. In discussing this approach, it is convenient to think of the scaled range variable X as a time-like variable, as the fourth moment equation is of parabolic form. The derivatives in the transverse direction can then be discretized, using standard finite difference approximations. This gives rise to a system of complex-valued coupled ordinary differential equations with independent variable X for the unknowns $m_{4,i,j} = m_4(X, \zeta_{1,i}, \zeta_{2,j}), i = 1, ..., N, j = 1, ..., N$, where N is typically between 20 and 50. This can be re-expressed as a real system by taking real and imaginary parts. Any standard variable step ODE integration package can then be used to solve the system of N^2 differential equations, e.g. a Runge-Kutta-Fehlberg method of order (4,5) has been used successfully in practice (e.g. in Macaskill and Ewart, 1995), as the equations are not stiff. The mesh variation with range is determined by the solution of auxiliary differential equations which can be formulated in a number of ways, but which are essentially set up to ensure that the grid points are clustered where the fourth moment changes most quickly. (In practice, this means that high resolution is needed on the axes $\zeta_1 = 0, \zeta_2 = 0$ and particularly near the origin $\zeta_1 = \zeta_2 = 0$, which happen to be the regions of practical interest, because as discussed above the fourth moment reduces to the correlation of intensity on the axes, and at the origin to the second moment of intensity; these are the quantities most easily measured.)

The point source initial condition, which corresponds to the case encountered in ocean propagation, is the most demanding numerically. Reeve et al (1990) tranformed to cylindrical coordinates to deal with this problem: this changes the form of the differential operator in the moment equation, but the method of lines can still be applied. This approach, in a simplified form, has been used by Macaskill and Ewart (1995), to make predictions of the intensity fluctuations for the lower path at MATE. No allowance was made there for the weak deterministic refraction present. In Figure 4, the real and imaginary parts of the fourth moment for the 2kHz experiment over the MATE lower path, are plotted against ζ_1 and ζ_2 and also against the grid indices i and j. It is clear that a non-uniform grid is essential.

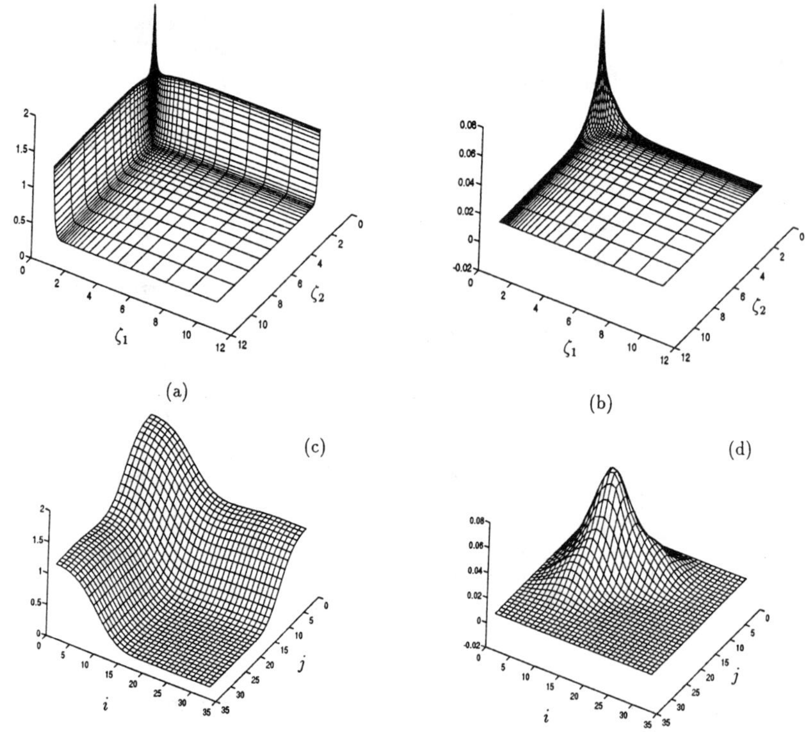

Figure 4. The predicted fourth moment at MATE for the 2kHz case. The real and imaginary parts are shown in (a) and (b) respectively. The real and imaginary parts are plotted against the grid indices i and j in (c) and (d).

To deal with more general initial conditions, one can take a Fourier transform in Z, as described in Tur (1985); the problem is then essentially no more difficult but requires more calculation. Moreover, one can reduce the problem of the fourth moment for waves of different acoustic frequencies to a form equivalent to (4.12), using the transformation of variables given in Uscinski and Macaskill (1985), for example. At this stage, no attempt has been made to incorporate oceanic surface and bottom conditions into numerical solutions of the fourth moment equation, or to include refractive effects in a numerical solution of the fourth moment equation. Treatment of decorrelation in the horizontal, which requires re-introduction of the other transverse variable y, and hence an increase in the dimensionality of the problem, has not been attempted. In addition, no attempt has been made so far to solve the moment equation over the long ranges often encountered in the ocean at lower acoustic frequencies. Furthermore, the form (4.12) above is not appropriate near ray turning points, as discussed earlier, and so should be replaced (at least) by an integro-differential form. These limitations mean that at least at this stage the direct numerical solution of the fourth moment equation is probably restricted to relatively short ranges, with scattering effects dominant and with only weak deterministic refraction.

A more general numerical approach is to return to the stochastic wave equation (2.3) and use Monte Carlo methods directly. Such methods have been employed successfully in three-dimensional atmospheric propagation of light through turbulence by Martin and Flatté (1990).

In the context of ocean acoustics, direct numerical treatment of the parabolic equation (2.3) in the absence of any stochastic variation has been extensively studied using a number of different techniques, of which the most ubiquitous is the split-step Fourier technique of Tappert and Hardin (1974) (see also the review paper of Tappert, 1977). Flatté and Tappert (1975) were the first to apply this technique to the stochastic problem of sound propagation through internal waves. The basic idea is to solve (2.3) for each of an ensemble of realizations of the random process corresponding to the sound-speed fluctuations and then finally to take averages over the realizations, to give a numerical estimate of the various moments of the propagating field. Approximating the refractive index as in (3.1) in terms of the deterministic sound speed variation $\hat{c}(z)/c_0$ (where we assume a range-independent profile for the average sound-speed) and the random fluctuations of sound speed $\delta c(x,z)/c_0$, and as previously neglecting the dependence on y, we have

$$i\frac{\partial}{\partial x}\Psi = -\frac{1}{2k_0}\left(\frac{\partial}{\partial z}\right)^2\Psi + k_0\frac{\hat{c}(z)}{c_0}\Psi + k_0\frac{\delta c(x,z)}{c_0}\Psi. \tag{4.14}$$

The split-step algorithm is similar in concept to the the splitting techniques widely used in computational fluid mechanics (Canuto et al, 1988) in the sense that diffraction on the one hand, and the refractive and scattering effects of the medium on the other, are treated separately in each range step. As described above in the physical interpretation of the path-integral formulation, we imagine the medium split into a collection of slabs, corresponding to range steps, each of which has thickness greater than or equal to the horizontal correlation scale L_p. The refractive and scattering effects of the medium in each slab are imagined to act like a single phase-changing screen. We can see this by ignoring the diffraction terms in (4.14)

$$i\frac{\partial}{\partial x}\Psi = k_0\frac{\hat{c}(z)}{c_0}\Psi + k_0\frac{\delta c(x,z)}{c_0}\Psi. \tag{4.15}$$

and then integrating from x to $x+\Delta x$ to find

$$\Psi(x+\Delta x) = \Psi(x)\exp\int_x^{x+\Delta x}(k_0\frac{\hat{c}(z)}{c_0} + k_0\frac{\delta c(x,z)}{c_0})dx' \tag{4.16}$$

or

$$\Psi(x+\Delta x) = \Psi(x)\exp(k_0\frac{\hat{c}(z)}{c_0}\Delta x + k_0\int_x^{x+\Delta x}\frac{\delta c(x,z)}{c_0}dx'), \tag{4.17}$$

where the term in the exponential gives the (random) change in phase introduced by the medium over the slab $(x, x+\Delta x)$.

Thus in (4.17) in order to account for the stochastic term, for each range step an independent realization of the process corresponding to the integral of $\delta c(x,z)$ over the interval $(x, x+\Delta x)$ is required. As the medium is assumed to have Gaussian statistics, it is enough to calculate the appropriate correlation function, and then any standard procedure can be used to generate the realizations of the required process (see e.g. Macaskill and Ewart, 1984a). The correlation $\rho(x_1 - x_2, z_1 - z_2)$ we need is given by

$$\rho(x_1 - x_2, z_1 - z_2) = \int_{x_1}^{x_1+\Delta x_1}\int_{x_2}^{x_2+\Delta x_2}\langle\frac{\delta c(x_1', z_1)}{c_0}\frac{\delta c(x_2', z_2)}{c_0}\rangle dx_1' dx_2' \tag{4.18}$$

or, on switching to sum and difference variables $x = (x_1' + x_2')$, $\xi = x_1' - x_2'$, and $z = (z_1 + z_2)$, $\zeta = z_1 - z_2$ we find

$$\rho(\xi,\zeta) = \frac{\Delta x}{c_0^2}\int_{-\infty}^{\infty} R(\xi,\zeta)d\xi, \tag{4.19}$$

where $R(\xi,\zeta) = \langle\delta c(x_1,z_1)\delta c(x_2,z_2)\rangle$ is just the correlation function for the sound-speed fluctuations, which have been assumed to be statistically homogeneous.

In (4.19) we have extended the limits of integration for the difference variable to $\pm\infty$ on the basis that the step size is greater than the horizontal integral scale L_p, and that the contribution added by the change of limits is therefore zero. Thus we find that the appropriate correlation function for our algorithm, $\rho(\xi, \zeta)$ is in fact just the *transverse* correlation function, or the projection of the full correlation function on to the vertical. We can then determine the diffraction over the interval $(x, x + \Delta x)$ using Fourier techniques as

$$\Psi(x + \Delta x) = \boldsymbol{F}^{-1}[\exp(-\frac{i\nu^2}{2k_0}\Delta x)\boldsymbol{F}\Psi(x)] \tag{4.20}$$

where it has been assumed that Ψ is periodic in z and F represents a discrete Fourier transform between z and ν, which will usually be determined using FFT (Fast Fourier Transform) techniques. Equations (4.17) and (4.20) taken together give an algorithm for advancing the field through one range step. We note that in the limit as the range steps become small, and with the phase-changing screens statistically independent, we regain the path-integral formulation, with δ-correlation of the medium in the range direction. In practice, one needs to ensure that each independent screen only gives a weak modulation of the field in order to obtain accurate solutions. In addition, one can allow the screens to be correlated, and thus obtain solutions for the case where the Markov approximation is not valid.

Intensity conservation for arbitrary initial conditions is equivalent to unitarity of the matrix that relates the initial conditions to the propagated field. The split-step method preserves this unitarity to round-off accuracy. That is because the individual operations of multiplying by a function of magnitude 1 (in either position or wavenumber space) and FFTing are unitary operations. Similar split-step techniques are used in nonlinear dynamics to preserve the symplectic structure of Hamiltonian flows (which generalizes the unitarity of the linear problem).

Realistic calculations of this sort, with correlations of the sound-speed fluctuations modeled as due to internal waves, and with fluctuations correlated in range, are described in Jackson and Ewart (1994), where an absorbing layer is used above and below the water column to remove energy that is not constrained to remain in the sound channel. For very long range propagation, over 1000 km, at a frequency of 250 Hz, Colosi, Flatté and Bracher (1994) used simulation of a modified (wide-angle) parabolic equation (Berman et al, 1992), again with range-correlated fluctuations, to make comparisons with the results of the SLICE89 experiment (Duda et al, 1992). In these simulations Colosi et al used a reflecting ocean surface and an absorbing ocean bottom. Excellent agreement of direct simulation with experimental data has been demonstrated in the problem of atmospheric scattering of laser light by Flatté et al (1993).

What are the advantages and disadvantages of such direct simulation procedures when compared with theoretical approaches, such as the analytic or even numerical treatment of the moment equations? Clearly, a simulation allows us to see what is taking place in detail, and allows realistic representation of boundary conditions and range-dependent fluctuations of the medium. However, particularly at high acoustic frequencies, very high resolution is required in order to obtain accurate results, as the detailed fluctuations of the propagating field must be resolved. In addition to this, averages over many realizations must be used to limit the errors due to statistical fluctuations. Furthermore, in order to deal with pulse propagation, one must carry out a full simulation for each of a suitable number of single frequency components corresponding to the Fourier decomposition of the pulse, before weighting by a suitable filter function corresponding to the pulse shape and then transforming back to the time domain (see, e.g. Colosi et al, 1994).

By contrast the moment equations have solutions that are smooth and deterministic, but as discussed earlier the precise form of the equation used must be appropriate (i.e. do we use the isotropic form or the integro-differential form?) and as the order of the moment being treated increases, so does the dimensionality of the problem. This last feature is a particular difficulty for numerical solutions. Certainly, the treatment of moments of order higher than fourth by either analytic or numerical techniques appears fairly hopeless at this stage, whereas higher order moments can be obtained from simulations, and one can explore the probability distribution of intensity in this way, or indeed any statistical quantity of interest.

5. The pdf of intensity

The determination of the one-point probability distribution of intensity remains as one of the major unsolved problems of wave propagation through random media, at least from the theoretical viewpoint. The development of powerful simulation techniques, based on representing the medium as a sequence of phase-changing screens, means that we have a much better understanding of the distribution of intensity and its evolution with range, but as yet we have no real predictive ability, particularly in the intermediate regions of strong focusing, even for the idealized case of an isotropic random medium.

We have the following understanding of the evolution of the form of the probability distribution in range, when the scattering is strong. For short ranges, as we have seen above, the Rytov approximation is appropriate, and correspondingly, the intensity has a log-normal distribution. For extreme ranges, the energy reaching any receiver has been spread out in the vertical by much more than L_v for most of the range, and the mean squared phase fluctuation is large, i.e. $\langle|\Phi|^2\rangle \gg 1$. Thus, different interfering parts of that energy approaching the receiver in different directions have undergone statistically independent large phase shifts and become independent mean zero variables. The central limit theorem can then be used to determine the statistics. In particular, the complex amplitude has a two-dimensional Gaussian distribution, corresponding to a Rayleigh distribution for its magnitude and an exponential distribution for its intensity. This *saturated* case corresponds to the scintillation index, or variance of intensity, taking the value one, so that the second moment of the intensity is two. Ewart (1989) has shown that in the ocean, this saturated state can never be reached in practice, for if the scattering and the range are large enough so that in principle one would expect this situation to hold, then the sound absorption will be so strong that there will be no significant signal. Nevertheless, the concept of saturation is useful, particularly as one can look for corrections to this state. The exponential distribution of intensity corresponds to the moments

$$m_n = \langle I^n\rangle/\langle I\rangle^n = n! \tag{5.1}$$

As indicated in (5.1) we write $m_n = \langle I^n\rangle/\langle I\rangle^n$ henceforth.

Corrections to this result were obtained by path integral methods by Zavorotny et al (1977) and Dashen (1979) and by Gochelashvili and Shishov (1978):

$$\langle I^n\rangle/\langle I\rangle^n = n!\left[1 + \frac{1}{2}n(n-1)\gamma\right] \tag{5.2}$$

where γ depends on the spectrum of the medium. This was also shown to hold for the case where there is a deterministic sound channel, with a suitable interpretation of γ in terms of the basic scattering parameters, in Flatté et al (1983).

Dashen (1984) extended this result for the uniform medium case, although the essential argument should also apply to the ocean case. For the uniform case, $V_0 = 0$, so that the phase (action) is

$$S = \left(\int \left[\frac{1}{2}\left(\frac{dz}{dx}\right)^2 - \mu(x, z(x))\right] dx\right) \tag{5.3}$$

Using the path-integral representation of equation (4.9) for the complex field,

$$\Psi = \int \mathcal{D}\text{paths } \exp[ik_0 S] \tag{5.4}$$

leads to

$$\langle I^n \rangle = \int \mathcal{D}^{2n}\text{paths } \exp[ik_0(\sum_{\Psi} S - \sum_{\Psi^*} S)] \tag{5.5}$$

where the phase terms from Ψ and its complex conjugate have opposite signs. Dashen (1984) then approximates equation (5.5) by considering individual paths for Ψ and Ψ^* that are close to one another, but where typically any two such pairs of paths are well separated compared to the correlation scale, as would be expected over long ranges.

Adding all such contributions, and making corrections for any pairs of paths that are in fact close, gives the form

$$m_n = n! \exp\left[\binom{n}{2}\gamma f_2(\gamma) + \binom{n}{3}\gamma^2 f_3(\gamma) + \binom{n}{4}\gamma^3 f_4(\gamma) + \cdots\right] \quad . \tag{5.6}$$

Dashen then makes an analogy between interactions of the n pairs of paths and interactions of n particles via two body potentials; using this analogy equation (5.6) corresponds to the cluster expansion for the free energy, as determined in statistical mechanics texts.

Here the $f(\gamma)$'s can in principle be calculated from a knowledge of the characteristics of the medium, but Dashen gives no details for the actual calculation of these quantities. However, in a recent series of papers, (Dashen and Wang, 1993, Wang and Dashen, 1993, Dashen et al, 1993), for media with pure power-law spectra, a detailed asymptotic prescription for calculating the $f(\gamma)$'s is provided, and good agreement is obtained with the results of direct numerical simulation.

An alternative model has been proposed for the near-saturated case, the so-called K-distribution, with moments

$$m_n = n!\gamma^n \frac{\Gamma(n + 1/\gamma)}{\Gamma(1/\gamma)} \tag{5.7}$$

This distribution arises from modeling the scattering as a two-dimensional random walk in the complex amplitude plane (Jakeman and Pusey, 1978). Here the scattering is assumed strong enough so that the phase is uniformly distributed, and a step in the random walk step has length equal to the magnitude of the amplitude change that the field undergoes in a scattering event. For an infinite number of steps, or scatters, the exponential distribution is obtained, but for a finite number of steps, there are a number of possibilities, one of which is equation (5.7). As noted by Dashen (1984), this happens to reduce to the form of equation (5.2) in the limit as the second moment approaches two, (i.e. as $\gamma \to 0$ in the saturated regime) and this probably explains the success of this model in predicting the moments of the propagating field in a range of experimental studies involving scattering in extended random media. Note, however, that the K-distribution is only a one-parameter distribution, and that it does not reduce to the log-normal distribution for short range and/or weak scattering; indeed it is not defined for second moments less than two.

One can use the *form* of equation (5.6) to construct a heuristic description of the intensity pdf at moderate to large range. If it is required that (5.6) should be exact for the second and third moments, then

$$\gamma f_2(\gamma) = \log m_2/2, \quad \gamma^2 f_3(\gamma) = \log\left[4m_3/3m_2^3\right].$$

If we then ignore terms of $O(\gamma^3)$ and higher order in equation (5.6), we have an approximate expression for all higher moments of intensity in terms of the second and third moments of intensity. These two moments then play the rôle of parameters in the appropriate intensity distribution, and may of course be found from measurements. Alternatively, analytic and numerical evaluations of the second moment are already available, as described earlier; the third moment however remains relatively intractable.

As mentioned above, the log-normal distribution is appropriate for very short ranges. Furutsu (1976) has generalized this to the form

$$(5.8) \quad m_n = \exp\left[\binom{n}{2}\log m_2 + \binom{n}{3}\gamma^2\left(\frac{m_3}{m_2^3}\right) + \binom{n}{4}\log\left(\frac{m_4 m_2^6}{m_3^4}\right) + \cdots\right]$$

using a cluster expansion; Furutsu gives expressions for m_2 and m_3 under various approximations. In any case, one can use (5.8) in a heuristic way, by ignoring all but the first two terms in the exponential; if one does this one has a log-normal distribution for short range. Unfortunately, but as would be expected, (5.8) does not reduce to the exponential distribution in the saturation limit. Similarly, the Dashen description (5.6) does not give the log-normal result at short range. Thus the descriptions (5.8) and (5.6) are complementary; comparisons with data in the ocean case as in Macaskill and Ewart (1984b) show that both are quite useful.

The problem then remains to find a complete analytic description, valid for all ranges. A large number of possible candidate distributions have been proposed. For example, Ewart and Percival (1986) and Ewart

(1989) show that a three-parameter generalized gamma distribution gives a good fit to experimental data and to the results of numerical simulations corresponding to the ocean propagation case; since they do not impose a value for the mean intensity a priori, the number of free parameters is equivalent to the two-parameter heuristics described above. For simulations of optical propagation in the atmosphere, Flatté et al (1994) propose an exponential distribution convolved with a log-normal distribution as a good model fit when the inner scale of the medium fluctuations is small, and a log-normal distribution convolved with a K-distribution when the inner scale is not small. However, none of these papers takes the fundamental approach of starting from the wave equation, assuming some model of the medium, and then predicting the distribution of intensity. This general problem remains to be solved, but at least we now have a phenomenological form with which to compare theoretical results.

We note however, that analytic progress has been made in one very special case. For a phase screen with an exponential correlation function, Buckley (1986) has derived an exact Fokker-Planck equation for the characteristic function for the real and imaginary parts of the field. This was solved numerically by Roberts (1986) in a companion paper, giving rise to a distribution that numerically was indistinguishable from a Rice distribution.

6. Discussion

We started with the fundamentals of sound propagation through ocean fluctuations, and found that the assumptions and approximations that are made have differing plausibility. In particular, the expansion around deterministic rays and the Markov approximation are expected to fail sometimes. We proceeded to the calculation of moments of the field, and found the methods to do these calculations for the low moments to be well developed when the approximations are valid. The situation with the probability density function of intensity is not so good. This distribution has a high-intensity tail due to focusing, and that tail is related to the high moments that are difficult to calculate. There is some theory for the probability density function, and a good phenomenological model.

Other statistics have not been evaluated at all. One can easily think of such interesting statistics: the average intensity at a depth z' conditional on the intensity at z describes the sharpness of a focus. The intensity averaged over a vertical aperture depends on larger scales in the medium than the unaveraged intensity, and can be correlated with the phase. Similarly, smaller scale fluctuations can be selected for the phase by taking phase differences over small spacings. In ray tracing, the intensity is the reciprocal of the ray spacing, so negative moments of the intensity might be of interest. (Both log-normal and generalized gamma distributions have the property that the reciprocal of the intensity obeys a distribution in the same family.)

We have, on several occasions, mentioned the MATE experiment of Ewart and collaborators. There were two deterministic rays connecting the source and receiver array in this experiment. The experiment was designed to look at the lower path; the oceanography was well measured. The standard set of approximations is valid for the lower path, and the fit of theory and measurement has been very successful. The upper path is just the opposite. There is good reason to distrust the standard approximations. The oceanography was not measured extensively, and there has not been a good match of theory and measurement. Simulations are expected to be reliable, so one could make models of the oceanography and simulate the acoustics. The theoretical expressions could then be compared with the simulations. (We don't expect the oceanography to be uniquely determined by the actual measurements, but reasonable possible descriptions can be obtained.)

This field is largely driven by applied issues. Many of the currently important applications are in shallow water, such as on the continental shelf. Such environments are likely to resemble the upper path at MATE much more than the lower path which is typical of the deep-ocean situations that have been investigated. Shallow water concerns also tend to emphasize the lesser-understood small-scale fluctuations (having more diffraction) rather than the better-understood large scale fluctuations that primarily cause phase fluctuations. Frequencies are often much higher, and turbulence as well as internal waves might be important.

Another application that stretches the boundaries of this subject is the long-range low-frequency propagation used for global warming investigations. As we have seen, ray tracing doesn't work very well for the deterministic problem over large distances, so we ought not to expand around rays for the stochastic problem. Moreover, much of the methodology is based on the absence of deterministic refraction, and for long ranges the deterministic refraction is dominant; along any ray there is another caustic every 25 km or so. Ocean mesoscale becomes one of the fluctuations to be included along with internal waves.

Thus wave propagation in random media as applied to the ocean has parts that are very well understood and other parts that are completely unexplored. There is room for both applications of existing theory, and development of new ideas to attack unsolved problems.

Acknowledgment. We have had innumerable stimulating conversations on this topic with Terry Ewart over the years. Steve Reynolds assisted with calculations and figures. Frank Henyey's work was supported primarily by internal funds from the Applied Physics Laboratory, and also by the Office of Naval Research.

References

[1] M.J. Beran and T.L. Ho, 1969: Propagation of the fourth-order coherence function in a random medium. J. Opt. Soc. Am., **59**, 1134–1138.

[2] M.J. Beran and J.J. McCoy, 1976: Propagation through an isotropic random medium: an integro-differential formulation. J. Math. Phys., **17**, 1186-1189.

[3] M.J. Beran and A.M. Whitman, 1975: Scattering of a finite beam in a random medium with a nonhomogeneous background. J. Math. Phys., **16**, 214-217.

[4] D.H. Berman, E.B. Wright and R.N Baer, 1992: An optimal PE-type wave equation. J. Acoust. Soc. Am., **86**, 228–233.

[5] I.M. Besieris and F.D. Tappert, 1973: Kinetic equations for the quantized motion of a particle in a randomly perturbed potential field. J. Math. Phys., **14**, 1829–1836.

[6] W.P. Brown, Jr., 1972: Fourth moment of a wave propagating in a random medium. J. Opt. Soc. Am., **62**, 966–971.

[7] R. Buckley, 1986: A Fokker-Planck equation for the probability distribution of the field scattered by a random phase screen. in *Wave Propagation and Scattering.*, ed. B.J. Uscinski, Clarendon Press, Oxford, 113–127.

[8] C. Canuto, M.Y. Hussaini, A. Quarteroni and T.A. Zang, 1988: *Spectral Methods in Fluid Dynamics.*, Springer-Verlag, Berlin, 222–225,

[9] L.A. Chernov, 1969: Equations for the statistical moments of the field in a randomly inhomogeneous medium.*Sov. Phys. Acoust.*, **15**, 511–517.

[10] J.L. Codona, D.B. Creamer, S.M. Flatté, R.G. Frehlich and F.S. Henyey, 1985: Average arrival time of wave pulses through continuous random media. *Phys. Rev. Lett.*, **55** 9–12.

[11] J.L. Codona, D.B. Creamer, S.M. Flatté, R.G. Frehlich and F.S. Henyey, 1986a: Moment equation and path-integral techniques for wave propagation in random media. *J. Math. Phys.*, **27**, 171–177.

[12] J.L. Codona, D.B. Creamer, S.M. Flatté, R.G. Frehlich and F.S. Henyey, 1986b: Solution for the fourth moment of waves propagating in random media. *Radio Science.*, **21** 929–948.

[13] J.A. Colosi, S.M. Flatté and C. Bracher, 1994: Internal-wave effects on 1000-km oceanic acoustic pulse propagation: simulation and comparison with experiment. *J. Acoust. Soc. Am.*, **96** 452–468.

[14] I.M. Dagkesamanskaya and I.V. Shishov, 1970: Strong intensity fluctuations during wave propagation in statistically homogeneous and isotropic media. *Radiophys. Quantum Electron.*, **13**, 9–11.

[15] R. Dashen, 1979: Path integrals for waves in random media. *J. Math. Phys.*, **20**, 894–920.

[16] R. Dashen, 1984: The distribution of intensity in a multiply scattering medium. *Opt. Lett.*, **10**, 110–112.

[17] R. Dashen and G.-Y. Wang, 1993: Intensity fluctuation for waves behind a phase screen: a new asymptotic scheme. *J. Opt. Soc. Am.*, **A, 10**, 1219–1225.

[18] R. Dashen, G.-Y. Wang, S.M. Flatté and C. Bracher, 1993: Moments of intensity and log intensity: new asymptotic results for waves in power-law media. *J. Opt. Soc. Am.*, **A, 10**, 1233–1242.

[19] Y.J.F. Desaubies, 1978: On the scattering of sound by internal waves in the ocean. *J. Acoust. Soc. Am.*, **64**, 1460–1469.

[20] E.A. Dorfi and L. O'C. Drury, 1987: Simple adaptive grids for 1-D initial value problems. *J. Comp. Phys.*, **69**, 175–195.

[21] T.F. Duda, S.M. Flatté, J.A. Colosi, B.D. Cornuelle, J.A. Hildebrand, W.S. Hodgkiss, P.F. Worcester, B.M Howe, J.A. Mercer and R.C. Spindel, 1992: Measured wave-front fluctuations in 1000-km pulse propagation in the Pacific Ocean. *J. Acoust. Soc. Am.*, **92** 939–955.

[22] T.E. Ewart, 1989z: A model for the intensity probability distribution for wave propagation in random media. *J. Acoust. Soc. Am.*, **86**, 1490-1498.

[23] T.E. Ewart and D.B. Percival, 1986: Forward scattered waves in random media – The probability distribution of intensity. J. Acoust. Soc. Am., **80**, 1745–1753.

[24] T.E. Ewart and S.A. Reynolds, 1984: The mid-ocean acoustic transmission experiment – MATE. *J. Acoust. Soc. Am.*, **75**, 785-802.

[25] T.E. Ewart and S.A. Reynolds, 1991: Experimental ocean acoustic field moments versus predictions. in *Ocean Variability and Acoustic Propagation.*, eds. J. Potter and A. Warn-Varnas, Kluwer Academic Publishers.

[26] T.E. Ewart and S.A. Reynolds, 1992: Ocean acoustic propagation measurements and wave propagation in random media. in *Wave Propagation in Random Media (Scintillation).*, V.I. Tatarski, A. Ishimaru and V.U. Zavorotny, Eds., SPIE and IOP, 100–123.

[27] W. Feller, 1971: *An Introduction to Probability Theory and Its Applications* Volume II, 2nd edn., John Wiley and Sons.

[28] R.P. Feynman and A.R. Hibbs, 1965: *Quantum Mechanics and Path Integrals*, McGraw-Hill, New York.

[29] S.M. Flatté, 1983: Wave propagation through random media: Contributions from ocean acoustics. Proc. of the IEEE, **71**, 1267–1294.

[30] S.M. Flatté, D.R. Bernstein and R. Dashen, 1983: Intensity moments by path integral techniques for wave propagation through random media, with application to sound in the ocean. *Phys. Fluids.*, **26**, 1701–1713.

[31] S.M. Flatté, C. Bracher and G.-Y. Wang, 1994: Probability-density functions of irradiance for waves in atmospheric turbulence calculated by numerical simulation. *J. Opt. Soc. Am.*, **A, 11**, 2080–2092.

[32] S.M. Flatté, R. Dashen, W.M. Munk, K.M. Watson and F. Zachariasen, 1979: *Sound Transmission through a Fluctuating Ocean.*, Cambridge University Press, Cambridge.

[33] S.M. Flatté and F. D. Tappert, 1975: Calculation of the effect of internal waves on oceanic sound transmission. *J. Acoust. Soc. Am.*, **58**, 1151-1159.

[34] S.M. Flatté, G.-Y. Wang and J. Martin, 1993: Irradiance variance of optical waves through atmospheric turbulence by numerical simulation and comparison with experiment. *J. Opt. Soc. Am.* **A, 10**, 2363–2370.

[35] S. Frankenthal, A.M. Whitman and M.J. Beran, 1984: Two-scale solutions for intensity fluctuations in strong scattering. *J. Opt. Soc. Am* **A 1**, 585–597.

[36] K. Furutsu, 1976: Theory of irradiance distribution function in turbulent media-cluster approximation. *J. Math. Phys.*, **17**, 1252–1263.

[37] K. Furutsu, 1988: Intensity correlation functions of lightwaves in a turbulent medium: an exact version of the two-scale method. *Appl. Opt.*, **27**, 2127–2144.

[38] K.S. Gochelashvili and V.I. Shishov, 1978: Strong fluctuations of laser radiation intensity in a turbulent atmosphere – the distribution function. *Sov. Phys. JETP.*, **47**, 1028 –1030.

[39] J. Gozani, 1985: Numerical solution of the fourth order coherence function of a plane wave propagating in a two-dimensional Kolmogorovian medium. *J. Opt. Soc. Am.*, **A, 12**, 2111–2126.

[40] J. Gozani, 1987: Improvement in the two-scale solution for wave propagation in a random medium. *Opt. Lett.*, **12**, 1–3.

[41] F.S. Henyey, 1991: Acoustic scattering from ocean microbubble plumes in the 100 Hz to 2kHz region. *J. Acoust. Soc. Am.*, **90**, 399–405.

[42] A. Ishimaru, 1978: *Propagation and Scattering in Random Media.*, Academic Press, New York.

[43] D.R. Jackson and T.E. Ewart, 1994: The effect of internal waves on matched-field processing. *J. Acoust. Soc. Am.*, **96**, 2945–2955.

[44] D.R. Jackson and K.B. Briggs, 1992: High-frequency bottom backscattering: Roughness versus sediment volume scattering. *J. Acoust. Soc. Am.*, **92**, 962–977.

[45] D.R. Jackson, D.P. Winebrenner and A. Ishimaru, 1986: Application of the composite roughness model to high-frequency bottom backscattering. *J. Acoust. Soc. Am.*, **80**, 1410–1422.

[46] E. Jakeman and J.G. McWhirter, 1977: Correlation-function dependence of the scintillation behind a deep random phase screen. *J. Phys. A: Math. Gen.*, **10**, 1599–1643.

[47] E. Jakeman and P.N. Pusey, 1978: Significance of K distributions in scattering experiments. *Phys. Rev. Lett.*, **40**, 546–550.

[48] L.D. Landau and L.M.Lifschitz, 1975: *The Classical Theory of Fields.*, 4th revised English edition, Pergamon Press.

[49] S.R. Leonard, D.E. Reeve and M.C. Cook, 1990: Solution of the fourth-moment equation by an adaptive grid method. *J. Mod. Optics.*, **37**, 965–975.

[50] S.R. Leonard and B.J. Uscinski, 1993: Accurate numerical solutions for the frequency cross correlation of intensity fluctuations in a random medium. *Appl. Opt.*, **32**, 2656–2663.

[51] M. Leontovich and V.A. Fock, 1946: Solution of the problem of propagation of electromagnetic waves along the earth's surface by the parabolic equation method. *Zh. Eksp. Teor. Fiz.*, **16**, 557–573.

[52] A.P. Lyons, A.L. Anderson and F.S. Dwan, 1994: Acoustic scattering from the seafloor: modeling and data comparison. *J. Acoust. Soc. Am.*, **95**, 2441–2451.

[53] C. Macaskill, 1983: An improved solution to the fourth moment equation for intensity fluctuations. *Proc. R. Soc. London.*, Ser. **A, 386**, 461–474.

[54] C. Macaskill, 1986: Scattering in a random medium with a mean refractive index profile. in *Wave Propagation and Scattering.*, ed. B.J. Uscinski, Clarendon Press, Oxford, 65–81.

[55] C. Macaskill and T.E. Ewart, 1984a: Computer simulation of two-dimensional random wave propagation. *IMA J. Appl. Math.*, **33**, 1–15.

[56] C. Macaskill and T.E. Ewart, 1984b: The probability distribution of intensity for acoustic propagation in a randomly varying ocean. *J. Acoust. Soc. Am.*, **76**, 1466–1473.

[57] C. Macaskill and T.E. Ewart, 1995: Numerical solution of the fourth moment equation for acoustic intensity correlations and comparison with MATE. submitted to *J. Acoust. Soc. Am.*

[58] C. Macaskill and B.J. Uscinski, 1981: Propagation in waveguides containing random irregularities: the second moment equation. *Proc. R. Soc. London.*, Ser. **A, 377**, 73–98.

[59] J.M. Martin and S.M. Flatté, 1988: Intensity images and statistics from numerical simulation of wave propagation in 3-D random media. *Appl. Opt.*, **27**, 2111–2126.

[60] J.M. Martin and S.M. Flatté, 1990: Simulation of point-source scintillation through three-dimensional random media. *J. Opt. Soc. Am.*, **A, 7**, 838–847.

[61] J.J. McCoy and M.J. Beran, 1979: Directional spectral spreading in randomly inhomogeneous media. *J. Acoust. Soc. Am.*, **66**, 1468-1481.

[62] R.P. Mercier, 1962: Diffraction by a screen causing large random phase fluctuations.*Proc. Camb. Phil. Soc.*, **A, 58**, 382–400.

[63] S.J. Miller and B.J. Uscinski, 1986: Frequency cross-correlation of intensity fluctuations. Limitations of multiple-scatter solutions. *Opt. Acta.*, **33**, 1341–1358.

[64] D. Palmer, 1978: An Introduction to the Application of Feynman Path Integrals to Sound Propagation in the Ocean. NRL report 8148.

[65] R. Pinkel and S. Anderson, 1992: Toward a statistical description of finescale strain in the thermocline. *J. Phys. Oceanogr.*, **22**, 773–795.

[66] A.M Prokhorov, F.V. Bunkin, K.S. Gochelashvili and V.I. Shishov, 1975: Laser irradiance propagation in turbulent media. *Proc. IEEE.*, **63**, 790–811.

[67] D.E. Reeve, S.R. Leonard, and M.C. Cook, 1990: Numerical solution of the fourth moment equation for a point source. *J. Mod. Optics.*, **37**, 965–975.

[68] D.L. Roberts, 1985: PhD thesis, Statistical Properties of Waves Diffracted by a Random Phase Screen, University of Cambridge.

[69] D.L. Roberts 1986: The probability distribution of intensity fluctuations due to a random phase screen with an exponential autocorrelation function. in *Wave Propagation and Scattering.*, ed. B.J. Uscinski, Clarendon Press, Oxford, 113–127.

[70] S.M. Rytov, Y.A. Kravtsov, and V.I. Tatarski, 1988: *Principles of Statistical Radiophysics.*, vols. 1–4, Springer, Berlin.

[71] L.S. Schulman, 1981: *Techniques and Applications of Path Integration.*, Wiley-Interscience.

[72] V.I. Shishov, 1968: Theory of wave propagation in random media. *Radiophys. Quantum Electron.*, **11**, 500–505.

[73] D. Sornette, 1989: Acoustic waves in random media. II. Coherent effects and strong disorder regime. *Acustica.*, **67**, 251–65.

[74] M. Spivack and B.J. Uscinski, 1988: Accurate numerical solution of the fourth moment equation for very large values of Γ. *J. Mod. Optics.*, **35**, 1741–1755.

[75] M. Spivack and B.J. Uscinski, 1989: The split-step solution in random wave propagation. *J. Comp. App. Math.*, **27**, 349–361.

[76] F.D. Tappert, 1977: The Parabolic Approximation Method. in *Wave Propagation and Underwater Acoustics.*, J.B. Keller and J.S. Papadakis, eds., Lect. Notes in Phys. **70**, Springer-Verlag, Berlin.

[77] F.D. Tappert and R.H. Hardin, 1974: *Proceedings of the Eighth International Congress on Acoustics.*, Vol.II, Goldcrest, London, 452.

[78] V.I. Tatarski, 1969: Light propagation in a medium with random refractive index inhomogeneities in the Markov random process approximation. *Sov. Phys. JETP.*, **29**, 1133–1138.

[79] V.I. Tatarski, 1971: *The Effects of the Turbulent Atmosphere on Wave Propagation.*, (U.S. Dept. of Commerce, National Technical Information Service, Springfield, VA).

[80] V.I. Tatarski, A. Ishimaru and V.U. Zavorotny, Eds., 1992: *Wave Propagation in Random Media (Scintillation)*, SPIE and IOP.

[81] E.I. Thorsos, 1990: Acoustic scattering from a Pierson-Moskowitz sea surface *J. Acoust. Soc. Am.*, **88**, 335–349.

[82] M. Tur, 1982: Numerical solutions for the fourth moment of a plane wave propagating in a random medium. *J. Opt. Soc. Am.*, **72**, 1683–1691.

[83] M. Tur, 1985: Numerical solutions for the fourth moment of a finite beam propagating in a random medium. *J. Opt. Soc. Am.*, **A, 2**, 2161–2170.

[84] B.J. Uscinski, 1977: *The Elements of Wave Propagation in Random Media.*, McGraw-Hill, New York.

[85] B.J. Uscinski, 1982: Intensity fluctuations in a multiple scattering medium. Solution of the fourth moment equation. *Proc. R. Soc. London.*, Ser. **A, 380**, 137–169.

[86] B.J. Uscinski, 1985: Analytical solution of the fourth-moment equation and interpretation as a set of phase screens. *J. Opt. Soc. Am.*, **A, 2**, 2077–2091.

[87] B.J. Uscinski, 1986: Acoustic scattering by ocean irregularities. Aspects of the inverse problem. *J. Acoust. Soc. Am.*, **79**, 347–355.

[88] B.J. Uscinski and C. Macaskill, 1985: Frequency cross-correlation of intensity fluctuations in multiple scattering. *Optica Acta.*, **32**, 71-89.

[89] N.G. Van Kampen, 1981: *Stochastic Processes in Physics and Chemistry.*, North Holland, Amsterdam.

[90] G.-Y. Wan and R. Dashen, 1993: Intensity moment for waves in random media: three-order standard asymptotic calculation. *J. Opt. Soc. Am.*, **A, 10**, 1226–1232.

[91] G.B. Whitham, 1974: *Linear and Nonlinear Waves.*, Wiley, New York.

[92] P.F. Worcester, R.C. Spindel and B.M. Howe, 1985: Reciprocal acoustic transmissions: instrumentation for mesoscale monitoring of ocean currents. *IEEE J. Ocean. Eng.*, **OE-10**, 123–137

[93] V. U. Zavarotny, V.I. Tatarski and V.I. Klyatskin, 1977: Strong fluctuations of the intensity of electromagnetic waves in randomly inhomogeneous backgrounds. *Sov. Phys. JETP.*, **46**, 252–260.

FRANK S. HENYEY
Applied Physics Laboratory
University of Washington
1013 NE 40th Street
Seattle, Washington, 98105

CHARLES MACASKILL
School of Mathematics and Statistics
University of Sydney, NSW
Australia 2006

frank@apl.washington.edu

Stochastic modeling of turbulent flows

J.R. Herring

Abstract

We discuss applications of the statistical theory of turbulence to problems related to oceanographic flows. We begin by considering inviscid flows, for which there is an exact formalism for the underlying probability distribution function for the vorticity field (a solution to Liouville's equation). Such a solution, although remote from real flows, has been used to gain insights into certain features of the latter. We review some of these results, in particular the case of barotropic flow over a random topographic relief, noting briefly some successes and failures. Our remarks here are confined largely to the mean flow in a closed basin. We then describe stochastic models in which the nonlinear terms are modeled as a random stirring supplemented by eddy viscosity, focusing attention on the case of homogeneous flows with no mean circulation. Two methods are discussed: (1) Langevin modeling, in which the turbulence is activated by a white-noise random force, and (2) a modeling in which the temporal aspects of the force are allowed to be determined self consistently, according to the context of the problem. Here, the random force represents the model's rendering of pressure and advective nonlinearities in the equations of motion. From the perspective of the Langevin model, we note how such theories make contact with simpler heuristic ideas of energy transfer, and discuss how they yield inertial range distribution of eddy scales-sizes. We then apply the second modeling to flow over random topography, noting how part of the vorticity field is locked to the topographic relief, with no temporal fluctuations, in the stationary state. We stress that both types of modeling prescribe generalized eddy viscosities, whose forms are determined self consistently in terms of the flow's energy spectrum and molecular-dissipation properties. Finally, we note the failure of this method to properly incorporate isolated, intense, vortex structures.

1. Introduction

Ocean currents and eddies are clearly turbulent; thus, a stastical description of their scale–size distribution and associated transport properties may be sought in terms of the language and concepts of non–equilibrium statistical mechanics. The latter discipline finds its best known representative in the renormalized perturbation techniques leading to the Direct Interaction Approximation of Kraichnan (1959), and its Lagrangian extensions (Kraichnan (1964), Kaneda (1981)). Such a point of view naturally has its drawbacks. First, only simple averages (ensemble means, denoted here by $\langle\cdot\rangle$) are easily obtained from such theories. Secondly, the theoretical foundations of these theories seem as yet rather fragile, in that the existence of intense structures is at present beyond their pale (McWilliams, (1984), Herring and McWilliams (1985)). Nonetheless, stochastic models have elucidated certain dynamics of large–scale flows (such as inverse cascading aspects of quasi –geostrophic flows (Kraichnan (1967), and ocean basin guyers (Griffa and Salmon (1987), Holloway (1992)). Their use of computer resources is modest in comparison with ocean modeling *via* direct numerical simulations. Moreover, such methods also offer promise of providing a logical basis for large eddy simulations (LES) (Kraichnan (1976), Chollet and Lesieur (1982), Sadourny and Basdevant (1985)), a method in which the effects of scales beyond the resolution capacity of the computer are represented statistically in terms of well–defined eddy viscosity and conductivity concepts. In this paper, we will review some oceanographic research of which the starting point is the statistical theory of fluid flows. We start with some general underlying concepts, the most fundamental of which is the probability density function, $\mathcal{P}(\psi_1, \psi_2, \cdots, \psi_n, \cdots, t)$ that the dynamical degrees of freedom, $\{\psi_1, \psi_2, \cdots, \psi_n, \cdots\}$ attain values specified by the arguments of $\mathcal{P}$. Here, the degrees of freedom, ψ_n, may be thought of as a collection of velocity and temperature fields, $(\boldsymbol{u}(x), T(\boldsymbol{x}))$, and the index n is a labeling of the set of points $\boldsymbol{x}$.

Given the equations of motion for $\psi(t)$, which we write as

$$d\psi_i/dt = F_i(\psi_1, \psi_2, \cdots, \psi_n, \cdots, t) \tag{1.1}$$

we may infer for $\mathcal{P}$ the evolution equation (Edwards, (1964)),

$$\partial\mathcal{P}/\partial t = \sum_i \partial\{F_i\mathcal{P}\}/\partial\psi_i \tag{1.2}$$

Notice that in (1.2), ψ_i is a degree of freedom, and is not a function of time, as is ψ_i in (1.1). Technically, we should use different symbols here, but usually the context will suffice to make clear which interpretation is intended (we shall use $\hat{\psi}_i(t)$ to indicate solutions to (1.1) if there seems a danger of confusion.) In fact, the $\psi(t)$ in (1.1) are just the characteristics of (1.2), in the sense that a solution of (1.2) is,

$$\mathcal{P} = \langle \prod_i \delta(\psi_i - \hat{\psi}_i(t)) \rangle_{t=0} \tag{1.3}$$

The notation $\langle \cdot \rangle_{t=0}$ means an average over the ensemble of initial conditions for $\hat{\psi}_i$. The forcing function F_i may be further separated into viscous, non–linear, and stochastic components:

$$F_i = -\sum_j \nu_{ij}\psi_j + f_i + \mathcal{F}_i(t) \tag{1.4}$$

where the first term represents viscous or conductive dissipation (ie: $\nabla^2\psi$), $f_i(\{\psi\})$ the non–linear force, and $\mathcal{F}_i$ is a stochastic forcing, which, for the moment, we take as white noise. In case the timescale of $\mathcal{F}_i$ is much shorter than the dynamical timescale, we have the theorem of Novikov (1963):

$$\partial\langle \mathcal{P} \rangle_{\mathcal{F}} = \rho\partial^2\langle \mathcal{P} \rangle_{\mathcal{F}}/\partial\psi^2 + \sum_i \partial F'_i \langle \mathcal{P} \rangle_{\mathcal{F}}/\partial\psi_i \tag{1.5}$$

where

$$\rho = \int_{-\infty}^{\infty} ds \langle \mathcal{F}(t)\mathcal{F}(s) \rangle_{\mathcal{F}}, \; F'_i = -\sum_j \nu_{ij}\psi_j + f_i \tag{1.6}$$

and we assume the random forcing is stationary. Notice that if the viscous force is of the form $\nu_n\psi_i$, and if we suppress f_i, then $\langle \mathcal{P} \rangle_{\mathcal{F}}$ becomes Gaussian.

At this point, we record the form of f_i

$$f_i = \sum_{jk} C_{ijk}\psi_j\psi_k \tag{1.7}$$

that is associated with advection and pressure effects. For example, for advection of the vorticity, $\zeta(\boldsymbol{x}, t)$, f_i may be written (perhaps somewhat awkwardly) as:

$$f(\boldsymbol{x}) = -\int_{\mathcal{D}} d\boldsymbol{x}' d\boldsymbol{x}'' \boldsymbol{u}(x')\zeta(\boldsymbol{x}'')\nabla_{\boldsymbol{x}''}\delta(\boldsymbol{x}'' - \boldsymbol{x}')\delta(\boldsymbol{x}' - \boldsymbol{x}) \tag{1.8}$$

$$= \boldsymbol{u}(\boldsymbol{x}) \cdot \nabla\zeta(\boldsymbol{x}) \tag{1.9}$$

Finally, we note that for simple two–dimensional turbulence, for which both kinetic energy and vorticity are conserved, we may use the relationship between $\boldsymbol{u}$ and ζ to write the inviscid equations for $\zeta(\boldsymbol{x}, t)$ as:

$$\partial\zeta(\boldsymbol{x},t)/\partial t = -\boldsymbol{u}\cdot\nabla\zeta = \int d\boldsymbol{x}d\boldsymbol{x}'C(\boldsymbol{x},\boldsymbol{x}',\boldsymbol{x}'')\zeta(\boldsymbol{x}')\zeta(\boldsymbol{x}'') \tag{1.10}$$

where,

$$C(\boldsymbol{x},\boldsymbol{x}',\boldsymbol{x}'') = \{-\partial_{y'}\partial_{x''} + \partial_{x'}\partial_{y''}\}G(\boldsymbol{x}\mid\boldsymbol{x}')\delta(\boldsymbol{x}-\boldsymbol{x}'') \tag{1.11}$$

and

$$\nabla^2 G(\boldsymbol{x}\mid x') = \delta(\boldsymbol{x}-\boldsymbol{x}') \tag{1.12}$$

Here, (1.11) is a full articulation of (1.7).[1] In general, inviscid fluid dynamics dictates global conservation laws for flows in a finite domain. For incompressible flows, such conservation laws include those quadratic invariants responsible for conservation of total momentum, vorticity, energy, helicity (the volume integral of $\boldsymbol{u}\cdot\nabla\times\boldsymbol{u}$), and, in two dimensions, squared vorticity. For flow in a closed domain, the total momentum is zero. For quasi–two dimensional flows of interest here, the helicity vanishes. Thus, such inviscid, integral constraints have the general form:

$$I^1 = \sum_i C_i\psi_i \tag{1.13}$$

$$I_n^2 \equiv \sum_{ij} \mathcal{A}_{ij}^n\psi_i\psi_j,\ (n=1,2) \tag{1.14}$$

We call I_1^2 energy, and I_2^2 the enstrophy[2]. Conservation laws such as(1.14) correspond to a set of constraints on the form of f_i:

$$\sum_i C_i f_i = 0 \tag{1.15}$$

---

[1]The Fourier representations of (1.10), (1.11) and (1.12) are perhaps more familiar. These are obtained by introducing $G(\boldsymbol{x}\mid\boldsymbol{x}') = (1/2\pi)^2 \int d\boldsymbol{p}p^{-2}\exp(i\boldsymbol{p}\cdot(\boldsymbol{x}-\boldsymbol{x}'))$, and $\delta(\boldsymbol{x}-\boldsymbol{x}') = (1/2\pi)^2\int d\boldsymbol{q}\exp(-\boldsymbol{q}\cdot(\boldsymbol{x}-\boldsymbol{x}'))$ into (1.11) and forming from (1.10) the equation of motion for $\zeta(\boldsymbol{k}) \equiv \int d\boldsymbol{x}\exp(i\boldsymbol{k}\cdot\boldsymbol{x})\zeta(\boldsymbol{x})$. There results $d\zeta(\boldsymbol{k})/dt = \sum_{\boldsymbol{p},\boldsymbol{q}} C(\boldsymbol{k},\boldsymbol{p},\boldsymbol{q})\zeta(\boldsymbol{p})\zeta(\boldsymbol{q})$,with $C(\boldsymbol{k},\boldsymbol{p},\boldsymbol{q}) = \delta(\boldsymbol{k}-\boldsymbol{p}-\boldsymbol{q})(p_x q_y - p_y q_x)/p^2$.

[2]These conservation laws correspond to symmetries of $C(\boldsymbol{x},\boldsymbol{x}',\boldsymbol{x}'')$: $C(\boldsymbol{x},\boldsymbol{x}',\boldsymbol{x}'') = -C(\boldsymbol{x}'',\boldsymbol{x}',\boldsymbol{x})$, and $\nabla_{\boldsymbol{x}}^{-2}C(\boldsymbol{x},\boldsymbol{x}',\boldsymbol{x}'') = -\nabla_{\boldsymbol{x}'}^{-2}C(\boldsymbol{x}',\boldsymbol{x},\boldsymbol{x}'')$.

$$\sum_{ij} \mathcal{A}^n_{ij} \psi_j f_i = 0, \ n = (1,2) \tag{1.16}$$

A time–independent solution of (1.2) for a system in which F_i is given by (1.7), with constraints (1.13) and (1.14), is any function, E, whose sole argument is any function of the total vorticity, energy, and enstrophy. We shall be particularly interested in the case in which this function is a linear combination, which we denote,

$$\mathcal{H} = \alpha I^1 + \beta I^1_2 + \gamma I^2_2 \tag{1.17}$$

The proof that $\mathcal{P} = E(\mathcal{H})$ satisfies (1.2) follows from direct substitution. In (1.17), (α, β, γ) are Lagrange multipliers, whose values are determined by the condition that $\int \delta\psi I_n(\{\psi\})$ have preassigned values.

Of particular interest is the case for which the function E is an exponential, the *thermal equilibrium* distribution:

$$\mathcal{P}^{therm}(\{\psi\}) = \mathcal{C}\exp(-\mathcal{H}\{\psi\}) \tag{1.18}$$

Here, $\mathcal{C}$ is a normalizing constant such that $\int \delta\psi \mathcal{P}(\{\psi\}) = 1$, $\delta\psi \equiv \prod_{i=1}^{\infty} d\psi_i$. We remark that (1.18) has another interpretation: given a system with constraints (1.13), the most probable distribution of $\{\psi\}$ is (1.16) (Orszag (1970), (1974)). The formal proof proceeds by extremalizing the entropy, $Z \equiv \ln \int \delta\psi \mathcal{P}^{therm}(\{\psi\})$.

We should stress that thermal equilibrium considerations omit entirely dissipative effects associated with viscosity and conductivity. The latter have the effect of selectively dissipating the small scales, which in themselves may not be of interest. But the dissipation scales induce a cascade of energy from the large, essentially inviscid scales toward small scales, and such a cascade is entirely missing from(1.18). To see this, we form the equation of motion for $d\langle\psi_i^2/dt\rangle$. From (1.1), (1.4), and (1.7), we see that the essential term that transfers energy (ψ_i^2) out of mode i is $\sim \langle\psi_i\psi_j\psi_k\rangle \equiv \mathcal{T}_{ijk}$. But this third–order moment is, according to (1.16), zero, by symmetry. The question then is the relevance of thermal equilibrium issues to real flows.

Two points may be made in this connection. First, any theory of turbulence which proposes a formula for $\mathcal{T}_{ijk}$ in terms of $\langle\psi_i^2\rangle$ (and auxiliary quantities) ought to satisfy (1.18), if dissipative effects are turned off. In this sense, equilibrium is a test of a theory. The second and related point is that in regions of phase space remote from the dissipative degrees of freedom, the distribution of variance (energy and enstrophy) among the degrees of freedom may be expected to be equipartitioned. For example, an ensemble of flows whose initial statistics are anisotropic should tend towards isotropy, in the absence of anisotropic forcing, and in regions of scale sizes where dissipative effects are small. It remains to be seen if thermal equilibrium can yield more quantitative information. The answer to this question depends on the system investigated, and in particular, the number, if there are inviscid constraints such as in (1.16). In three dimensions, there is only one such constant, the total energy. This leads to the well–known result for the energy spectrum, $E(k) \sim k^2$ (Lee, (1952)). There would seem to be scant conclusions to be drawn from this, except isotropy. Nevertheless, this profile at small k may figure in a vital way in determining the decay of total energy (Saffman (1967)). For quasigeostrophic flow, the number of quadratic constants is larger (≥ 3, depending on the context), and the quantitative conclusions richer. The basic reason for the more interesting domain in near two–dimensional flows is that the additional constraints dramatically steepen the spectra at small scales, mitigating the ultraviolet catastrophe of $E(k) \sim k^2$. We will discuss two examples of the application of thermal equilibrium flows: two–layer quasigeostrophic flow, and barotropic flow over a topographic relief.

In the first example, the two–layer representation of the quasi-geostrophic flow, we will, for simplicity, take the layers of equal depth. Also, for simplicity, we take the ensemble mean fields to be zero. The equations of motion assert that each layer's potential vorticity is conserved following fluid particles. The potential vorticity consists of two parts: the vorticity ($\zeta_i = -(\partial_x^2 + \partial_y^2)\psi_i(x,y)$), with ψ_i the stream function for layer $i = (1,2)$, and the vortex stretching term, $(-1)^i R^{-2}(\psi_1 - \psi_2)$. Here, R is the (external) Rossby radius of deformation. In discussing this problem, it is useful to discriminate between scales of motion which are larger or smaller that R in their lateral extent. For the present purposes, we estimate the scale of $\psi(x,y)$ as $\ell \equiv \sqrt{\langle\psi^2(x,y)\rangle/\langle(\nabla^2\psi^2\rangle}$, where we recall that $\langle\cdot\rangle$ has the effect of averaging over the (x,y) periodic domain. The inviscid system has three quadratic constants of motion: the total energy $\langle(\nabla\psi_1)^2 + (\nabla\psi_2)^2 + R^{-2}(\psi_1 - \psi_2)^2\rangle$, and the enstrophy of each layer, $\langle\{\nabla^2\psi_i + (-1)^i R^{-2}(\psi_1 - \psi_2)\}^2\rangle$. The thermal equipartition solution to this problem (Salmon *et al.* (1976)) is nearly barotropic $((\psi_1 = \psi_2)$ at scales larger that the Rossby radius, and is a random, equal–parts mixture of barotropic and baroclinic motion at values of ℓ smaller than R $(\langle\psi_1\psi_2\rangle \sim 0)$. Qualitatively, we may understand this behavior in terms of a steepest descent estimation of the integral of (1.18) needed to evaluate second moments of $\{\psi_1,\ \psi_2\}$. At large scales, $\mathcal{H}$ is minimized by a barotropic stratification, since the potential energy, $\sim (\psi_1 - \psi_2)^2$ is then the dominant contributor to $\mathcal{H}$. At small scales, where the gradients are dominant, $\mathcal{H}$ is minimized by independent but equal stream functions. The transition zone $\ell \simeq R$, where the correlation between layers decreases from near unity to zero requires a computation, which has been made by Salmon *et al.* (1976). Their results seem plausible when compared with direct numerical simulations and observations.

We now discuss in more detail the problem of barotropic flow over an arbitrary topographic relief, $h(\boldsymbol{x})$. The latter we take as non–distributed with respect to the ensemble of flows (the same $h(\boldsymbol{x})$ for each member of the ensemble). Let ζ be the potential vorticity for barotropic flow above a topography, whose relief is $h(x, y)$. Then,

$$\partial\zeta(\boldsymbol{x}, t)/\partial t = -\boldsymbol{u} \cdot \nabla\{\zeta + h(\boldsymbol{x})\} \tag{1.19}$$

The three conservation laws associated with (1.19) are:

$$I_1 = \int_{\mathcal{D}} d\boldsymbol{x}\zeta \tag{1.20}$$

$$I_2 = \int_{\mathcal{D}} d\boldsymbol{x}(\zeta - h)^2 \tag{1.21}$$

and

$$I_3 = \int_{\mathcal{D}} d\boldsymbol{x}\psi\zeta \tag{1.22}$$

where $\zeta = -\nabla^2\psi$ is the stream function, and $\mathcal{D}$ is the closed domain of the flow. The thermal equilibrium prescription for computing $\langle\zeta(\boldsymbol{x})\rangle$ is then, from (1.18):

$$\langle\zeta\rangle = \frac{\int \delta\zeta\, \zeta \exp(-\mathcal{H})}{\int \delta\zeta \exp(-\mathcal{H})} \tag{1.23}$$

where

$$\mathcal{H} = \sum_i B_i\zeta_j + \sum_{ij} A_{ij}\zeta_i\zeta_j, \; \delta\zeta \equiv \prod_i d\zeta_i \tag{1.24}$$

and,

$$B_i = (\gamma - 2\alpha h_i), \; A_{ij} = \alpha\delta_{ij} - \beta G_{ij} \tag{1.25}$$

Here, $-G_{ij}$ corresponds to the inverse of the Laplacian ∇^2 in (1.12). Then

$$\langle \zeta_n \rangle = \partial/\partial B_n \{ \ln \int \delta\zeta \exp(-B\zeta - \zeta A \zeta) \} \tag{1.26}$$

$$= \partial/\partial B_n \ln(Z). \tag{1.27}$$

To proceed, we introduce a transformation of basis $\zeta = \alpha\xi$, such that $A_{ij}\zeta_i\zeta_j = A_{ij}\alpha_{in}\alpha_{jm}\xi_n\xi_m = \lambda_n\xi_n^2$ (summation convention invoked). Here, λ_n are the eigenvalues of the symmetric matrix A. Then,

$$Z = \int \delta\xi \exp\{-\sum[\sqrt{\lambda_n}\xi_n + C_n/(2\sqrt{\lambda_n})]^2 + \sum C_n^2/(4\lambda_n)\} \tag{1.28}$$

$$= \exp(\sum C_n^2/(2\lambda_n)^2)/\sqrt{|A|}, \quad C_n = \sum_m \alpha_{nm} B_m \tag{1.29}$$

and,

$$\ln(Z) = \sum (B_i \alpha_{im} B_j \alpha_{jm})/(4\lambda_m) - \tfrac{1}{2}\ln(|A|)$$
$$= \tfrac{1}{4}(BA^{-1}B) - \tfrac{1}{2}\ln(|A|) \tag{1.30}$$

Hence,

$$\langle \zeta_n \rangle = \tfrac{1}{4}\{A^{-1}B + BA^{-1}\}_n = \tfrac{1}{2}\{A^{-1}B\}_n \tag{1.31}$$

Interpreting (1.31) in terms of its spatial representation and using (1.25) gives:

$$\langle \zeta \rangle(\boldsymbol{x}) = \tfrac{1}{2}\nabla^2(\alpha\nabla^2 - \beta)^{-1}\{\gamma - 2\alpha h\} \tag{1.32}$$

so that the stream function satisfies:

$$(\alpha\nabla^2 - \beta)\langle \psi(\boldsymbol{x}) \rangle = \alpha h(\boldsymbol{x}) - \tfrac{1}{2}\gamma \tag{1.33}$$

Finally, we note that two–point covariances $\langle \zeta(\boldsymbol{x})\zeta(\boldsymbol{x}') \rangle \equiv Z(\boldsymbol{x} \mid \boldsymbol{x}')$ follow from (1.30) and (1.18) as:

$$\langle \zeta(\boldsymbol{x})\zeta(\boldsymbol{x}') \rangle = \partial \ln(Z)/\partial A_{ij} = \mathcal{G}(\boldsymbol{x} \mid \boldsymbol{x}') + \langle \zeta(\boldsymbol{x}) \rangle \langle \zeta(\boldsymbol{x}') \rangle \tag{1.34}$$

where, $\langle \zeta(\boldsymbol{x}) \rangle$ is given by (1.32), and $\mathcal{G}(\boldsymbol{x} \mid \boldsymbol{x}')$ satisfies:

$$\nabla^2(\alpha\nabla^2 - \beta)\mathcal{G}(\boldsymbol{x} \mid \boldsymbol{x}') = -\delta(\boldsymbol{x} \mid \boldsymbol{x}') \tag{1.35}$$

In the homogeneous context ($\langle\psi(\boldsymbol{x})\rangle = 0$), the stream function power spectrum is,

$$\Psi(\boldsymbol{k}) = 1/(\alpha k^4 + \beta k^2), \tag{1.36}$$

as proposed by Kraichnan (1975). The power spectrum $\Psi(k)$ seems reasonable, but that for $Z(k)$ becomes unrealistically independent of k as $k \to \infty$.

The form of $\langle\psi(\boldsymbol{x})\rangle$ predicted by (1.33) for a closed basin seem plausible, and related to other deterministic models. Here, we should mention two studies for which (1.33) plays a pivotal role: that of Griffa and Salmon (1989), and that of Holloway (1992). The first of these is a study of closed basin barotropic circulation on a β–plane. This problem is equivalent to (1.19), if we take

$$h(x, y) = \varpi(y - y_0) \tag{1.37}$$

where, ϖ is the (constant) "beta" term, the northward derivative of the Coriolis force. Griffa and Salmon first demonstrate that a numerically simulated ocean basin flow without friction or wind stress evolves toward a solution of (1.33). In this case, the central region of flow is described by $-\beta\langle\psi(\boldsymbol{x})\rangle = \alpha\langle h(\boldsymbol{x})\rangle$, while the $\alpha\nabla^2\langle\psi(\boldsymbol{x})\rangle$ permits the flow to obey slip boundaries in the coastal region. Such flows are quasi–steady, and are designated by them as Fofonoff flows (although the latter is strictly steady (Fofonoff (1954)). For a wind–driven flow with bottom–drag, the flow is quite similar, provided the wind stress exerts a torque which balances the bottom–drag torque.[3] If the wind opposes the Fofonoff flow, then flow becomes turbulent, with a much smaller mean flow. The connection of the forced–dissipative flow with (1.33) is more tenuous.

The study by Holloway examines the relevance of (1.33) to real ocean circulations, utilizing observed $h(\boldsymbol{x})$. He proposes a simplified picture, based on ignoring the ∇^2 term in (1.33), and finds the (North Atlantic) circulation so produced (called an unprejudiced circulation) plausible, provided regions of direct forcing (such as the Gulf Stream) are avoided. Holloway suggests that (1.33) may be a useful zeroth–order circulation, in the sense that a correct representation of eddy viscosities should impel the system toward (1.33), rather than towards a state of rest.

Equation (1.10) has an infinity of constants of motion, (an integral of any function of the vorticity, ζ), but the formalism developed here only uses

---

[3]It can be demonstrated that the equipartitioning solution (1.18) survives in the forced–dissipative case, provided there is a mode–by–mode balance of forcing and dissipation.

linear and quadratic constraints. Recently, methods have been developed that are able to incorporate these additional constraints into the statistical mechanics (Weichman (1993), Miller *et al.* (1992)). Such methods have been applied to Jupiter's Red Spot, and its application to Oceanographic problems would seem a useful project. The structure of the mean field equations for $\langle\psi\rangle$ specifies the mean vorticity as a function of the mean stream function.

2. Langevin Modeling

We now turn to methods which are suitable for real, dissipative flows. Our interest here is in obtaining variance information, and we will focus on simple problems in which the ensemble mean flow is absent, and the flow may be regarded as homogeneous with impunity.[4] The principle item of interest here is to obtain accurate information about energy spectra with reasonable economy in computation. If such methods are found, incorporating flows with (ensemble) mean fields presents only technical problems. Over the past quarter of a century, the Langevin equation has proved a useful basis for generating quantitative information at the covariance level. The basic physical *ansatz* is that the nonlinear force (as in (1.1)–(1.7)) acts, to first order, as a random forcing of ψ_i. This starting point, that f_i is nearly Gaussian, despite its dependence on the dynamical variables ψ_i, would seem credible only if the number of degrees of freedom in (1.7) is large. But such a replacement neglects those higher–order correlations necessary to maintain the conservation laws (1.14): the total variance, $\sum\langle\psi_i^2\rangle$ would increase with time. Thus, if we argue that a Langevin equation is a plausible zeroth–order approximation, we must simultaneously augment the molecular dissipation of (1.1)–(1.4) in such a way that conservation principles are observed by the Langevin system, insofar as possible[5]. Thus, the prescription is to replace

$$d\psi_i/dt = \sum_{jk} C_{ijk}\psi_j\psi_k - \nu_i\psi_i \tag{2.1}$$

with

$$d\psi_i/dt = \sigma(t)\sum_{jk} C_{ijk}\sqrt{\theta_{ijk}}\chi_j\chi_k - \mu_i\psi_i \tag{2.2}$$

---

[4]For homogeneous flows, the Fourier representation renders $\langle\psi_i\psi_j\rangle = \delta_{ij}\langle\psi_i^2\rangle$, where (i,j) stand for $\boldsymbol{k}_i$ and $\boldsymbol{k}_j$. We use this simplification in the remainder of this section.

[5]Langevin models maintain conservation principles only in ensemble mean.

Here, χ_i are Gaussian fields contrived to have the same covariances as ψ_i, and $\sigma(t)$ is a white–noise, stationary force, $\langle\sigma(t)\sigma(t')\rangle = \delta(t-t')$. The basic idea in (2.2) is to chose θ_{ijk} and μ_i so that the variances of (2.2), $\Psi_i(t)$, approximate those of (2.1) (which we designate $\hat{\Psi}_i$). To see how this may be effected, we advance (2.1) forward in time from an initial state of Gaussian chaos: $\mathcal{P}(\psi_1, \psi_2, \cdots) \sim \exp(-\sum_i a_i\psi_i^2)$. To order dt we find for $\hat{\Psi}_i \equiv \langle\psi_i^2\rangle$

(2.3)

$$(\frac{1}{2}\partial_t + 2\nu_i)\hat{\Psi}_i(dt) = 2dt\sum_{jk} C_{ijk}^2\hat{\Psi}_j(0)\hat{\Psi}_k(0) + 4dt\sum_{jk} C_{ijk}C_{jki}\hat{\Psi}_i(0)\hat{\Psi}_k(0)$$

while from (2.2) we have (exactly, at any time):

$$(\frac{1}{2}\partial_t + 2\mu_i)\Psi_i(t) = 2\sum_{jk} C_{ijk}^2\Psi_j(t)\Psi_k(t)\theta_{ijk} \tag{2.4}$$

where we have used the constraint $\langle\chi_i\chi_j\rangle = \delta_{ij}\Psi_i(t)$. Comparing (2.3) to (2.4) we see that requiring $\hat{\Psi}_i = \Psi_i$ implies (to order dt^2)

$$\mu_i = \nu_i - 4dt\sum_{jk} C_{ijk}C_{jik}\Psi_k\theta_{ijk} \tag{2.5}$$

and

$$\theta_{ijk} = dt \tag{2.6}$$

Imagine now that the procedure for evaluating (2.4) carried out to arbitrary order in dt. Through second order this recipe presents no problems. But beyond that, such a perturbation analysis of (2.1) will contain a vast number of terms. Only a sub–group will (at any order) have the form of the right–hand side of (2.4), (with suitable extensions of (2.6)). Another sub–group will have the form of (2.5) times $\Psi_i(t)$. Since at orders higher that dt^2 the perturbation method contains terms *not* of the form (2.4), we must admit that at best a judicious selection of factors θ_{ijk} will make (2.4) correspond to certain well defined perturbation schemes, whose validity must be judged on other grounds.

Finally, note that if the damping term μ_i in (2.4) is to fulfill conservation principles (1.14), it must satisfy,

$$\mu_i = -4\sum_{jk} C_{ijk}C_{jik}\theta_{ijk}\Psi_k \tag{2.7}$$

So far, equation (2.4) contains considerable arbitrariness in that the factor θ_{ijk} is as yet unspecified. This quantity has dimensions of time, and physically is the duration of the interactions among modes (i, j, k).

One approach to determine θ_{ijk} is to assume that covariances are accurately given by some form of renormalized perturbation theory, such as the Direct Interaction Approximation (DIA) (Kraichnan (1959)), and then determine θ by a comparison of (2.4) to such a theory. Of course, the DIA is not of the Langevin form, so the determination of θ is only approximate. To see how this may be done, we note that the DIA may be written in the form (2.4), with θ replaced by Θ, where:

$$\Theta_{ijk} = \int_0^t dsG_i(t \mid s)\{\Psi_j(t \mid s)/\Psi_j(t \mid t)\}\{\Psi_k(t \mid s)/\Psi_k(t \mid t)\} \tag{2.8}$$

Here, $\Psi_i(t \mid s) \equiv \langle \psi_i(t)\psi_i(s) \rangle$, and $G_i(t \mid s)$ is the response of mode $\psi_i(t)$ at a small perturbation at time s. With respect to the "two–time" quantities $\Psi_i(t, t')$, we first note the thermal equilibrium relation:

$$\Psi_i^{therm}(t \mid s) = \Psi_i^{therm}(s \mid s) G_i^{therm}(t \mid s) \tag{2.9}$$

By thermal equilibrium, we mean that the distribution of modes ψ_i satisfies (1.18). Equation (2.9) is known as the "fluctuation dissipation relation", and is frequently used even for dissipative flows. Its extension to non–stationary flows may be justified by an alternate perturbation theory, (Herring, (1965), (1966)), in which (2.9) emerges as a consistent *a priori* estimate of $\Psi_i(t \mid s)$ given the value of $\Psi(s \mid s)$. In the DIA, $G_i(t \mid s)$ satisfies:

$$G_i(t \mid s) = \delta\Psi_i(t, t')/\delta\Psi_i(t', t'), \tag{2.10}$$

so that (2.9) may be satisfied. The DIA equation for $\Psi_i(t, t')$ is,

$$\partial_t \Psi_i(t, t') = 2C_{ijk}^2 \int_0^{t'} G_i(t', s)\Psi_j(t, s)\Psi_k(t, s)$$

$$-4C_{ijk}C_{jki} \int_{t'}^{t} \Psi_i(s, t')G_j(t, s)\Psi_k(t, s) \tag{2.11}$$

To complete the determination of θ_{ijk}, we parameterize $G(t, s)$ by

$$G_i(t, s) = \exp(-\int_s^t ds'\tilde{\mu}_i(s')) \tag{2.12}$$

and use (2.9) (incorrectly, out of thermal equilibrium). This allows the identification of Θ_{ijk} in (2.8) with θ_{ijk} in (2.2). Using (2.12) and (2.8), we find,

$$\frac{d\theta_{ijk}}{dt} = 1 - (\tilde{\mu}_i + \tilde{\mu}_j + \tilde{\mu}_k)\theta_{ijk} \tag{2.13}$$

The $\tilde{\mu}_i$ may be evaluated by integrating the equation of motion for $G_i(t,t')$ ((2.10) applied to (2.11)) over t, and using (2.12) as an approximation for $G_i(t,t')$. For k larger than the peak in the energy spectrum $E(k) \equiv 2\pi k^3\Psi(k)$, we find (Kraichnan (1959), (Herring (1975)):

$$\hat{\mu}(k) \sim k\sqrt{\int_0^k dpE(p)} \tag{2.14}$$

Notice that (2.14) implies a dependence of $\hat{\mu}$ on large– scale sweeping: an addition of a *random* uniform translation, $\boldsymbol{u}_0\delta(\boldsymbol{k})$ to the velocity comprising $E(k)$ adds $k\sqrt{\langle \boldsymbol{u}_0^2\rangle}$ to (2.14). Although such dependence on large–scale sweeping seems appropriate for G_i, it is not for Ψ_i (or more precisely, the velocity field belonging to Ψ), since small–scale spectra are unaware of the Galilean frame in which they move. The problem stems from adapting the Eulerian frame to formulating the problem, and can be avoided by a formulation in an entirely Lagrangian frame. The latter may approximately be realized by a modification of the characteristic time–scale $\hat{\mu}(k)$ (that of $G_i(t \mid s)$) to a Lagrangian time–scale:

$$\eta(k) \sim \sqrt{\int_0^k dp\; p^2E(p)} \tag{2.15}$$

Equation (2.15) should then be used in place of $\hat{\mu}$ to compute θ_{ijk}. Such methods have been used in a number of studies to predict energy spectra for quasigeostrophic flows, and we will review certain of these shortly. Before such a summary, however, we should indicate in general terms what violence the Markov modeling as embodied in (2.2) does to the dynamics of the basic equations. We recall that the first step was to replace the actual nonlinear force by a random stirring. Broadly speaking, the effect of the random forcing is to destroy in the model any effects of structures that Navier–Stokes may possess. In this sense, such models are "structureless," and their deviations from reality may give some indication of the effects of structures on covariance information.

From the historical perspective, stochastic models were developed *after* results of the analytic theories (such as the DIA) were known, at least in their broad outline. One reason for their development was to assure that a particular analytic theory was realizable: they were "proof by construction" that the covariance of an analytical theory is physically realizable, a point of some insecurity among theorists in the early days (1955–1964). Finally, it was realized that the stochastic models could be used to generate perturbation expansions in their own right, with the analytic theory serving as the unperturbed state (Phythian (1969), Kraichnan (1971), Herring and Kraichnan (1972)). Stochastic models may also be used as Monte Carlo simulations in complex flows, for which analytic theories may present equations too formidable for practical application. This approach has not been much used, but may be of computational advantage for certain flows, as has been found by Kaneda (1992).

The system ((2.4), (2.7)), with θ given by (2.15), is called an Eddy Damped Quasinormal Markovian Approximation (EDQNM), and a systematic review of their application to various problems in quasigeostrophic flows may be found in Lesieur (1990) . Of course, such studies rely heavily on numerical methods for solution, and such may not be a suitable basis for gaining insight into the underlying physics. One useful procedure to reveal the physics is to assume that the energy spectrum is strongly peaked at large scales, and approximate the wavenumber integrals in (2.4) by assuming that the wavenumber belonging to either j or k (but not both) is small[6]. This leads to (see Lesieur (1990), p. 278):

$$\partial E(k)/\partial t = k^{-2}\partial_k\{k^3\partial_k\{kE(k)\eta(k)\}\} - 2\nu k^2 E(k) \tag{2.16}$$

where we recall (2.15) for $\eta(k)$. Thus, the energy (or enstrophy) cascade to small scales is diffusive; energy passes from one scale to adjacent scales. The rate at which this happens, $\eta(k)$, is essentially the *rms* strain. Such a modeling of turbulent transfer was first proposed by Leith (1967), on heuristic grounds. Equation (2.16) contains both the inverse cascade of energy (for the case in which a small–scale forcing term is added to its right–hand side), and forward enstrophy cascades. The former leads to the well–known $k^{-5/3}$ inverse cascade range, while the latter leads to a $k^{-3}(\ln(k/k_i))^{-1/3}$ enstrophy range (Kraichnan (1971)).

These methods may be readily extended to anisotropic turbulence, such as occurs with β–plane flows (Holloway and Hendershott (1977)). If we examine the rate equations for anisotropic turbulence at small scales, using

---

[6]Note that according to (1.10)–(1.11), $\sum_{jk} C_{ijk}C_{jki}\Psi_j\Psi_k = \int d\boldsymbol{p}d\boldsymbol{q}\delta(\boldsymbol{p}+\boldsymbol{q}-\boldsymbol{k})\cdot B(\boldsymbol{k},\boldsymbol{p},\boldsymbol{q})\Psi(\boldsymbol{p})\Psi(\boldsymbol{q})$. What is proposed is to evaluate this integral by assuming $\boldsymbol{p}$ to be small, (hence, $\sim \boldsymbol{q} \sim \boldsymbol{k}$), and expand $\Psi(\boldsymbol{k}-\boldsymbol{p})$ about $\boldsymbol{p}=\boldsymbol{0}$, retaining only the lead, non–vanishing terms.

the diffusion expansion described above, a rather simple heuristic emerges (Herring, 1975)): at small scales, k, the level of anisotropy at a given time is determined by a balance between its rate of production by large–scale straining (strain rate times the small–scale isotropic component of the spectrum), and its rate of destruction, by isotropization at scales roughly comparable to k. The isotropizing rate is essentially $\eta(k)$, as given by (2.15), times the small–scale anisotropy. Thus, if $E(k)$ falls off faster that k^{-3}, the small scales retain the anisotropy of the large scales.

3. Non–Markovian Stochastic Models

Although Markovian models are easy to implement either analytically or numerically, they are unable to cope with systems that have disparate time scales, as would be encountered for example, in turbulence coexisting with waves. A simple example is barotropic flow over random topography, for which a component of the vorticity is locked to topography. We discuss here briefly the application of the DIA to this case, starting from the point of view of a non–Markovian model of the DIA, as developed in (Kraichnan (1971), and (Herring and Kraichnan (1972)). We take as the basic dynamical variable the Fourier transform of the vorticity field as given by (1.19), which we write as:

$$\partial\zeta_{\boldsymbol{k}}/\partial t = F_{\boldsymbol{k}}(\zeta, h) - \nu k^2 \zeta_{\boldsymbol{k}} \tag{3.1}$$

where $F_{\boldsymbol{k}}$ is the Fourier transform of the right–hand side of (1.19), whose Fourier transform is written as:

$$F_{\boldsymbol{k}} = \sum_{\boldsymbol{k}=\boldsymbol{p}+\boldsymbol{q}} \{C_{\boldsymbol{k},\boldsymbol{p},\boldsymbol{q}}\zeta_{\boldsymbol{p}}\zeta_{\boldsymbol{q}} + E_{\boldsymbol{k},\boldsymbol{p},\boldsymbol{q}}\zeta_{\boldsymbol{p}}h_{\boldsymbol{q}}\} \tag{3.2}$$

To emphasize the idea that (3.1) couples a fast mode ($\zeta(\boldsymbol{x}, t)$ with an (infinitely) slower component, $h(\boldsymbol{x})$ we also write,

$$\partial h_{\boldsymbol{k}}/\partial t = 0 \tag{3.3}$$

We shall first write down the DIA stochastic model for (3.1) and then justify it. It is

$$\partial\zeta_{\boldsymbol{k}}/\partial t = -\int_{t'}^{t} ds(\rho_{\boldsymbol{k}}(t,s)\zeta_{\boldsymbol{k}}(s) + \hat{\rho}_{\boldsymbol{k}}(t,s)h_{\boldsymbol{k}}) + F_{\boldsymbol{k}}(\chi, h) - \nu k^2\zeta_{\boldsymbol{k}} \tag{3.4}$$

Here, $h_{\boldsymbol{k}}$ denotes the Fourier transform of $h(\boldsymbol{x})$, and $\chi_{\boldsymbol{k}}$ is a Gaussian variable, as in (2.2). The visco–elastic terms, $\rho_{\boldsymbol{k}}$ and $\hat{\rho}_{\boldsymbol{k}}$ act to preserve energy and enstrophy conservation under the replacement of $F_{\boldsymbol{k}}(\xi, h)$ by $F_{\boldsymbol{k}}(\chi, h)$. To see how this may be implemented, we form the equation of motion for $2Z_{\boldsymbol{k}}(t,t) = \{\langle \zeta_{\boldsymbol{k}}(t')\partial_t \zeta_{\boldsymbol{k}}(t) + \zeta_{\boldsymbol{k}}(t)\partial_{t'}\zeta_{\boldsymbol{k}}(t')\rangle\}_{t\to t'}$. From (3.4):

(3.5)
$$\partial_t Z_{\boldsymbol{k}}(t,t) = -2\int_0^t ds[\rho_{\boldsymbol{k}}(t,s)Z_{\boldsymbol{k}}(t,s) + \hat{\rho}_{\boldsymbol{k}}(t,s)X(t') + \langle \zeta_{\boldsymbol{k}}(t)F_{\boldsymbol{k}}(\chi,h,t)\rangle$$

and for $X_{\boldsymbol{k}}(t) \equiv \langle h_{\boldsymbol{k}}\zeta_{\boldsymbol{k}}(t)\rangle$

$$\partial_t X_{\boldsymbol{k}} = -\int_0^t ds[\rho_{\boldsymbol{k}}(t,s)X_{\boldsymbol{k}}(s) + \hat{\rho}(t,s)H_{\boldsymbol{k}}] + \langle h_{\boldsymbol{k}}F_{\boldsymbol{k}}(\chi,h)\rangle \tag{3.6}$$

In these equations $H_{\boldsymbol{k}} \equiv \langle h_{\boldsymbol{k}}h_{-k}\rangle$. To evaluate $\langle \zeta_{\boldsymbol{k}}(t)F_{\boldsymbol{k}}(\chi,h,t)\rangle$, we note that from (3.4)

$$\begin{pmatrix}\zeta\\ h\end{pmatrix}(t) = \int_0^t ds \begin{pmatrix} G^{11}(t,s) & G^{12}(t,s)\\ G^{21}(t,s) & G^{22}(t,s)\end{pmatrix}\begin{pmatrix}F(\chi,h)\\ 0\end{pmatrix} \tag{3.7}$$

where $G^{ij}_{\boldsymbol{k}}(t,s)$ is:

$$\partial_t \begin{pmatrix} G^{11}(t,t') & G^{12}(t,t')\\ G^{21}(t,t') & G^{22}(t,t')\end{pmatrix} + \int_{t'}^t ds \begin{pmatrix}\rho(t,s) & \hat{\rho}(t,s)\\ 0 & 0\end{pmatrix}\begin{pmatrix} G^{11}(s,t') & G^{12}(s,t')\\ G^{21}(s,t') & G^{22}(s,t')\end{pmatrix}$$

$$= \begin{pmatrix}1 & 0\\ 0 & 1\end{pmatrix}\delta(t,t') \tag{3.8}$$

In (3.7) and (3.8), we have omitted the $\boldsymbol{k}$ subscripts for simplicity. Introducing (3.7) into (3.5) permits the evaluation of $\langle \zeta_{\boldsymbol{k}}F_{\boldsymbol{k}}(\chi,h)\rangle$ as:

$$\langle \zeta_{\boldsymbol{k}}F_{\boldsymbol{k}}(\chi,h)\rangle = \int_0^t ds G^{11}_{\boldsymbol{k}}(t,s)\langle F_{\boldsymbol{k}}(\chi,h,s)F_{\boldsymbol{k}}(\chi,h,t)\rangle \tag{3.9}$$

Notice that the right hand side ensemble average is, according to (3.2), and (3.4) a fourth moment of Gaussian variables, ($\xi_{\boldsymbol{k}}$, $h_{\boldsymbol{k}}$), and hence becomes a bilinear function of $Z_{\boldsymbol{k}}$, $X_{\boldsymbol{k}}$, and $H_{\boldsymbol{k}}$. The visco–elastic operators $\rho_{\boldsymbol{k}}(t,s)$, $\hat{\rho}_{\boldsymbol{k}}(t,s)$ may now be determined by the constraint that for $\nu = 0$ the energy $((1/2)\sum_{\boldsymbol{k}} Z_{\boldsymbol{k}}/k^2)$ and enstrophy $(\sum_{\boldsymbol{k}}(Z_{\boldsymbol{k}} + 2X_{\boldsymbol{k}} + H_{\boldsymbol{k}})$ are conserved. We shall not write these equations here. $\hat{\rho}(t,s)$ stems entirely from the interaction of topography with vorticity, and characterizes topographic drag on mode $\zeta_{\boldsymbol{k}}$. The realizability of $Z_{\boldsymbol{k}}(t,t') \equiv \langle \zeta_{\boldsymbol{k}}(t)\zeta_{-\boldsymbol{k}}(t')\rangle$ is fully guaranteed by (3.4). The full two–time form of $Z_{\boldsymbol{k}}(t,t')$ follows from (3.4) by multiplying (3.4) by $\zeta_{-\boldsymbol{k}}(t')$ and ensemble averaging.

We now simplify the notation by denoting the convolution $\int_0^t dsx(t,s) \cdot y(s,t')$ as $x * y$. Where useful, we denote the operator acting on $\boldsymbol{G}(t,t')$ in (3.8) as $\{\boldsymbol{G}\}^{-1}$. Then, (3.5) becomes more simply,

$$\{G^{11}_{\boldsymbol{k}}\}^{-1} * Z_{\boldsymbol{k}} = G^{11} * \langle FF^{\dagger}\rangle - \hat{\rho}_{\boldsymbol{k}} * X^{\dagger} \tag{3.10}$$

Where † denotes the transpose in (t,t'). Similarly, with the equation for $X(t)$

$$\{G^{12}_{\boldsymbol{k}}\}^{-1} * X_{\boldsymbol{k}} = -\hat{\rho} * H_{\boldsymbol{k}} \tag{3.11}$$

It is of interest to examine the stationary state form of these equations (assuming either inviscid flow, or an additional forcing, compatible with stationary turbulence). In that case, $X(t,t') = X(\tau \equiv t-t')$ and the equations for the Fourier transforms

$$\tilde{X}(\omega) \equiv \int_{-\infty}^{\infty} d\tau X(\tau) \tag{3.12}$$

are of the form:

$$i\omega \tilde{Z}_{\boldsymbol{k}}(\omega) = -\tilde{\rho}_{\boldsymbol{k}}(\omega)\tilde{X}_{\boldsymbol{k}}(\omega) - \tilde{\hat{\rho}}_{\boldsymbol{k}}(\omega)X_{\boldsymbol{k}}(\omega) + G^{11}_{\boldsymbol{k}}(\omega)\langle FF^{\dagger}\rangle(\omega) \tag{3.13}$$

and

$$0 = -[\tilde{\rho}_{\boldsymbol{k}}(\omega)X_{\boldsymbol{k}} + \tilde{\hat{\rho}}_{\boldsymbol{k}}(\omega)H_{\boldsymbol{k}}]\delta(\omega) + \langle h_{\boldsymbol{k}}F_{\boldsymbol{k}}(\chi,h) \tag{3.14}$$

Here we have used the fact that $\tilde{X}_{\boldsymbol{k}}$ and $\tilde{H}_{\boldsymbol{k}} \sim \delta(\omega)$, for stationary turbulence. It follows from (3.13) and (3.14) that $Z_{\boldsymbol{k}}$ must have a static component, $Z^{static}_{\boldsymbol{k}}(\tau)$ independent of τ:

$$\hat{\rho}_{\boldsymbol{k}}(0)Z^{\text{static}}_{\boldsymbol{k}} = -\tilde{\hat{\rho}}_{\boldsymbol{k}}(0)H_{\boldsymbol{k}} \tag{3.15}$$

We recall that $\rho_{\boldsymbol{k}}(0)$ and $\hat{\rho}_{\boldsymbol{k}}(0)$ are $\int d\tau \exp(i\omega\tau)(\rho_{\boldsymbol{k}}(\tau),\ \hat{\rho}_{\boldsymbol{k}}(\tau)$ evaluated at $\omega = 0$.

For regions of scale–size for which dissipation plays little role, it may be shown that $\tilde{\rho}(0)_{\boldsymbol{k}} \simeq \tilde{\hat{\rho}}(0)_{\boldsymbol{k}}$, (Holloway, (1976), Herring (1975)) so that $Z_{\boldsymbol{k}} \sim= H_{\boldsymbol{k}}$, and $X_{\boldsymbol{k}} = -H_{\boldsymbol{k}}$. This result is plausible if we argue that if $\boldsymbol{u}$ is large (in some nominal sense) then $\partial_t\zeta$ in (1.19) is nominally small only if $\zeta = -h$, in detail. What the statistical theory gives is a way to quantify how strong this correlation is, and how it depends on the flow context. A close examination of (3.4), and the subsequent equations show that there is always the possibility of a static solution to the case of stationary turbulence, in the DIA, even for $h(\boldsymbol{x}) = 0$. By static solution, we mean one in which $\tilde{Z}_{\boldsymbol{k}}(\omega) \sim \delta(\omega)$. Of course, such would require a static forcing; one which has no time dependence, but which varies from ensemble to ensemble. Although these static solutions to DIA are possible, studies of two–dimensional turbulence, and also convection (Herring, (1969)), suggest that they are not stable, even at quite small Reynolds number. But solutions to the DIA show a tendency for $Z_{\boldsymbol{k}}(\tau)$ to have a much longer timescale than $G(\tau)$, especially at small scales, where viscosity is important. Such is at variance with the fluctuation dissipation relation (2.9). In practice this has been found to make little difference, except for the present problem.

We should mention that (3.10) and (3.11) are similar to Markovian equations proposed by Holloway (1976, 1986). Roughly speaking, to obtain the latter, we parameterize the (t, t') arguments as $\exp(-\eta(k) \mid t - t' \mid)$, and assume that the relaxation of tipple moments is much faster than that of $Z_{\boldsymbol{k}}(t, t)$. Holloway compared his Markovian theory with DNS and found quite satisfactory results. The topographic drag (as contained in (3.4)–(3.10)) was also generalized for realistic anisotropic flow (on a β–plane), and compared to other results.

4. Concluding Comments

This review has focused on methods which assume that the full distribution function for the fluid vorticity field is, in some sense, close to multivariate Gaussian. We note that the inviscid equipartitioned distribution is itself multivariate Gaussian (in vorticity), and perhaps it is this fact, together with the tentative agreement of its predictions with simulations and observations, that has encouraged its generalization to forced dissipative flows. But the inviscid solution fails at small scales, where a divergence of total enstrophy is encountered. The latter is attributable to the absence of dissipative effects. Langevin models are able to give reasonable small–scale spectra, but of course have unrealistic lagged covariances. The latter problem is less severe for the stochastic models of Sec. 3, which have self–consistently–determined lagged covariances. Yet all these methods are unable to incorporate structures; indeed, their replacement of actual non–linearities with Gaussianly composed force fields would suggest their inadequacies in this respect. Space–filling structures are not the problem, since they may not produce strong non–Gaussian signals. It is only if structures become isolated that measures of non Gaussianity become significant. Such is the case for decaying homogeneous two–dimensional turbulence (McWilliams (1985)), as well as homogeneous quasigeostrophic flow (McWilliams *et al.* (1994)). But the presence of random topography, random forcing, and Rossby waves that exists in real applications may well disrupt the formation of isolated vortices and lead to a chaotic flow more commensurate with stochastic modeling. At least this has been found to be so in the study of scalar mixing by Bartello and Holloway (1991).

References

[1] Bartello, P., and G. Holloway, 1991: Passive scalar transport in β–plane turbulence. *J. Fluid Mech.*, **223**, 521-536.

[2] Chollet, J.P., and M. Lesieur, 1981: Parameterization of small scales of three-dimensional isotropic turbulence utilizing spectral closure. *J. Atmos. Sci.*, **38**, 2747-2757.

[3] Edwards, S. F., 1964: The theoretical dynamics of homogeneous turbulence. *J. Fluid Mech.*, **18**, 239-273.

[4] Fofonoff, N. P., 1954: Steady flow in a frictionless homogeneous ocean. *J. Mar. Res.*, **13**, 254–262.

[5] Griffa, A., and R. Salmon, 1989: Steady flow in a frictionless homogeneous ocean. *J. Marine Research.*, **47**, 457–492.

[6] Herring, J. R., 1969: Statistical Theory of Thermal Convection at Large Prandtl Number. *Phys. Fluids.*, **12**, 39-52.

[7] Herring, J. R., and R. H. Kraichnan, 1972: Comparison of some approximations for isotropic turbulence. *Lecture Notes in Physics,* M. Rosenblatt and C. Van Atta, Eds., Springer– Verlag, **12**.

[8] Herring, J. R., and J. C. McWilliams, 1985: Comparison of direct numerical simulation of two-dimensional turbulence with two-point closure: The effects of intermittency. *J. Fluid Mech.*, **158**, 229–242.

[9] Herring, J. R., 1965: Self-consistent field approach to turbulence theory. *Phys. Fluids.*, **8**, 2219–2225.

[10] Herring, J. R., 1966: Self-consistent field approach to non– stationary turbulence *Phys. Fluids.*, **9**, 2106–2110.

[11] Herring, J. R., 1975: Theory of two-dimensional anisotropic turbulence. *J. Atmos. Sci.*, **32**, 2254–2271.

[12] Holloway, G. 1976: Statistical hydromechanics: application in mesoscale ocean circulation. PhD thesis, Univ. Calif. San Diego.

[13] Holloway, G. and M. C. Hendershott, 1977: Stochastic modeling for non–linear Rossby waves. *J. Fluid Mech.*, **82**, 747–765.

[14] Holloway, G., 1986: Eddies, waves, circulation and mixing: statistical geofluid mechanics. *Ann. Rev. Fluid Mech.*, **82**, 747–765.

[15] Holloway, G., 1992: Representing Topographic Stress for Large–Scale Ocean Models. *J. Phys. Ocean.*, **22**, 1033–1046.

[16] Kaneda, Y., 1981: Renormalized expansions in the theory of turbulence

with the use of the Lagrangian position function. *J. Fluid Mech.*, **107**, 131–145.

[17] Kaneda, Y., 1992: Application of Monte Carlo Method to the Lagrangian Renormalized Approximation. *Research Trends in Physics: Chaos and Transport in Fluids and Plasmas*, W. Horton, Y. Ichikawa, I. Prigogine, and G. Zaslavsky, Eds., La Jolla International School of Physics, The Institute for Advanced Studies.

[18] Kraichnan, R. H., 1959: The structure of isotropic turbulence at very high Reynolds numbers. *J. Fluid Mech.*, **5**, 497–543.

[19] Kraichnan, R. H., 1964: Lagrangian-history closure approximation for turbulence. *Phys. Fluids.*, **8**, 575– 598.

[20] Kraichnan, R. H., 1967: Inertial Range in Two-Dimensional Turbulence. *Phys. Fluids.*, **10**, 1417-1423.

[21] Kraichnan, R. H., 1971: Inertial–range transfer in two and three dimensional turbulence. *J. Fluid Mech.*, **47**, 525 –535.

[22] Kraichnan, R. H., 1975: Statistical dynamics of two dimensional turbulence. *J. Fluid Mech.*, **67**, 155– 175.

[23] Kraichnan, R. H., 1976: Eddy viscosity in two- and three– dimensional turbulence. *J. Atmos. Sci.*, **33**, 1521– 1536.

[24] Lee, T. D., 1952: On some statistical properties of hydrodynamical and magnetohydrodynamical fields. *Q. Appl. Math.*, **10**, 69.

[25] Leith, C.E., 1967: Diffusion approximation to the inertial energy transfer in isotropic turbulence. *Phys. Fluids.*, **10**, 1409-1416.

[26] Lesieur, M., 1990: *Turbulence in Fluids* Dordrecht.

[27] McWilliams, J.C, 1984: The emergence of isolated vortices in turbulent flows. *J. Fluid Mech.*, **146**, 21-43. Novikov, E. A., 1963: Method of random forces in turbulence theory. *Zh. Exper. Teor. Fiz.*, **44**, 2159–2168.

[28] Miller, J., P. B. Weichman, and M. C. Cross 1991: Statistical mechanics, Euler's equation, and Jupiter's Red Spot. *Phys. Rev. A.*, **45**, 2328–2359.

[29] Orszag, S. A., 1970: The statistical theory of turbulence, *Unpublished manuscript*, book.

[30] Orszag, S. A., 1974: Statistical theory of turbulence. *Les Houches Summer School in Physics.*, (R. Balian and J.-L. Peabe), Eds., Gordon and Breach.

[31] Phythian, R., 1969: Self-consistent perturbation series for stationary homogeneous turbulence. *J. Phys.* A, **2**, 181.

[32] Sadourny, R., and C. Basdevant, 1985: Parameterization of subgrid scale barotropic and baroclinic eddies in quasi geostrophic models: anticipated potential method. *J. Atmos. Sci.*, **42**, 1353–1363.

[33] Saffman, P. G., 1967: Note on the decay of homogeneous turbulence. *Phys. Fluids.*, **10**, 1349.

[34] Salmon, R., G. Holloway, and M. C. Hendershott, 1978: The equilibrium statistical mechanics of simple quasi–geostrophic models. *J. Fluid Mech.*, **75**, 691–703.

[35] Weichman, P. B., 1993: Statistical Mechanics, Euler's Equation, and Jupiter's Red Spot in *Nonlinear waves and Weak Turbulence with Applications in Oceanography and Condensed Matter Physics.* N. Fitsmaurice, D. Guarie F. McCaughan, and W.A. Woyczyński Eds., Birkhäuser, Boston. 240–277.

National Center for Atmospheric Research
Box 3000
Boulder, COLORADO
80307

herring@oak.mmm.ucar.edu

Neptune effect: statistical mechanical forcing of ocean circulation

Greg Holloway

Abstract

Beginning from a particular concern about the role of pressure-topography correlations in coastal dynamics, for which the whimsical term "Neptune effect" arose, a broader and more powerful context unfolded. It has become clear that this effect has strength comparable to wind or thermohaline forcing, yet it is absent or grossly corrupted in even the most powerful computer models. A novel approach considers idealized statistical mechanics, then takes account that applied forces such as wind or thermohaline will draw the state of the ocean away from a higher entropy configuration. Nonequilibrium tendencies which arise due to departure from higher entropy serve to parameterize unresolved eddy effects. Introduction into conventional ocean models leads to large differences which markedly improve model skill measured against a global inventory of long term current meter records.

1. Brief history of "Neptune effect"

Sometimes an idea starts in a limited context with a good deal of labor. Later one may see the broader context, as an idea becomes easier to understand and more encompassing. "Neptune effect" started in one such small way with a lot of bother. It's not yet clear where the idea may lead. This article includes four parts. The first, without detail, recalls the earlier limited context in which the term "Neptune effect" emerged. The second provides the extension, which is a simplification that yields practical application. The third discusses consequences relative to what is known about actual ocean circulation. Then an appendix comments upon a stochastic model representation.

Early concern was for "form drag", also called "topographic drag" or "mountain torque". One imagines large scale flow driven over a bumpy bottom surface. Interaction of the large scale flow with the bumps produces meanders and eddies in the flow. Presumably eddies may be present for other reasons as well. The eddying motion implies horizontal pressure differences, whether geostrophic or not. Correlations between pressure anomalies and the bumps can result in large exchanges of momentum (angular momentum from global perspective) between overlying fluid and underlying earth. How to estimate that exchange?

A short calculation shows the magnitude of the forces involved and the subtlety of the dynamics needed to produce such forces. Pressure p on a bottom of slope ∇H transfers horizontal momentum $p \nabla H$ per unit time per unit horizontal area, thus a " form stress". Let p' be a typical amplitude of pressure deviation (from horizontal mean), and h' a representative height of bump. If v' is a typical deep flow speed (above frictional benthic boundary layer) and l' a characteristic horizontal scale for variation of v', then geostrophy provides an estimate $p' = \rho l' f v'$, with ρ =density. Ageostrophic contribution to p' may be neglected in this rough estimation of amplitudes. ∇H can be estimated as h'/l' using the same l' since only the deviations of p correlated with deviations of ∇H will contribute net transfer (integrated over larger areas). Simple product of amplitudes shows that a scaling for $p\nabla H$ is then $\rho f v' h'$. Consider some numbers. Deep flows in the open ocean may have amplitudes $v' \sim .03$ m/s, while deep ocean "bumps" may be of heights 300m (over horizontal scales of hundreds of km such as may be represented in a global ocean model). With $\rho \sim 10^3$ kg/m^3, $f \sim 10^{-4}s^{-1}$, a simple product of amplitudes yields$\rho f v' h' \sim$ 1N/m^2. This is an order of magnitude larger than typical mean wind stresses thought to dominate the driving of mean ocean circulation. Clearly the form stress cannot be so large (in the mean), and clearly I have erred by paying no attention to the correlation coeffcient between v and h deviations. What we here appreciate is that subtle eddy-topography interactions which could produce correlation coefficients only of magnitude 0.1 can produce men forcing fully as large as the wind driving of oceans.

We may be alarmed when large scale ocean models, or analytical theories of circulation, omit this eddy-topography forcing altogether. Less clear is the situation for fine resolution (so-called "eddy resolving") ocean models which might, in principle, explictly resolve the dynamics which would lead to form stress. What is not known is how fine must be the resolution before such a model will perform faithfully. Alternatively, one may hope some parameterization scheme can represent the effect.

An approach was formulated by extending theory from quasi-geostrophic turbulence over topography (Herring, 1977; Holloway, 1978) to include the role of imposed mean (uniform) flow in a statistically homogenous setting. Despite such idealizations as barotropic quasigeostrophy and statistical homogeneity, results (Holloway, 1987, hereafter H87) were laborious and not readily applicable to more realistic circumstances. Details aren't needn't here. Only one aspect of the result is germane: It was found that if the environment supported large scale wave propagation (planetary or topographic Rossby waves), then form stress should exhibit a bias in the direction of wave phase propagation. This could contradict a simple assumption that "form drag", as a drag, should oppose the mean flow. Depending on the sense of mean flow relative to wave propagation, the form stress could very well propel the mean flow. Indeed, in absence of mean flow, the result was clearly that form stress would generate a mean flow. Indeed, in absense of mean flow, the result was clearly that form stress would generate a mean flow. Thus in particular, no simple drag parameterization (not even if unsymmetrical with respect to the sense of wave propagation) could represent the phenomen.

The aforementioned results led to a change of terminology. Whereas it had been usual to speak of topographic drag or form drag, one now encounters "topographic stress", emphasizing that the sense may not be opposed to mean flow.

And "Neptune effect". This bit of whimsy has aquired a life of its own since appearing in a newsletter article (Holloway, 1986) that anticipated later results. The context was coastal circulation, commenting on circumstances where applied windstress and longshore pressure gradient (so well as that can be known) should appear to drive flows in one direction, while observed flows can be persistently backwards to their apparent driving. How can that be? It must be King Neptune, prodding with his trident (evidenced in bottom roughness). This was cartooned in the newsletter.

2. A broader, simpler view leads to application

Details were not given in the preceding section because they are labored and not readily shown to yield a simple, robust parameterization which might serve practical application. Progress resulted from considering the broader context as well as taking a number of (as yet unsubstantiated) leaps. Broadly, the problem we address is one of nonequilibrium statistical mechanics: a system with many interacting degrees of freedom, subject to external forcing. The interacting degrees of freedom include the macroscopic motion in the ocean, ranging from sub-centimeter to planetary radius. The number of such degrees of freedom is in order of magnitude like Avogadro's number! External forcing includes wind, thermal and freshwater fluxes, tidal potential, the swimming of fish, etc., as well as effects of viscosity and conduction which we regard as "external" to the degrees of freedom involving macroscopic motion.

Methods in statistical mechanics address two classes of problems: equilibrium and nonequilibrium. In contrast with the nonequilibrium circumstance just described, equilibrium statistical mechanics considers a system of many degrees of freedom which has been isolated from external influences for a sufficient length of time that any memory of external influences (initial conditions) has been erased. Only overall integrals of the motion, such as total energy, remain. A probability distribution for possible states of the system can be derived by maximizing entropy

$$S = -\int P ln P dY \tag{1}$$

subject to whatever integrals of the motion one has been able to identify. In (1), dY is a volume element in a phase space for which multi-dimensional $\mathbf{Y}$ represents the collection of coordinates (degrees of freedom) that completely define the state of the fluid. $P(\mathbf{Y})dY$ is the probability of realizing a state within neighborhood dY of $\mathbf{Y}$. Certain restrictions apply. The phase space must be such that motion $\mathbf{Y}\circ$ satisfies the Liouville property $\partial\mathbf{Y}\circ/\partial\mathbf{Y} = 0$. Linear transformations of coordinates will affect an additive constant, not listed in (1), but not affect the distribution at maximum entropy.

Allure of the equilibrium problem is that it may be more tractable. In some circumstances for a sufficiently isolated system, it may be the appropriate problem. Unhappily the circumstance of actual oceans, subject to external forcing and internal dissipation, precludes direct application of these equilibrium ideas. Nonetheless let us recall briefly a result from Salmon et al. (1976, hereafter SHH) who considered two-layer, unforced, inviscid, quasigeostrophic flow above topography. Projecting motion onto a finite (however large) number of basis functions, SHH solved for equilibrium P(Y). Among their results it was found that for scales of motion larger than the internal deformation radius, the equilibrium flow was nearly barotropic. For the case of barotropic flow over topography, an expected mean flow was obtained as

$$(\alpha_1/\alpha_2 - \nabla^2) < \Psi >= h \tag{2}$$

where Ψ is barotropic streamfunction, h is bottom topography expressed as a fractional change of depth of fluid times Coriolis parameter, ∇^2 is two-dimensional Laplacian, and α_1 and α_2 are Lagrange multipliers which serve to constrain entropy maximization subject to total energy and enstrophy (square of potential vorticity).

The question must be asked if a result such as (2) is only academically interesting or if it may have something to do with actual ocean circulation. Manifestly the actual ocean is a forced-dissipative open system, compelling the more laborious (but yet uncertain) methods of nonequilibrium statistical mechanics. What to do?

An important link between the equilibrium and nonequilibrium accounts was made by Carnevale et al. (1981, hereafter CFS) who demonstrated explictly that an entire class of nonequilbrium theories, including that employed by H87, satisfy a so-called H-theorem. This is the statement that interactions among the internal degrees of freedom provide only non-negative contributions to overall temporal evolution of entropy S. In absence of external forcing and imposed dissipation, theory such as H87 leads to SHH. Complications for H87 have to do with calculating rates so that one can evaluate the relative competition between entropy-generating internal interactions and entropy-limiting external inputs.

Let us only borrow this conceptual idea of entropy-generating tendencies competing with entropy-limiting inputs. Conventional ocean theories and models already deal with the inputs: wind, thermal, freshwater, and assigned dissipation, on account of which the state of the ocean is pulled away from an equilibrium such as (2). Missing from our theories and models is an entropy-generating tendency which should arise in consequence of the ocean state being displaced from equilibrium. Let y_i be the expectation for the i-th component of $\mathbf{Y}$; then, even far from maximum entropy eqilibrium, we expect a conjugate force $\mathbf{\Lambda}_{ij}\partial S/\partial y_j$ to act on y_i. The difficulty is that far from equilibrium we've no confident means to estimate either $\partial S/\partial y_j$ or the coupling coefficients $\mathbf{\Lambda}_{ij}$

We are obliged to attempt some guesses. Our only excuse for the clumsiness of these guesses is to consider the clumsiness of the alternative, namely to ignore (effectively, set to zero) the entropy-generating terms. If $\mathbf{Y}$ were near maximum entropy, say $\mathbf{Y}*$, then we could say that the gradient $\partial S/\partial \mathbf{Y}$ would be linear in $\mathbf{Y} * - \mathbf{Y}$ and the tensor of coupling coefficients $\mathbf{\Lambda}$ could be evaluated at $\mathbf{Y}*$. Although assumption that a restoring force is linear in the departure from $\mathbf{Y}*$ cannot be justified for large departures, until we have greater sophistication to evaluate $\partial S/\partial \mathbf{Y}$ far from $\mathbf{Y}*$, the linear assumption is simplest and no other is supported. Likewise we assume from simplicity that the $\mathbf{\Lambda}$ are diagonal, i.e. that the departure of any y_i from y_i* produces a force only on y_i proportional to $y_i * - y_i$. Again, we've no supporting argument at this time for any more sophisticated assumption. As well, we may be seeking only some initial feedback, making some trial assumptions just to see if consequences are obviously nuts. If we find instead some agreeable results, this may encourage further effort along these lines.

Practically, what we have done (Alvarez *et al.*, 1994; Eby and Holloway, 1994, hereafter EH) is to adopt a very conventional ocean model (the "GFDL Ocean Model" as described by Pacanowski *et al.*, 1991) set up in a very conventional way. Alvarez *et al.* performed studies of circulation in the Mediterranean while EH evaluated global ocean circulation after 1000 year integration. In each study, two cases were compared. A "control" was run in an entirely conventional manner. Then the model was modified to take account of entropy-generating tendencies ("Neptune") which should compete with imposed forcing. The two outcomes were compared to see (a) if significant differences appeared and (b) if it was possible to see if one was "better".

Modification to an ocean model consisted of two steps. First one must approximately characterize $\mathbf{Y}*$. Then one must introduce tendencies given by $\mathbf{\Lambda}(\mathbf{Y}* - \mathbf{Y})$.

Characterization of $\mathbf{Y}*$ follows (2). Define $L^{-2} = \alpha_1/\alpha_2$. Although α_1 and α_2 would be given precise values for the idealized case considered by SHH, their values will not be well known far from equilibrium under forcing and dissipation. We are not even assured that $\alpha_1/\alpha_2 > 0$, although evaluations under circumstances of SHH tend to yield a positive ratio when topography is present. We expect L is a real length. I *speculate* that L is a length associated with the higher wavenumber range of mesoscale eddy vorticity, reflecting the strong generation of entropy should variance be added at these scales. This is no more than order of magnitude guesstimation. Km to tens of km might be guesses, whereas cm or earth radii should be excessive. Even in a limited domain such as the Mediterranean, and certainly for the global domain, we are inclined to compute on grids coarser than the first deformation radius. (Our idea is to *represent* effects of eddies without the enormous cost of attempting to resolve them.) Coarseness helps us because we need only consider the baratrropic aspect from SHH. Future extensions to smaller scales will need to address baroclinic considerations. (As ever, we recognize that the actual ocean is highly baroclinic even on the largest scales. This is in consequence of the nature of imposed forcing that we envision competing with entrophy-generating tendencies which, themselves, would drive toward barotrophy at scales larger that first radius.) Coarseness gives us a further advantage. If the length scales on which we care to view solutions to (2) are larger that L, as we soppase our grid spacing will be, then we may omit ∇^2 in (2), yielding only $< \Psi >= L^2 h$. This may be the *simplest* "theory" of ocean circulation ever!

A further ambiguity arises because (2) is based on quasigeostrophy; hence h reflects only small fractional change of depth of fluid. It is ambiguous in (2) whether ψ is a velocity or transport streamfunction. As we mean to apply the idea to a model encompassing the full depth of ocean variation, we proceed on basis of simplicity until more sophistication is warranted. Taking account that the depth H of the actual ocean will vary on length scales shorter than scale of variation of Coriolis parameter f, I've adopted a transport streamfunction $\Psi* = -fL^2H$, the sign following definition of H as total depth rather than elevation above a reference depth. Thus a depth-independent velocity field $\mathbf{U}*$ is obtained as $H\mathbf{U}* = z \times \nabla\Psi*$. [This is an oversimplification which "works" under the clumsiness of the conventional model used by EH. More sophisticated models that permit continuous shoaling to quite shallow depths will get into trouble with unbounded $\mathbf{U}*$ in shallow water. A "fix", albeit clumsy, is to write $\Psi* = -fL^2\frac{H^2}{(H+H_0)}$ with H_0 a fittable depth scale chosen to avoid shallow water catastrophe.]

What remains is to introduce a tendency toward $\mathbf{U}*$ into the ocean model. One could, for simplicity, attach a linear restoration $\Lambda(\mathbf{U}*-u)$ to the momentum equation where Λ is some chosen inverse time scale and u is the space-time-dependent 3-D velocity field. Although we did some early experiments in this way, there are a couple of practical disadvantages. Over large areas with not-too-strong $\mathbf{U}*$, the result is an overall damping which may produce undesired loss of energy. As well, one is obliged to choose a parameter Λ (restoring inverse timescale). Alternatively, EH addressed the lateral eddy viscosity operator $A\nabla^2u$ which is already present in the GFDL model and which will be needed (in some form) to prevent grid-scale accumulation of velocity variance. *Ad hoc* fudge factor A will still need be chosen in any case. A plausible remark is that rate of restoring toward higher entropy may be more rapid at shorter length scales, perhaps more in accord with $A\nabla^2$ than under simple Λ. Clearly this is a point (among others!) that will invite further research. What EH did was to rewrite the eddy visvosity as $A\nabla^2(u - \mathbf{U}*)$, thereby expressing internal eddy tendency toward higher entrophy (approximately $\mathbf{U}*$) rather than conventional $A\nabla^2$ centered about a state of rest.

3. Does it matter (really) ?

EH, like Alvarez *et al.*, ran twin experiments. One case ("control") left the original model unmodified. The second case ("topostress") included Neptune effect as described above. Within the limited confines of the Mediterranean, Alvarez et al. selected parameter $L = 4$ km. For their global study, EH chose a latitude-dependent L increasing from 3 km at the pole to 12 km at the equator, reflecting some increase of eddy size from polar to low latitudes. The reader is referred to those publications for further detail. Here we illustrate only briefly from EH.

Figure 1 shows the $U*$ field obtained by EH for use in their coarse (1.9 deg grid) global model.

Figure 2 compares flow fields at 1805 m depth after 800 years integration under steady forcing (mean wind from Hellerman and Rosenstien (1983), surface temperature and salinity relaxation to mean Levitus (1982)). The top panel is "control"; lower panel "topostress". Differences between the panels are apparent. Topostress exhibits a stronger southward flow more tightly confined to the western margin of the North Atlantic as well as poleward flows along the eastern margins in both the Atlantic and Pacific. These and many other impacts are more fully described by EH, and by Alvarez *et al.* in the case of the Mediterranean.

When viewing a figure such as Figure 2, two questions arise. First, are there significant differences? Second, are differences of a sense that things are improved? We ask the first question in order to apprise the value of further study of Neptune effect. If resulting differences were slight, one might better invest effort in other matters. We see that differences are not slight; "right" or "wrong", it seems important to give attention to this matter. Presumably we are "wrong" in the line of development leading to Figure 2. The issue is to measure if we become "less wrong".

One can seek to measure skill against various direct observations. For example, the temperature (T) and salinity (S) fields throughout the world oceans have been more extensively observed than other oceanic fields. Nonconservative tracers such as oxygen and nutrients have been widely observed. Transient tracers show "age" of water since exposed to the sea surface. Current meters and surface and mid-depth drifters reveal aspects of flow. Other technologies include electromagnetics, acoustic tomography and satellite altimetry. Among these sundry observations, how to measure possible impact of Neptune effect?

Initial efforts based on T and S proved not so useful, and won't be discussed here in detail. It seems that models drift away from observed T and S for a host of reasons due in parts to poorly known surface forcings (especially in freshwater, hence S) and to poorly represented interior processes of eddy stirring-mixing-transport and convection. "Control" and "topostress" fare about equally poorly against observed T and S. We recognize though that the main differences between "control" and "topostress" occur in nearly depth-independent flow, hence will not be expressed by differences of horizontal (thermal wind) gradients in T and S. From inverse modelling perspective, Neptune most affects the reference level

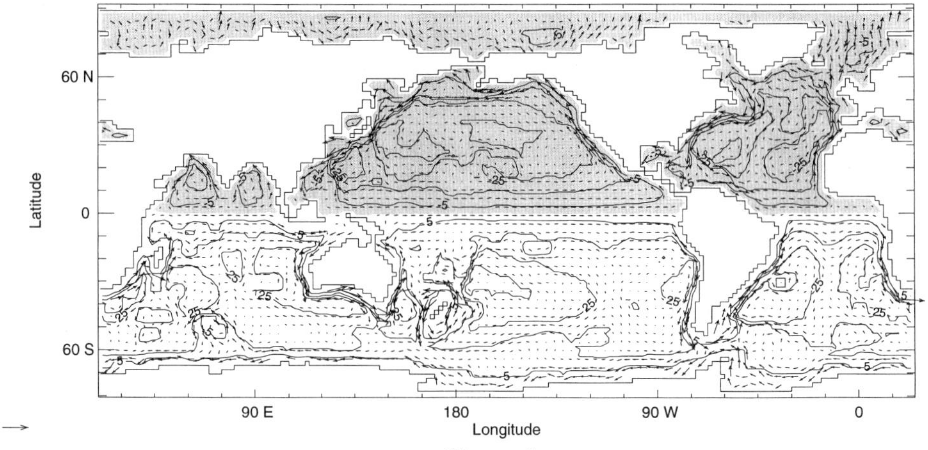

Figure 1

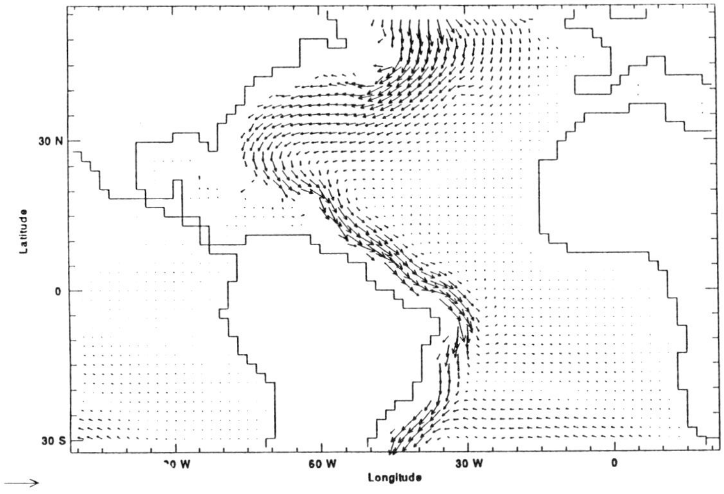

Figure 2a

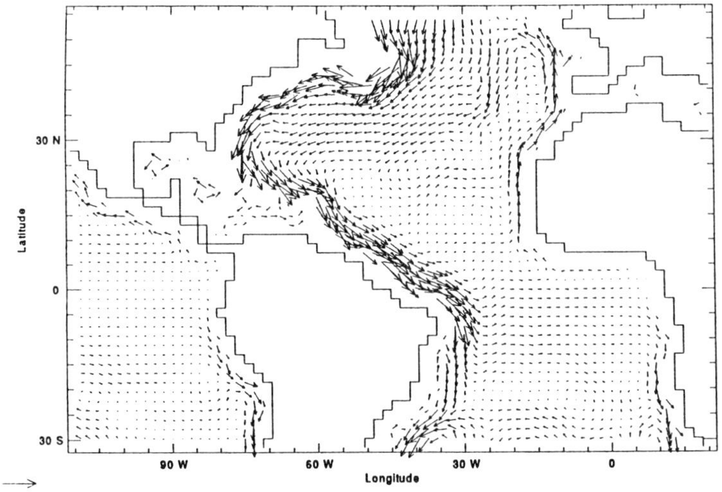

Figure 2b

determination. We also appreciate that a model which might somehow get T and S just right could have a circulation quite wrong.

Time and resource have not to date permitted exploration of Neptune effect against transient tracers, electromagnetics, tomography and altimetry. Only preliminary effort has been made against drifters. Principle effort has focussed on direct velocity determination from current meters. Nearly 2000 records have been obtained from colleagues worldwide based on criteria of minimum duration 100 days and instrument depth greater than 100 m. The effort is ongoing as the database expands.

Figure2 Let d_i be the observed mean flow at current meter i, where "mean" will mean time average over the duration at i. Let ${\sigma_i}^2$ be the total variance of fluctuations of velocity about d_i. Let m_i be the model flow interpolated to the position of current meter i. Let d be the collection of d_i, m be the collection of m_i, and V be a diagonal matrix of entries ${\sigma_i}^2$ normalized such that trace $V = 1$. A natural measure of model error is the kinetic energy of the difference field, weighted by V^{-1} : $(d-m)\cdot V^{-1}\cdot(d-m)$. It is convenient to combine this with the data kinetic energy $d \cdot V^{-1} \cdot d$, to form a skill

$$E = \frac{[d \cdot V^{-1} \cdot d - (d-m) \cdot V^{-1} \cdot (d-m)]}{[d \cdot V^{-1} + (d-m) \cdot V^{-1} \cdot (d-m)]}$$

such that a perfect model ($m = d$) yields $E = +1$ while an awful model with $|d - m| >> |d|$ yields $E = -1$. If a model produced m which was simply randomly unrelated to d, its skill would take the value

$$F = \frac{-m \cdot V^{-1} \cdot m}{[2d \cdot V^{-1} \cdot d + m \cdot V^{-1} \cdot m]}$$

Thus the achieved skill of the model is E-F. As well it may be interesting to inquire about the skill of a model only to get direction of flow right. The natural measure for this skill is the weighted inner product of unit vectors $d/|d|$ and $m/|m|$ labelled

$$D = \frac{d \cdot V^{-1} \cdot m}{|d||m|}$$

where again a perfect model achieves $D = +1$, an awful model with velocities everywhere backwards yields $D = -1$ while random m yield $D = 0$.

Skills D,E, and F were calculated for the "control" and "topostress" cases from EH, samples from which are seen in Figure 2. Results are tabulated below:

	D	E	F	E-F
control	0.1035	0.0030	-0.0649	0.0679
topstress	0.2833	0.0930	-0.0998	0.1928

All of the numbers may appear disappointingly small. This is easy to appreciate though when one considers a model grid spacing near 200 km while a current meter, effectively at a point, will be quite affected by very nearby topography (among other things). Even if a current meter is averaged over 1000 days or more, that time average may be quite unrepresentative of the 200 km spatial average seen by the model. This will reflect both the persistent effect of nearby topography as well as model infidelity. We may hope that a sufficiently large number of current meters will reveal significant measures even when each individual current meter is unrepresentative. To assess this, the above skills were recalculated in ten independent trials each using a random selection of approximately one half of the global current meter inventory. The mean skills achieved across these ten trials were consistent with the global skills, while standard deviation across the ten trials indicated one standard deviation of D at about .07 to .08 and of E-F at about .03 -.04.

Both "control" and "topostress" achieve significant skills in both D and E-F. What may be interesting is that statistical mechanical considerations which are quite "foreign" to traditional ocean modelling are here indicated to provide an increment in skill greater than all the skill achieved *hitherto* (i.e. as "control"). So that seems kind of exciting. However, some cautions apply. The skills reported above are from a particular model formulation executed on a particular grid spacing. Whether this enhancement of skill may occur also for other models at different resolutions remains to be tested.

Acknowledgements . Work reported in this chapter rests substantially upon efforts by Michael Eby and Tessa Sou, supported by the Office of Naval Research. Preparation of this chapter has been supported by the National Science Foundation.

Appendix: stochastic model representation

Although this volume addresses stochastic processes, relation of the present chapter to stochastic processes *per se* may not be clear. However, there is a close connection worth noting. We previously discussed equilibrium and nonequilibrium statistical mechanics and their relation via an H-theorem. For the nonequilibrium eddy-topography problem, early studies from turbulence perspective were cited (Herring, 1977; Holloway, 1978). Given statistically homogenous, barotropic, quasigeostrophic flow, the deterministic equation of motion is

$$(3) \qquad \partial\zeta_k/\partial_t + \sigma A_{kpq}(\zeta_p\zeta_q + \zeta_p h_q) + \upsilon_k\zeta_k = F_k$$

where ζ_k is a coefficient of expansion of the relative vorticity $\zeta = z \cdot \nabla x u$ on horizontal Fourier bases $\exp(ik \cdot x)$, summation is over p and q such that $k + p + q = 0, A_{kpq}$ are coefficients dependent upon k, p and q, h_q is a coefficient of Fourier expansion of topography $h(x)$, υ_k expresses a scale-dependent dissipation, and F_k is a Fourier coefficient of any externally applied torque. An aim of turbulence theory is to achieve approximate equations for evolution of moments $< \zeta_k\zeta_{-k} >$ and $< \zeta_k h_{-k} >$, given a statistical prescription of initial conditions and/or F_k. While such equations are only hoped to approximate (3), they can be shown to be exact for another equation set

$$(4) \qquad \partial\zeta_k/\partial t + \gamma_k h_k + (\eta_k + \upsilon_k)\partial_k = f_k + F_k$$

where γ_k and η_k are expressions dependent upon all the moments $< \zeta_k\zeta_{-k} >$ and $< \zeta_k h_{-k} >$, and f_k is a stochastic forcing also prescribed from moments $< \zeta_k\zeta_{-k} >$ and $< \zeta_k h_{-k} >$. Thus (4) is a stochastic model representation of (3) with details given by Holloway (1978). From (3) we can only speculate that in absence of υ_k and F_k the evolution is toward (2). From (4) the H-theorem shown by CFS assures evolution toward (2). However, we use such a stochastic model (or the companion set of moment equations) far from absolute equilibrium (2), where $\eta_k + \upsilon_k$ may be seen as a renormalization of the linear dissipation operator υ_k while $f_k + F_k$ renormalizes the forcing F_k. Of special interest is that an entirely new term occurs in (4) for which there was no precedent in (3). This is $\gamma_k h_k$. The stochastic model reveals that this is *the term* which drives the flow to non-zero vorticity-topography correlation, hence for given h leads to generation of mean flow from random eddies.

References

[1] Alvarez, A., J. Tintore, G. Holloway, M. Eby and J. M. Beckers, 1994: Effect of topographic stress on the circulation in the western Mediterranean. *J. Geophys Res.*, **99**, 16053-64.

[2] Carnevale, G. F., U. Frisch and R. Salmon, 1981: H-theorems in statistical fluid dynamics. *J. Phys. A.*, **14**, 1701-1718.

[3] Eby, M. and G. Holloway, 1994: Sensitivity of a large scale ocean model to a parameterization of topographic stress. *J. Phys. Oceanogr.*, **24**, 2577-2588.

[4] Hellerman, S. and M. Rosenstein, 1983: Normal monthly wind stress over the World Ocean with error estimates. *J. Phys. Oceanogr.*, **13**, 1093-1104.

[5] Herring, J. R., 1977: Two-dimensional topographic turbulence. *J. Atmos. Sci.*, **34**, 1731-50.

[6] Holloway, G., 1978: A spectral theory of nonlinear barotropic motion above irregular topography. *J. Phys. Oceanogr.*, **8**, 414-427.

[7] Holloway, G., 1986: A shelf wave / topographic pump drives mean coastal circulation. Parts 1 & 2, Ocean Modelling No's 68 & 69.

[8] Holloway, G., 1987: Systematic forcing of large-scale geophysical flows by eddy-topography interaction. *J. Fluid Mech.*, **184**, 463-476.

[9] Holloway, G., 1992: Representing topographic stress for large scale ocean models. *J. Phys. Oceanogr.*, **22**, 1033-46.

[10] Levitus, S., 1982: Climatological Atlas of the World Ocean. NOAA Prof. Paper 13, Washington, D. C.

[11] Pacanowski, R., K. Dixon and A. Rosati, 1991: The GFDL Modular Ocean Model Users Guide v. 1.0. Geophysical Fluid Dynamics Lab/NOAA, Princeton, NJ.

[12] Salmon, R., G. Holloway and M. C. Hendershott, 1976: The equilibrium statistical mechanics of simple quasi-geostrophic models. *J. Fluid Mech.*, **75** , 691-703.

Institute of Ocean Sciences
Sidney, V8L 4B2
CANADA

zounds@ios.bc.ca

Short-Time Correlation Approximations for Diffusing Tracers in Random Velocity Fields: A Functional Approach

V.I. Klyatskin, W.A. Woyczynski and D. Gurarie

Contents

Abstract

The paper studies statistical characteristics of the passive tracer concentration field, and of its spatial gradient, in random velocity fields. Those include mean values, correlation functions as well as probability distributions. The functional approach is used. Influence of the mean flow (on the example of linear shear flow) and of the molecular diffusion coefficient on the statistical characteristics is analysed. Most of our analysis is conducted in the framework of the delta-correlated (in time) approximation and conditions for its applicabillity are established. We also consider approximations taking account of the positive but finite correlation radius, such as the telegraph process approximation and the diffusion approximation.

The authors made an effort to provide a broad perspective of and the background for the issues under consideration. To make the material accessible to physical scientists they tried to avoid use of overly abstract mathematical structures and conduct their presentation in the language and an idiom common in the physical and applied mathematics literature.

1. Introduction

A study of passive tracer (or passive scalar in another terminology) transport in random velocity flows is one of the classical topics in statistical fluid mechanics. Its applications range from questions of environmental pollutant diffusion in a turbulent atmosphere, to problems of advection of the heat and the salinity in oceanic currents,[1–5], and from diffusion in porous media[6], to questions of the large scale mass distribution in the late stages of the Universe.[7,8]

The problem has been studied since the end of the Fifties, beginning with pioneering papers[9–10]. Later on, numerous researchers (see e.g. [11–15]) obtained many equations describing statistical characteristics of the passive tracer field, both in Eulerian and in Lagrangian descriptions. This process continues vigorously at present time.

It is relatively easy to write down equations for the statistical characteristics of the passive tracer field in the so-called *delta-correlated approximation* for the random velocity field (see, e.g., [16–19]). There, in the Lagrangian picture, a particle behaves like an ordinary Brownian particle. More difficult problem is to account, approximately, for finite *time correlation radius* of the velocity field.

In this paper, on the basis of a uniform *functional approach*, we study the problem of passive tracer diffusion in random velocity fields. Both the general set-up, and approximate methods permitting efficient numerical computations are considered. Applicability conditions for the latter are formulated.

First, we consider in detail the delta-correlated (in time) approximation of random velocity field. Here a relatively complete picture is obtained. Then we proceed to more general approximations, such as the so-called telegraph process and the diffusion approximations, and we observe that some important features of the delta-correlated case remain valid for them.

The paper analyzes both the average concentration field and its correlation function which, in our approximations, are described by linear equations due to linearity of the original equations. In absence of the molecular diffusion, we study probability distribution of the magnitude of the tracer concentration gradient (which turns out to be lognormal), statistical characteristics of the contours of constant concentration (which in the diffusion process assume a "fractal" character), and the impact of the mean flow gradient on statistical characteristics of the tracer concentration.

In the process, we also find conditions under which the influence of the molecular diffusion could be neglected. For the diffusion of tracer with the constant mean gradient we show that one-point probability densities of concentration are Gaussian and that , as $t \to \infty$, the stationary probability distribution exists. It turns out that the variance of tracer concentration fluctuations depends on the molecular diffusion coefficient in a logarithmic fashion.

Finally, we discuss the so-called telegraph process approximation and the diffusion approximation which permit us to account for the finite correlation radius of the velocity field and, consequently, go beyond the delta-correlation case.

All the approximations considered in this paper can be viewed as short-time correlation approximations. Physically, the assumption is that the influence of velocity field's fluctuations on statistical characteristics of tracer concentration is small on time scales of the order of its temporal correlation radius. A more systematic mathematical formulation of this problem is possible along the lines of[20].

The present paper serves two purposes. On the one hand it reviews the existing results and methods in Sections 2, 3, 4.1, 5.1,and 5.2. On the other hand, we bring in some new results and ideas in Sections 4.2-4.6, 5.3, and 5.4. The unifying theme is a functional approach to the whole complex of short-time correlation problems for randomly diffusing tracers.

The number of publications in the area grows rapidly. So our list of references includes mostly selected papers published in recent years; many of them, however, contain the detailed references to earlier work.

2. Evolution of passive tracer concentration.

In this section we formulate the dynamical problem in the Lagrangian and Eulerian descriptions, establish their connection, and prepare the background for the statistical analysis of mean concentration and its correlation on the one hand, and the probability distribution of the tracer concentration and its spatial gradient on the other.

The first equation

$$(1) \qquad \left(\frac{\partial}{\partial t} + \boldsymbol{V}(\boldsymbol{r},t)\frac{\partial}{\partial \boldsymbol{r}}\right) q(\boldsymbol{r},t) = \kappa \frac{\partial^2}{\partial \boldsymbol{r}^2} q(\boldsymbol{r},t), \qquad q(\boldsymbol{r},0) = q_0(\boldsymbol{r}),$$

describes scalar fields $q = q(\boldsymbol{r},t)$ representing such quantities as temperature, salinity, etc., of special interest in oceanography. In the case of incompressible flows (div $\boldsymbol{V} = 0$) to be considered later on, it also covers concentrations of "matter", like air pollutants and oil droplets in an oil slick.

The second equation

(2) $$\left(\frac{\partial}{\partial t} + \boldsymbol{V}(\boldsymbol{r},t)\frac{\partial}{\partial \boldsymbol{r}}\right) p_i(\boldsymbol{r},t) = -\frac{\partial V_k(\boldsymbol{r},t)}{\partial r_i} p_k(\boldsymbol{r},t) + \kappa\frac{\partial^2}{\partial \boldsymbol{r}^2} p_i(\boldsymbol{r},t),$$

$$\boldsymbol{p}(\boldsymbol{r},0) = \boldsymbol{p}_0(\boldsymbol{r}) = \frac{\partial}{\partial \boldsymbol{r}} q_0(\boldsymbol{r}),$$

describes the spatial gradient $\boldsymbol{p}(\boldsymbol{r},t) = \partial q(\boldsymbol{r},t)/\partial \boldsymbol{r}$ of the concentration field $q(\boldsymbol{r},t)$. Throughout the paper we shall conform to the usual summation convention of summing over repeated indices. We make one exception in (21) when the Furutsu-Novikov formula is introduced for the first time.

For incompressible flows, equation (1) without the diffusion term has the form of a conservation law, the quantity

$$Q = \int d\boldsymbol{r}\, q(\boldsymbol{r},t) = \int d\boldsymbol{r}\, q_0(\boldsymbol{r})$$

being conserved. In both equations κ denotes the "molecular" diffusion coefficient.

The velocity field $\boldsymbol{V}$ is assumed to be random with finite expectations, decomposable into the mean component

$$\boldsymbol{v}(\boldsymbol{r},t) = \langle \boldsymbol{V}(\boldsymbol{r},t)\rangle,$$

and the random fluctuation

$$\boldsymbol{F}(\boldsymbol{r},t) = \boldsymbol{V}(\boldsymbol{r},t) - \boldsymbol{v}(\boldsymbol{r},t).$$

Although equations (1-2) are linear, solutions q and $\boldsymbol{p}$ depend in a complicated, implicit nonlinear way on the velocity field $\boldsymbol{V}$. Observe also that evolution (1) of q implies a similar relation for higher moments $q^n(\boldsymbol{r},t)$:

(1′) $$\left(\frac{\partial}{\partial t} + \boldsymbol{V}(\boldsymbol{r},t)\cdot\frac{\partial}{\partial \boldsymbol{r}}\right) q^n(\boldsymbol{r},t)$$

$$= \kappa\frac{\partial^2}{\partial \boldsymbol{r}^2} q^n(\boldsymbol{r},t) - \kappa n(n-1) q^{n-2}(\boldsymbol{r},t)\boldsymbol{p}^2(\boldsymbol{r},t),$$

The latter, however, couples nonlinearly the moments of q to gradient $\boldsymbol{p}$ given by (2). Equations (1-2) give the *Eulerian description* of the system.

A direct study of the probability distribution of $q(\boldsymbol{r},t)$ is not possible as (1) contains the second order (diffusion) term in $\boldsymbol{r}$. However, one could write a variational Hopf equation[(21)] for the characteristic functional of

q, whose study would correspond to the statistical analysis of solutions of (1) in an infinite dimensional functional space.[1,16,17] However, at present there are no satisfactory methods for the analysis and solution of these variational equations.

Next, we observe that the effect of molecular diffusion $\kappa\Delta q$ in equation (1) could be modeled by introducing an auxiliary (independent of $\boldsymbol{V}$) delta-correlated Gaussian vector process $\boldsymbol{A}(t) = (A_i(t))$, with zero mean and covariance matrix

$$\left\langle A_i(t)A_j(t')\right\rangle = 2\kappa\delta_{ij}\delta(t-t'), \quad i,j = 1,2,3,$$

and replacing deterministic (for a fixed realization of velocity field $\boldsymbol{V}$) equation with the stochastic equation

$$(3) \quad \left(\frac{\partial}{\partial t} + \boldsymbol{V}(\boldsymbol{r},t)\cdot\frac{\partial}{\partial \boldsymbol{r}}\right)\tilde{q}(\boldsymbol{r},t) = -\boldsymbol{A}(t)\cdot\frac{\partial}{\partial \boldsymbol{r}}\tilde{q}(\boldsymbol{r},t), \qquad \tilde{q}(\boldsymbol{r},0) = q_0(\boldsymbol{r}).$$

Then, the ansemble average of $\tilde{q}$ with respect to $\boldsymbol{A}$ gives precisely the deterministic solution of equation (1),[22] i.e.,

$$(4) \qquad q(\boldsymbol{r},t) = \left\langle \tilde{q}(\boldsymbol{r},t)\right\rangle_{\boldsymbol{A}}.$$

Formula (4) is, essentially, a path integral representation of a solution of (1), and it is convenient to apply whenever solutions of equation (3) can be written explicitly.[23,24]

The first order partial differential equation (3) is solved by the standard *method of characteristics* (or passage to particle dynamics), which gives

$$(5) \qquad \begin{aligned} \frac{d}{dt}\boldsymbol{r}(t) &= \boldsymbol{V}(\boldsymbol{r}(t),t) + \boldsymbol{A}(t), \quad & \boldsymbol{r}(0) = \boldsymbol{\xi}, \\ \frac{d}{dt}\tilde{q}(t) &= 0, & \tilde{q}(0) = q_0(\boldsymbol{\xi}). \end{aligned}$$

The solutions

$$\boldsymbol{r}(t) = \boldsymbol{r}(t|\boldsymbol{\xi}), \qquad \tilde{q}(t) = \tilde{q}(t|\boldsymbol{\xi}),$$

of (5) depend on the characteristic initial parameter $\boldsymbol{\xi}$. We shall use a vertical bar to separate the time variable t from the initial parameter $\boldsymbol{\xi}$. This is the *Lagrangian description* of the passive tracer transport.

Eliminating parameter $\boldsymbol{\xi}$ in system (5), via function $\boldsymbol{\xi}(\boldsymbol{r},t)$ inverse to $\boldsymbol{r} = \boldsymbol{r}(t|\boldsymbol{\xi})$, will bring back the *Eulerian description* of the evolving tracer concentration

$$\tilde{q}(\boldsymbol{r},t) = \tilde{q}\left(t \,\middle|\, \boldsymbol{\xi}(\boldsymbol{r},t)\right).$$

Observe, that in this case $\tilde{q}(t|\boldsymbol{\xi}) = q_0(\boldsymbol{\xi})$, so the tracer concentration is preserved along characteristics.

To find the statistics of solutions of (5) let us introduce a generalized (indicator of particle's position) function

$$\tilde{\Phi}(\boldsymbol{r};t) = \delta(\boldsymbol{r}(t) - \boldsymbol{r}),$$

satisfying the Liouville's equation[(16–18)]

$$\left(\frac{\partial}{\partial t} + \boldsymbol{V}(\boldsymbol{r},t)\frac{\partial}{\partial \boldsymbol{r}}\right)\tilde{\Phi}(\boldsymbol{r},t) = -\boldsymbol{A}(t)\cdot\frac{\partial}{\partial \boldsymbol{r}}\tilde{\Phi}(\boldsymbol{r},t), \qquad \tilde{\Phi}(\boldsymbol{r},0) = \delta(\boldsymbol{\xi} - \boldsymbol{r}).$$

Once again, averaging the Liouville's equation over the $\boldsymbol{A}$-ensemble, we get the stochastic equation

$$(6) \qquad \left(\frac{\partial}{\partial t} + \boldsymbol{V}(\boldsymbol{r},t)\frac{\partial}{\partial \boldsymbol{r}}\right)\phi(\boldsymbol{r};t) = \kappa\frac{\partial^2}{\partial \boldsymbol{r}^2}\phi(\boldsymbol{r};t), \qquad \phi(\boldsymbol{r};0) = \delta(\boldsymbol{\xi} - \boldsymbol{r})$$

for the mean $\phi(\boldsymbol{r};t) = \langle\tilde{\Phi}(\boldsymbol{r},t)\rangle_{\boldsymbol{A}}$. Let us note the role of $\phi(\boldsymbol{r};t) = \phi(\boldsymbol{r};t|\boldsymbol{\xi})$ as the probability density of $\boldsymbol{r}(t)$ (for a fixed realization of $\boldsymbol{V}$) on the one hand, and the Green's function of (1) on the other.

Equation (6) coincides with (1) but the initial conditions are different. That reflects the different mathematical contents of their solutions: equation (6) describes the probability density of particle's position while (1) describes the tracer concentration in the Eulerian picture. But their behavior is identical if the initial tracer concentration is the point source, that is

$$q(\boldsymbol{r};0) = \delta(\boldsymbol{\xi} - \boldsymbol{r}).$$

Hence, solution of (1) with an arbitrary initial data q_0 is accomplished by a (generalized) convolution of the Green's function ϕ with $q_0(\boldsymbol{\xi})$.

Transition from the Lagrangian to the Eulerian description, i.e. from solutions of equation (5) to solutions of (1) is provided by the Jacobian of map $\boldsymbol{\xi} \mapsto \boldsymbol{r}$, i.e.

$$J(t|\boldsymbol{\xi}) = \mathrm{Det}\left\|\frac{\partial r_i}{\partial \xi_j}\right\|, \qquad J(0|\boldsymbol{\xi}) = 1,$$

which becomes trivial for incompressible flows, where $J(t|\boldsymbol{\xi}) \equiv 1$.

The basic linear stochastic equation (1) can be used to compute the mean concentration $\langle q(\boldsymbol{r},t)\rangle$, or the one-point Lagrangian probability density $\phi(\boldsymbol{r},t)$. To compute correlations (or higher moments) one has to consider the product concentration field $\tilde{\Gamma}(\boldsymbol{r}_1,\boldsymbol{r}_2,t) = q(\boldsymbol{r}_1,t)q(\boldsymbol{r}_2,t)$ that obeys a linear equation

$$(7) \qquad \begin{aligned} \left(\frac{\partial}{\partial t} + \boldsymbol{V}(\boldsymbol{r}_1,t)\frac{\partial}{\partial \boldsymbol{r}_1} + \boldsymbol{V}(\boldsymbol{r}_2,t)\frac{\partial}{\partial \boldsymbol{r}_2}\right)\tilde{\Gamma}(\boldsymbol{r}_1,\boldsymbol{r}_2,t) & \\ = \kappa\left(\frac{\partial^2}{\partial {\boldsymbol{r}_1}^2} + \frac{\partial^2}{\partial {\boldsymbol{r}_2}^2}\right)\tilde{\Gamma}(\boldsymbol{r}_1,\boldsymbol{r}_2,t), & \\ \tilde{\Gamma}(\boldsymbol{r}_1,\boldsymbol{r}_2,0) = q_0(\boldsymbol{r}_1)q_0(\boldsymbol{r}_2), & \end{aligned}$$

then average it with respect to the velocity ensemble $\boldsymbol{V}$.

We would like to recast the diffusion equation (7) in the evolutionary form (3) with random delta-correlated Gaussian sources. However, the product $\tilde{q}(\boldsymbol{r}_1,t)\tilde{q}(\boldsymbol{r}_2,t)$ of two stochastic solutions of (3) averaged over the $\boldsymbol{A}$-ensemble does not yield solutions of equation (7) as

$$\left\langle \tilde{q}(\boldsymbol{r}_1,t)\tilde{q}(\boldsymbol{r}_2,t)\right\rangle_{\boldsymbol{A}} \neq \left\langle \tilde{q}(\boldsymbol{r}_1;t)\right\rangle_{\boldsymbol{A}}\left\langle \tilde{q}(\boldsymbol{r}_2;t)\right\rangle_{\boldsymbol{A}}.$$

To interpret the solution of (7) as a probability density we shall model the continuous concentration field q by a system of *particles.* The k-th particle will be described by the ordinary stochastic differential equation

$$\text{(8)} \qquad \frac{d}{dt}\boldsymbol{r}^{(k)}(t) = \boldsymbol{V}(\boldsymbol{r}^{(k)}(t),t) + \boldsymbol{A}^{(k)}(t), \qquad \boldsymbol{r}^{(k)}(0) = \boldsymbol{\xi}^{(k)},$$

where stochastic processes $\boldsymbol{A}^{(k)}(t)$, $k = 1,2,\ldots$, are statistically independent [3].

Next we introduce, as above, a generalized random (indicator) function

$$\tilde{\varphi}(\boldsymbol{r}_1,\boldsymbol{r}_2,t) = \delta\left(\boldsymbol{r}^{(1)}(t) - \boldsymbol{r}_1\right)\delta\left(\boldsymbol{r}^{(2)}(t) - \boldsymbol{r}_2\right),$$

which determines the joint probability distribution for two particles. Function $\tilde{\varphi}$ satisfies a Liouville-type equation

$$\left(\frac{\partial}{\partial t} + \boldsymbol{V}(\boldsymbol{r}_1,t)\frac{\partial}{\partial \boldsymbol{r}_1} + \boldsymbol{V}(\boldsymbol{r}_2,t)\frac{\partial}{\partial \boldsymbol{r}_2}\right)\tilde{\varphi}(\boldsymbol{r}_1,\boldsymbol{r}_2,t)$$

$$= -\left(\boldsymbol{A}^{(1)}(t)\cdot\frac{\partial}{\partial \boldsymbol{r}_1} + \boldsymbol{A}^{(2)}(t)\cdot\frac{\partial}{\partial \boldsymbol{r}_2}\right)\tilde{\varphi}.$$

Averaging it over ensembles $\boldsymbol{A}^{(k)}(t)$, $k = 1,2$, we get equation

$$\text{(9)} \qquad \left(\frac{\partial}{\partial t} + \boldsymbol{V}(\boldsymbol{r}_1,t)\cdot\frac{\partial}{\partial \boldsymbol{r}_1} + \boldsymbol{V}(\boldsymbol{r}_2,t)\cdot\frac{\partial}{\partial \boldsymbol{r}_2}\right)\varphi(\boldsymbol{r}_1,\boldsymbol{r}_2,t)$$

$$= \kappa\left(\frac{\partial^2}{\partial \boldsymbol{r}_1^2} + \frac{\partial^2}{\partial \boldsymbol{r}_2^2}\right)\varphi(\boldsymbol{r}_1,\boldsymbol{r}_2,t),$$

$$\varphi(\boldsymbol{r}_1,\boldsymbol{r}_2,0) = \delta\left(\boldsymbol{\xi}^{(1)} - \boldsymbol{r}_1\right)\delta\left(\boldsymbol{\xi}^{(2)} - \boldsymbol{r}_2\right),$$

describing evolution of a joint probability density

$$\varphi(\boldsymbol{r}_1,\boldsymbol{r}_2,t) = \langle\tilde{\varphi}(\boldsymbol{r}_1,\boldsymbol{r}_2,t)\rangle_{\boldsymbol{A}^{(1)}(t);\boldsymbol{A}^{(2)}(t)}$$

---

[3] The stochastic processes $\boldsymbol{A}^{(k)}(t)$ could result from molecular forces affecting tracer particles, and their mutual independence reflects a short range of molecular interactions.

of two particles. The latter equation is identical to (7).

Our next goal is to average the above linear stochastic equations (1-2), (7), and (9), over the ensembles of velocity field fluctuations $\boldsymbol{F}(\boldsymbol{r},t)$, to obtain explicit formulas for the evolution of the mean concentration $\langle q(\boldsymbol{r},t)\rangle_{\boldsymbol{F}}$, the two point correlations

$$\Gamma(\boldsymbol{r}_1,\boldsymbol{r}_2,t) = \left\langle \tilde{\Gamma}(\boldsymbol{r}_1,\boldsymbol{r}_2,t)\right\rangle_{\boldsymbol{F}} = \left\langle q(\boldsymbol{r}_1,t)q(\boldsymbol{r}_2,t)\right\rangle_{\boldsymbol{F}},$$

and other statistical characteristics.

We shall also analyze the probability distributions of solutions of equation (1-2) and explain the role of the molecular diffusion coefficient κ, assumed to be sufficiently small. One point of interest is the limiting behavior of solutions as $\kappa \to 0$. Here, we require the initial tracer concentration $q_0(\boldsymbol{r})$ and its gradient $\boldsymbol{p}_0(\boldsymbol{r})$ to be large-scale (the precise meaning will be explained later). Then, one can drop terms containing κ in (1-2), and consider a simplified problem described by equations

$$(10) \qquad \left(\frac{\partial}{\partial t} + \boldsymbol{V}(\boldsymbol{r},t)\cdot\frac{\partial}{\partial \boldsymbol{r}}\right) q(\boldsymbol{r},t) = 0, \qquad q(\boldsymbol{r},0) = q_0(\boldsymbol{r}),$$

$$(11) \qquad \left(\frac{\partial}{\partial t} + \boldsymbol{V}(\boldsymbol{r},t)\cdot\frac{\partial}{\partial \boldsymbol{r}}\right) p_i(\boldsymbol{r},t) = -\frac{\partial V_k(\boldsymbol{r},t)}{\partial r_i} p_k(\boldsymbol{r},t),$$

$$\boldsymbol{p}(\boldsymbol{r},0) = \boldsymbol{p}_0(\boldsymbol{r}) = \frac{\partial}{\partial \boldsymbol{r}} q_0(\boldsymbol{r}).$$

Let us introduce another generalized (indicator) function

$$(12) \qquad \Phi_{t,\boldsymbol{r}}(q,\boldsymbol{p}) = \delta(q(\boldsymbol{r},t)-q)\delta(\boldsymbol{p}(\boldsymbol{r},t)-\boldsymbol{p})$$

that determines the joint one-point probability distribution of fields q and $\boldsymbol{p}$ at a given spatial point in the Eulerian coordinates, and more general function

$$(12') \qquad \Phi_{t,\boldsymbol{r}_1,\boldsymbol{r}_2}(q_1,\boldsymbol{p}_1;q_2,\boldsymbol{p}_2)$$

$$= \delta(q(\boldsymbol{r}_1,t)-q_1)\delta(\boldsymbol{p}(\boldsymbol{r}_1,t)-\boldsymbol{p}_1)\delta(q(\boldsymbol{r}_2,t)-q_2)\delta(\boldsymbol{p}(\boldsymbol{r}_2,t)-\boldsymbol{p}_2)$$

$$\equiv \Phi_{t,\boldsymbol{r}_1}(q_1,\boldsymbol{p}_1)\Phi_{t,\boldsymbol{r}_2}(q_2,\boldsymbol{p}_2).$$

The latter determines the joint two-point distribution of the tracer concentration field and its gradient, and contains additional information compared to (12). In particular, it permits the analysis of various functionals constructed from fields q and $\boldsymbol{p}$.

Based on equations (10) and (11), one can easily obtain dynamic evolution of functions (12) and (12'). In particular, $\Phi_{t,\boldsymbol{r}}(q,\boldsymbol{p})$ satisfies the Liouville equation[16,17]

$$(13) \quad \left(\frac{\partial}{\partial t} + \boldsymbol{V}(\boldsymbol{r},t)\cdot\frac{\partial}{\partial \boldsymbol{r}}\right) \Phi_{t,\boldsymbol{r}}(q,\boldsymbol{p}) = \frac{\partial V_k(\boldsymbol{r},t)}{\partial r_i}\frac{\partial}{\partial p_i}\Big(p_k\Phi_{t,\boldsymbol{r}}(q,\boldsymbol{p})\Big),$$

with the initial condition

$$\Phi_{0,\boldsymbol{r}}(q,\boldsymbol{p}) = \delta(q_0(\boldsymbol{r}) - q)\delta(\boldsymbol{p}_0(\boldsymbol{r}) - \boldsymbol{p}).$$

In the zero diffusion case, the method of characteristics (particle dynamics) yields an equation

$$\frac{d\boldsymbol{r}(t|\boldsymbol{\xi})}{dt} = \boldsymbol{V}(\boldsymbol{r},t), \qquad \boldsymbol{r}(0|\boldsymbol{\xi}) = \boldsymbol{\xi}. \tag{14}$$

Then, equations (10) and (11) take the form

$$\frac{d}{dt}q(t|\boldsymbol{\xi}) = 0, \qquad q(0|\boldsymbol{\xi}) = q_0(\boldsymbol{\xi}), \tag{14'}$$

$$\frac{d}{dt}p_i(t|\boldsymbol{\xi}) = -\frac{\partial V_k(\boldsymbol{r},t)}{\partial r_i}p_k(t|\boldsymbol{\xi}), \qquad p_i(0|\boldsymbol{\xi}) = \frac{q_0(\boldsymbol{\xi})}{\partial \xi_i},$$

along the characteristic curve. Clearly, equations (14) and (14') form a closed system in the Lagrangian description. Solution q remains constant along characteristics, so

$$q(t|\boldsymbol{\xi}) \equiv q_0(\boldsymbol{\xi}).$$

Introducing one more generalized (indicator) function

$$\Phi_t(\boldsymbol{r},q,\boldsymbol{p}|\boldsymbol{\xi}) = \delta(\boldsymbol{r}(t|\boldsymbol{\xi}) - \boldsymbol{r})\delta(q(t|\boldsymbol{\xi}) - q)\delta(\boldsymbol{p}(t|\boldsymbol{\xi}) - \boldsymbol{p}), \tag{15}$$

that determines the joint density of particle distribution, we can write a similar Liouville equation

$$\left(\frac{\partial}{\partial t} + \boldsymbol{V}(\boldsymbol{r},t)\cdot\frac{\partial}{\partial \boldsymbol{r}}\right)\Phi_t(\boldsymbol{r},q,\boldsymbol{p}|\boldsymbol{\xi}) = \frac{\partial V_k(\boldsymbol{r},t)}{\partial r_i}\frac{\partial}{\partial p_i}\Big(p_k\Phi_t(\boldsymbol{r},q,\boldsymbol{p}|\boldsymbol{\xi})\Big), \tag{15'}$$

with the initial condition

$$\Phi_0(\boldsymbol{r},q,\boldsymbol{p}|\boldsymbol{\xi}) = \delta(\boldsymbol{\xi} - \boldsymbol{r})\delta(q_0(\boldsymbol{\xi}) - q)\delta(\frac{\partial q_0(\boldsymbol{\xi})}{\partial \boldsymbol{\xi}} - \boldsymbol{p}), \tag{15''}$$

Once again, two problems (13) and (15') differ only by the initial conditions. This coincidence reflects the connection between the Lagrangian and the Eulerian description. Indeed, taking into account the incompressibility of the flow, function (15) can be rewritten in the form

$$\Phi_t(\boldsymbol{r},q,\boldsymbol{p}|\boldsymbol{\xi}) = \delta(\boldsymbol{\xi}(\boldsymbol{r},t) - \boldsymbol{\xi})\Phi_{t,\boldsymbol{r}}(q,\boldsymbol{p}).$$

Subsequent integration over the characteristic initial parameter $\boldsymbol{\xi}$, yields the essential relation between the Eulerian and the Lagrangian densities:

$$\Phi_{t,\boldsymbol{r}}(q,\boldsymbol{p}) = \int d\boldsymbol{\xi}\,\Phi_t(\boldsymbol{r},q,\boldsymbol{p}|\boldsymbol{\xi}). \tag{16}$$

Since parameter $\boldsymbol{\xi}$ enters into (15') only through the intial condition (15"), clearly, the equations for the Eulerian and the Lagrangian densities should coincide.

Let us also observe that variable q in equations (13) and (15') enters only through the initial conditions. For that reason, multiplying those by q^n and integrating over q and $\boldsymbol{p}$ we get a dynamic evolution equation for quantity q^n, that coincides with (13) and (15'). The latter property is connected with the conservation of q along characteristics.

To recapitulate, quantities of interest to us, Eulerian $q^n(\boldsymbol{r},t)$ and $\Phi_{t,\boldsymbol{r}}(q,\boldsymbol{p})$, obey the same dynamic equations as the corresponding Lagrangian probability densities $\Phi_t(\boldsymbol{r}|\boldsymbol{\xi})$ and $\Phi_t(\boldsymbol{r},q,\boldsymbol{p}|\boldsymbol{\xi})$.

In a similar manner one can introduce a two-particle system

$$
\begin{aligned}
\frac{d\boldsymbol{r}_1(t)}{dt} &= \boldsymbol{V}(\boldsymbol{r}_1,t), \qquad \boldsymbol{r}_1(0)=\boldsymbol{\xi}_1,\\
\frac{dp_{1i}(t)}{dt} &= -\frac{\partial V_k(\boldsymbol{r}_1,t)}{\partial r_{1i}}p_{1k}(t), \qquad p_{1i}(0)=\frac{\partial q_0(\boldsymbol{\xi}_1)}{\partial \xi_{1i}},\\
\frac{d\boldsymbol{r}_2(t)}{dt} &= \boldsymbol{V}(\boldsymbol{r}_2,t), \qquad \boldsymbol{r}_2(0)=\boldsymbol{\xi}_2,\\
\frac{dp_{2i}(t)}{dt} &= -\frac{\partial V_k(\boldsymbol{r}_2,t)}{\partial r_{2i}}p_{2k}(t), \qquad p_{2i}(0)=\frac{\partial q_0(\boldsymbol{\xi}_2)}{\partial \xi_{2i}}.
\end{aligned} \tag{17}
$$

The corresponding two-particle Lagrangian density will then be described by the same equation as the two-point Eulerian density (12'). All (Eulerian and Lagrangian) densities above obey stochastic partial differential equations with the randomness introduced through the velocity fluctuations $\boldsymbol{F}$. Their ensemble averaging yields the evolution of the probability densities

$$
P_{t,\boldsymbol{r}}(q,\boldsymbol{p}) = \langle \Phi_{t,\boldsymbol{r}}(q,\boldsymbol{p})\rangle_{\boldsymbol{F}}, \qquad P_t(\boldsymbol{r},q,\boldsymbol{p}|\boldsymbol{\xi}) = \langle \Phi_t(\boldsymbol{r},q,\boldsymbol{p}|\boldsymbol{\xi})\rangle_{\boldsymbol{F}}, \tag{18}
$$

in both the Eulerian, and the Lagrangian descriptions.

3. Statistical averaging

The averaging procedure mentioned in Section 2 can be implemented in a number of cases. For instance, averaging equations (1) over the $\boldsymbol{F}$-ensemble yields an evolution of the mean field, where random velocities $\boldsymbol{F}$ are coupled to random solution $q = q[\boldsymbol{F}]$, itself a functional of $\boldsymbol{F}$, through the fluctuation term

$$
\left\langle \boldsymbol{F}\cdot\frac{\partial}{\partial \boldsymbol{r}}q\right\rangle. \tag{19}
$$

So, to get the effective mean field evolution one needs to decouple the cross-correlation term (19). The decoupling methods strongly depend on the nature of random field $\boldsymbol{F}$.

In the Gaussian case[4] decoupling exploits the so-called *Furutsu-Novikov* formula,(25,26) (see the Appendix to the present paper). Namely, given a zero-mean Gaussian random vector field $\boldsymbol{F} = (F_i)$, any functional $R[\boldsymbol{F}]$ satisfies

$$(20)\ \Big\langle F_i(\boldsymbol{r},t)R[\boldsymbol{F}]\Big\rangle = \int d\boldsymbol{r}' \int dt' \sum_j \Big\langle F_i(\boldsymbol{r},t)F_j(\boldsymbol{r}',t)\Big\rangle \Big\langle \frac{\delta R[\boldsymbol{F}]}{\delta F_j(\boldsymbol{r}',t')}\Big\rangle.$$

So, the cross-term decouples into a superposition of products of the correlation coefficients and the mean variational derivatives of R. Applying the Furutsu-Novikov formula (20) to the cross-term (19) of equation (1) we get that

$$(21)\quad \left(\frac{\partial}{\partial t} + \boldsymbol{v}(\boldsymbol{r},t)\cdot\frac{\partial}{\partial \boldsymbol{r}}\right)\langle q(\boldsymbol{r},t)\rangle$$
$$+\int dt' \int d\boldsymbol{r}' \sum_j B_{ij}(\boldsymbol{r},t;\boldsymbol{r}',t')\frac{\partial}{\partial r_i}\Big\langle \frac{\delta q(\boldsymbol{r},t)}{\delta F_j(\boldsymbol{r}',t')}\Big\rangle = \kappa\frac{\partial^2}{\partial \boldsymbol{r}^2}\langle q(\boldsymbol{r},t)\rangle,$$

where $B_{ij}(\boldsymbol{r},t;\boldsymbol{r}',t') = \langle F_i(\boldsymbol{r},t)F_j(\boldsymbol{r}',t')\rangle$ is the space-time correlation of $\boldsymbol{F}$. Although equation (21) is exact for any zero-mean Gaussian field $\boldsymbol{F}$, it is not closed since the evolution of the mean field is coupled to the mean variational derivative with respect to $\boldsymbol{F}$. The variational derivative $\delta q/\delta \boldsymbol{F}$, itself solves a stochastic differential equation

$$\frac{\partial}{\partial t}\left(\frac{\delta q(\boldsymbol{r},t)}{\delta F_j(\boldsymbol{r}',t')}\right)+[\boldsymbol{v}(\boldsymbol{r},t)+\boldsymbol{F}(\boldsymbol{r},t)]\cdot\frac{\partial}{\partial \boldsymbol{r}}\left(\frac{\delta q(\boldsymbol{r},t)}{\delta F_j(\boldsymbol{r}',t')}\right) = \kappa\frac{\partial^2}{\partial \boldsymbol{r}^2}\left(\frac{\delta q(\boldsymbol{r},t)}{\delta F_j(\boldsymbol{r}',t')}\right)$$
(22)

obtained by varying (1) in $\boldsymbol{F}$, and satisfies the initial condition

$$\frac{\delta q(\boldsymbol{r},t)}{\delta F_j(\boldsymbol{r}',t')}\bigg|_{t\to t'+0} = -\delta(\boldsymbol{r}-\boldsymbol{r}')\frac{\partial}{\partial r_j}q(\boldsymbol{r},t').$$

Notice, that the above procedure is essentially equivalent to finding the Green's function for the original problem (1). Taking the ensemble average of (22) and then applying to it the Furutsu-Novikov formula would produce higher order variational derivatives $\langle \delta^2 q/\delta F_i \delta F_j\rangle$, etc. Such system would require a suitable closure, hypothesis that could be rigorously implemented only in certain cases. Two of these cases, namely the delta-correlated in time random fields $\boldsymbol{F}(\boldsymbol{r},t)$, and $\boldsymbol{F}(\boldsymbol{r},t)$ which are the so-called telegraph-type, will be discussed below.

[4] Flows with non-Gaussian fluctuations were also considered in other papers.(27,28)

4. Delta-correlated random field approximation

4.1. Mean tracer concentration and its correlation function. Let $\boldsymbol{F}(\boldsymbol{r},t)$ be a zero-mean Gaussian, delta-correlated random field with covariance structure

$$B_{ij}(\boldsymbol{r},t;\boldsymbol{r}',t') = \langle F_i(\boldsymbol{r},t)F_j(\boldsymbol{r}',t')\rangle = 2B^{\text{eff}}_{ij}(\boldsymbol{r};\boldsymbol{r}',t)\delta(t-t'). \tag{23}$$

In this case, the variational derivative in (21) is expressed through $\delta q(\boldsymbol{r};t)/\delta F_j(\mathbf{r}';t')$ at $t \to t'$, i.e. through the initial condition of (22), and as a result we get that

$$\left(\frac{\partial}{\partial t} + \boldsymbol{v}(\boldsymbol{r},t)\frac{\partial}{\partial \boldsymbol{r}}\right)\langle q(\boldsymbol{r},t)\rangle$$

$$-2\int_0^t dt' \int d\boldsymbol{r}' B^{\text{eff}}_{ij}(\boldsymbol{r};\boldsymbol{r}',t)\delta(t-t')\frac{\partial}{\partial r_j}\delta(\boldsymbol{r}-\boldsymbol{r}')\frac{\partial}{\partial r_i}\langle q(\boldsymbol{r};t)\rangle = \kappa\frac{\partial^2}{\partial \boldsymbol{r}^2}\langle q(\boldsymbol{r};t)\rangle .$$

Integration in variables t' and $\boldsymbol{r}'$ gives a closed-form equation

$$\left(\frac{\partial}{\partial t} + \boldsymbol{v}(r,t)\frac{\partial}{\partial \boldsymbol{r}}\right)\langle q(\boldsymbol{r},t)\rangle = \frac{\partial}{\partial r_i}\left[B^{\text{eff}}_{ij}(\boldsymbol{r};\boldsymbol{r},t)\frac{\partial}{\partial r_j} + \kappa\frac{\partial}{\partial r_i}\right]\langle q(\boldsymbol{r},t)\rangle .$$

So, the spatial variance $B^{\text{eff}}(\boldsymbol{r};\boldsymbol{r},t)$ of $\boldsymbol{F}$ becomes the effective diffusion coefficient for the mean concentration.

If the velocity fluctuation field $\boldsymbol{F}(\boldsymbol{r},t)$ is homogeneous and isotropic in space and stationary in time, then

$$B^{\text{eff}}_{ij}(\boldsymbol{r};\boldsymbol{r}',t) = B^{\text{eff}}_{ij}(|\boldsymbol{r}-\boldsymbol{r}'|),$$

$$B^{\text{eff}}_{ij}(\boldsymbol{r}-\boldsymbol{r}') = \frac{1}{2}\int_{-\infty}^{\infty} dt' B_{ij}(\boldsymbol{r}-\boldsymbol{r}',t-t'),$$

and

$$B^{\text{eff}}_{ij}(0) = \delta_{ij}D_1,$$

where

$$D_1 = \frac{1}{N}B^{\text{eff}}_{ii}(0), \tag{23'}$$

and N (= 2 or 3) denotes the dimension of space. Hence, we get

$$\left(\frac{\partial}{\partial t} + \boldsymbol{v}(r,t)\frac{\partial}{\partial \boldsymbol{r}}\right)\langle q(\boldsymbol{r},t)\rangle = \left(D_1+\kappa\right)\frac{\partial^2}{\partial \boldsymbol{r}^2}\langle q(\boldsymbol{r},t)\rangle. \tag{24}$$

Similarly, one obtains equation

$$\left(\frac{\partial}{\partial t} + \boldsymbol{v}(\boldsymbol{r}_1,t)\frac{\partial}{\partial \boldsymbol{r}_1} + \boldsymbol{v}(\boldsymbol{r}_2,t)\frac{\partial}{\partial \boldsymbol{r}_2}\right)\Gamma(\boldsymbol{r}_1,\boldsymbol{r}_2,t)$$

$$= \int_0^t dt' \int dr' \left[B_{ij}(\boldsymbol{r}_1, t; \boldsymbol{r}', t') \frac{\partial}{\partial r_{1i}} \right.$$

$$\left. + B_{ij}(\boldsymbol{r}_2, t; \boldsymbol{r}', t') \frac{\partial}{\partial r_{2i}} \right] \left\langle \frac{\delta q(\boldsymbol{r}_1, t) q(\boldsymbol{r}_2, t)}{\delta F_j(\mathbf{r}', t')} \right\rangle + \kappa \left[\frac{\partial^2}{\partial \boldsymbol{r}_1^2} + \frac{\partial^2}{\partial \boldsymbol{r}_2^2} \right] \Gamma(\boldsymbol{r}_1, \boldsymbol{r}_2, t)$$

for the correlation function $\Gamma(\boldsymbol{r}_1, \boldsymbol{r}_2, t) = \langle q(\boldsymbol{r}_1, t) q(\boldsymbol{r}_2, t) \rangle_{\boldsymbol{F}}$. In the case of a homogeneous, isotropic and δ-correlated velocity fluctuation field $\boldsymbol{F}(\boldsymbol{r}, t)$, the equation takes the form

$$\left(\frac{\partial}{\partial t} + \boldsymbol{v}(\boldsymbol{r}_1, t) \frac{\partial}{\partial \boldsymbol{r}_1} + \boldsymbol{v}(\boldsymbol{r}_2, t) \frac{\partial}{\partial \boldsymbol{r}_2} \right) \Gamma(\boldsymbol{r}_1, \boldsymbol{r}_2, t) = \tag{25}$$

$$\left(\left(D_1 + \kappa \right) \left(\frac{\partial^2}{\partial \boldsymbol{r}_1^2} + \frac{\partial^2}{\partial \boldsymbol{r}_2^2} \right) + 2 \sum_{i,j} B_{ij}^{\text{eff}}(\boldsymbol{r}_1 - \boldsymbol{r}_2) \frac{\partial^2}{\partial r_{1i} \partial r_{2j}} \right) \Gamma(\boldsymbol{r}_1, \boldsymbol{r}_2, t).$$

If, in particular, the mean flow $\boldsymbol{v}(\boldsymbol{r}, t) = 0$ and the initial tracer concentration $q(\boldsymbol{r}, 0) = q_0(\boldsymbol{r})$ is a homogeneous random field, then the random field $q(\mathbf{r}, t)$ will also be homogeneous and isotropic. Hence,

$$\Gamma(\boldsymbol{r}_1, \boldsymbol{r}_2, t) = \Gamma(\boldsymbol{r}_1 - \boldsymbol{r}_2, t),$$

and equation (25) simplifies to

$$\frac{\partial}{\partial t} \Gamma(\boldsymbol{r}, t) = 2 \left(\kappa + D_1 \right) \frac{\partial^2}{\partial \boldsymbol{r}^2} \Gamma(\boldsymbol{r}, t) - 2 \sum_{i,j} B_{ij}^{\text{eff}}(\boldsymbol{r}) \frac{\partial^2}{\partial r_i \partial r_j} \Gamma(\boldsymbol{r}, t), \tag{26}$$

for $\boldsymbol{r} = \boldsymbol{r}_1 - \boldsymbol{r}_2$. Alternatively, we get

$$\frac{\partial}{\partial t} \Gamma(\boldsymbol{r}, t) = 2\kappa \frac{\partial^2}{\partial \boldsymbol{r}^2} \Gamma(\boldsymbol{r}, t) + 2 \frac{\partial^2}{\partial r_i \partial r_j} D_{ij}(\boldsymbol{r}) \Gamma(\boldsymbol{r}, t), \tag{26'}$$

where

$$D_{ij}(\boldsymbol{r}) = B_{ij}^{\text{eff}}(0) - B_{ij}^{\text{eff}}(\boldsymbol{r}) \tag{27}$$

is the matrix-valued structure function of field $\boldsymbol{F}$.

Let us remark, that equations (24) and (25) have the Fokker-Planck form for the one-particle and two-particle probability densities of the Langrangian coordinates. Furthermore, the Lagrangian description of problem (8) yields a Markov process. Additionally, equation (26) (or (26')) describes relative diffusion of two particles. For sufficiently small initial distances between two particles ($r_0 \ll l_0$, where l_0 is the spatial radius of correlation of the fluctuation field $\boldsymbol{F}$) function $D_{ij}(\boldsymbol{r})$ can be expanded in the Taylor series and in the first approximation

$$D_{ij}(\boldsymbol{r}) = -\frac{1}{2} \frac{\partial^2 B_{ij}^{eff}(\boldsymbol{r})}{\partial r_k \partial r_l} \bigg|_{\boldsymbol{r}=0} r_k r_l. \tag{27'}$$

Now, let us introduce the spectral density of energy of the flow by the formula

$$B_{ij}^{eff}(\boldsymbol{r}) = \int d\boldsymbol{k} E(k)(\delta_{ij} - \frac{k_i k_j}{\boldsymbol{k}^2})e^{i\boldsymbol{k}\boldsymbol{r}}. \tag{28}$$

Then

$$-\frac{\partial^2 B_{ij}^{eff}(\boldsymbol{r})}{\partial r_k \partial r_l}\Big|_{\boldsymbol{r}=0} = D_2\Big\{(N+1)\delta_{ij}\delta_{k\ell} - \delta_{ik}\delta_{jl} - \delta_{jk}\delta_{il}\Big\}, \tag{29}$$

where

$$D_2 = \frac{1}{N(N+2)}\int d\boldsymbol{k}\,\boldsymbol{k}^2 E(k). \tag{30}$$

Note that the introduced earlier quantity D_1 is also determined by the spectral density $E(\kappa)$ via equality

$$D_1 = \frac{N-1}{N}\int d\boldsymbol{k}\, E(k). \tag{30'}$$

In this case the diffusion tensor (27') simplifies and can be written in the form

$$D_{ij}(\boldsymbol{r}) = \frac{1}{2}D_2\{(N+1)\boldsymbol{r}^2\delta_{ij} - 2r_i r_j\}. \tag{31}$$

Substituting (31) into (26'), multiplying both sides of the obtained equation by $\boldsymbol{r}^2$ and integrating over $\boldsymbol{r}$, we obtain the equation

$$\frac{d}{dt}\langle \boldsymbol{r}^2(t)\rangle = 4\kappa N + 2(N+2)(N-1)D_2\langle \boldsymbol{r}^2(t)\rangle \tag{32}$$

for variance $\langle \boldsymbol{r}^2\rangle$ (the mean $\langle \boldsymbol{r}(t)\rangle$ is conserved). Its solution has the structure

$$\langle \boldsymbol{r}^2\rangle = r_0^2 e^{2(N+2)(N-1)D_2 t} + \frac{2\kappa N}{(N+2)(N-1)D_2}\Big\{e^{2(N+2)(N-1)D_2 t} - 1\Big\}. \tag{33}$$

It is clear from (74) that under condition

$$\kappa \ll D_2 r_0^2 \tag{34}$$

the effects of molecular diffusion on particle are not essential and the last term in (33) can be omitted. In this case the solution becomes an exponentially growing function in time

$$\langle \boldsymbol{r}^2\rangle = r_0^2 e^{2(N+2)(N-1)D_2 t}. \tag{33'}$$

Expression (33') is valid whenever expansion (27') is, that is for the time range

$$D_2 t \ll \frac{1}{(N+2)(N-1)}\ln\frac{l_0}{r_0}. \tag{35}$$

Note that the influence of the molecular diffusion for the above one-particle probability density, according to (24), can be neglected if condition

$$(34') \qquad \kappa \ll D_1$$

is satisfied. Approximation (27') is, however, not valid for the turbulent fluid flow. That case has been studied elsewhere.[29]

So far, we considered the mean concentration of the tracer and its correlation function which are described in the closed form due to the linearity of the basic equation (1). If one considers higher moment functions of the tracer concentration described by equation (1') then we do not get a closed form description. Indeed, averaging (1') over the $\boldsymbol{F}$-ensemble we obtain equation

$$(36) \qquad (\frac{\partial}{\partial t} + \boldsymbol{v}(\boldsymbol{r},t)\frac{\partial}{\partial \boldsymbol{r}})\langle q^n(\boldsymbol{r},t)\rangle$$
$$= (D_1+\kappa)\frac{\partial^2}{\partial \boldsymbol{r}^2}\langle q^n(\boldsymbol{r},t)\rangle - \kappa n(n-1)\langle q^{n-2}(\boldsymbol{r},t)\boldsymbol{p}^2(\boldsymbol{r},t)\rangle,$$

whose right-hand side contains an unknown function—covariance of the concentration field and its spatial gradient. In order to understand better the structure of the tracer gradient field one can, in the first approximation, neglect effects of the molecular diffusion, i.e. consider the stochastic system (10-11).

4.2. Fine structure of passive tracer fluctuations in random velocity fields. In this subsection we look at subtler charcateristics of tracer fluctuations than the means and correlations considered in subsection 4.1, like the joint probability density of the tracer concentration and its gradient. Averaging Liouville equation (13) over the ensemble of realizations of fluctuation field $\boldsymbol{F}$ and using the Furutsu-Novikov formula and the formula (13')

$$\frac{\delta}{\delta u_j(\boldsymbol{r}',t-0)}\Phi_{t,\boldsymbol{r}}(q,\boldsymbol{p}) = \left\{-\delta(\boldsymbol{r}-\boldsymbol{r}')\frac{\partial}{\partial r_j} + \frac{\partial\delta(\boldsymbol{r}-\boldsymbol{r}')}{\partial r_i}\frac{\partial}{\partial p_i}p_j\right\}\Phi_{t,\boldsymbol{r}}(q,\boldsymbol{p}),$$

for the variational derivative obtained from (13), we get for the one-point joint probability density of fields $q(\boldsymbol{r},t)$ and $\boldsymbol{p}(\boldsymbol{r},t)$ the Fokker -Planck equation

$$(37) \quad \left\{\frac{\partial}{\partial t} + \boldsymbol{v}(\boldsymbol{r},t)\frac{\partial}{\partial \boldsymbol{r}} - \frac{\partial \boldsymbol{p}\boldsymbol{v}(\boldsymbol{r},t)}{\partial \boldsymbol{r}}\frac{\partial}{\partial \boldsymbol{p}}\right\}P_{t,\boldsymbol{r}}(q,\boldsymbol{p}) = D_1\frac{\partial^2}{\partial \boldsymbol{r}^2}P_{t,\boldsymbol{r}}(q,\boldsymbol{p}) +$$
$$+D_2\left\{(N+1)\frac{\partial^2}{\partial \boldsymbol{p}^2}\boldsymbol{p}^2 - 2\frac{\partial}{\partial \boldsymbol{p}}\boldsymbol{p} - 2\left(\frac{\partial}{\partial \boldsymbol{p}}\boldsymbol{p}\right)^2\right\}P_{t,\boldsymbol{r}}(q,\boldsymbol{p})$$

with initial condition

$$(37') \qquad P_{0,\boldsymbol{r}}(q,\boldsymbol{p}) = \delta(q_0(\boldsymbol{r})-q)\delta(\frac{\partial}{\partial \boldsymbol{r}}q_0(\boldsymbol{r})-\boldsymbol{p}).$$

Constants D_1 and D_2 introduced in (30) and (30') become, that way, the new diffusion coefficients for the Fokker-Planck equation in $\boldsymbol{r}$- and $\boldsymbol{p}$-spaces, respectively.

Equation (37) can be written in an operator form

$$\frac{\partial}{\partial t}P_{t,\boldsymbol{r}}(q,\boldsymbol{p}) = \hat{L}(\boldsymbol{r},t)P_{t,\boldsymbol{r}}(q,\boldsymbol{p}) + \hat{M}(\boldsymbol{r},\boldsymbol{p},t)P_{t,\boldsymbol{r}}(q,\boldsymbol{p}), \tag{38}$$

where operators $\hat{L}$ and $\hat{M}$ are defined by formulas

$$\hat{L}(\boldsymbol{r},t) = -\boldsymbol{v}(\boldsymbol{r},t)\frac{\partial}{\partial \boldsymbol{r}} + D_1\frac{\partial^2}{\partial \boldsymbol{r}^2},$$

(38')

$$\hat{M}(\boldsymbol{r},\boldsymbol{p},t) = \frac{\partial \boldsymbol{p}\boldsymbol{v}(\boldsymbol{r},t)}{\partial \boldsymbol{r}}\frac{\partial}{\partial \boldsymbol{p}} + D_2\left\{(N+1)\frac{\partial^2}{\partial \boldsymbol{p}^2}\boldsymbol{p}^2 - 2\frac{\partial}{\partial \boldsymbol{p}}\boldsymbol{p} - 2\left(\frac{\partial}{\partial \boldsymbol{p}}\boldsymbol{p}\right)^2\right\}.$$

As was discussed earlier, operator $\hat{L}(\boldsymbol{r},t)$ defines the spatial diffusion of the Lagrangian particle, while $\hat{M}(\boldsymbol{r},\boldsymbol{p},t)$ defines the diffusion of the tracer gradient and its correlation with the particle's position. In the simplest case of zero mean flow ($\boldsymbol{v}=0$), or in the case of shear flow with constant gradient $\partial \boldsymbol{p}\boldsymbol{v}(\boldsymbol{r},t)/\partial \boldsymbol{r}$, the operators $\hat{L}$ and $\hat{M}$ commute, which reflects statistical independence of diffusions in the position $\boldsymbol{r}$-space and gradient $\boldsymbol{p}$-space.

Also, notice that for spatially homogeneous and isotropic Gaussian velocity fluctuations, the corresponding diffusion operators are also isotropic. This fact will give us additional information on the fluctuations of the spatial gradient of the tracer concentration field that we shall now outline.

Consider the case of the zero mean flow ($\boldsymbol{v}=0$). Then, the solution of equation (38) has the structure corresponding to the averaged equation(16)

$$P_{t,\boldsymbol{r}}(q,\boldsymbol{p}) = \int d\boldsymbol{\xi}P_t(\boldsymbol{r}|\boldsymbol{\xi})P_t(q,\boldsymbol{p}|\boldsymbol{\xi}). \tag{39}$$

Here, $P_t(\boldsymbol{r}|\boldsymbol{\xi})$ is the probability density of the Lagrangian coordinate of the particle described by equation

$$\frac{\partial}{\partial t}P_t(\boldsymbol{r}|\boldsymbol{\xi}) = D_1\frac{\partial^2}{\partial \boldsymbol{r}^2}P_t(\boldsymbol{r}|\boldsymbol{\xi}), \qquad P_0(\boldsymbol{r}|\boldsymbol{\xi}) = \delta(\boldsymbol{r}-\boldsymbol{\xi}), \tag{40}$$

and $P_t(q,\boldsymbol{p}|\boldsymbol{\xi})$ is the joint probability density of the tracer concentration and its spatial gradient, satisfying equation

$$\frac{\partial}{\partial t}P_t(q,\boldsymbol{p}|\boldsymbol{\xi}) = D_2\left\{(N+1)\frac{\partial^2}{\partial \boldsymbol{p}^2}\boldsymbol{p}^2 - 2\frac{\partial}{\partial \boldsymbol{p}}\boldsymbol{p} - 2\left(\frac{\partial}{\partial \boldsymbol{p}}\boldsymbol{p}\right)^2\right\}P_t(q,\boldsymbol{p}|\boldsymbol{\xi}) \tag{41}$$

$$P_0(q,\boldsymbol{p}|\boldsymbol{\xi}) = \delta(q_0(\boldsymbol{\xi})-q)\delta(\boldsymbol{p}_0(\boldsymbol{\xi})-\boldsymbol{p}).$$

Specifically, in the 3-D case solution of equation (40) is

$$(40') \quad P_t(\boldsymbol{r}|\boldsymbol{\xi}) = \exp\left[D_1 t \frac{\partial^2}{\partial \boldsymbol{r}^2}\right] \delta(\boldsymbol{r}-\boldsymbol{\xi}) = (4\pi D_1 t)^{-3/2} \exp\left[-\frac{(\boldsymbol{r}-\boldsymbol{\xi})^2}{4D_1 t}\right].$$

It corresponds to the Brownian motion of the particle with parameters

$$\langle \boldsymbol{r}(t|\boldsymbol{\xi})\rangle = \boldsymbol{\xi}, \qquad \sigma_{ij}^2(t) = 2D_1\delta_{ij}t.$$

From (41) we derive the following moment equation for gradient $\boldsymbol{p}(t|\boldsymbol{\xi})$:

$$(42) \qquad \frac{\partial}{\partial t}\langle |\boldsymbol{p}(t|\boldsymbol{\xi})|^n\rangle = D_2 n(N+n)(N-1)\langle |\boldsymbol{p}(t(\boldsymbol{\xi})|^n\rangle,$$

$$(42') \qquad \frac{\partial}{\partial t}\langle \boldsymbol{p}|\boldsymbol{p}(t|\boldsymbol{\xi})|^n\rangle = D_2 n(N+n+2)(N-1)\langle \boldsymbol{p}|\boldsymbol{p}|^n\rangle,$$

and, in particular,

$$\frac{\partial}{\partial t}\langle \boldsymbol{p}(t|\boldsymbol{\xi})\rangle = 0, \qquad \text{i.e.} \qquad \langle \boldsymbol{p}(t|\boldsymbol{\xi})\rangle = \boldsymbol{p}_0(\boldsymbol{\xi}).$$

So the moment functions grow exponentially in time, with the exception of the conserved quantity $\langle \boldsymbol{p}(t|\boldsymbol{\xi})\rangle$. If, in addition, the initial tracer concentration $q_0(\boldsymbol{r})$ is spherically symmetric (3-D case) or radially symmmetric (2-D case), then in the Eulerian representation, we obtain from (39) that

$$\langle \boldsymbol{p}(\boldsymbol{r},t)|\boldsymbol{p}(\boldsymbol{r},t)|^n\rangle = 0.$$

However, arbitrarily small deviations from the rotational symmetry, make the above quantity grow unboundedly at large times t, so we encounter the *stochastic instability*.

Notice, that in the Eulerian representation, (39) implies that the exponential time-growth of moments $\langle |\boldsymbol{p}(\boldsymbol{r},t)|^n\rangle$ and $\langle \boldsymbol{p}(\boldsymbol{r},t)|\boldsymbol{p}(\boldsymbol{r},t)|^n\rangle$ is accompanied by their spatial dissipation with the diffusion rate D_1.

Equation (42) also implies that the normalized quantity $|\boldsymbol{p}(t|\boldsymbol{\xi})|/|\boldsymbol{p}_0(\boldsymbol{\xi})|$ has a lognormal probability distribution, i.e.,

$$\chi(t) = \ln\frac{|\boldsymbol{p}(t|\boldsymbol{\xi})|}{|\boldsymbol{p}_0(\boldsymbol{\xi})|}$$

is Gaussian with parameters

$$(43) \qquad \langle \chi(t)\rangle = D_2 N(N-1)t, \qquad \sigma_\chi^2(t) = 2D_2(N-1)t.$$

Properties of the lognormal distribution were studied in detail in another paper[(30)] (see also[(31)]), where it was shown that the typical realization of process $|\boldsymbol{p}(t)|$ has an exponential growth

$$|\boldsymbol{p}(t|\boldsymbol{\xi})| \sim |\boldsymbol{p}_0(\boldsymbol{\xi})|\exp[D_2 N(N-1)t],$$

accompanied by large excursions relative to the above exponential curve. In addition, there exist several lower estimates for the quantity $\chi(t)$. Note that this situation is fundamentally different from the one-dimensional problem (where the fluid flow is always compressible). There, the gradient conserves its sign and a typical realization of the process is an exponentially decaying curve.(19,32) It is worth mentioning that the lognormal distribution of the norm of the tracer gradient, first proposed by Gurvich and Yaglom(33), agrees well with experimental atmospheric data.(34,35)

We also get from (41) the evolution equation

$$(44)\ \frac{\partial}{\partial t}\langle p_i(t|\boldsymbol{\xi})p_j(t|\boldsymbol{\xi})\rangle = -4D_2\langle p_i(t|\boldsymbol{\xi})p_j(t|\boldsymbol{\xi})\rangle + 2D_2(N+1)\delta_{ij}\langle \boldsymbol{p}^2(t)|\boldsymbol{\xi}\rangle,$$

for the covariance $\langle p_i(t|\boldsymbol{\xi})p_j(t|\boldsymbol{\xi})\rangle$. Clearly, cross-terms $i \neq j$ of the correlation of different components of gradient $\boldsymbol{p}(t|\boldsymbol{\xi})$ converge rapidly (exponentially) to zero. So, for large time values $D_2 t \gg 1/4$ vector $\boldsymbol{p}(t|\boldsymbol{\xi})$ undergoes full statistical isotropisation independent of the initial conditions.

Equation (41) is written in Cartesian coordinates in the $\boldsymbol{p}$ space. Using isotropy of the diffusion operator, that equation could be simplified by passage to the spherical (3-D case) or polar (2-D case) coordinates. So, for example, in the 2-D case, replacing the Euclidean coordinates (p_x, p_y) by the standard polar (p, φ),

$$(45)\qquad p_x = p\cos\varphi, \qquad p_y = p\sin\varphi, \qquad (|\boldsymbol{p}| = p)$$

and taking into account the relation

$$P_t(q, p, \varphi) = pP_t(q, \boldsymbol{p}|\boldsymbol{\xi}),$$

between the probability densities $P_t(q,p,\varphi)$ and $P_t(q,\boldsymbol{p}|\boldsymbol{\xi})$, we obtain the following equation for P_t:(32)

$$(46)\qquad \frac{\partial}{\partial t}P_t(q,p,\varphi) = D_2\left\{\frac{\partial}{\partial p}p\frac{\partial}{\partial p}p - 2\frac{\partial}{\partial p}p + 3\frac{\partial^2}{\partial \varphi^2}\right\}P_t(q,p,\varphi).$$

The change of variables (45) in the Fokker-Planck equation (41) corresponds to the introduction of two scalar random functions $p(t)$ and $\varphi(t)$ instead of a single vector random function $\boldsymbol{p}(t|\boldsymbol{\xi})$. So equation (46) becomes the Fokker-Planck equation for their joint one-time probability density. It is clear from (46) that the stochastic processes $p(t)$ and $\varphi(t)$ are statistically independent, and that process $p(t)$ has a lognormal distribution while $\varphi(t)$ is Gaussian.

Equation (46) is statistically equivalent to a system of stochastic differential equations

$$\frac{d}{dt}p(t) = 2D_2 p(t) + z_1(t)p(t), \quad p(0) = p_0,$$

(47) $$\frac{d}{dt}\varphi(t) = z_2(t), \quad \varphi(0) = \varphi_0,$$

where processes $z_1(t)$ and $z_2(t)$ are statistically independent white noise processes with parameters

(48) $$\langle z_i(t)\rangle = 0, \quad \langle z_1(t)z_1(t')\rangle = 2D_2\delta(t-t'), \ \langle z_2(t)z_2(t')\rangle = 6D_2\delta(t-t').$$

Hence, it follows that the quantity $\chi(t) = \ln(p(t)/p_0)$is statistically equivalent to the solution of the stochastic equation

(47′) $$\frac{d}{dt}\chi(t) = 2D_2 + z_1(t), \quad \chi(0) = 0.$$

Another derivation is possible directly from equation (46). Quantity $\chi(t)$ has Gaussian distribution with parameters

$$\langle\chi(t)\rangle = 2D_2 t, \qquad \sigma_\chi^2 = 2D_2 t,$$

which, naturally, agrees with formula (43) for $N = 2$.

Here we have limited out attention to moments of the tracer gradient (42-42'). Paper(32) concentrates on a geometric interpretation of these quantities and we will analyse them from that perspective below.

4.3. Geometric interpretation of the fine structure. In this subsection, as in(32), we will restrict our attention to the case of two dimensional fluid flows and study the mean value (in a geometric sense)

(49) $$A_n(t) = \int d\boldsymbol{r}|\boldsymbol{p}(\boldsymbol{r},t)|^{n+1}\delta[q(\boldsymbol{r},t)-q] = \oint |p(\boldsymbol{r},t)|^n d\ell$$

of the n-th power of norm of the tracer concentration gradient along the level curve

$$q(\boldsymbol{r},t) = q$$

of the concentration field itself. If q is is interpreted as temperature then the corresponding curves are isotherms. In particular, for $n = 0$, formula (49) gives the length of the contour(36–40)

(49′) $$A_0(t) = \ell(t) = \int d\boldsymbol{r}|\boldsymbol{p}(\boldsymbol{r},t)|\delta[q(\boldsymbol{r},t)-q] = \oint d\ell.$$

The mean value of the tracer concentration gradient over the planar domain bound by the level curve is given by expression

(49″) $$\boldsymbol{A}(t) = \int_S \nabla q(\boldsymbol{r},t)dS = q\int d\boldsymbol{r}\,\boldsymbol{p}(\boldsymbol{r},t)\delta[q(\boldsymbol{r},t)-q] = q\oint \frac{\boldsymbol{p}(\boldsymbol{r},t)}{|\boldsymbol{p}(\boldsymbol{r},t)|}d\ell.$$

Equation (49-49") can be rewritten in terms of the distribution function $\Phi_{t,\boldsymbol{r}}(q,\boldsymbol{p})$ as follows:

(50) $$A_n(t) = \int d\boldsymbol{r}\int d\boldsymbol{p}\,p^{n+1}\Phi_{t,r}(q,\boldsymbol{p}), \quad \boldsymbol{A}(t) = q\int d\boldsymbol{r}\int d\boldsymbol{p}\,\boldsymbol{p}\Phi_{t,r}(q,\boldsymbol{p}).$$

Consequently, their averages are determined by one-point probability density $P_{t,\boldsymbol{r}}(q,\boldsymbol{p})$ via formulas

$$\langle A_n(t)\rangle = \int d\boldsymbol{r}\int d\boldsymbol{p}\, p^{n+1} P_{t,\boldsymbol{r}}(q,\boldsymbol{p}), \qquad \langle \boldsymbol{A}(t)\rangle = q\int d\boldsymbol{r}\int d\boldsymbol{p}\,\boldsymbol{p}P_{t,\boldsymbol{r}}(q,\boldsymbol{p}). \tag{51}$$

Substituting (39) for $P_{t,\boldsymbol{r}}(q,\boldsymbol{p})$ and taking into account that $\int d\boldsymbol{r}\, P_t(\boldsymbol{r}|\boldsymbol{\xi}) = 1$, we express the quantities of interest

$$\langle A_n(t)\rangle = \int d\boldsymbol{\xi}\int d\boldsymbol{p}\, p^{n+1} P_t(q,\boldsymbol{p}|\boldsymbol{\xi}), \qquad \langle \boldsymbol{A}(t)\rangle = q\int d\boldsymbol{\xi}\int d\boldsymbol{p}\,\boldsymbol{p}P_t(q,\boldsymbol{p}|\xi), \tag{51'}$$

in terms of Lagrangian probability density. In the 2-D case, those satisfy equation (41)

$$\frac{\partial}{\partial t}P_t(q,\boldsymbol{p}|\boldsymbol{\xi}) = D_2\left\{3\frac{\partial^2}{\partial \boldsymbol{p}^2}\boldsymbol{p}^2 - 2\frac{\partial}{\partial \boldsymbol{p}}\boldsymbol{p} - 2(\frac{\partial}{\partial \boldsymbol{p}}\boldsymbol{p})^2\right\}P_t(q,\boldsymbol{p}|\xi),$$

$$P_0(q,\boldsymbol{p}|\boldsymbol{\xi}) = \delta(q_0(\boldsymbol{\xi}) - q)\delta(\boldsymbol{p}_0(\boldsymbol{\xi}) - \boldsymbol{p}), \qquad \boldsymbol{p}_0(\xi) = \frac{\partial q_0(\boldsymbol{\xi})}{\partial \boldsymbol{\xi}}. \tag{52}$$

Differentiating expression (51') with respect to time and applying equality (42) with $N=2$, we obtain differential equations for the mean values

$$\frac{d}{dt}\langle A_n(t)\rangle = (n+1)(n+3)D_2\langle A_n(t)\rangle, \quad \langle A_n(0)\rangle = A_n(0),$$

$$\frac{d}{dt}\langle \boldsymbol{A}(t)\rangle = 0, \qquad \langle \boldsymbol{A}(0)\rangle = \boldsymbol{A}(0). \tag{53}$$

Consequently, their solutions correspond to exponentially growing in time functions

$$\langle \ell(t)\rangle = \ell_0 e^{3D_2 t}, \quad \langle A_n(t)\rangle = A_n(0)e^{(1+n)(n+3)D_2 t}, \tag{54}$$

and the mean concentration gradient, averaged over the area, is conserved, i.e. $\langle \boldsymbol{A}(t)\rangle = \boldsymbol{A}(0)$.

Let us note that in the 3-D case, (50) become integrals over the level surfaces $q(\boldsymbol{r},t) = q$ and, respectively, the regions bounded by these surfaces.

The exponential growth of (54) indicates strong increase of roughness of the level curves of tracer concentration with time, that creates a "fractal" structure, and the situation is similar for the mean tracer gradient.[(41)] Examples of numerical modeling of this phenomenon can be found in dissertation[(42)].

It follows from (50) that higher moments of random quantities $\ell(t)$ and $A_n(t)$ are connected with multi-point (or multi-particle) probability distributions. Saichev and Woyczynski[(32)] have shown that it is possible to evaluate the second moment of quantity $A_1(t) = \oint |p(r,t)d\ell$ — the contour mean of the total tracer concentration gradient. Indeed, it satisfies equation

$$(55)\langle A_1^2(t)\rangle = \int d\boldsymbol{r}_1 \int d\boldsymbol{p}_1 \int d\boldsymbol{r}_2 \int d\boldsymbol{p}_2\, \boldsymbol{p}_1^2\boldsymbol{p}_2^2 \langle \Phi_{t,\boldsymbol{r}_1}(q,\boldsymbol{p}_1)\Phi_{t,\boldsymbol{r}_2}(q,\boldsymbol{p}_2)\rangle.$$

Differentiating (55) with respect to time, using the dynamic equation (13) (in absence of the mean flow), the Furutsu-Novikov formula and equation (13') for the variational derivative, we get

$$(56)\qquad \frac{d}{dt}\langle A_1^2(t)\rangle = \int d\boldsymbol{r} \int d\boldsymbol{p} \int d\boldsymbol{r}_1 \int d\boldsymbol{p}_1\, \boldsymbol{p}^2\boldsymbol{p}_1^2 \times$$

$$\times\left\{\hat{M}(\boldsymbol{p}) + \hat{M}(\boldsymbol{p}_1) + \hat{N}(\boldsymbol{r}-\boldsymbol{r}_1;\boldsymbol{p},\boldsymbol{p}_1)\right\}\left\langle \Phi_{t,\boldsymbol{r}}(q,\boldsymbol{p})\Phi_{t,\boldsymbol{r}_1}(q,\boldsymbol{p}_1)\right\rangle.$$

Here, operators

$$\hat{M}(\boldsymbol{p}) = D_2\left\{3\frac{\partial^2}{\partial \boldsymbol{p}^2}\boldsymbol{p}^2 - 2\frac{\partial}{\partial \boldsymbol{p}}\boldsymbol{p} - 2(\frac{\partial}{\partial \boldsymbol{p}}\boldsymbol{p})^2\right\},$$

(57)

$$\hat{N}(\boldsymbol{r}-\ \boldsymbol{r}_1;\boldsymbol{p},\boldsymbol{p}_1) = -\frac{\partial^2 B_{kj}(\boldsymbol{r}-\boldsymbol{r}_1)}{\partial r_\ell \partial r_i}\left\{\frac{\partial}{\partial p_{1i}}p_{1k}\frac{\partial}{\partial p_\ell}p_j + \frac{\partial}{\partial p_i}p_k\frac{\partial}{\partial p_{1\ell}}p_{1j}\right\}.$$

In general, equation (56) can not be closed because its right-hand side contains the covariance function $B_{ij}(\boldsymbol{r}-\boldsymbol{r}_1)$ of the velocity fluctuations field $\boldsymbol{F}(\boldsymbol{r},t)$. However, if the initial domain of the tracer concentration has sufficiently small characteristic size r_0 ($r_0 \ll l_0$, where l_0 is the spatial velocity correlation radius), then functions B_{ij} can be expanded in the Taylor series in $\boldsymbol{r}$. This expansion, however, would be valid only for a limited time range (35), that is time t for which the domain size of the concentration density at t does not exceed l_0. As a result, operator $\hat{N}(\boldsymbol{r}-\boldsymbol{r}_1;\boldsymbol{p},\boldsymbol{p}_1)$ simplifies and, via (29), takes the form

$$\hat{N}(0;\boldsymbol{p},\boldsymbol{p}_1) = 2D_2\left\{3\frac{\partial^2}{\partial \boldsymbol{p}\partial \boldsymbol{p}_1}(\boldsymbol{p}\boldsymbol{p}_1) - (\frac{\partial}{\partial \boldsymbol{p}_1}\boldsymbol{p}_1)(\frac{\partial}{\partial \boldsymbol{p}}\boldsymbol{p}) - \frac{\partial}{\partial \boldsymbol{p}_1}(\boldsymbol{p}_1\frac{\partial}{\partial \boldsymbol{p}})\boldsymbol{p}\right\}.$$

(58)

Integrating (56) by parts, we obtain an equation for $\langle A_1^2(t)\rangle$ of the form

$$(59)\qquad \frac{d}{dt}\langle A_1^2(t)\rangle = 8\,D_2\langle A_1^2(t)\rangle + 16\,D_2\langle\Psi(t)\rangle.$$

Here we introduced a new function

$$(60)\quad \langle\Psi(t)\rangle = \int d\boldsymbol{r} \int d\boldsymbol{p} \int d\boldsymbol{r}_1 \int d\boldsymbol{p}_1 (\boldsymbol{p}\boldsymbol{p}_1)^2 \langle \Phi_{t,\boldsymbol{r}}(q,\boldsymbol{p})\Phi_{t,\boldsymbol{r}_1}(q,\boldsymbol{p}_1)\rangle.$$

Differentiating (60) with respect to time and repeating the above procedure, we obtain equation

$$(61)\qquad \frac{d}{dt}\langle\Psi(t)\rangle = 16\,D_2\langle\Psi(t)\rangle + 8D_2\langle A_1^2(t)\rangle.$$

The system of equations (59) and (61) is already closed. It is easily integrable and we get

$$\langle A_1^2(t)\rangle - \langle \Psi(t)\rangle = A_1^2(0) - \Psi(0) = const.$$

For a radially symmetric initial tracer concentration it becomes

$$\langle A_1^2(t)\rangle - \langle \Psi(t)\rangle = \frac{1}{2}A_1^2(0). \tag{62}$$

Consequently, $\langle A_1^2(t)\rangle$ itself is described by a first order (in time) differential equation

$$\frac{d}{dt}\langle A_1^2(t)\rangle = 24\, D_2\langle A_1^2(t)\rangle - 8\, D_2 A_1^2(0), \tag{63}$$

whose solution

$$\langle A_1^2(t)\rangle = \frac{1}{3}\langle A_1^2(0)(1+2e^{24D_2t}). \tag{63'}$$

Similarly, one can study higher moments $\langle A_1^n(t)\rangle$, that obey a closed system of n first order equations in time. However, in contrast to (59) and (61), the latter can not be reduced to a first order equation for $\langle A_1^n(t)\rangle$, since it has no first integrals. Saichev and Woyczynski[(32)] obtained equation for probability density of the normalized quantity $u = A_1(t)/A_1(0)$, namely

$$\frac{\partial}{\partial t}P_t(u) = 4D_2\frac{\partial}{\partial u}(u^2-1)\frac{\partial}{\partial u}P_t(u), \quad P_0(u) = \delta(u-1). \tag{64}$$

Equations for quantities $\langle u\rangle$ and $\langle u^2\rangle$ which follow from (64), coincide with equations (53) and (63). However, equation (64) fails to correctly describe higher moments of u. Distribution from (64) corresponds to the additional assumption

$$\int d\boldsymbol{r}\int d\boldsymbol{p}\int d\boldsymbol{r}_1\int d\boldsymbol{p}_1 \Big\langle [\boldsymbol{p}^2\boldsymbol{p}_1^2 - (\boldsymbol{p}\boldsymbol{p}_1)^2]\Phi_{t,\boldsymbol{r}}(q,\boldsymbol{p}) \times \Phi_{t,\boldsymbol{r}_1}(q,\boldsymbol{p}_1)e^{i\kappa A_1(t)}\Big\rangle = \frac{1}{2}\langle e^{i\kappa A_1(t)}\rangle, \tag{65}$$

where $\langle e^{i\kappa A_1(t)}\rangle$ denotes the characteristic function of the random variable $A_1(t)$. Equation (65) is satisfied in the diffusionless case $\kappa = 0$ but not in general. However, the probability distribution of large values of u corresponds to small values of κ. Consequently, equation (65) correctly predicts the probability distribution of large values of u.

Derivation of equations (63) and (65) was heavily based on the Taylor expansion of the covariance function $B_{ij}(\boldsymbol{r})$ in the neighborhood of zero. The latter is valid (without further assumptions) only for a sufficiently small initial contour and a bounded time interval. We have observed before (in Section 4.1) that the relative diffusion of two particles leads to their exponential separation and that condition (35) guarantees the validity of the Taylor expansion of B. Given that the rate of growth in time of $\langle A_1^2(t)\rangle$ of (63) significantly exceeds that of $\langle \boldsymbol{r}^2(t)\rangle$, one can conclude that for the time-range (35) quantity $\langle A_1^2(t)\rangle$ grows exponentially along with other (higher) moments.

Let us observe that the similar analysis can not be carried out for other quantities, like the contour length because, unlike $A_1(t)$, their moment functions are described by an infinite system of the first order differential equations in time.

In this subsection, we have provided a detailed analysis of statistical characteristics of tracer concentration in random velocity fields and, in absence of the mean flow, of its derivative. In presence of the mean flow, even for the deterministic case one can observe both, the steepening of the gradient of tracer concentration and the deformation of the contour of constant concentration. The presence of even small fluctuations of the velocity field quickly accelerates these processes. We will illustrate this phenomenon by the example of the simplest two-dimensional flow with linear shear.

4.4. Planar-parallel flow with linear shear. In this subsection we shall consider a simple shear mean flow

$$v_x = \alpha y, \qquad v_y = 0,$$

and analyze the effect of the shear on the statistics of tracer concentration.

In view of (37), the probability distribution $P_{t,\boldsymbol{r}}(q,\boldsymbol{p})$ is described by equation ($N=2$)

$$(66)\qquad \frac{\partial}{\partial t}P_{t,\boldsymbol{r}}(q,\boldsymbol{p}) = \left\{-\alpha y\frac{\partial}{\partial x} + D_1\frac{\partial^2}{\partial \boldsymbol{r}^2}\right\}P_{t,\boldsymbol{r}}(q,\boldsymbol{p})$$

$$+\left\{\alpha p_1\frac{\partial}{\partial p_2} + D_2\left[3\frac{\partial^2}{\partial \boldsymbol{p}^2}\boldsymbol{p}^2 - 2\frac{\partial}{\partial \boldsymbol{p}}\boldsymbol{p} - 2\left(\frac{\partial}{\partial \boldsymbol{p}}\boldsymbol{p}\right)^2\right]\right\}P_{t,\boldsymbol{r}}(q,\boldsymbol{p}),$$

where $(p_1,p_2) = (p_x,p_y)$. As before, the effective spatial diffusion of particles will be described by operator

$$\hat{L}(\boldsymbol{r}) = -\alpha y\frac{\partial}{\partial x} + D_1\frac{\partial^2}{\partial \boldsymbol{r}^2}.$$

The corresponding probability distribution is Gaussian with parameters[(43)]

$$\langle x(t)\rangle = x_0 + \alpha y_0 t, \quad \langle y(t)\rangle = y_0,$$

$$(67)\qquad \sigma_{xx}^2(t) = 2D_1t(1+\frac{1}{3}\alpha^2t^2), \quad \sigma_{yy}^2 = 2D_1t, \quad \sigma_{xy}^2 = \alpha D_1t^2.$$

Operator

$$\hat{M}(\boldsymbol{p}) = \alpha p_1 \frac{\partial}{\partial p_2} + D_2 \left[3 \frac{\partial^2}{\partial \boldsymbol{p}^2} \boldsymbol{p}^2 - 2 \frac{\partial}{\partial \boldsymbol{p}} \boldsymbol{p} - 2 \left(\frac{\partial}{\partial \boldsymbol{p}} \boldsymbol{p} \right)^2 \right]$$

describes the diffusion of the tracer concentration gradient and is now anisotropic. In this case the mean value of vector $\boldsymbol{p}$ is not conserved. It gives the solution

$$\langle p_1(t) \rangle = p_1(0), \quad \langle p_2(t) \rangle = p_2(0) - \alpha p_1(0) t \tag{68}$$

of the problem in the case of zero velocity fluctuations.

Consider second moments of vector $\boldsymbol{p}$ and write the Lagrangian equations for them:

$$\begin{aligned} \frac{d}{dt} \langle p_1^2 \rangle &= 2 D_2 \langle p_1^2 \rangle + 6 D_2 \langle p_2^2 \rangle, \\ \frac{d}{dt} \langle p_2^2 \rangle &= 2 D_2 \langle p_2^2 \rangle + 6 D_2 \langle p_1^2 \rangle - 2\alpha \langle p_1 p_2 \rangle, \\ \frac{d}{dt} \langle p_1 p_2 \rangle &= -4 D_2 \langle p_1 p_2 \rangle - \alpha \langle p_1^2 \rangle. \end{aligned} \tag{69}$$

Alternatively,

$$\begin{aligned} \frac{d}{dt} \langle \boldsymbol{p}^2 \rangle &= 8 D_2 \langle \boldsymbol{p}^2 \rangle - 2\alpha \langle p_1 p_2 \rangle, \\ \frac{d}{dt} \langle p_1 p_2 \rangle &= -4 D_2 \langle p_1 p_2 \rangle - \alpha \langle p_1^2 \rangle \\ \frac{d}{dt} \langle p_1^2 \rangle &= -4 D_2 \langle p_1^2 \rangle + 6 D_2 \langle \boldsymbol{p}^2 \rangle. \end{aligned} \tag{70}$$

Solving (70) in the exponential form $e^{\lambda t}$ we get the characteristic equation for λ

$$(\lambda + 4D_2)^2 (\lambda - 8D_2) = 12 \alpha^2 D_2, \tag{71}$$

whose roots, essentially, depend on α / D_2.

For small $\alpha / D_2 \ll 1$, the roots of equation (71) are, approximately,

$$\lambda_1 = 8 D_2 + \frac{1}{12} \frac{\alpha^2}{D_2}, \quad \lambda_2 = -4 D_2 + i|\alpha|, \quad \lambda_3 = -4 D_2 - i|\alpha|. \tag{72}$$

Hence, in the time range $D_2 t \gg 1/4$, solution of the problem is completely controlled by random factors. That means that random velocity fluctuations fully dominate in problems with weak gradients of the mean field.

In the case of large $\alpha / D_2 \gg 1$, characteristic equation (71) has approximate roots

$$\lambda_1 = (12 \alpha^2 D_2)^{1/3}, \quad \lambda_2 = (12 \alpha^2 D_2)^{1/3} e^{i(2/3)\pi}, \quad \lambda_3 = (12 \alpha^2 D_2)^{1/3} e^{-i(2/3)\pi}. \tag{73}$$

Since the real parts of λ_2 and λ_3 are negative, for $(12\alpha^2 D_2)^{1/3} t \gg 1$, solutions are asymptotic to

$$(74) \qquad \langle p^2(t) \rangle \sim \exp\{(12\alpha^2 D_2)^{1/3} t\},$$

so even small velocity fluctuations have significant effect in presence of sufficiently strong mean gradient of the flow.

A more detailed structure of fluctuations of the tracer gradient can be obtained in the polar coordinates representation. In this case, unlike (46), one obtains the Fokker-Planck equation

$$\frac{\partial}{\partial t} P_t(q,p,\varphi) = \frac{\alpha}{2} \left\{ \frac{\partial}{\partial p} p \sin 2\varphi + \frac{\partial}{\partial \varphi}(1 + \cos 2\varphi) \right\} P_t(q,p,\varphi) +$$

$$(75) \qquad + D_2 \left\{ \frac{\partial}{\partial p} p \frac{\partial}{\partial p} p - 2 \frac{\partial}{\partial p} p + 3 \frac{\partial^2}{\partial \varphi^2} \right\} P_t(q,p,\varphi)$$

for the probability density $P_t(q,p,\varphi)$. Equation (75) is statistically equivalent to a stochastic system

$$\frac{dp(t)}{dt} = -\frac{\alpha}{2} p \sin 2\varphi + 2D_2 p + z_1(t) p,$$

$$(76) \qquad \frac{d\varphi(t)}{dt} = -\frac{\alpha}{2}(1 + \cos 2\varphi) + z_2(t),$$

where $z_1(t)$ and $z_2(t)$ are independent white noise processes with parameters

$$\langle z_1(t) z_1(t') \rangle = 2D_2 \delta(t - t'), \langle z_2(t) z_2(t') \rangle = 6D_2 \delta(t - t').$$

In this case quantity $p(t)$ has an exponential form

$$p(t) = e^{\chi(t)},$$

with process $\chi(t)$ solving a stochastic equation

$$(77) \qquad \frac{d}{dt} \chi(t) = -\frac{\alpha}{2} \sin 2\varphi + 2D_2 + z_1(t).$$

That allows a closed form solution for angle φ. As for random function $\chi(t)$ it is no longer Gaussian. A comparison of the stochastic system (76) with (70) shows that in the case $\alpha / D_2 \gg 1$, the cross-correlations of $p(t)$ with $\varphi(t)$ become crucial.

4.5. Effects of the molecular diffusion. We shall start with an observation made earlier: as time passes the tracer concentration field acquires more chaotic structure and its spatial gradient steepens. In addition, finer and finer scales are created. This tendency will be checked at the level of molecular diffusion, so such dynamical picture would be valid only for a limited time interval. Our goal is to estimate the length of this time interval. To that end we shall utilize exact equation (36) in the absence of the mean flow:

$$(78)\quad \frac{\partial}{\partial t}\langle q^n \boldsymbol{r},t)\rangle = (D_1+\kappa)\frac{\partial^2}{\partial \boldsymbol{r}^2}\langle q^n(\boldsymbol{r},t)\rangle - \kappa n(n-1)\langle q^{n-2}(\boldsymbol{r},t)\boldsymbol{p}^2(\boldsymbol{r},t)\rangle.$$

For small $\kappa \ll D_1$, solution of (78) can be written in the form

$$\langle q^n(\boldsymbol{r},t)\rangle = \exp\left[tD_1\frac{\partial^2}{\partial \boldsymbol{r}^2}\right] q_0^n(\boldsymbol{r}) -$$

$$(79)\quad -\kappa n(n-1)\int_0^t d\tau \exp\left[(t-\tau)D_1\frac{\partial^2}{\partial \boldsymbol{r}^2}\right]\langle q^{n-2}(\boldsymbol{r},\tau)\boldsymbol{p}^2(\boldsymbol{r},\tau)\rangle.$$

To evaluate the last term in (79) exploit equation (38) with the zero mean flow. Then we get

$$\frac{\partial}{\partial t}\langle q^{n-2}(\boldsymbol{r},t)\boldsymbol{p}^2(\boldsymbol{r},t)\rangle = D_1\frac{\partial^2}{\partial \boldsymbol{r}^2}\langle q^{n-2}(\boldsymbol{r},t)\boldsymbol{p}^2(\boldsymbol{r},t)\rangle +$$

$$(80)\quad 2D_2(N+2)(N-1)\langle q^{n-2}(\boldsymbol{r},t)\boldsymbol{p}^2(\boldsymbol{r},t)\rangle,$$

for $\langle q^{n-2}(\boldsymbol{r},t)\boldsymbol{p}^2(\boldsymbol{r},t)\rangle$. Its solution is

$$\langle q^{n-2}(\boldsymbol{r},t)\boldsymbol{p}^2(\boldsymbol{r},t)\rangle = \exp\left[2D_2(N+2)(N-1)t + D_1 t\frac{\partial^2}{\partial \boldsymbol{r}^2}\right] q_0^{n-2}(\boldsymbol{r})\boldsymbol{p}_0^2(\boldsymbol{r}).$$

(81)

Substituting (81) into (79) and carrying out integration with respect to τ, we get

$$(82)\quad \langle q^n(\boldsymbol{r},t)\rangle = \exp\left[tD_1\frac{\partial}{\partial \boldsymbol{r}^2}\right]\{q_0^n(\boldsymbol{r})$$

$$-\kappa\frac{n(n-1)}{2D_2(N+2)(N-1)}[e^{2D_2(N+2)(N-1)t}-1]q_0^{n-2}(\boldsymbol{r})p_0^2(\boldsymbol{r})\Big\}.$$

Formula (82) shows that the molecular diffusion becomes insignificant under two conditions. One of them limits the initial spatial domain of tracer from below:

$$(83)\quad 2(N+2)(N-1)D_2 r_0^2 \gg \kappa n(n-1).$$

Here, $\boldsymbol{r}_0$ is the characteristic size of initial tracer concentration $q_0(\boldsymbol{r})$ (compare with formula (34)). The other condition limits the time range by

$$(84)\quad D_2 t \ll \frac{1}{2(N+2)(N-1)}\ln\frac{D_2 r_0^2}{\kappa n^2}.$$

Notice that the time domain (84) decreases with the growth of power n.

4.6. Fluctuations of passive tracer with mean concentration gradient. In the present subsection, we follow(44) and discuss evolution of statistical characteristics of passive tracer in the presence of nonzero mean concentration gradient and absence of the mean flow. In other words, we study equations (1-2) with the initial conditions

$$q_0(\boldsymbol{r}) = \boldsymbol{G}\boldsymbol{r}, \quad \boldsymbol{p}_0(\boldsymbol{r}) = \boldsymbol{G} = (G_1, G_2).$$

The problem has recently attracted considerable attention, both from the theoretical and experimental side.(45–50,63) The quoted papers used numerical modeling and phenomenological models to analyze behavior of the stationary ($t \to \infty$) probability density of the tracer gradient. They observed among other the appearance of distributions with "slowly decaying tails" of exponential type. Note, that (63) has demonstrated that the stationary probability density of the scalar concentration also has the "slowly decaying tails". Here, we shall turn attention to the concentration itself rather than its gradient, and study both time-dependent and stationary regimes.

Representing concentration $q(\boldsymbol{r},t)$ in the form

$$q(\boldsymbol{r},t) = \boldsymbol{G}\boldsymbol{r} + \tilde{q}(\boldsymbol{r},t), \tag{85}$$

we obtain equation

$$\left(\frac{\partial}{\partial t} + \boldsymbol{F}\frac{\partial}{\partial \boldsymbol{r}}\right)\tilde{q}(\boldsymbol{r},t) = -\boldsymbol{G}\boldsymbol{F}(\boldsymbol{r},t) + \kappa\frac{\partial^2}{\partial \boldsymbol{r}^2}\tilde{q}(\boldsymbol{r},t), \quad \tilde{q}(\boldsymbol{r},0) = 0, \tag{86}$$

for the fluctuating component $\tilde{q}$ of the tracer concentration. Statistical spatial homogeneity of field $\tilde{q}(\boldsymbol{r},t)$ makes analysis of equation (86) simpler than the problem (1). In particular, statistical moments $\langle \tilde{q}^n(\boldsymbol{r},t)\rangle$ are independent of $\boldsymbol{r}$, and satisfy equations

$$\frac{d}{dt}\langle \tilde{q}^n(\boldsymbol{r},t)\rangle = -n\boldsymbol{G}\Big\langle \boldsymbol{F}(\boldsymbol{r},t)\tilde{q}^{n-1}(\boldsymbol{r},t)\Big\rangle - \kappa n(n-1)\Big\langle \tilde{q}^{n-2}(\boldsymbol{r},t)\,\tilde{\boldsymbol{p}}^2(\boldsymbol{r},t)\Big\rangle, \tag{87}$$

where

$$\tilde{\boldsymbol{p}}(\boldsymbol{r},t) = \frac{\partial}{\partial \boldsymbol{r}}\tilde{q}(\boldsymbol{r},t) = \boldsymbol{p}(\boldsymbol{r},t) - \boldsymbol{G}$$

denotes the fluctuating part of the concentration gradient.

We apply once again the Furutsu-Novikov formula, and the relation

$$\frac{\delta\tilde{q}(\boldsymbol{r},t)}{\delta F_j(\boldsymbol{r}',t-0)} = -\delta(\boldsymbol{r}-\boldsymbol{r}')\frac{\partial}{\partial r_j}\tilde{q}(\boldsymbol{r},t) - G_j\delta(\boldsymbol{r}-\boldsymbol{r}') \tag{88}$$

for the variational derivative, to split covariances $\langle \boldsymbol{F}\tilde{q}^{n-1}\rangle$. The resulting equation

$$\frac{d}{dt}\langle \tilde{q}^n(\boldsymbol{r},t)\rangle = n(n-1)D_1\boldsymbol{G}^2\langle \tilde{q}^{n-2}(\boldsymbol{r},t)\rangle - \kappa n(n-1)\Big\langle \tilde{q}^{n-2}(\boldsymbol{r},t)\tilde{\boldsymbol{p}}^2(\boldsymbol{r},t)\Big\rangle \tag{89}$$

yields, in the stationary regime ($t \to \infty$), the following relation for the time-independent statistical characteristics:

$$\langle \tilde{q}^{n-2}\tilde{\boldsymbol{p}}^2 \rangle = \frac{D_1 \boldsymbol{G}^2}{\kappa} \langle \tilde{q}^{n-2} \rangle. \tag{90}$$

In particular, $n = 2$ gives the stationary value of the second moment of the gradient

$$\langle \tilde{\boldsymbol{p}}^2(\boldsymbol{r},t) \rangle = \langle [\tilde{q}(\boldsymbol{r},t)]^2 \rangle = D_1 \boldsymbol{G}^2/\kappa. \tag{90$'$}$$

Consequently, equation (90) can be rewritten in the form

$$\langle \tilde{q}^{n-2}\tilde{\boldsymbol{p}}^2 \rangle = \langle \tilde{\boldsymbol{p}}^2 \rangle \langle \tilde{q}^{n-2} \rangle, \tag{90$''$}$$

i.e., quantities $\tilde{q}(\boldsymbol{r},t)$ and $\tilde{\boldsymbol{p}}^2(\boldsymbol{r},t)$ are statistically uncorrelated in the stationary regime. So, equation (89) can be rewritten in the form

$$\frac{d}{dt}\langle q^n(\boldsymbol{r},t) \rangle = n(n-1) D_1 \boldsymbol{G}^2 \langle f(\boldsymbol{r},t) \tilde{q}^{n-2}(\boldsymbol{r},t) \rangle, \tag{91}$$

with function

$$f(\boldsymbol{r},t) = 1 - \frac{\kappa}{D_1 \boldsymbol{G}^2} \tilde{\boldsymbol{p}}^2(\vec{r},t).$$

Equation (91) implies (note that $\langle \tilde{q}(\boldsymbol{r},t) \rangle = 0$) that the variance of the concentration field $\tilde{q}(\boldsymbol{r},t)$

$$\langle \tilde{q}^2(\boldsymbol{r},t) \rangle = 2 D_1 \boldsymbol{G}^2 \int_0^t d\tau\, f(\boldsymbol{r},\tau). \tag{92}$$

In absence of the molecular diffusion, function $f(\tau) = 1$, and

$$\langle \tilde{q}^2(\boldsymbol{r},t) \rangle = 2 D_1 \boldsymbol{G}^2 t. \tag{93}$$

In this case, the one-point distribution of the field $\tilde{q}(\boldsymbol{r},t)$ is Gaussian and $\tilde{q}$ is uncorrelated with its spatial derivatives. In general, solution (93) is valid only for sufficiently small time.

Notice that the covariance function $\Gamma(\boldsymbol{r},t) = \langle \tilde{q}(\boldsymbol{r}_1,t)\tilde{q}(\boldsymbol{r}_2,t) \rangle$, $\boldsymbol{r} = \boldsymbol{r}_1 - \boldsymbol{r}_2$, of tracer fluctuations satisfies equation

$$\frac{\partial}{\partial t}\Gamma(\boldsymbol{r},t) = 2\Big[B_{ij}^{eff}(0) - B_{ij}^{eff}(\boldsymbol{r}) + \kappa\delta_{ij}\Big] \frac{\partial^2}{\partial r_i \partial r_j}\Gamma(\boldsymbol{r},t) + 2G_iG_jB_{ij}^{eff}(\boldsymbol{r}).$$

Hence, its stationary (limit) values $\Gamma(\boldsymbol{r}) = \lim_{t\to\infty}\Gamma(\boldsymbol{r},t)$ are described by equation

$$G_iG_jB_{ij}^{eff}(\boldsymbol{r}) = -\Big[B_{ij}^{eff}(0) - B_{ij}^{eff}(\boldsymbol{r}) + \kappa\delta_{ij}\Big] \frac{\partial^2}{\partial r_i \partial r_j}\Gamma(\boldsymbol{r}). \tag{90$'''$}$$

For $\boldsymbol{r} = 0$, we derive once again (90'). Differentiating (90''') twice with respect to $\boldsymbol{r}$ at $\boldsymbol{r} = 0$, we get

$$\kappa^2 \Big\langle \Big(\frac{\partial^2}{\partial \boldsymbol{r}^2} \tilde{q} \Big)^2 \Big\rangle = (N+2)(N-1)) D_2 (D_1 + \kappa) \boldsymbol{G}^2.$$

In order to pass to the limit $\langle \tilde{q}^2 \rangle |_{t \to \infty}$ in (92) we need information about the time evolution of the second moment $\langle \tilde{\boldsymbol{p}}^2(\boldsymbol{r}, t) \rangle$. That information follows from (2) without the diffusion term. Although such approximation for the equation (2) is good only for the finite time interval[(64)], the knowledge of the exact value (90') will give us an opportunity to estimate the stationary value of the variance of the concentration of passive scalar in presence of molecular diffusion.

We proceed by introducing a generalized (indicator) function

$$\Phi_{t,\boldsymbol{r}}(\boldsymbol{p}) = \delta(\boldsymbol{p}(\boldsymbol{r}, t) - \boldsymbol{p}),$$

that solves the Liouville equation

$$\left(\frac{\partial}{\partial t} + \boldsymbol{F}(\boldsymbol{r}, t) \frac{\partial}{\partial \boldsymbol{r}} \right) \Phi_{t,\boldsymbol{r}}(\boldsymbol{p}) = \frac{\partial}{\partial p_i} \frac{\partial F_k(\boldsymbol{r}, t)}{\partial r_i} p_k \Phi_{t,\boldsymbol{r}}(\boldsymbol{p}) \tag{94}$$

with the initial condition

$$\Phi_{0,\boldsymbol{r}}(\boldsymbol{p}) = \delta(\boldsymbol{G} - \boldsymbol{p}). \tag{95}$$

Then we take $\boldsymbol{F}$-ensemble average of (94), apply the Furutsu-Novikov formula along with the delta-correlated approximation (23), use the variational derivative formula

$$\frac{\delta}{\delta F_j(\boldsymbol{r}', t-0)} \Phi_{t,\boldsymbol{r}}(\boldsymbol{p}) = \{ -\delta(\boldsymbol{r} - \boldsymbol{r}') \frac{\partial}{\partial r_j} + \frac{\partial}{\partial p_i} \frac{\partial \delta(\boldsymbol{r} - \boldsymbol{r}')}{\partial r_i} p_j \} \Phi_{t,\boldsymbol{r}}(\boldsymbol{p}),$$

a consequence of (94), and exploit the statistical homogeneity of the field $\boldsymbol{p}(\boldsymbol{r}, t)$. All that results in Fokker-Planck equation

$$\frac{\partial}{\partial t} P_t(\boldsymbol{p}) = D_2 \Big((N+1) \frac{\partial^2}{\partial \boldsymbol{p}^2} \boldsymbol{p}^2 - 2 \frac{\partial}{\partial \boldsymbol{p}} \boldsymbol{p} - 2 (\frac{\partial}{\partial \boldsymbol{p}} \boldsymbol{p})^2 \Big) P_t(\boldsymbol{p}),$$

$$P_0(\boldsymbol{p}) = \delta(\boldsymbol{G} - \boldsymbol{p}), \tag{96}$$

for the probability density $P_t(\boldsymbol{p}) = \langle \Phi_{t,\boldsymbol{r}}(\boldsymbol{p}) \rangle$ of the passive scalar concentration's gradient.

The lognormal probability distribution of the concentration gradient has, apparently, been first discovered in[(64)] (although equations like (96) do not appear there), and its properties can be found, for example, in[(30)]. As we indicated earlier, the validity of the lognormal approximation of the probability density (in absence of the molecular diffusion) and, in particular, of the exponential growth of (97) in time, extends only for a finite time interval. However, the exact knowledge of (90') allows us to approximate by the exponential law (97) the time behavior of the quantity $\langle\tilde{\boldsymbol{p}}^2(\boldsymbol{r},t)\rangle$ in presence of the molecular diffusion, up to time

$$T_0 \sim \frac{1}{2D_2(N+2)(N-1)} \ln \frac{D_1+\kappa}{\kappa}, \tag{97}$$

when it attains its stationary value $\langle\tilde{\boldsymbol{p}}^2\rangle$ described by equation (90'). In other words, for $t < T_0$, we use equation (2) and for $t > T_0$, equation (90'). Hence, $\int_0^\infty dt f(t) \sim T_0$ and, by (92), we get the stationary variance of field $\tilde{q}(\boldsymbol{r},t)$

$$\lim_{t\to\infty} \langle\tilde{q}^2(\boldsymbol{r},t)\rangle \sim \frac{1}{(N+2)(N-1)} \frac{D_1}{D_2} \boldsymbol{G}^2 \ln \frac{D_1+\kappa}{\kappa}. \tag{98}$$

Taking into account that

$$D_1 \sim \sigma_u^2 t_0 \qquad \text{and} \qquad D_1/D_2 \sim l_0^2,$$

where σ_u^2 is the variance of velocity fluctuations, and t_0 and l_0 are, respectively, its temporal and spatial radii of correlations, it follows form (98-99) that time T_0, in view of its logarithmic dependence on parameter κ, can not be too large. In addition,

$$\langle\tilde{q}^2\rangle \sim \boldsymbol{G}^2 l_0^2 \ln \frac{\sigma_u^2 t_0}{\kappa}, \quad \kappa \ll \sigma_u^2 t_0.$$

Earlier, we have observed that the concentration of passive scalar is uncorrelated with the square of its gradient, both in the stationary regime and in the initial time interval (in which the molecular diffusion effects are not important). If this were true in the whole time interval, we would have obtained from equation (91) that the one-point distribution of the field $\tilde{q}$ was Gaussian. However, as follows from[(63)] (see also[(65)]), this is not the case and the existence od such correlation is the most important factor in forming of the stationary regime for fluctuations of $\tilde{q}$.

5. Other approximations

In the previous section, we provided a detailed statistical analysis of passive tracer transport in the delta-correlated approximation for the velocity fluctuations $\boldsymbol{F}$. Our next goal is to relax the delta-correlation assumption. Different approximation techniques could be utilized to that end. We shall demonstrate a few of them, including successive iterations, telegrapher process, and diffusion approximations. For the sake of comparison, we shall start with the delta-correlated approximation discussed at length in Section 4.

5.1. Delta-correlated field approximation revisited. Let $\boldsymbol{F}(\boldsymbol{r},t)$ be a zero-mean Gaussian field with correlation matrix $B_{ij}(\boldsymbol{r},t;\boldsymbol{r}',t)$. The delta-correlated approximation of field $\boldsymbol{F}(\boldsymbol{r},t)$ corresponds to approximation

$$B_{ij}(\boldsymbol{r},t;\boldsymbol{r}',t') \approx 2B_{ij}^{\text{eff}}(\boldsymbol{r},\boldsymbol{r}',t)\delta(t-t'),$$

where

$$2B_{ij}^{\text{eff}}(\boldsymbol{r},\boldsymbol{r}',t) = \int_{-\infty}^{\infty} dt'\, B_{ij}(\boldsymbol{r},t;\boldsymbol{r}',t')$$

and the quantity B^{eff} is proportional to the time correlation radius t_0. In this case, the problem was shown in the previous section to allow a closed form representation. The condition of applicability of such approximation is that the temporal correlation radius t_0 of fluctuations $\boldsymbol{F}$ is much smaller than any other temporal scale arising in the problem.(19)

For problems considered in Subsection 4.1 and 4.2, in the absence of mean flow, the temporal scales connected with the statistics of the fluctuating component of the velocity field include

$$r_0^2/D_1, \qquad l_0^2/D_1, \qquad \text{and} \qquad 1/D_2.$$

Here, r_0 is the characteristic dimension of the initial tracer concentration $q_0(\boldsymbol{r})$, and l_0 is the spatial correlation radius of the velocity field. Other temporal scales, related to the coefficient of molecular diffusion, are

$$r_0^2/\kappa, \qquad \text{and} \qquad l_0^2/\kappa.$$

In presence of the mean flow, new temporal scales appear. Thus, for the mean flow considered in Subsection 4.4, such additional scale is quantity $1/\alpha$. The problem of fluctuating passive tracer with mean concentration gradient, considered in Subsection 4.6, will be revisited in Section 5.4.

5.2. The method of successive iterations. Let again $\boldsymbol{F}(\boldsymbol{r},t)$ be a zero-mean Gaussian field with correlation matrix $B_{ij}(\boldsymbol{r},t;\boldsymbol{r}',t')$. It appears as coefficients in the exact equation (21) that also contains variational derivative $\delta q/\delta F_j$. The latter, in turn, is described by the stochastic equation (22). Averaging equation (22) over the ensemble we get another exact equation that contains the second variation $\delta^2 q/\delta F_i \delta F_j$, and so on. The method of successive iterations asks for the closure of the resulting system at the n-th level, that is for $\delta^n/\delta F\ldots\delta F$. The usual way to proceed is to make a suitable δ-correlation assumption.(51,17,18) So, the method consists of successive improvements of the functional dependence of q on $\boldsymbol{F}$. In some cases, one can argue on physical grounds that the n-th order corrections to $q[\boldsymbol{F}]$ give negligible contribution to the solution, hence, could be dropped.

Let us remark that in the case of zero mean flow $\boldsymbol{v}$ ($\boldsymbol{v}=0$), and rapidly decaying space-time correlation functions $B_{ij}(\boldsymbol{r},t)$, the delta-correlated approximation predicts the correct asymptotic solutions as $t\to\infty$. This result was also confirmed by the numerical simulation.[52] If, however, correlation functions $B_{ij}(\boldsymbol{r},t)$ exhibit a more complex structure as in turbulent velocity fields[1]), or in the presence of nonzero mean flow, the delta-correlated approximation is clearly inadequate.

For practical purposes, it is often sufficient to close the mean field equation (22) using the delta-correlation hypothesis at the second step. Although such closures are widely used in some areas, e.g., plasma, ionosphere physics, and, particularly, magnetohydrodynamics,[53] there is no satisfactory mathematical justification of such closure and the validity of the resulting approximation of q.

5.3. Telegraph process approximation. In general, a Gaussian Markov process $X(t)$, i.e. a Gaussian process with correlation function $a^2e^{-2\nu(t-t')}$, can be obtained as a limit[17,18]

$$X(t) = \lim_{n\to\infty} \sum_1^n Z_k(t),$$

where $Z_k(t)$ are so-called telegraph processes (to be defined below) with cross-correlations

$$\langle Z_i(t)Z_j(t')\rangle = \delta_{ij}\frac{a^2}{n}e^{-2\nu|t-t'|}.$$

The telegraph processes considered as building blocks of general Gaussian Markov processes justify the approximation considered in this subsection.

However, before we embark on the exposition of this approximation, let us consider an exactly solvable model where the velocity fluctuations $\boldsymbol{F}(\boldsymbol{r},t)$ are assumed to be of the form[19]

$$\boldsymbol{F}(\boldsymbol{r},t) = \boldsymbol{g}(\boldsymbol{r},t)Z(t).$$

Here, $\boldsymbol{g}(\boldsymbol{r},t)$ represents a deterministic factor, while $Z(t)$ is the telegraph process defined as follows:

$$Z(t) = A(-1)^{N(0,t)},$$

where A is a random variable taking values $\pm a$ with probability 1/2, and $N(s,t), s < t$, is the increment of the Poisson process (independent of A). In other words, $N(s,t)$ are integer-valued random variables satisfying the following conditions:

- $N(t_1,t_3) = N(t_1,t_2) + N(t_2,t_3)$, for any $t_1 < t_2 < t_3$,
- $N(t_1,t_2)$ and $N(t_2,t_3)$ are statistically independent for $t_1 < t_2 < t_3$,
- probability distribution of $N(t_1,t_2)$ is given by the formula

$$P\Big(N(t_1,t_2) = m\Big) = \frac{(\nu|t_1-t_2|)^m}{m!}e^{-\nu|t_1-t_2|}.$$

In particular, the distribution is completely determined by the mean number

$$\langle N(t_1, t_2)\rangle = \nu|t_1 - t_2|,$$

of jumps over the time interval $[t_1, t_2]$.

Process $Z(t)$ is stationary with correlation function

$$\langle Z(t_1)Z(t_2)\rangle = a^2 e^{-2\nu|t_1-t_2|},$$

and the correlation radius $l_0 = 1/2\nu$.

As before, we approach the passive tracer evolution problem by averaging equation (1) over the ensemble $\{Z(t)\}$ which this time gives

$$\left(\frac{\partial}{\partial t} + \boldsymbol{v}(\boldsymbol{r},t)\cdot\frac{\partial}{\partial \boldsymbol{r}}\right)\langle q(\boldsymbol{r},t)\rangle_Z + a^2\boldsymbol{g}(\boldsymbol{r},t)\cdot\frac{\partial}{\partial \boldsymbol{r}}\Psi(\boldsymbol{r},t) = \kappa\frac{\partial^2}{\partial \boldsymbol{r}^2}\langle q(\boldsymbol{r},t)\rangle_Z, \tag{99}$$

where

$$\Psi(\boldsymbol{r},t) = \langle Z(t)q(\boldsymbol{r},t)\rangle_Z$$

is the new cross-correlation function, which is shown to satisfy equation

$$\left(\frac{\partial}{\partial t} + 2\nu\right)\Psi(\boldsymbol{r},t) + \boldsymbol{v}(\boldsymbol{r},t)\cdot\frac{\partial}{\partial \boldsymbol{r}}\Psi(\boldsymbol{r},t) + a^2\boldsymbol{g}(\boldsymbol{r},t)\cdot\frac{\partial}{\partial \boldsymbol{r}}\langle q(\boldsymbol{r},t)\rangle_Z = \kappa\frac{\partial^2}{\partial \boldsymbol{r}^2}\Psi(\boldsymbol{r},t). \tag{100}$$

The latter plays the role of the Furutsu-Novikov formula in the delta-correlated case.

The derivation of (100) is based on the following *differentiation rule*

$$\left(\frac{d}{dt} + 2\nu\right)\langle Z(t)R_t[Z(\tau)]\rangle = \left\langle Z(t)\frac{d}{dt}R_t[Z(\tau)]\right\rangle,$$

valid for the correlation function of the telegraph process $Z(t)$ and an arbitrary functional $R_t[Z(\tau)]$, with $\tau \le t$.[(17,18,54)] Equations (99-100) form a closed system for an unknown mean field $\langle q\rangle$ and correlation Ψ. *Remark 1.* As $\nu, a^2 \to \infty$ so that $a^2/2\nu = const$, the (apropriate rescaled interpolation of the) telegraph process converges weakly to a Gaussian δ-correlated process. Hence, in this limit,

$$\Psi(\boldsymbol{r},t) = -\frac{a^2}{2\nu}\boldsymbol{g}(\boldsymbol{r},t)\cdot\frac{\partial}{\partial \boldsymbol{r}}\langle q(\boldsymbol{r},t)\rangle_Z,$$

we arrive at a single equation for the mean field $\langle q(\boldsymbol{r},t)\rangle_Z$, which corresponds to the Fokker-Planck equation.

The effectiveness of the above procedure has been demonstrated above for velocity field fluctuations of the form $\boldsymbol{F}(\boldsymbol{r},t)=\boldsymbol{g}(\boldsymbol{r},t)Z(t)$, where $Z(t)$ was the telegraph process but the underlying method is more general, and we shall now review equation (1) with the same assumption on $Z(t)$ but from a somewhat different perspective.[17,18] Once again, after averaging equation (1) we get

$$(101)\qquad \left(\frac{\partial}{\partial t}+\boldsymbol{v}(\boldsymbol{r},t)\cdot\frac{\partial}{\partial \boldsymbol{r}}\right)\langle q(\boldsymbol{r},t)\rangle + a^2\boldsymbol{g}(\boldsymbol{r},t)\cdot\frac{\partial}{\partial \boldsymbol{r}}\langle Z(t)q(\boldsymbol{r},t)\rangle = \kappa\frac{\partial^2}{\partial \boldsymbol{r}^2}\langle q(\boldsymbol{r},t)\rangle.$$

To decouple the cross-correlation term $\langle Z(t)q(\boldsymbol{r},t)\rangle$, we shall use the identity (see Appendix)

$$(102)\qquad \langle Z(t)R_t[Z(\tau)]\rangle = a^2\int_0^t dt'\, e^{-2\nu(t-t')}\left\langle \frac{\delta}{\delta Z(t')}\tilde{R}_t[t',Z(\tau)]\right\rangle,$$

where

$$\tilde{R}_t[t',Z(\tau)] = R_t[Z(\tau)\Theta(t'-\tau)],$$

and where Θ denotes the Heavyside step-function. The decoupling formula (102) closely resembles the Furutsu-Novikov formula (20) for Gaussian fields with exponential correlation functions. The only difference is that the right-hand side of (102) contains a truncated functional $\tilde{R}_t$ on interval $[t',t]$, rather than $R_t[Z(\tau)]$ itself. More precisely,

$$\tilde{R}_t[t',Z(\tau)] = \begin{cases} R_t[Z(\tau)], & \text{if } \tau<t'; \\ R_t[0], & \text{if } \tau>t'. \end{cases}$$

Thus, we get from (102) that

$$(103)\qquad \left(\frac{\partial}{\partial t}+\boldsymbol{v}(r,t)\frac{\partial}{\partial \boldsymbol{r}}\right)\langle q(\boldsymbol{r},t)\rangle + a^2\boldsymbol{g}(\boldsymbol{r},t)\frac{\partial}{\partial \mathbf{r}}\int_0^t dt'\, e^{-2\nu(t-t')}\left\langle \frac{\delta}{\delta Z(t')}\tilde{q}[t',Z(\tau)]\right\rangle = \kappa\frac{\partial^2}{\partial \boldsymbol{r}^2}\langle q(\boldsymbol{r},t)\rangle.$$

Functional $\tilde{q}[t',Z(\tau)]$ is described by equation (1) with the velocity fluctuation field

$$\boldsymbol{F}(\boldsymbol{r},t)=\boldsymbol{g}(\boldsymbol{r},t)Z(t)\Theta(t'-t).$$

Hence, for $t>t'$, we get equation

$$(104)\qquad \left(\frac{\partial}{\partial t}+\boldsymbol{v}(\boldsymbol{r},t)\cdot\frac{\partial}{\partial \boldsymbol{r}}\right)\tilde{q}(\boldsymbol{r},t)=\kappa\frac{\partial^2}{\partial \boldsymbol{r}^2}\tilde{q}(\boldsymbol{r},t),\quad \tilde{q}(\boldsymbol{r},t)\Big|_{t=t'}=q(\boldsymbol{r},t').$$

Taking the variational derivative of equation (104) with respect to $Z(t')$ we get, again for $t > t'$, that

(105) $$\left(\frac{\partial}{\partial t} + \boldsymbol{v}(\boldsymbol{r},t)\cdot\frac{\partial}{\partial \boldsymbol{r}}\right)\frac{\delta \tilde{q}}{\delta Z(t')} = \kappa \frac{\partial^2}{\partial \boldsymbol{r}^2}\frac{\delta \tilde{q}}{\delta Z(t')},$$

with the initial condition

(106) $$\left.\frac{\delta \tilde{q}}{\delta Z(t')}\right|_{t=t'} = -\boldsymbol{g}(\boldsymbol{r},t')\cdot\frac{\partial}{\partial \boldsymbol{r}}q(\boldsymbol{r},t'),$$

i.e., equation of the type of equation (22) but without the fluctuating velocity component. If we introduce function

$$\Psi(\boldsymbol{r},t) = a^2\int_0^t dt'\, e^{-2\nu(t-t')}\left\langle\frac{\delta}{\delta Z(t')}\tilde{q}[t', Z(\tau)]\right\rangle,$$

then, in view of (104), Ψ is easily seen to be a solution of equation (100). Thus, if the velocity field has fluctuations of the telegraph process-type, then we are able to obtain a closed system of equations (99-100).

Remark 2. If the velocity fluctuations field is of the form $\boldsymbol{F}(\boldsymbol{r},t) = \boldsymbol{g}(\boldsymbol{r},t)Z(t)$ where $Z(t)$ is a Gaussian Markov process (rather than the telegraph process with the same parameters a and ν), then the stochastic equation for the variational derivative $\delta q/\delta Z$ takes the form

(107′)
$$\left(\frac{\partial}{\partial t} + \boldsymbol{v}(\boldsymbol{r},t)\cdot\frac{\partial}{\partial}\right)\frac{\delta q}{\delta Z(t')} + \boldsymbol{F}(\boldsymbol{r},t)\cdot\frac{\partial}{\partial \boldsymbol{r}}\frac{\delta q}{\delta Z(t')} = \kappa\frac{\partial^2}{\partial \boldsymbol{r}^2}\frac{\delta q}{\delta Z(t')}, \qquad t' < t.$$

It resembles equation (105), but it has an additional random advection term $\boldsymbol{F}\cdot\partial/\partial\boldsymbol{r}$ on the left-hand side. If we drop it, we'll get back the stochastic equation for the telegraph process. So, in that sense, the telegraph process could be thought of as an approximation of the Gaussian Markov process (see also comment at the beginning of this subsection).

Now, assume that the fluctuations field $\boldsymbol{F}(\boldsymbol{r},t)$ is Gaussian, with correlation function B_{ij}, and that it is characterized, as a function of difference $t - t'$, by the correlation radius t_0. Taking into account the range of integration of t' in (103), we find that the principal range of integration of $\delta q/\delta F_j$ is of order t_0. If we make a, physically justified, assumption that on such scales the velocity fluctuation component $\boldsymbol{F}$ does not enter into $\delta q/\delta F_j$ (i.e. the latter remains functionally independent of $\boldsymbol{F}$ in the sense that equation for $\delta q/\delta F_j$ is deterministic), then we can drop the fluctuating term of (22) and arrive at a closed-form description called the *telegraph process approximation.* In this case we get a system of coupled equations for $\langle q\rangle$ and $\delta q/\delta F_j$:

(107) $$\left(\frac{\partial}{\partial t} + \boldsymbol{v}(\boldsymbol{r},t)\cdot\frac{\partial}{\partial \boldsymbol{r}}\right)\langle q(\boldsymbol{r},t)\rangle$$

$$+\int_0^t dt' \int d\boldsymbol{r}' \sum_j B_{ij}(\boldsymbol{r},t;\boldsymbol{r}',t') \frac{\partial}{\partial r_i} \left\langle \frac{\delta q(\boldsymbol{r},t)}{\delta F_j(\boldsymbol{r}',t')} \right\rangle = \kappa \frac{\partial^2}{\partial \boldsymbol{r}^2} \langle q(\boldsymbol{r},t)\rangle,$$

$$\left(\frac{\partial}{\partial t} + \boldsymbol{v}(\boldsymbol{r},t)\cdot\frac{\partial}{\partial \boldsymbol{r}}\right) \left\langle \frac{\delta q(\boldsymbol{r},t)}{\delta F_j(\boldsymbol{r}',t')} \right\rangle = \kappa \frac{\partial^2}{\partial \boldsymbol{r}^2} \left\langle \frac{\delta q(\boldsymbol{r},t)}{\delta F_j(\boldsymbol{r}',t')} \right\rangle,$$

with initial condition

$$\left\langle \frac{\delta q(\boldsymbol{r},t)}{\delta F_j(\boldsymbol{r}',t')} \right\rangle \Bigg|_{t\to t'+0} = -\delta(\boldsymbol{r}-\boldsymbol{r}') \frac{\partial}{\partial r_j} \langle q(\boldsymbol{r},t')\rangle.$$

We have already mentioned that determining the variational derivative is essentially equivalent to finding the Green function of the original problem (1) in absence of the velocity fluctuations. Let us also remark that system (107) is equivalent to the one obtained in[55], but our approach is very different. Also, a well known equation that appears in[14] can be derived from system (107) in the case where both, the molecular diffusion coefficient and the mean velocity field vanish. Of course, system (107) is itself too complicated for the direct analysis, so one would like to simplify it further.

5.4. The diffusion approximation. Here, in addition to the previous assumptions of the telegraph process approximation made at the end of previous subsection, we assume that the velocity fluctuations $\boldsymbol{F}$ do not affect the dynamics of q on scales of the order of t_0. Hence, the dynamics of passive tracer at these temporal scales can be approximately described by equation

$$\left(\frac{\partial}{\partial t} + \boldsymbol{v}(\boldsymbol{r},t)\cdot\frac{\partial}{\partial \boldsymbol{r}}\right) q(\boldsymbol{r},t) = \kappa \frac{\partial^2}{\partial \boldsymbol{r}^2} q(\boldsymbol{r},t),$$

$$q(\boldsymbol{r},t)|_{t=t'} = q(\boldsymbol{r},t').$$

The latter is deterministic and randomness enters here only through the initial condition. The *diffusion approximation* will give an additional relation between quantities $q(\boldsymbol{r},t)$ and $q(\mathbf{r},t')$, and will allow us to get a closed form first order equation in t for the mean field $\langle q(\boldsymbol{r},t)\rangle$. In this approximation, field $q(\boldsymbol{r},t)$is Markovian at large time scales $t \gg t_0$.[56]

Let us also remark that the diffusion approximation becomes exact in the case of linear equation with an additive noise. Here, the variational derivative coincides with the Green's function of the deterministic equation.

As an example we will consider a two-dimensional diffusion of a particle by the Gaussian incompressible velocity field with the parallel mean flow. The Lagrangian coordinates $\boldsymbol{r} = (x,y)$ of such particle obey equation

$$\frac{d\boldsymbol{r}(t)}{dt} = v(y)\boldsymbol{e} + \boldsymbol{F}(\boldsymbol{r},t) \quad \boldsymbol{r}(0) = \boldsymbol{\xi}, \tag{108}$$

where $\boldsymbol{e} = (1,0)$ is the unit vector along the flow. The fluctuating field $\boldsymbol{F}$ is characterized by its covariance matrix

$$\Big\langle F_i(\boldsymbol{r},t)F_j(\boldsymbol{r}',t')\Big\rangle = B_{ij}(\boldsymbol{r}-\boldsymbol{r}';t-t').$$

Introducing the usual generalized (indicator) function

$$\Phi_t(\boldsymbol{r}) = \delta(\boldsymbol{r}(t) - \boldsymbol{r}),$$

we obtain for it the Liouville equation

$$\left(\frac{\partial}{\partial t} + v(y)\frac{\partial}{\partial x}\right)\Phi_t(\boldsymbol{r}) = -\boldsymbol{F}(\boldsymbol{r},t)\frac{\partial}{\partial \boldsymbol{r}}\Phi_t(\boldsymbol{r}). \tag{109}$$

Those were shown on several occasions (Section 2) to describe a dynamic evolution for the tracer transport in the Eulerian coordinates when the molecular diffusion is absent. The special case of equation (109) with a linear shear was considered in the delta-correlated approximation in Subsection 4.4.

Averaging equation (109) over the $\boldsymbol{F}$-ensemble, we obtain equation for probability density (or mean tracer concentration) $P_t(\boldsymbol{r}) = \langle\Phi_t(\boldsymbol{r})\rangle$ of the particle coordinate of the form

$$\left(\frac{\partial}{\partial t} + v(y)\frac{\partial}{\partial x}\right)P_t(\boldsymbol{r}) \tag{110}$$

$$= -\int_0^t dt' \int d\boldsymbol{r}' B_{ij}(\boldsymbol{r}-\boldsymbol{r}', t-t')\frac{\partial}{\partial r_i}\left\langle \frac{\delta}{\delta F_j(\boldsymbol{r}',t')}\Phi_t(\boldsymbol{r})\right\rangle.$$

In the diffusion approximation the variational derivative $\delta\Phi_t/\delta F_j$ satisfies equation

$$\left(\frac{\partial}{\partial t} + v(y)\frac{\partial}{\partial x}\right)\frac{\delta}{\delta F_j(\boldsymbol{r}',t')}\Phi_t(\boldsymbol{r}) = 0. \tag{111}$$

Given the initial condition

$$\frac{\delta}{\delta F_j(\boldsymbol{r}',t')}\Phi_t(\boldsymbol{r})\Big|_{t=t'} = -\delta(\boldsymbol{r}-\boldsymbol{r}')\frac{\partial}{\partial r_j}\Phi_{t'}(\boldsymbol{r}),$$

solution of equation (111) becomes

$$\frac{\delta}{\delta F_j(\boldsymbol{r}',t')}\Phi_t(\boldsymbol{r}) = -\exp\left[-(t-t')v(y)\boldsymbol{e}\frac{\partial}{\partial \boldsymbol{r}}\right]\delta(\boldsymbol{r}-\boldsymbol{r}')\frac{\partial}{\partial r_j}\Phi_{t'}(\boldsymbol{r}). \tag{112}$$

For times scales of the order of t_0, function $\Phi_t(\boldsymbol{r})$ is described by (109) but without the fluctuation term on the right-hand side, that is by the equation

$$\left(\frac{\partial}{\partial t} + v(y)\frac{\partial}{\partial x}\right)\Phi_t(\boldsymbol{r}) = 0, \qquad \Phi_t(\boldsymbol{r})\Big|_{t=t'} = \Phi_{t'}(\boldsymbol{r}). \tag{113}$$

Its solution has the form

$$\Phi_t(\boldsymbol{r}) = \exp\left[-(t-t')v(y)\boldsymbol{e}\frac{\partial}{\partial \boldsymbol{r}}\right]\Phi_{t'}(\boldsymbol{r}). \tag{114}$$

Consequently, quantity $\Phi_{t'}(\boldsymbol{r})$ on the right-hand side of (112) can be expressed via $\Phi_t(\boldsymbol{r})$ as follows:

$$\Phi_{t'}(\boldsymbol{r}) = \exp\left[(t-t')v(y)\boldsymbol{e}\frac{\partial}{\partial \boldsymbol{r}}\right]\Phi_t(\boldsymbol{r}). \tag{115'}$$

Substituting (115') into (112) we obtain the following expression for the mean value of the variational derivative

(115)
$$\left\langle \frac{\delta}{\delta F_j(\boldsymbol{r}',t')}\Phi_t(\boldsymbol{r})\right\rangle$$
$$= -\exp\left[-(t-t')v(y)\boldsymbol{e}\frac{\partial}{\partial \boldsymbol{r}}\right]\delta(\boldsymbol{r}-\boldsymbol{r}')\frac{\partial}{\partial r_j}\exp\left[(t-t')v(y)\boldsymbol{e}\frac{\partial}{\partial \boldsymbol{r}}\right]P_t(\boldsymbol{r}).$$

The latter is then substituted in (110) and the integration is carried out with respect to $\boldsymbol{r}'$. The result is a closed form equation for the probability density $P_t(\boldsymbol{r})$

(116)
$$\left(\frac{\partial}{\partial t}+v(y)\frac{\partial}{\partial x}\right)P_t(\boldsymbol{r})$$
$$= \frac{\partial}{\partial r_i}\int_0^t d\tau\, B_{ij}(\tau v(y)\boldsymbol{e},\tau)\exp\left[-\tau v(y)\boldsymbol{e}\frac{\partial}{\partial \boldsymbol{r}}\right]\frac{\partial}{\partial r_j}\exp\left[\tau v(y)\boldsymbol{e}\frac{\partial}{\partial \boldsymbol{r}}\right]P_t(\boldsymbol{r}),$$

which can be rewritten in the final form

(117)
$$\left(\frac{\partial}{\partial t}+v(y)\frac{\partial}{\partial x}\right)P_t(\boldsymbol{r}) = \frac{\partial}{\partial r_i}\left\{D_{ij}^{(1)}(\boldsymbol{r},t)\frac{\partial}{\partial r_j}+D_{i2}^{(2)}(\boldsymbol{r},t)\frac{\partial}{\partial x}\right\}P_t(\boldsymbol{r}),$$

with the diffusion coefficients $D^{(k)}$ determined by

(118)
$$D_{ij}^{(1)}(\boldsymbol{r},t) = \int_0^t d\tau\, B_{ij}(\tau v(y)\boldsymbol{e},\tau),$$
$$D_{i2}^{(2)}(\boldsymbol{r},t) = \int_0^t d\tau\,\tau B_{i2}(\tau v(y)\boldsymbol{e},\tau)\frac{dv(y)}{dy}.$$

Equation (117) is valid uniformly in time t, including small times $t \le t_0$, and could be formally called a *generalized Fokker-Planck equation.* However, the multi-time probability does not factor, which means the stochastic process $\boldsymbol{r}(t)$ does not have the Markov property.[56]

For large times $t \gg t_0$ the limits of integration in (118) could be replaced by infinite limits. That will produce a Markov approximation of the stochastic process $\boldsymbol{r}(t)$.

As $t_0 \to 0$ the coefficients $D_{i2}^{(2)} \to 0$, while the coefficients

$$D_{ij}^{(1)}(\boldsymbol{r},t) = \int_0^t d\tau\, B_{ij}(0,\tau) = B_{ij}^{\text{eff}}(0),$$

which corresponds to the above delta-correlated case. In the absence of the mean flow ($v(y) = 0$) the diffusion approximation essentially coincides with the delta-correlated approximation.

As a second example we consider the diffusion approximation for fluctuations of passive tracer with mean concentration gradient and provide conditions of applicability of the delta-correlated approximation of subsection 4.6. In this case we have an exact equation

$$(119) \quad \frac{d}{dt}\langle \tilde{q}^n(\boldsymbol{r},t)\rangle = -n(n-1)G_i \int d\boldsymbol{r}' \int_0^t dt'\, B_{ij}(\boldsymbol{r}-\boldsymbol{r}',t-t') \times \left\langle \tilde{q}^{n-2}(\boldsymbol{r},t)\frac{\delta\tilde{q}(\boldsymbol{r},t)}{\delta F_j(\boldsymbol{r},t')}\right\rangle - \kappa n(n-1)\langle \tilde{q}^{n-2}(\boldsymbol{r},t)\tilde{\boldsymbol{p}}^2(\boldsymbol{r},t)\rangle.$$

Since the principal contribution in the integral with respect to time on the right-hand side of (119) comes from the domain of integration where $t - t' \leq t_0$, one can neglect the fluctuation terms in the corresponding equation for the variational derivative. As a result, we get that

$$(120) \quad \frac{\delta\tilde{q}(\boldsymbol{r},t)}{\delta F_j(\boldsymbol{r}',t')} = -e^{\kappa(t-t')\Delta}\Big(\delta(\boldsymbol{r}-\boldsymbol{r}')\frac{\partial}{\partial r_j}\tilde{q}(\boldsymbol{r},t') - G_j\delta(\boldsymbol{r}-\boldsymbol{r}')\Big),$$

where Δ is the Laplacian. In the same time interval,

$$\tilde{q}(\boldsymbol{r},t') = e^{-\kappa(t-t')\Delta}\tilde{q}(\boldsymbol{r},t).$$

In view of the spatial homogeneity, the first term on the right-hand side of (121) makes no contribution, and we get equation

$$(121) \quad \frac{d}{dt}\langle \tilde{q}^n(t)\rangle = -\kappa n(n-1)\langle \tilde{q}^{n-2}(t)\tilde{\boldsymbol{p}}^2(t)\rangle + n(n-1)\boldsymbol{G}^2 D_1(t,\kappa)\langle \tilde{q}^{n-2}(t)\rangle,$$

where

$$(122) \quad D_1(t,\kappa) = \frac{N-1}{N}\int_0^t d\tau \int d\boldsymbol{k}\, E(k,\tau)e^{-\kappa\tau \boldsymbol{k}^2}.$$

Hence, the condition of applicability of the diffusion approximation in this case is

$$(123) \quad D_1(t,\kappa)\boldsymbol{G}^2 t_0 \ll 1.$$

For the delta-correlated approximation, in addition to (124), one assumes that

$$(124) \quad \kappa t_0/l_0^2 \ll 1, \quad \text{and} \quad t \gg t_0.$$

The coefficient $D_1(t,\kappa) \sim D_1$ of (23'), and condition (124) can be rewritten in the form

$$(125) \quad \sigma_u^2 \boldsymbol{G}^2 t_0^2 \ll 1,$$

where σ_u^2 is the variance of the velocity fluctuation field. Also, recall that in the derivation of the stationary value of $\langle \tilde{q}^2 \rangle$ (see (107)), we needed information about the evolution of random field $\boldsymbol{p}$. This was provided by formula (112) the validity thereof depended on the condition $D_2 t_0 \ll 1$ where D_2 was defined in (29). Equivalently

$$\sigma_u^2 t_0^2 \ll l_0^2. \tag{126}$$

Thus, finally, we have shown that (125-127) constitute applicability conditions of the delta-correlated approximation of Section 4.6. They demand both, the fluctuations of the velocity field, and the molecular diffusion coefficient to be small. These requirements are met, for instance, in many geophysical systems.

6. Conclusions

Utilizing a unified functional approach, the paper discusses the statistical description of passive tracer diffusion in Gaussian random velocity fields. It provides both, the general setup of the problem and several methods of approximate analysis.

In the general case the problem is very complicated. It contains many parameters given by the mean flow, statistical properties of the fluctuating component of the flow, and molecular diffusion rate. To understand their combined effect one can not limit oneself to analysis of the mean tracer concentration or its correlation function. Rather, one needs to study the statistics of the problem at the level of joint probability densities of the tracer concentration and of its spatial gradient. Even in the simplest case: zero mean flow, no molecular diffusion, the delta-correlated approximation, the picture of time evolution of the tracer fluctuations is still rather complicated. The initially smooth distribution of tracer becomes more and more spatially chaotic and disordered, its spatial gradients show strong blow-up, and the lines of constant concentration assume fractal shape. In addition, this process goes in the direction of decreasing spatial scales, so that for sufficiently large times one needs to take into account molecular diffusion. In this case, however, we can not describe the precise statistical properties of probability densities. So, one has to work out various approximate schemes. The first efforts of this kind can be found in several papers.[(57–60)] The quantitative description of effects of mean flow (even in the simplest case of paraller shear flow) and specific models of velocity field fluctuations (for example, correlation tensor of the fluctuating component) can be analyzed on the basis of numerical solutions of the corresponding Fokker-Planck equations and computer simulations.

The functional approach presented in the paper is based essentially on the assumption of finite time correlation radius of the velocity field. The conditions of applicability of different approximation schemes are expressed in terms of the correlation radius. Our results do not apply to fields with large or infinite correlation radius, such as *stationary (in time) random velocity field.* The latter case has not been studied in complete generality yet, although some statistical problems of this type have been considered.[61–62]

Appendix:

Correlation splitting for Gaussian and telegraph processes.

A1. Furutsu-Novikov formula. Let $\boldsymbol{F} = (F_i)$ be a (vector or scalar) random Gaussian field with zero mean $\langle \boldsymbol{F}(\boldsymbol{x})\rangle = 0$ and correlation-function

$$B_{ij}(\boldsymbol{x};\boldsymbol{x}') = \langle F_i(\boldsymbol{x})F_j(\boldsymbol{x}')\rangle.$$

For any functional $R[\boldsymbol{F}]$ one has

$$\langle F_i(\boldsymbol{x})R[\boldsymbol{F}]\rangle = \int d\boldsymbol{x}' \sum_j \langle F_i(\boldsymbol{x})F_j(\boldsymbol{x}')\rangle \Big\langle \frac{\delta R[\boldsymbol{F}]}{\delta F_j(\boldsymbol{x}')}\Big\rangle. \tag{127}$$

To derive (127), we shall follow the functional method of Klyatskin.[16–18] The argument consists of several steps and exploits the Gaussianity of $\boldsymbol{F}$, and its characteristic functional

$$\Phi(\boldsymbol{\lambda}) = \langle e^{i\int \boldsymbol{\lambda}\cdot\boldsymbol{F}}\rangle.$$

Step 1. The cross-correlation term could be written as

$$\begin{aligned}\langle F_i(\boldsymbol{x})R[\boldsymbol{F}]\rangle &= \langle F_i(\boldsymbol{x})R[\boldsymbol{F}+\boldsymbol{\eta}]\rangle|_{\boldsymbol{\eta}=0} \\ &= \Big\langle F_i(\boldsymbol{x})\exp\Big(\int d\boldsymbol{x}'\boldsymbol{F}(\boldsymbol{x}')\frac{\delta}{\delta\boldsymbol{\eta}(\boldsymbol{x}')}\Big)\Big\rangle R[\boldsymbol{\eta}]\,\Big|_{\boldsymbol{\eta}=0},\end{aligned} \tag{128}$$

where $\boldsymbol{\eta}(\boldsymbol{x})$ is an arbitrary deterministic function. Formula

$$R[\boldsymbol{F}+\boldsymbol{\eta}] = \exp\Big(\int d\boldsymbol{x}'\boldsymbol{F}(\boldsymbol{x}')\frac{\delta}{\delta\boldsymbol{\eta}(\boldsymbol{x}')}\Big)R[\boldsymbol{\eta}] \tag{129}$$

is a variational (infinite dimensional) analog of the well-known relation for functions in n variables:

$$R(\boldsymbol{x}+\boldsymbol{u}) = \exp\left[\sum_j u_j\frac{\partial}{\partial x_j}\right]R(\boldsymbol{x}),$$

where $\boldsymbol{u} = (u_1,\ldots,u_n)$. **Step 2.** Our goal is to split the cross-correlation

(128) into ensemble averages of F_i and $R[\boldsymbol{F}]$. To this end we rewrite the right-hand side of (128) in the form that involved the statistical mean of R

$$\frac{\langle F_i(\boldsymbol{x})\exp(\int \boldsymbol{F}\frac{\delta}{\delta\boldsymbol{\eta}})\rangle}{\langle\exp(\int \boldsymbol{F}\frac{\delta}{\delta\boldsymbol{\eta}})\rangle}\left\langle \exp(\int \boldsymbol{F}\frac{\delta}{\delta\boldsymbol{\eta}})\right\rangle R[\boldsymbol{\eta}] = \frac{\langle F_i(\boldsymbol{x})\exp(\int \boldsymbol{F}\frac{\delta}{\delta\boldsymbol{\eta}})\rangle}{\langle\exp\int \boldsymbol{F}\frac{\delta}{\delta\boldsymbol{\eta}}\rangle}\langle R[\boldsymbol{F}+\boldsymbol{\eta}]\rangle$$

(130)

The fraction $\langle\ldots\rangle/\langle\ldots\rangle$ in the left-hand side of (130) represents the ratio of two variational (differential) operators, averaged over the ensemble. We can write it as an 'operator function' $\boldsymbol{\Omega}\,(\delta/i\delta\boldsymbol{\eta})$, where

$$\boldsymbol{\Omega}(\boldsymbol{\lambda}) = \frac{\langle \boldsymbol{F}(\boldsymbol{x})e^{i\int \boldsymbol{F}\boldsymbol{\lambda}}\rangle}{\langle e^{i\int \boldsymbol{F}\boldsymbol{\lambda}}\rangle} = \frac{\delta}{i\delta\boldsymbol{\lambda}}\ln\left\langle e^{i\int \boldsymbol{F}\boldsymbol{\lambda}}\right\rangle = \frac{\delta}{i\delta\boldsymbol{\lambda}}\ln\Phi[\boldsymbol{\lambda}].$$

Step 3. It remains to recall that the Gaussianity of $\boldsymbol{F}$ implies that

$$\Phi[\boldsymbol{\lambda}] = \exp\left(-\frac{1}{2}\boldsymbol{B}\boldsymbol{\lambda}\cdot\boldsymbol{\lambda}\right).$$

Hence

$$\frac{\delta}{i\delta\boldsymbol{\lambda}}\log\Phi[\boldsymbol{\lambda}] = i\boldsymbol{B}\boldsymbol{\lambda},$$

and a formal substitution $i\boldsymbol{\lambda}\to\delta/\delta\boldsymbol{\eta}$ yields

$$\boldsymbol{\Omega}\left(\frac{\delta}{i\delta\boldsymbol{\eta}}\right) = \boldsymbol{B}\frac{\delta}{\delta\boldsymbol{\eta}}.$$

The latter, upon insertion into (130), yields the desired result

$$\begin{aligned}\langle F_i(\boldsymbol{x})R[\boldsymbol{F}]\rangle &= \int d\boldsymbol{x}'B(\boldsymbol{x};\boldsymbol{x}')\frac{\delta}{\delta\boldsymbol{\eta}(\boldsymbol{x}')}\langle R[\boldsymbol{F}+\boldsymbol{\eta}]\rangle\bigg|_{\boldsymbol{\eta}=0}\\ &= \int d\boldsymbol{x}'B(\boldsymbol{x};\boldsymbol{x}')\langle\frac{\delta}{\delta\boldsymbol{F}(\boldsymbol{x}')}R[\boldsymbol{F}]\rangle.\end{aligned}$$

Note that for a non-Gaussian field $\boldsymbol{F}$ the expansion of the logarithm of the characteristic functional $\ln\Phi[\boldsymbol{\lambda}]$ determines the cumulant functions of the field:

$$\ln\Phi[\boldsymbol{\lambda}] = \sum_{n=1}^{\infty}\frac{i^n}{n!}\int\cdots\int d\boldsymbol{x}_1\ldots d\boldsymbol{x}_n\,\kappa_n^{i_1,\ldots,i_n}(\boldsymbol{x}_1,\ldots,\boldsymbol{x}_n)\lambda_{i_1}(\boldsymbol{x}_1)\ldots\lambda_{i_n}(\boldsymbol{x}_n),$$

which gives the following generalization of formula (120):[(16–18)]

$$\langle F_l(\boldsymbol{x})R[\boldsymbol{F}]\rangle = \sum_{n=1}^{\infty}\frac{1}{(n-1)!}\int\cdots\int d\boldsymbol{x}_1\ldots d\boldsymbol{x}_{n-1}\,\kappa_n^{i_1,\ldots,i_{n-1},l}(\boldsymbol{x}_1,\ldots,\boldsymbol{x}_{n-1};\boldsymbol{x})$$

$$\times \left\langle \frac{\delta^{n-1}}{\delta F_{i_1}(\boldsymbol{x}_1)\ldots\delta F_{i_{n-1}}(\boldsymbol{x}_{n-1})} R[\boldsymbol{F}] \right\rangle .$$

A2. The telegraph process. For the telegraph process $Z(t)$ defined in Section 5 we can explicitly compute a number of useful statistical characteristics. In particular, the first moment $\langle Z(t)\rangle = 0$, and for correlations we get that, for $t_1 > t_2$,

$$\langle Z(t_1)Z(t_2)\rangle = a^2\langle(-1)^{N(0,t_1)+N(0,t_2)}\rangle = a^2 e^{-2\nu(t_1-t_2)}.$$

More generally, the k-point correlator for $t_k < t_{k-1} < \ldots < t_1$ is equal to

$$M_k(t_1,\ldots t_k) = \langle Z(t_1)Z(t_2)\ldots Z(t_k)\rangle = a^2 e^{-2\nu(t_1-t_2)} M_{k-2}(t_3,\ldots,t_n).$$

For a Gaussian process the latter would be expressed through the sum of all pair correlations. Furthermore, the characteristic functional

$$\Phi_t[v] = 1 - a^2 \int_0^t dt_1 \int_0^{t_1} dt_2 e^{-2\nu(t_1-t_2)} v(t_1)v(t_2)\Phi_{t_2}[v].$$

Hence, we get the following equations

$$\text{(131)} \qquad \frac{d}{dt}\Phi_t[v] = -a^2 v(t) \int_0^t dt_1 e^{-2\nu(t-t_1)} v(t_1)\Phi_{t_1}[v], \quad \Phi_0 = 1,$$

and

$$\text{(132)} \qquad \frac{d^2}{dt^2}\Phi_t + \left[2\nu - \frac{d}{dt}\log v(t)\right]\frac{d}{dt}\Phi_t + a^2 v^2(t)\Phi_t = 0$$

$$\Phi_0 = 1; \quad \frac{d}{dt}\Phi\Big|_{t=0} = 0.$$

A3. Correlation splitting. Now, we shall apply the above formulae to split the cross-correlation of telegraph process $Z(t)$ with a functional $R_t[Z]$. Some steps will follow the Gaussian case above. For $\tau < t$, write

$$\langle Z(t)R_t[Z+\eta]\rangle = \left\langle Z(t)\exp\left[\int_0^t d\tau Z(\tau)\frac{\delta}{\delta\eta(\tau)}\right]\right\rangle R_t[\eta] = \Psi_t\left[\left(\frac{\delta}{i\delta\eta}\right)\right]R_t[\eta],$$

where functional Ψ_t is expressed through the characteristic functional as follows:

$$\text{(133)} \qquad \Psi_t[v] = \left\langle Z(t)\exp\left[i\int_0^t d\tau Z(\tau)v(\tau)\right]\right\rangle$$

$$= \frac{1}{iv(t)}\frac{d}{dt}\left\langle \exp\left[i\int_0^t d\tau Z(\tau)v(\tau)\right]\right\rangle = \frac{1}{iv(t)}\frac{d}{dt}\Phi_t[v].$$

Now, we substitute differential equation (131) for Φ_t, and write (133) as

$$\Psi_t[v] = ia^2 \int_0^t dt_1 e^{-2\nu(t-t_1)} v(t_1)\Phi_{t_1}[v].$$

In other words

$$\Psi_t[v] = ia^2 \int_0^t dt_1 e^{-2\nu(t-t_1)} v(t_1)\Big\langle \exp\left[i\int_0^{t_1} d\tau Z(\tau)v(\tau)\right]\Big\rangle.$$

Hence, it follows that

$$\langle Z(t)R_t[Z+\eta]\rangle = ia^2 \int_0^t dt_1 e^{-2\nu(t-t_1)} \frac{\delta}{i\delta\eta(t_1)}\Big\langle \exp\left[\int_0^{t_1} d\tau Z(\tau)\frac{\delta}{\delta\eta(\tau)}\right]\Big\rangle R_t[\eta]$$

$$= a^2 \int_0^t dt_1 e^{-2\nu(t-t_1)} \frac{\delta}{\delta\eta(t_1)}\Big\langle \exp\left[\int_0^{t} d\tau Z(\tau)\Theta(t_1-\tau)\frac{\delta}{\delta\eta(\tau)}\right]\Big\rangle R_t[\eta] \quad (134)$$

where $\Theta(t)$ is the Heaviside step-function. The right-hand side of (135) can be rewritten in the form

$$a^2 \int_0^t dt_1 e^{-2\nu(t-t_1)} \frac{\delta}{\delta\eta(t_1)}\langle R_t[\eta(\tau)+Z(\tau)\Theta(t_1-\tau)]\rangle.$$

Finally, a substitution $\eta = 0$ yields the requisite decoupling formula for the cross-correlation:(17,18)

$$\langle Z(t)R[Z]\rangle = a^2 \int_0^t dt_1 e^{-2\nu(t-t_1)} \Big\langle \frac{\delta}{\delta Z(t_1)} R_t[Z(\tau)\Theta(t_1-\tau+0)]\Big\rangle.$$

Acknowledgments: This work was supported, in part, by a grant from the Office of Naval Research, grant MBPOO from the International Science Foundation and Project 94-05-16151 from the Russian Fund for Fundamental Research. The authors also thank Professor Alexander I. Saichev for numerous discussions.

References

[1] A.S. Monin and A.M. Yaglom, *Statistical Fluid Mechanics*, (MIT Press, Cambridge, Mass., 1980).

[2] G.T. Csanady, *Turbulent Diffusion in the Environment*, (D.Reidel Publ. Co, Dordrecht, 1980).

[3] A. Okubo, *Diffusion and Ecological Problems: Mathematical Models*, (Springer-Verlag, N.Y., 1980).

[4] M. Lesieur, *Turbulence in Fluids*, (Kluwer, Boston, 1990).

[5] W. McComb, *The Physics of Fluid Turbulence*, (Clarendon Press, Oxford, 1990).

[6] G. Dagan, Theory of solute transport by groundwater, *Ann. Rev. Fluid Mech.*, **19**: 183(1987).

[7] S.F. Shandarin and Ya.B. Zel'dovich, Turbulence, intermittency, structures in a self-gravitating medium: the large scale structure of the Universe, *Rev. Modern Phys* **61**: 185 (1989).

[8] S. Gurbatov, A. Malakhov and A. Saichev, *Nonlinear random waves and turbulence in nondispersive media: waves, rays and particles*, (Manchester U Press, Cambridge, 1991).

[9] G.K. Batchelor, Small-scale variation of convected quantities like temperature in turbulent fluid. 1. General discussion and the case of small conductivity, *J. Fluid Mech.*, **5**: 113(1959).

[10] G.K. Batchelor, I.D. Howells and A.A. Townsend, Small-scale variation of convected quantities like temperature in turbulent fluid. 2. The case of large condictivity, *J. Fluid Mech.*, **5**: 134(1959).

[11] P.H. Roberts, Analytical theory of turbulent diffusion, *J. Fluid Mech.*, **11**: 257(1961).

[12] R.H. Kraichnan, Small scale structure of scalar field convected by turbulence, *Phys. Fluids* **11**: 945 (1968).

[13] R.H. Kraichnan, Diffusion by a random velocity field, *Phys. Fluids* **13**: 22 (1970).

[14] P.G. Saffman, Application of the Wiener-Hermite expansion to the diffusion of passive scalar in a homogeneous turbulent flow, *Phys. Fluids* **12**(9): 1786 (1972).

[15] D. McLaughlin, G. Papanicolaou and O.R. Pironneau, Convection of microstructures and related problems, *SIAM J. Appl. Math.*, **45**: 780(1985).

[16] V.I. Klyatskin, *Statistical Depsription of Dynamical Systems with Fluctuating Parameters*, (Nauka, Moscow, 1975, in Russian).

[17] V.I. Klyatskin, *Ondes at equations stochastiques dans les milieus aleatoirement non homogenes*, (Editions de Physique, Besancon-Cedex, 1985, in French).

[18] V.I. Klyatskin, *Stochastic Equations and Waves in Random Media*, (Nauka, Moscow, 1980, in Russian).

[19] V.I. Klyatskin, Statistical description of the diffusion of tracers in random velocity fields, *Physics-Uspekhi*, **37**(5): (1994).

[20] L. Piterbarg, Short-correlation approximation in models of turbulent diffusion, in *Stochastic Models in Geosystems*, (IMA Volumes, Springer-Verlag, N.Y.,1995), to appear.

[21] E. Hopf, Statistical hydrodynamics and functional calculus, *J. Ration Mech. Anal.* **1**: 87(1953).

[22] S. Chandrasekhar, Stochastic problems in physics and astronomy, *Rev. Modern Phys.* **15**: 1(1943).

[23] M. Avellaneda and A. Majda, Mathematical models with exact renormalization for turbulent transport, *Comm. Math. Phys.* **131**: 381(1990).

[24] A.J. Majda, The random uniform shear layer: An explicit example of turbulent diffusion with broad tail probability distribution, *Phys. Fluids A* **5**(8): 1963(1993).

[25] K. Furutsu, On the statistical theory of electromagnetic waves in a fluctuating media, *J. Res. NBS*, **D-67**: 303(1963).

[26] E.A. Novikov, Functionals and the random-force method in turbulence theory, *Sov. Phys. JETP*, **20**(5): 1290(1964).

[27] V.R. Chechetkin, V.S. Lutovinov and A.A. Samokhin, On the diffusion of passive impurities in random flow, *Physica A* **175**: 87(1991).

[28] A.A. Samokhin and V.R. Chechetkin V.R., Diffusion of passive admixtures in a turbulent fluid, *Izvestia Atmosph.Oceanic Physics* **27**(6): 434 (1991).

[29] V.S. Lutovinov and V.R. Chechetkin, The Komogorov-Obukhov spectrum for the paired correlation function of passive tracers in a turbulent fluid, *Izvestia Atmosph. Oceanic Physics* **25**(3): 195 (1989).

[30] V.I. Klyatskin and A.I. Saichev, Statistical and dynamical localization of plane waves in randomly layered media, *Sov. Phys. Usp.* **35**(3): 231 (1992).

[31] V.I. Klyatskin and W.A. Woyczynski, *Dynamical and statistical characteristics of geophysical fields and waves and related boundary-value problems, Stochastic models in geosystems, (IMA Volumes,* Springer-Verlag, N.Y.1995), to appear.

[32] A.I. Saichev and W.A. Woyczynski, Probability distributions of passive tracers in randomly moving media, in *Stochastic Models in Geosystems,* (IMA Volumes, Springer-Verlag, N.Y. 1995), to appear.

[33] A.S. Gurvich and A.M. Yaglom, Breakdown of eddies and probability distributions for small-scale turbulence, *Phys. Fluids Suppl.* **10**(9): 559(1967).

[34] A.R. Kerstein and W.T. Ashurst, Lognormality of gradients of diffusive scalars in gomogeneous, two-dimensional mixing systems, *Phys. Fluids* **27**(12): 2819(1984).

[35] W.J.A. Dahm and K.A. Buch, Lognormality of the scalar dissipation pdf in turbulent flows, *Phys. Fluids* **A1**(7): 1290(1989).

[36] S.D. Rice, Mathematical analysis of random noise, *Bell. Syst. Tech. J.*, **23**: 282, (1944).

[37] S.D. Rice, Mathematical analysis of random noise, *Bell. Syst. Tech. J.*, **24**: 46, (1945).

[38] M.S. Longuet-Higgins, The statistical analysis of a random moving surface, *Philos. Trans. R. Soc. London, Ser.* **A249**: 321(1957).

[39] M.S. Longuet-Higgins, Statistical properties of an isotropic random surface, *Philos. Trans. R. Soc. London, Ser.* **A250**: 157(1957).

[40] P. Swerling, Statistical properties of the countours of random surfaces, *IRE Trans. Inf. Theory*, **IT-8**: 315(1962).

[41] M.B. Isichenko, Percolation, statistical topography, and transport in random media, *Rev. Modern Phys.,* **64**(4): 961 (1992).

[42] C. Zirbel, Stochastic flows: dispersion of a mass distribution and Lagrangian observations of a random field (Ph.D. Dissertation, Princeton University, 1993).

[43] E. Zambianchi and A. Griffa, Effects of finite scales of turbulence on disperions estimates, *J. Marine Res.* **52**: 129 (1994).

[44] V.I. Klyatskin and W.A. Woyczynski, Fluctuations of passive scalar with nonzero mean concentration gradient in random velocity fields, *Phys. Rev. Lett.* (submitted).

[45] A. Pumir, B. Shraiman and E. Siggia, Exponential tails and random advection, *Phys. Rev. Lett.*, **66**(23): 2984(1991).

[46] J. Gollub, J. Clarke, M. Gharib, B. Lane and O. Mesquita, Fluctuations and transport in a stirred fluid with a mean gradient, *Phys. Rev. Lett.*, **67**(25): 3507(1991).

[47] M. Holzer and A. Pumir, Simple models of non-Gaussian statistics for a turbulently advected passive scalar, *Phys. Rev.*, **E47**(1): 202(1993).

[48] A. Pumir, A numerical study of the mixing of a passive scalar in three dimensions in the presence of a mean gradient, *Phys. Fluids*, **A6**(6): 2118(1994).

[49] M. Holzer and E. Siggia, Turbulent mixing of a passive scalar, *Phys. Fluids* **6**(5): 1820(1994).

[50] A. Kerstein and P.A. McMurtry, Mean-field theories of random advection, *Phys. Rev.*, **E49**(1): 474(1994).

[51] V.I. Klyatskin and V.I. Tatarskii, A new method of successive approximations in the problem of the propagation of waves in a medium having random large-scale inhomogeneities, *Radiophysics and Quantum Electronics* **14**: 1100 (1971).

[52] A. Careta, F. Sagues, L. Ramirez-Piscina and J.M. Sancho, Effective diffusion in a stochastic velocity field, *J. Stat. Phys.* **71**: 235 (1993).

[53] S.I. Vainstein, F.F. Ruzmaikin and Ya.B. Zel'dovich, *Turbulent Dynamo in Astrophysics*, (Nauka, Moscow, 1980, in Russian).

[54] V.E. Shapiro and V.M. Loginov, *Dynamical Systems under Random Influences*, (Nauka, Novosibirsk, 1983, in Russian).

[55] J.T. Lipscomb, A.L. Frenkel and D. Ter Haar, On the convection of a passive scalar by a turbulent Gaussian velocity field , *J. Statistical Phys.* **63**: 305 (1991).

[56] V.I. Klyatskin, Approximations by delta-correlated random processes and diffusive approximation in stochastic problems, *Lectures in Appl. Math*, **27**: 447 (1991).

[57] Ya.G. Sinai and V. Yakhot, Limiting probability distributions of a passive scalar in a random velocity field, *Phys.Rev. Lett.* **63**: 1962 (1989).

[58] H. Chen, S. Chen and R.H, Kraichnan, Probability distribution of a stochastically advected scalar field, *Phys. Rev. Lett.* **63**: 2657(1989).

[59] Y. Kimura and R.H. Kraichnan, Statistics of an advected passive scalar, *Phys. Fluids A* **5**: 2264 (1993).

[60] F. Gao, An analytical solution for the scalar probability density function in homogeneous turbulence, *Phys. Fluids A* **3**: 511 (1991).

[61] M. Avellaneda and A. Majda, An integral representation and bounds on the effective diffusivity in passive advection by laminar and turbulent flows, *Commun. Math. Phys.*, **138**: 339(1991).

[62] A.J. Majda, Random shearing direction models for isotropic turbulent diffusion, *J. Stat. Phys.* **75**(516): 1153(1994).

[63] B.I. Shraiman and E.D. Siggia, Lagrangian path integrals and fluctuations in random flow, *Phys. Rev. E* **49**(4): 2912(1994).

[64] R.H. Kraichnan, Convection of a passive scalar by a quasi-uniform random straining field, *J. Fluid Mech.* **64**(4): 737(1974).

[65] R.H. Kraichnan, Anomalous scaling of a randomly advected passive scalar, *Phys. Rev. Letters* **72**(7): 1016(1974).

[66] A. Bershadskii, Topological and fractal properties of turbulent passive scalar fluctuations at small scales *J. Stat. Phys.* **77**: 909(1994).

[65] S.A. Molchanov and L.I. Piterbarg, Heat transport in random flows, *Russian J. Math. Phys.* **1**(3): 353(1974).

V.I. KLYATSKIN
Institute of Atmospheric Physics
Russian Academy of Sciences
Moscow 109017, Russia

Pacific Oceanological Institute
Russian Academy of Sciences
Vladivostok 690041, Russia

W.A. WOYCZYNSKI
D. GURARIE
Department of Statistics and
Center for Stochastic
and Chaotic Processes
in Science and Technology
Case Western Reserve University
Cleveland, OH 44106

Particles, vortex dynamics and stochastic partial differential equations

Peter Kotelenez

Abstract

The derivation of quasilinear stochastic partial differential equations (SPDE's) for mass distributions and their generalizations is reviewed in Section 2. Special emphasis is given to the vorticity distribution and its macroscopic limit in a 2D-fluid. In Section 4 and 5 bilinear SPDE's on weighted Hilbert spaces are derived from the underlying particle system. Moreover, it is shown that spatially homogeneous initial conditions imply that the solution is also spatially homogeneous. I.e., (non-Gaussian!!) homogeneous random fields are derived from an underlying particle system using SPDE methods.

1. Introduction

The irregular movement of particles immersed into a fluid is often referred to as Brownian motion. The rigorous foundation to the study of this phenomenon was laid by Einstein, Smoluchowski, Langevin, Ornstein, Uhlenbeck, Wiener, Levy, and Kolmogorov (see Nelson [22]). Physically there are two main approaches to study Brownian motion. The first is by introducing a random force and applying Newton's law $f = ma$ to get a stochastic ordinary differential equation (SODE) for the positions and momenta of the particles. The other approach uses a Hamiltonian formalism to derive the distribution of the positions and momenta (cf. Lebowitz and Rubin [20], where these equations have been derived without priori introduction of a stochastic term).

The object of this paper is to extend the first approach by introducing more complicated and, in our view, more realistic random forces. Tradi-

This research was supported by NSF grant DMS92-11438

tionally the random force acting on the position of a particle is assumed to be Gaussian white noise (in t), and the Gaussian white noises associated with different particles are assumed to be independent. Alternatively, this random force may be finite sum of independent Gaussian white noises (in t) coupled by some state dependent diffusion coefficient (cf. [Dawson and Vaillancourt [6] for the latter model). The latter version is one way of introducing correlations into the fluctuation forces. Another way of defining correlated fluctuation forces is by using the convolution of space-time white noise with an appropriate correlation function. This was done by Kotelenez [14], [15], and in this paper we will extend this approach.

One of the main advantages of our approach is the construction of smooth random fields representing a "mass" distribution as solutions of stochastic partial differential equations (SPDE's). These SPDE's are derived from stochastic ordinary differential equations (SODE's) for the positions of finitely many point particles (cf. 2.12). Following the widely accepted generalization of particle methods to fluid mechanics, in particular, for the calculation of the vorticity distribution of an incompressible 2D-fluid, we obtain the SODE (3.10) for the positions of N point vortices. We briefly comment of the difference between (3.10) and the traditional SODE's for point vortices: (cf. Chorin [3, 4]).

Let $\beta^1(t), \ldots, \beta^N(t)$ be i.i.d. standard $\Re^2$-valued Brownian motions. $q^i(t)$ is the position of the ith point vortex of time t, a_i its intensity, ν the viscosity, and K_δ is a smoothed version of the two-dimensional Biot-Savart kernel (cf. Section 3). Then the classical SODE is given by

$$dq^i(t) = \sum_{j=1}^{N} a_j K_\delta(q^i - q^j)dt + \sqrt{2\nu}d\beta^i(t), \qquad i = 1, \ldots, N$$

The smoothness of K_δ implies that (1.1) has a unique Itô-solution if the initial conditions $q^i(0)$, $i = 1, \ldots, N$, are adapted to an appropriate filtration and a.s. finite. The vorticity determined by (1.1) is given by the empirical process

$$\mathcal{Y}_N(t) := \sum_{i=1}^{N} a_i \delta_{q^i(t)},$$

where δ_q is the point-measure concentrated in q. It can be shown (Kotelenez [14], (1.11)) that a (formal) SPDE which would have (1.2) as a solution cannot have smooth random fields as solutions even if the initial conditions are smooth.

The main difference of our approach is the replacement of $\beta^i(t)$ by the convolution of the Brownian sheet, or, equivalently, of infinitely many i.i.d.

$\Re^2$-valued Brownian motions (cf. (2.13) and (3.10)) with a smooth kernel $\hat{\Gamma}_\varepsilon$ which defines the correlations of the fluctuation "forces". In particular, the fluctuation "force" depends on the position of the particle (resp. point vortex), and the correlation of the fluctuation forces can be modeled to depend on the distance of the particles. In the case of just one particle (resp. point vortex) and the choice of a Gaussian kernel (cf. Example 2.1) (3.10) becomes equivalent to (1.1) by Levy's theorem. The same holds if one adds an external force acting on this particle. Thus for one particle our model coincides with the Einstein-Smoluchowski model (cf. Nelson [22]). However, for several particles, the correlations among the fluctuation forces built into our model allow the derivativation of smooth random fields for the vorticity (resp. mass) distribution as follows: We consider the empirical process $\mathcal{X}_N(t)$ determined by (3.10) (resp. (2.12)) as a map (or flow) from point measure valued initial conditions ($\mathcal{X}_N(0) = \sum_{i=0}^N a_i \delta_{r^i(0)}$) into measures at t (or, more general, into measure valued functions of t)—which, of course, happen to be point measures. We can extend this map by continuity from point measure valued initial conditions. Now if those initial conditions have a density with respect to the Lebesque measure, the image at time t will also have a density with respect to Lebesque measure. Moreover, this map is the solution of an SPDE. In particular, for the vorticity distribution this implies that (3.10) "generates" smooth distributions, whereas (1.1) does not. So our general approach is a follows:

SPDE's are derived as a mezoscopic model of mass distributions and its generalizations (such as vorticity) from an underlying microscopic model. The microscopic model is given by SODE's for the positions of N (possibly interacting) particles living on $\Re^d (d \geq 1)$. Section 2 and most of Section 3 are a review of the author's derivation of quasilinear SPDE's with mass conservation; the main results in Section 2 are existence (Theorems 2.5, 2.6), smoothness (Theorem 2.7) and uniqueness (Theorem 2.9). The key result for the smoothness is the Itô formula (2.24). In Section 3 we restrict ourselves to $d = 2$ and the vortex dynamics of an incompressible viscous $2D$-fluid. We derive the deterministic (or macroscopic) Navier-Stokes equation for the vorticity distribution as the macroscopic limit of our SPDE (as $N \to \infty$ and ε, the correlation length of the "fluctuation forces", tends to 0). Moreover, we indicate how to obtain the Euler equation as the limit of our SPDE's if the correlation length $\varepsilon = f(\nu)$, where ν is the viscosity and $\frac{\nu}{f(\nu)} \to 0$.

Let us now introduce some Hilbert spaces. For $p \geq 1$ let $L_p(\Re^d, dr)$ be the space of Borel measurable real valued functions on $\Re^d$ such that

$\int |f|^p(r)dr < \infty$, where $|\cdot|$ is the Euclidean norm on $\Re^d$. We set

$$(\boldsymbol{H}_0, \langle\cdot,\cdot\rangle_0) := (L_2(\Re^d, dr), \langle\cdot,\cdot\rangle_0),$$

where $\langle\cdot,\cdot\rangle_0$ is the standard L_2- scalar product on $\boldsymbol{H}_0$, and $|\cdot|_0$ will denote its associated norm.

In Section 4 the Markov property of our quasilinear SPDE's is proved on weighted L_2-spaces $\boldsymbol{H}_{0,\lambda}$ provided that the SPDE has a unique solution on $\boldsymbol{H}_{0,\lambda}$ satisfying an apriori estimate (4.1). (By the results of Section 2 all assumptions in Section 4 hold for bilinear SPDE's, if we take $\lambda(r) \equiv 1$, i.e., $\boldsymbol{H}_{0,\lambda} = \boldsymbol{H}_0$). The proof of Theorem 4.2 is an adaptation of Dynkin's proof ([9]) and its generalization by Arnold, Curtain and Kotelenez [2] to our setting. In Section 5 we derive for the special case of bilinear SPDE's spatially (strictly) homogeneous solutions on the weighted Hilbert space $\boldsymbol{H}_{0,\Phi}$, which is defined as follows: $\varphi\colon \Re_+ \to \Re_+$, it is at least twice continuously differentiable and all derivatives are bounded. Moreover,

$$\varphi(x) = \begin{cases} x^2, & \text{if } x \le \frac{1}{2}, \\ x, & \text{if } x > 1. \end{cases}$$

Then the weight function $\Phi(r), r \in \Re^d$, is given by

$$\Phi(r) := \exp(-\varphi(|r|)).$$

$f \in \boldsymbol{H}_{0,\Phi}$, if it is Borel measurable from $\Re^d$ into $\Re$ and

$$|f|^2_{0,\Phi} := \int f^2(r)\Phi(r)dr < \infty.$$

$|\cdot|_{0,\Phi}$ is the norm on $\boldsymbol{H}_{0,\Phi}$ and $\langle\cdot,\cdot\rangle_{0,\Phi}$ its associated scalar product.

The derivation of spatially (strictly) homogeneous random field solutions depends on a generalization of Itô's formula (2.24) to $||X(t)||^2_{0,\Phi}$ in the bilinear case. Since this generalization is straight forward but requires some lengthy calculations we refer to Kotelenez [16] for details. Our derivation of spatially (strictly) homogeneous random field solutions for bilinear SPDE's (which are non-Gaussian!) proves the usefulness of SPDE methods (based on the particle approach)–in the area of random fields, where to our knowledge most results have been on Gaussian fields (cf., e.g., Adler [1], Pitt [24]). It should be mentioned that the particle approach to SPDE's was first developed by Dawson [5] (creating the new area of superprocesses). Also the recent paper by Dawson and Vaillancourt [6] uses a particle approach, where quasilinear SPDE's are derived via a martingale problem approach.

2. Quasilinear stochastic partial differential equations arising in interacting particle systems

The proofs of the following statements (if not given here) can be found in Kotelenez [14], [15].

Let N point particles be distributed over $\Re^d$, where d is arbitrary and fixed. The position of the i-th particle at time t will be denoted by $r^i(t)$, which is by assumption in $\Re^d$ for all i. We will assume that by some kind of adiabatic elimination we can neglect the momenta (e.g. Kotelenez and Wang [18]). Thus we consider the position of the N-particle system at time t as a point in $\Re^{dN}$, i.e.,

$$(r^1(t), \dots, r^N(t)) \in \Re^{dN}, \tag{2.1}$$

and a description of its time evolution will be a microscopic model for the particle distribution. Thus we derive ordinary differential equations (ODE's) or stochastic ordinary differential equations (SODE's) for (2.1) from mechanical principles (cf. e.g., Lebowitz and Rubin [20], Il'in and Khasminskii [12], Nelson [22], Kotelenez and Wang [18] etc.). We will call such equations (describing the position of each particle, usually in dependence on the position of the other particles) *microscopic equations.* Now let the mass of the i-th particle be a_i. Then the empirical mass distribution at time t of the N-particle system is given by

$$\mathcal{X}_N(t) := \sum_{i=1}^{N} a_i \delta_{r^i(t))}, \tag{2.2}$$

where δ_r is the point measure, concentrated in $r \in \Re^d$. In other words, $\mathcal{X}_N(t)$ is a measure process and for $A \subset \Re^d$, $\mathcal{X}_N(t, A) = \sum_{i=1}^N a_i 1_{\{r^i(t) \in A\}}$ describes the mass in A at t irrespective of which particular particles are in A at t, i.e., $\mathcal{X}_N(t)$ reduces information given by (2.1) to the relevant one as far as the mass distribution is concerned. On the other hand, if we agree to call a mass distribution *macroscopic* if the particle structure cannot be seen and stochastic fluctuations are absent then we may call a description of the time evolution of (2.2) a *mezoscopic model*, in particular, if stochastic fluctuations are present in (2.1), which we will always assume. Our goal is to derive a stochastic partial differential equation (SPDE) for (2.2), which will be called a *mezoscopic equation.* Under physically and mathematically reasonable assumption on (2.1) this SPDE will be extendible from sums of weighted point measures to measures having densities with respect to the Lebesque measure dr on $\Re^d$.

To include the case of randomly moving point vortices in a $2D$-fluid as well as other possible signed measure valued cases (cf. Fife [10]) we will allow the a_i in (2.2) to be positive and negative and just call them positive and negative "masses". So let $a^+ > 0$ be the total "positive" mass of (2.2), $a^- \geq 0$ be the total negative mass and $a := a^+ + a^-$. These quantities will be fixed throughout the paper. Set

$$\boldsymbol{M}_a := \{\mu : \mu \text{ is a finite signed Borel measure on } \Re^d, \mu^\pm(\Re^d) = a^\pm\},$$

where μ^+ and μ^- is the Jordan decomposition of μ and for any Borel set $A \subset \Re^d$ and nonnegative numbers b^+, b^- $\mu^\pm(A) = b^\pm$ if and only if $\mu^+(A) = b^+$ and $\mu^-(A) = b^-$. First we define a metric on $\Re^d$ by

$$\rho(r,q) := (\bar{c}|r-q| \wedge 1), \tag{2.3}$$

where $r, q \in \Re^d$, $|r-q|$ is the Euclidean distance on $\Re^d$, $\bar{c}$ some positive constant and "$\wedge$" denotes "minimum". If $\mu, \tilde{\mu} \in \boldsymbol{M}_a$ we will call positive Borel measures $Q^\pm$ on $\Re^{2d}$ joint representations of $(\mu^+, \tilde{\mu}^+)$, resp $(\mu^-, \tilde{\mu}^-)$ if $Q^\pm(A \times \Re^d) = \mu^\pm(A)a^\pm$ and $Q^\pm(\Re^d \times B) = \tilde{\mu}^\pm(B)a^\pm$ for arbitrary Borel sets $A, B \subset \Re^d$. The set of all joint representations of $(\mu^+, \tilde{\mu}^+)$, resp. $(\mu^-, \tilde{\mu}^-)$ will be denoted by $C(\mu^+, \tilde{\mu}^+)$, resp. $C(\mu^-, \tilde{\mu}^-)$. For $\mu, \tilde{\mu} \in \boldsymbol{M}_a$ and $m = 1, 2$ set

$$\begin{aligned}\gamma_{m,a}(\mu,\tilde{\mu}) :=& [\inf_{Q^+ \in C(\mu^+,\tilde{\mu}^+)} \int\int Q^+(dr,dq)\varrho^m(r,q) \\ &+ \inf_{Q^- \in C(\mu^-,\tilde{\mu}^-)} \int\int Q^-(dr,dq)\varrho^m(r,q)]^{\frac{1}{m}},\end{aligned} \tag{2.4}$$

where the integration is taken over $\Re^d \times \Re^d$. (We will not indicate the integration domain when integrating over $\Re^d$.) By the boundedness of ϱ and the Cauchy-Schwarz inequality

$$\gamma_{1,a}(\mu,\tilde{\mu}) \geq \gamma_{2,a}^2(\mu,\tilde{\mu}) \geq \frac{1}{a^+ \vee a^-}\gamma_{1,a}^2(\mu,\tilde{\mu}), \tag{2.5}$$

where "$\vee$" denotes the maximum of two numbers.

After normalizing the measures by $\mu^\pm \rightarrow \frac{\mu^\pm}{a^\pm}$ (setting $\frac{\mu^-}{a^-} = 0$ if $a^- = 0$) the Kantorovich-Rubinstein theorem implies $\gamma_{2,a}(\mu,\tilde{\mu}) = 0$ if and only if $\mu^+ = \tilde{\mu}^+$ and $\mu^- = \tilde{\mu}^-$ (Dudley [8], Ch. 11). The triangle inequality for $\gamma_{2,a}(\mu,\tilde{\mu})$ follows as for the Wasserstein metric (which is $\gamma_1(\frac{\mu^+}{a^+}, \frac{\tilde{\mu}^+}{a^+})$ for $\frac{\mu^+}{a^+}$ and $\frac{\tilde{\mu}^+}{a^+}$). Hence $\gamma_{2,a}$ is a metric on $\boldsymbol{M}_a$, and $\boldsymbol{M}_a$ endowed with $\gamma_{2,a}$ will be denoted by $(\boldsymbol{M}_a, \gamma_{2,a})$. By (2.5), the Prohorov and the Kantorovich-Rubinstein theorems $(\boldsymbol{M}_a, \gamma_{2,a})$ is complete (cf. Dudley [8], 11.5.5 and

11.8.2). Moreover, as in the Wasserstein case (cf. De Acosta [7], Appendix, Lemma 4) we obtain that the set of linear combinations of signed point measures from $\boldsymbol{M}_a$ is dense in $(\boldsymbol{M}_a, \gamma_{2,a})$. For $f \in C(\Re^d, \Re)$, the space of real valued continuous functions on $\Re^d$, we set

$$|||f|||_L := \sup_{r,q\in\Re^d, r\neq q} \{\frac{|f(r)-f(q)|}{\varrho(r,q)}\}.$$

(2.5) and the Kantorovich-Rubinstein theorem imply

$$\gamma_{2,a}^2(\mu,\tilde{\mu}) \geq \frac{1}{2(a^+ \vee a^-)} \sup_{|||f|||_L \leq 1} |< \mu - \tilde{\mu}, f >^2|. \tag{2.6}$$

Next we introduce the stochastic set-up. $(\Omega, \mathcal{F}, \mathcal{F}_t, P)$ is a stochastic basis with right continuous filtration. All our stochastic processes are assumed to live on Ω and to be $\mathcal{F}_t$-adapted (including all initial conditions in stochastic ordinary differential equations (SODE's) and stochastic partial differential equations (SPDE's). Moreover, the processes are assumed to be $dP \otimes dt$-measurable, where dt is the Lebesgue measure on $[0,\infty)$. Let $w_\ell(r,t)$ be i.i.d. real valued Brownian sheets on $\Re^d \times \Re_+$, $\ell = 1,\dots,d$ (cf. Walsh [26] and Kotelenez [13]) with mean zero and variance $t|A|$, where A is a Borel set in $\Re^d$ with finite Lebesgue measure $|A|$. Adaptedness for $w_\ell(r,t)$ means that $\int_A w_\ell(dp,t)$ is adapted for any Borel set $A \subset \Re^d$ with $|A| < \infty$. Set $w(p,t) := (w_1(p,t),\dots,w_d(p,t))^T$, where "$T$" denotes the transpose.

In Section 5 we will also need stochastic equations driven by the shifted Brownian sheet $w_h(r,t) := w(r+h,t), h \in \Re^d$. w_h is itself a Brownian sheet and therefore all results in Sections 2-4 derived for equations driven by w also hold for those equations, if w is replaced by w_h.

For $m \in \mathcal{N}$ let $C_b^m(\Re^d, \Re)$ be the space of m times continuously differentiable bounded real valued functions on $\Re^d$, where all derivatives up to order m are bounded, $C_0^m(\Re^d,\Re)$ is the subspace of $C_b^m(\Re^d,\Re)$ whose elements vanish at infinity. If $f \in C_b^m(\Re^d,\Re)$ we set

$$|||f|||_m := \max_{\substack{\ell_1+\cdot+\ell_d=|\ell| \\ |\ell|\leq m}} \sup_{r\in\Re^d} |\partial^{|\ell|}_{\ell_1,\dots\ell_d} f(r)|,$$

where $\partial^{|\ell|}_{\ell_1\dots\ell_d} f(r) = \frac{\partial^{\ell_1+\cdot+\ell_d}}{(\partial r_{\ell_1})^{\ell_1}\dots(\partial r_{\ell_q})^{\ell_d}} f(r)$, and r_{ℓ_i} is the ℓ_i-th coordinate of r. If we take only one partial derivative, say with respect to r_ℓ, we will just write "∂_ℓ".

We will describe the interaction of a specific particle with the other particles and with the surrounding random medium through smooth kernels. The contribution of the interaction of this particle with n particles to its motion will be weighted by positive numbers $p_n \geq 0$, $q_n \geq 0$, $n \in \mathcal{N} \cup \{0\}$, such that

$$(2.7) \qquad \begin{cases} c_a := & \sum_{n=1}^{\infty} p_n n a^n (1 + \frac{1}{a^+} + \frac{1}{a^-} \cdot 1_{\{a^- > 0\}}) < \infty, \\ \tilde{c}_a^2 := & (\sum_{n=0}^{\infty} q_n (n+1)(a^n + a^{n-1} + 1) \\ & \qquad \times (1 + \frac{1}{a^+} + \frac{1}{a^-} \cdot 1_{\{a^- > 0\}}))^2 < \infty, \end{cases}$$

where $1_{\{a^- > 0\}} = 1$ if $a^- > 0$ and $\frac{1}{a^-} 1_{\{a^- > 0\}} := 0$ if $a^- = 0$. If $f \in L_1(\Re^d, dr)$ and $\mu \in \boldsymbol{M}_a$, we set

$$f * \mu^{*n}(r) := \begin{cases} \int f(r-p)\mu^{*n}(dp) := & \\ \int \ldots \int f(r - (p_1 + \ldots + p_n))\mu(dp_1)\ldots\mu(dp_n), & \text{if } n \geq 1, \\ f(r), & \text{if } n = 0. \end{cases}$$

The particle-particle interaction is governed by a sequence of kernels $K_n = (K_{1n}, \ldots, K_{dn}) : \Re^d \to \Re^d$ such that

$$(2.8) \quad \begin{aligned} &(i) K_{\ell n} \in C_b^1(\Re^d, \Re) \cap L_1(\Re^d, dr) \text{ for } n \in \mathcal{N} \cup \{0\}, \ell = 1, \ldots, d, \\ &(ii) \sup_{n \in \mathcal{N} \cup \{0\}} \max_{\ell = 1, \ldots, d} \{|||K_{\ell n}|||_1 + |||K_{\ell n}|||_L + \int |K_{\ell n}|(r) dr\} =: c_K < \infty. \end{aligned}$$

Similarly, the particle-medium-particle interaction is governed by a sequence of kernels $\Gamma_n = (\Gamma_{k\ell n}) : \Re^d \to \mathcal{M}_{d \times d}$, the $d \times d$-matrices over $\Re$, such that for some $\bar{m} \geq \frac{d}{2} + 1$

$$(2.9) \quad \begin{aligned} &(i) \Gamma_{k\ell n} = \Gamma_{\ell k n} \in C_b^{\bar{m}}(\Re^d, \Re) \cap \boldsymbol{H}_0 \text{ for } n \in \mathcal{N} \cup \{0\}, k, \ell = 1, \ldots, d, \\ &(ii) \sup_{n \in \mathcal{N} \cup \{0\}} \max_{\ell = 1, \ldots, d} \{|||\Gamma_{k\ell n}|||_{\bar{m}}^2 + |||\Gamma_{k\ell n}|||_{L,0}^2 + ||\Gamma_{k\ell n}||_2^2\} =: c_\Gamma^2 < \infty, \end{aligned}$$

where, for suitable $f : \Re^d \to \Re$

$$|||f|||_{L,0}^2 := \sup_{r,q} \int \frac{|f(r-p) - f(q-p)|^2}{\rho^2(r,q)} dp,$$

$$||f||_2^2 := ||f||_0^2 + \sum_{\ell=1}^{d} ||\partial_\ell f||_0^2 + \sum_{k,\ell=1}^{d} ||\partial_{k\ell}^2 f||_0^2.$$

Example 2.1. Let $\varepsilon > 0$ and $\tilde{\Gamma}_\varepsilon(r-q) := \{\frac{1}{(2\pi\varepsilon)^{d/2}} \exp(-\frac{|r-q|^2}{2\varepsilon}\}^{\frac{1}{2}}$. Further, let $\rho(r,q) := (\frac{|r-q|}{\sqrt{8\varepsilon}} \wedge 1)$. (2.9) can be easily verified for

$$\Gamma_{k\ell n} := \tilde{\Gamma}_\varepsilon \delta_{k\ell}, \tag{2.10}$$

for all n where $\delta_{k\ell} = 1$, if $k = \ell$, and $= 0$, if $k \neq \ell$.

Let us now introduce the following abbreviations:

$$\left\{ \begin{array}{l} F(\mu, r) := \sum_{n=0}^{\infty} p_n K_n * \mu^{*n}(r), \\ \mathcal{J}(\mu, r) := \sum_{n=0}^{\infty} q_n \Gamma_n * \mu^{*n}(r), \end{array} \right. \tag{2.11}$$

where $\mu \in \boldsymbol{M}_a$. In what follows N is fixed, $a_i \in \Re, i = 1, \ldots, N, \sum_{a_i \geq 0} a_i = a^+, -\sum_{a_i < 0} a_i = a^-$. We will assume that the positions of our N-particle system satisfy the following SODE:

$$\begin{aligned} dr^i(t) =& F(\mathcal{X}_N(t), r^i(t))dt + \int \mathcal{J}(\mathcal{X}_N(t), r^i(t) - p) w(dp, dt) \\ & r^i(0) = r_0^i, i = 1, \ldots, N, \mathcal{X}_N(t) = \sum_{i=1}^N a_i \delta_{r^i(t)}. \end{aligned} \tag{2.12}$$

Remark 2.2. (i) Let $\{\tilde{\phi}_n\}_{n \in \mathcal{N}}$ be a complete orthonormal system (CONS) in $\boldsymbol{H}_0$ and define an $\mathcal{M}_{d \times d}$-valued function ϕ_n whose entries on the main diagonal are all $\tilde{\phi}_n$ and whose other entries are all 0. Then for any adapted processes $\mu(t)$ and $r(t)$ with values in $\boldsymbol{M}_a$ and $\boldsymbol{R}^d$, respectively,

$$\begin{aligned} & \int \mathcal{J}(\mu(t), r(t) - p) w(dp, dt) \\ & = \sum_{n=1}^{\infty} \int \mathcal{J}(\mu(t), r(t) - p) \phi_n(p) dp d\beta^n(t), \end{aligned} \tag{2.13}$$

where $\beta^n(t)$ are $\Re^d$-valued i.i.d. standard Wiener processes. The right hand side of (2.13) defines the increment $dM(\mu(t), r(t), t)$ of an $\Re^d$-valued square integrable continuous martingale. For the verification of this statement it is enough to show that the quadratic variation of the right hand side of (2.13) is finite. Let [M] denote the quadratic variation of $\Re$-, $\Re^d$- or $\boldsymbol{H}_0$-valued square integrable martingales M. It will be clear from the context which state space is underlying in the definition of [ ]. The mutual quadratic variation of any two one-dimensional components of $dM(\mu(t), r(t), t)$ is (formally) given by

(2.14)

$$\begin{aligned} &d[M_k(\mu(t), r(t), t), M_\ell(\mu(t), r(t), t)] \\ &\quad = \sum_{n=1}^{\infty}\sum_{j=1}^{d} \int \mathcal{J}_{kj}(\mu(t), r(t)-p)\tilde{\phi}_n(p)dp \int \mathcal{J}_{\ell j}(\mu(t), r(t)-p)\tilde{\phi}_n(p)dpdt \\ &\quad = \sum_{j=1}^{d} \int \mathcal{J}_{kj}(\mu(t), r(t)-p)\mathcal{J}_{\ell j}(\mu(t), r(t)-p)dpdt \\ &\quad = \sum_{j=1}^{d}\sum_{m,\tilde{m}} q_m q_{\tilde{m}} \int \Gamma_{kjm} * \mu^{*m}(t)(r(t)-p)\Gamma_{\ell j\tilde{m}} * \mu^{*\tilde{m}}(t)(r(t)-p)dpdt. \end{aligned}$$

Hence, by (2.7) and (2.9)

$$[M_k(\mu(t), r(t), t), M_k(\mu(t), r(t), t)] \le dc_\Gamma^2\tilde{c}_a^2 t. \tag{2.15}$$

This shows that (1.14) is rigorous and altogether it follows that we may view (2.12) as an (Itô) SODE driven by infinitely many i.i.d. Wiener processes.

(ii) In view of the aforementioned adiabatic elimination we see that the right hand side of (2.12) is the sum of the slowly varying (F) and rapidly varying ($\int \mathcal{J}(\cdot, \cdot - p)w(dp, \cdot)$) components of the forces acting on the i-th particle. The dependence of $\mathcal{J}$ on $\mathcal{X}_N(t)$ (the "mean field") reflects the fact that the action of the medium on the i-th particle itself depends on the position of the other particles. We easily check that (2.8) and (2.9) imply suitable Lipschitz conditions on the coefficients in (2.11). Suppose $\mu, \tilde{\mu} \in \boldsymbol{M}_a$ and $r, \tilde{r} \in \Re^d$. Then

$$|F(\mu, r) - F(\tilde{\mu}, \tilde{r})| \le c_K c_a\{\rho(r, \tilde{r}) + \sqrt{2}d\gamma_{2,a}(\mu, \tilde{\mu})\}. \tag{2.16}$$

and

$$\begin{aligned} &[\int (\mathcal{J}(\mu, r-p) - \mathcal{J}(\tilde{\mu}, \tilde{r}-p))w(dp, dt)] \\ &\qquad \le 2d^2\tilde{c}_a^2 c_\Gamma^2\{\rho^2(r, \tilde{r}) + a\gamma_{2,a}^2(\mu, \tilde{\mu})dt. \end{aligned} \tag{2.17}$$

Next we will derive the (mezoscopic) SPDE for the mass distribution $\mathcal{X}_N(t)$ associated with (2.12), assuming that (2.12) is solvable. Let $\mu \in \boldsymbol{M}_a$ and $r \in \Re^d$. Set

$$D_{k\ell}(\mu, r) := \frac{d}{dt}[M_k(\mu, r, t), M_\ell(\mu, r, t)]. \tag{2.18}$$

(cf. 2.14.) By assumptions (2.7) and (2.9) $D_{k\ell}(\mu) := D_{k\ell}(\mu,\cdot) \in C_b^2(\Re^d,\Re)$ and $\mathcal{J}_{k\ell}(\mu) := \mathcal{J}_{k\ell}(\mu,\cdot) \in C_b^2(\Re^d,\Re) \cap \boldsymbol{H}_0$. Similarly $F(\mu) := F(\mu,\cdot) \in C_b^1(\Re^d,\Re) \cap L_1(\Re^d,\Re)$ (cf. 2.11). Let ∇ be the gradient on $\Re^d$ and "$\cdot$" denote the scalar product on $\Re^d$. Consider the following quasilinear SPDE on $\boldsymbol{M}_a$

$$
(2.19) \qquad d\mathcal{X} = \{\frac{1}{2}\sum_{k,\ell=1}^{d} \partial_{k\ell}^2(D_{k\ell}(\mathcal{X})\mathcal{X}) - \nabla\cdot(\mathcal{X}F(\mathcal{X}))\}dt - \nabla\cdot(\mathcal{X}\int \mathcal{J}(\mathcal{X},\cdot - p)w(dp,dt)),
$$

$$
(2.20) \qquad \mathcal{X}^{\pm}(\Re^d,t) \overset{a.s.}{=} \mathcal{X}^{\pm}(\Re^d,0) = a^{\pm}(\text{conservation of mass}).
$$

$\int \mathcal{J}(\mathcal{X},\cdot - p)w(dp,dt)$ is treated as a density with respect to $\mathcal{X}$, i.e. $\mathcal{X}\int \mathcal{J}(\mathcal{X},\cdot - p)w(dp,dt)$ is the $\Re^d$- valued (signed) measure $\int \mathcal{J}(\mathcal{X},\cdot - p)w(dp,dt)\mathcal{X}(d\cdot)$. Similarly for $D_{k\ell}(\mathcal{X})\mathcal{X}$ and $\mathcal{X}F(\mathcal{X})$. A weak solution of (2.19) is by definition a continuous $\boldsymbol{M}_a$-valued adapted process $\mathcal{X}(t)$ which satisfies

$$
(2.21) \qquad d\langle\mathcal{X},\varphi\rangle = \{\frac{1}{2}\sum_{k,\ell=1}^{d} \langle\mathcal{X}, D_{k\ell}(\mathcal{X})\partial_{k\ell}^2\varphi\rangle + \langle\mathcal{X}, F(\mathcal{X})\cdot\nabla\varphi\rangle\}dt + \langle\mathcal{X}, \int \mathcal{J}(\mathcal{X},\cdot - p)w(dp,dt)\cdot\nabla\varphi\rangle
$$

where $\varphi \in C_0^3(\Re^d,\Re)$ and $\langle\cdot,\cdot\rangle$ is the duality between measures and elements from $C_b(\Re^d,\Re)$, which is an extension of $\langle\cdot,\cdot\rangle_0$. The restriction to $C_0^3(\Re^2,\Re)$ is motivated by the fact that for those φ $|||\partial_\ell\varphi|||_L$ and $|||\partial_{k\ell}^2\varphi|||_L < \infty$ for $k,\ell = 1,\dots,d$ and, therefore, the right hand side of (2.21) is defined for $\mathcal{X} \in (\boldsymbol{M}_a,\gamma_{2,a})$. The fact that φ vanishes at infinity allows to come from (2.21) to (2.19) through integration by parts, where in (2.19) the derivatives have to be interpreted in the distributional sense. That (2.19)/(2.20) is also the "correct" SPDE for the mass distribution associated with (2.12) follows from the following lemma.

Lemma 2.3. *Suppose (2.12) has a solution* $(r^1(t,r_0),\dots,r^N(t,r_0))$. *Then* $\mathcal{X}_N(t) := \sum_{i=1}^N a_i\delta_{r^i(t)}$ *is a weak solution of (2.19)-(2.20).*

Proof . (i) $\mathcal{X}_N(t)$ satisfies (2.21) by Itô's formula. (ii) The mass conservation (2.20) follows from the construction.

Theorem 2.4. *To each $\mathcal{F}_0$-adapted initial condition $r_N(0) \in \Re^{dN}$ (2.12) has a unique $\mathcal{F}_t$-adapted solution $r_N(\cdot, r_N(0)) \in C([0,\infty);\Re^{dN})$ a.s., which is an $\Re^{dN}$-valued Markov process.*

Let $\mathcal{X}_N(t)$ be the empirical process for (2.12) starting at $\mathcal{X}_N(0)$. Set

$$\mathcal{X}(t, \mathcal{X}_N(0)) := \mathcal{X}_N(t).$$

Furthermore, set

$$\boldsymbol{M}_{d,a} := \{\mu \in \boldsymbol{M}_a : \mu \text{ is a finite linear combination of point measures on } \Re^d\}.$$

We introduce the following metric spaces of $\boldsymbol{M}_{d,a}$, resp. $\boldsymbol{M}_a$-valued $\mathcal{F}_0$-measurable random variables and of continuous adapted $\boldsymbol{M}$-valued processes:

$$\begin{aligned}\tilde{\mathcal{M}}_{0,a} &:= L_2(\Omega; \boldsymbol{M}_{d,a})\\ \mathcal{M}_{0,a} &:= L_2(\Omega; \boldsymbol{M}_a)\\ \mathcal{M}_{[0,T],a} &:= L_2(\Omega; C([0,T]; \boldsymbol{M}_a)),\end{aligned}$$

where the metric on the first two spaces is given by $(E\gamma_2^2(\mu,\tilde{\mu}))^{\frac{1}{2}}$ for $\mu, \tilde{\mu} \in \mathcal{M}_{0,a}$ and on the last one by $(E\sup_{0\le t\le T}\gamma_2^2(\mu_t,\tilde{\mu}_t))^{\frac{1}{2}}$ for $\mu_t, \tilde{\mu}_t \in \mathcal{M}_{[0,T],a}$. Note that $\mathcal{M}_{0,a}$ and $\mathcal{M}_{[0,T],a}$ are complete, since $\boldsymbol{M}_a$ is complete.

Theorem 2.5. *The map $\mathcal{X}_N(0) \mapsto \mathcal{X}(\cdot, \mathcal{X}_N(0))$ from $\tilde{\mathcal{M}}_{0,a}$ into $\mathcal{M}_{[0,T],a}$ extends uniquely to a map $\mathcal{X}_0 \mapsto \mathcal{X}(\cdot, \mathcal{X}_0)$ from $\mathcal{M}_{0,a}$ into $\mathcal{M}_{[0,T],a}$. Moreover, for any $\mathcal{X}_0, \mathcal{Y}_0 \in \mathcal{M}_{0,a}$*

$$E \sup_{0\le t\le T} \gamma_{2,a}^2(\mathcal{X}(t,\mathcal{X}_0), \mathcal{X}(t,\mathcal{Y}_0)) \le \bar{c}_{T,a,\Gamma,K} E\gamma_{2,a}^2(\mathcal{X}_0,\mathcal{Y}_0), \tag{2.22}$$

where $\bar{c}_{T,a,\Gamma,K} < \infty$.

Theorem 2.6. *For any $\varphi \in C_b^3(\Re^d, \Re)$ and $\mathcal{X}_0 \in \mathcal{M}_{0,a}$ $\langle \mathcal{X}(t,\mathcal{X}_0), \varphi\rangle$ satisfies (2.21), $0 < t < \infty$.*

Next we consider smoothness. Set for $\mathcal{Y} \in \mathcal{M}_{[0,T],a}$, $s \in [0,T]$

$$\hat{\mathcal{L}}(\hat{D})(\mathcal{Y}(s), r) : \lim_{q\to r} \sum_{k,\ell=1}^{d} \partial_{k\ell,r}^2 \hat{D}_{k\ell}(s,r,q). \tag{2.23}$$

Next we consider the quasilinear SPDE (2.19)/(2.20) with its weak solution from Theorems 3.4 and 3.5.

Theorem 2.7. *Suppose* $\mathcal{X}(0) =: X_0 \in \boldsymbol{H}_0$ *and* $E\|X_0\|_0^{4n} < \infty$ *for some* $n \geq 1$. *Then*
(I) $\mathcal{X}(t) =: X(t) \in \boldsymbol{H}_0$ *a.s. for all* $t \geq 0$;
(II)

$$\begin{aligned} \|X(t)\|_0^{2n} = \|X_0\|_0^{2n} &+ n\int_0^t \|X(s)\|_0^{2(n-1)}\langle X^2(s), \hat{\mathcal{L}}(\hat{D})(X(s))\rangle_0 ds \\ &- n\int_0^t \|X(s)\|_0^{2(n-1)}\langle X^2(s), \nabla\cdot F(X(s))\rangle_0 ds \\ &- n\int_0^t \|X(s)\|_0^{2(n-1)}\langle X^2(s), \nabla\cdot \int \mathcal{J}(X(s),\cdot - p)w(dp,ds)\rangle_0 \\ &+ n(n-1)\int_0^t \|X(s)\|_0^{2(n-2)}\int\int X^2(s,r)X^2(s,q) \\ &\times (\sum_{k,\ell=1}^d \partial^2_{k,r,\ell,q}D_{k\ell}(X(s),r,q))drdqds, \end{aligned} \tag{2.24}$$

with

$$D_{k\ell}(X(s),r,q) := \sum_{j=1}^d \int \mathcal{J}_{kj}(X(s),r-p)\mathcal{J}_{\ell j}(X(s),q-p)dpds;$$

(III)

$$X(\cdot) \in C([0,\infty);\boldsymbol{H}_0) \text{ a.s.}; \tag{2.25}$$

(IV) *for any* $T > 0$

$$E\sup_{0\leq t\leq T}\|X(t)\|_0^{2n} \leq 2\exp(c_{2n,a,\Gamma,K}T)(E\|X_0\|^{4n})^{\frac{1}{2}}, \tag{2.26}$$

with $c_{2n,a,\Gamma,K} < \infty$.

Remark 2.8. If $n = 1$ in (2.24), it follows that our quasilinear SPDE (2.19)/(2.20) cannot be treated by the usual variational methods on $\boldsymbol{H}_0$ (cf. Pardoux [23] and the generalization of Pardoux's variational approach by Krylov and Rozovskii [19].

We obtain strong (Itô) uniqueness for the (mezoscopic) SPDE if the initial condition X_0 is in $L_2(\Re^d, dr)$.

Theorem 2.9. *Suppose* $X_0 \in \boldsymbol{H}_0$ *and* $E\|X_0\|_0^4 < \infty$. *Let* $X(\cdot, X_0)$ *be the weak solution of (2.19)/(2.20) from Theorems 2.5 and 2.6 starting at* X_0.

Let $Y(\cdot, X_0)$ be an arbitrary solution of (2.19)/(2.20) with $Y(0) = X_0$ such that $Y(\cdot, X_0) \in C([0,\infty); \boldsymbol{H}_0)$ a.s. Then a.s.

$$X(\cdot, X_0) \equiv Y(\cdot, X_0). \tag{2.27}$$

3. Vortex dynamics in a viscous 2D-fluid and associated stochastic partial differential equations

Now we assume $d = 2$. Our aim is to model the time evolution of the vorticity of a two-dimensional incompressible fluid. The restriction to two dimensions is natural in applications like oceanography, where the depth is considered to be negligible in comparison to its planar extension. Although for applications in oceanography one should include the action of the Coriolis force on the vorticity distribution we will neglect its contribution here since we want to be conceptual. Moreover, we believe that it is fairly easy to include the Coriolis contribution into our models, since it acts in the form of an external force on the system. Under the above assumptions we obtain a *macroscopic equation* for the distribution of vorticity in a two-dimensional fluid:

$$\left.\begin{aligned} &\frac{\partial}{\partial t}X(r,t) = \nu\triangle X(r,t) - \bigtriangledown\cdot(U(r,t)X(r,t)) \\ &X(r,t) = curlU(r,t) = \frac{\partial U_2}{\partial r_1} - \frac{\partial U_1}{\partial r_2} \\ &\bigtriangledown\cdot U \equiv 0.\end{aligned}\right\} \tag{3.1}$$

Here $U(r,t)$ is the velocity field, $r \in \Re^2, \nu \geq 0$ is the kinematic viscosity, (or in the oceanographic setting the eddy diffusion coefficient). $\triangle$ is the Laplacian, $\bigtriangledown$ the gradient, and "$\cdot$" denotes the scalar product on $\Re^2$. If $\nu > 0$ we obtain the Navier-Stokes equation for the vorticity. If the fluid is inviscid, i.e., $\nu = 0$, we obtain the Euler equation. Note that by the incompressibility condition $\bigtriangledown \cdot U = 0$ we obtain

$$U(r,t) = \int (\bigtriangledown^{\perp} g)(r-q)X(q,t)dq, \tag{3.2}$$

where $g(|r|) := \frac{1}{2\pi}\ell n(|r|)$ with $|r|^2 = r_1^2 + r_2^2$ and $\bigtriangledown^{\perp} = (\frac{-\partial}{\partial r_2}, \frac{\partial}{\partial r_1})^T$ with $''T''$ denoting the transpose. $\int()dq$ denotes integration over $\Re^2$ with respect to the Lebesgue measure. As a consequence we can obtain the velocity field

U, which satisfies the standard Navier-Stokes equation from the vorticity distribution.

There is an extensive literature on the (numerical) solution of (1.1) by the so-called (random) point vortex method (cf. Chorin [3], [4] Puckett [25], and the references therein). A theoretical model related to the point vortex model has been analyzed by Marchioro and Pulvirenti [21], which is a special case of the following more general model:

Let $0 < \delta \leq 1$ and $g_\delta(|r|) \equiv g(|r|)$ for $\delta \leq |r| \leq \frac{1}{\delta}$ and $g_\delta(|r|)$ for all $r \in \Re^2$ at least twice continuously differentiable with bounded derivatives up to order 2 and $|g'_\delta(s)| \leq |g'(s)|, |g''_\delta(s))| \leq |g''(s)|$ for $s > 0$. Set

$$K_\delta(r) := \nabla^\perp g_\delta(|r|).$$

We may assume without loss of generality $g'(0) = 0$, which implies $K_\delta(0) = 0$. Thus we have the smoothed NSE

$$\text{(3.3)} \qquad \begin{cases} \frac{\partial}{\partial t} X(r,t) & = \nu \triangle X(r,t) - \nabla \cdot (U_\delta(r,t) X(r,t)) \\ U_\delta(r,t) & := \int K_\delta(r-q) X(q,t) dq. \} \end{cases}$$

We will assume in (2.11) that $K_1 \equiv K_\delta$ and $K_n \equiv 0$ for $n \neq 1$. Moreover $\Gamma_0 = \hat{\Gamma}_\varepsilon$ and $\Gamma_n \equiv 0$ for $n \neq 0$, where

$$\text{(3.4)} \qquad \hat{\Gamma}_\varepsilon(r,p) := \begin{pmatrix} \tilde{\Gamma}_\varepsilon(r,q) & 0 \\ 0 & \tilde{\Gamma}_\varepsilon(r,q) \end{pmatrix}.$$

$\varepsilon > 0$ and the correlation functions $\tilde{\Gamma}_\varepsilon : \Re^4 \to \Re_+$ are just bounded Borel-measurable functions which are symmetric in $r, p \in \Re^2$ such that the following conditions are satisfied.

$$\text{(3.5)} \qquad \int \tilde{\Gamma}^2_\varepsilon(r,p) dp = 1,$$

and there is a finite positive constant c such that for any $r, q \in \Re^2$

$$\text{(3.6)} \qquad \int \tilde{\Gamma}_\varepsilon(r,p) \tilde{\Gamma}_\varepsilon(q,p) dp = c\sqrt{\varepsilon} \cdot \tilde{\Gamma}_{2\varepsilon}(r,q).$$

There are finite positive constants c, $c_\varepsilon := \bar{c}$ such that defining

$$\text{(3.7)} \qquad \rho(r,q) := (c_\varepsilon |r-q|) \wedge 1$$

we have

$$\int (\tilde{\Gamma}_\varepsilon(r,p) - \tilde{\Gamma}_\varepsilon(q,p))^2 \leq c\rho^2(r,q). \tag{3.8}$$

A particular correlation function, where (3.5) - (3.8) can be verified is given in Example 2.1.

Remark 3.1. (i) Apparently, one can get a more general class of $\tilde{\Gamma}_\varepsilon(r,p)$ satisfying (3.5) - (3.8) than in that example by taking $\tilde{\Gamma}_\varepsilon(r,q) := \sqrt{p(\varepsilon,r,q)}$ where $p(\varepsilon,r,q)$ is the transition density of an $\Re^2$-valued diffusion process (at time $t=\varepsilon$), whose generator is a strictly elliptic (second order) operator with smooth coefficients.
(ii) Assume there are two $\Re^2$-valued adapted stochastic processes $q^1(t)$ and $q^2(t)$. Then we obtain from (2.14) that the $\int_0^t \int \hat{\Gamma}_\varepsilon(q^i(s),p)w(dp,ds)$ are $\Re^2$-valued square integrable continuous martingales ($i=1,2$) and their mutual quadratic variation is given by

$$\begin{aligned}&\langle\langle \int_0^t \int \tilde{\Gamma}_\varepsilon(q^1(s),p)w_k(dp,ds), \int_0^t \int \tilde{\Gamma}_\varepsilon(q^2(s),p)w_\ell(dp,ds)\rangle\rangle \\ &\qquad = \int_0^t \int \tilde{\Gamma}_\varepsilon(q^1(s),p)\tilde{\Gamma}_\varepsilon(q^2(s),p)dpds \cdot \delta_{k,\ell},\end{aligned} \tag{3.9}$$

$k,\ell = 1,2$ with $\delta_{k,\ell}=1$ if $k=\ell$, and $=0$ otherwise. Moreover, assuming the set-up of Example 2.1 (2.10) implies that correlations are negligible if $|q_1(s)-q_2(s)|^2 >> \varepsilon$ and that they are observable if $|q_1(s)-q_2(s)|^2 \sim \varepsilon$.

Consider N point vortices with intensities $a_i \in \Re^2$ and let r^i be the position of the i-th vortex. Abbreviate $r_N := (r^1,\ldots,r^N) \in \Re^{2N}$. Assume that the positions satisfy the stochastic ordinary differential equation (SODE):

$$dr^i(t) := \sum_{j=1}^N a_j K_\delta(r^i - r^j)dt + \sqrt{2\nu}\int \hat{\Gamma}_\varepsilon(r^i,p)w(dp,dt), \tag{3.10}$$

$i=1,\ldots N$. Set

$$\mathcal{X}_N(t) := \sum_{i=1}^N a_i \delta_{r^i(t)}, \tag{3.11}$$

where $r^i(t)$ are the solutions of (3.10).

Specializing now the results of Section 2 to the vorticity distribution we obtain that the empirical process $\mathcal{X}_N(t)$ given by (3.11) has a unique extension $\mathcal{X}_{\varepsilon,\delta}(t)$ to $\boldsymbol{M}_a = \boldsymbol{M}_a(\Re^2)$ which satisfies the following stochastic Navier-Stokes equation:

$$(3.12) \qquad \begin{cases} d\mathcal{X}(t) = [\nu\triangle\mathcal{X} - \nabla \cdot \tilde{U}_\delta \mathcal{X})]dt \\ \qquad\qquad -\sqrt{2\nu}\,\nabla \cdot (\mathcal{X} \int \hat{\Gamma}_\varepsilon(\cdot, p) w(dp, dt) \\ \tilde{U}_\delta(r,t) := \int K_\delta(r-q)\mathcal{X}(dq,t) \end{cases}$$

$$(3.13) \qquad \mathcal{X}^{\pm}(\Re^2, t) \overset{a.s.}{=} \mathcal{X}^{\pm}(\Re^2, 0) = a^{\pm}(\text{conservation of vorticity}).$$

Existence, uniqueness and smoothness follows from (resp. as in) Section 2 (here our assumptions on $\hat{\Gamma}_\varepsilon$ are weaker than those on $\Gamma_n, n = 0, 1, \ldots$). We will now describe the macroscopic behaviour of $\mathcal{X}_\varepsilon(t, \mathcal{X}_N(0))$ as $N \to \infty$ and $\varepsilon \to 0$. Set

$$(3.14) \qquad \Lambda_N := \{r_N \in \Re^{2N} : \exists (i,j), 1 \le i < j \le N, \text{ such that } r^i = r^j\}.$$

Let $r_N(0)$ be the $\Re^{2N}$-valued initial condition in (3.10) and $\mathcal{X}_N(0) := \sum_{i=1}^N a_i \delta_{r^i(0)}$.

Theorem 3.2. *For each $N \in \mathcal{N}$ suppose $r_N(0) \notin \Lambda_N$ a.s. Let $\varphi \in C_b(\Re^2, \Re)$ and suppose $E\langle \mathcal{X}_N(0), \varphi\rangle \to \langle X(0), \varphi\rangle_0$ as $N \to \infty$. Then there is a sequence $\delta(N) \to 0$, as $N \to \infty$, such that for any $t > 0$*

$$(3.15) \qquad E\langle \mathcal{X}_{\varepsilon,\delta(N)}(t), \varphi\rangle \to \langle X(t), \varphi\rangle_0, \text{ as } \varepsilon \to 0 \text{ and } N \to \infty.$$

In (3.10) we considered the dynamics of point vortices in dependence of two parameters, namely ν, the kinematic viscosity, and ε, the correlation length of the fluctuation forces. However, it seems to be more natural to take the correlation length $\varepsilon = f(\nu)$. Let $X_\nu(t) := X_\nu(t, X_{0,\nu})$ be the solution of (3.12) for that choice of ε, where we assume $E\|X_0\|_0^2 < \infty$. Then Itô's formula (2.24) for $n = 1$ yields

$$\begin{aligned} \|X_\nu(t)\|_0^2 = \|X_{0,\nu}\|_0^2 &+ \int_0^t \frac{\nu}{f(\nu)} \|X_\nu(s)\|_0^2 ds \\ &- \int_0^t \langle X_\nu^2(s), (\nabla \cdot K_\delta) * X_\nu(s)\rangle_0 ds \\ &- \sqrt{2\nu}\int_0^t \langle X_\nu^2(s), \nabla \cdot \int \hat{\Gamma}_{f(\nu)}(\cdot - p) w(dp, ds)\rangle_0 \end{aligned}$$

Remark 3.3. Let $Y(t)$ be the solution of the Euler equation ((3.1) with $Y = 0$). Suppose
(i) $E\|X_{0,\nu} - Y_0\|_0^2 \to 0$, as $\nu \to 0$;
(ii) $\frac{\nu}{f(\nu)} \to 0$;
(iii) $\delta \to 0$.

Then for any $t > 0$ we expect

$$E\|X_\nu(t) - Y(t)\|_0^2 \to 0, \text{ as } \nu \to 0$$

(cf. Kotelenez [17])

4. The Markov property

Let $\boldsymbol{H} := \boldsymbol{H}_{0,\lambda}$, where $\lambda(x)$ is some (positive) weight function. The scalar product $\langle \cdot, \cdot \rangle$ on $\boldsymbol{H}$ is given by $\langle f, g\rangle = \int f(r)g(r)\lambda(r)dr =: \langle f, g\rangle_{0,\lambda}$ for $f, g \in \boldsymbol{H}$ and $\|f\| := \langle f, f\rangle^{\frac{1}{2}}$. If $\lambda(r) \equiv 1$, then $\boldsymbol{H}_{0,\lambda} = \boldsymbol{H}_0$. Next let $Y_s^i \in \boldsymbol{H}$ be $\mathcal{F}_s$-measurable with $E\|Y_s^i\|^2 < \infty$, $s \geq 0$, $i = 1, 2$. Suppose that (2.19) has a unique weak solution $X(t, s, Y_s)$ on $\boldsymbol{H}$ with $X(s, s, Y_s) = Y_s$ and that for $i = 1, 2$

$$(4.1) \qquad \begin{cases} E\|X(t, s, Y_s^i)\|^2 \leq c_t E\|Y_s^i\|^2, \\ E\|X(t, s, Y_s^1) - X(t, s, Y_s^2)\|^2 \leq c_t E\|Y_s^1 - Y_s^2\|^2 \end{cases}$$

where $c_t < \infty$ does not depend on Y_s.

Remark 4.1. The first inequality holds on $\boldsymbol{H} = \boldsymbol{H}_0$ by Theorems 2.6, 2.7, and 2.9. The second inequality follows from the first one in the bilinear case.

Let $\mathcal{B}$ be the Borel σ-algebra on $\boldsymbol{H}$.

Theorem 4.2. *Suppose that the conditions preceding Remark 4.1 are satisfied on* $\boldsymbol{H}$. *Then for* $0 \leq s$ *and (deterministic)* $\xi \in \boldsymbol{H}$ *there is an* $\boldsymbol{H}$*-valued process* $\tilde{X}(t, s, \xi)(t \geq 0)$ *such that (i)* $\tilde{X}(t, s, \xi)$ *is the weak solution of (2.19) starting at* s *in* ξ*; (ii) for fixed* $s \leq t$ $\tilde{X}(t, s, \xi, \omega)$ *is* $\mathcal{B} \otimes \mathcal{F}_t - \mathcal{B}$*-measurable; (iii) if* $X(t, 0, X_0)$ *is the weak solution of (2.19) starting at* 0 *in* X_0 *then:*

$$(4.2) \qquad X(t, 0, X_0) = \tilde{X}(t, s, X(s, 0, X_0)) \text{ a.s.}$$

Proof . Since $\boldsymbol{H}$ is separable we can decompose it into a countable family $\{B_k^n\}_{k\in\mathcal{N}}$ with $\sup_{\xi,\eta\in B_k^n} \|\xi - \eta\| \leq 3^{-n}$. Choose $\xi_k^n \in B_k^n$ and set $g_n : \boldsymbol{H} \to \boldsymbol{H}$ by $g_n(\xi) = \xi_k^n$ for any $\xi \in B_k^n$. If we denote for arbitrary $\xi \in \boldsymbol{H}$ by $\tilde{X}(t, s, g_n(\xi))$ be solution of (2.19) starting at s in

$g_k(\xi)$, then $(\xi,\omega) \mapsto \tilde{X}(t,s,g_n(\xi),\omega)$ is obviously $\mathcal{B}\otimes\mathcal{F}_t$ measurable. Moreover $E\|\tilde{X}(t,s,g_n(\xi)) - \tilde{X}(t,s,\xi)\|^2 \le c_t\|g_n(\xi)-\xi\|^2 \le c_t 3^{-n} \to 0$ (where $c_t < \infty$). By the usual argument it follows $\tilde{X}(t,s,\xi,\omega)$ is $\mathcal{B}\otimes\mathcal{F}_t$-measurable as a function of (ξ,ω). We then see that $\tilde{X}(t,s,g_n(X(s,0,X_0))$ solves (2.19) with $\tilde{X}(s,s,g_n(X(s,0,X_0)) = g_n(X(s,0,X_0))$. By uniqueness $X(t,0,X_0) = X(t,s,X(s,0,t_0))$.

$$
\begin{aligned}
E\|\tilde{X}(t,s,&g_n(X(s,0,X_0)) - X(t,s,X(s,0,X_0))\|^2\\
&\le c_t E\|g_n(X(s,0,X_0) - X(s,0,X_0)\|^2\\
&= c_t\sum_k E1_{\{\Omega_k^n\}}\|g_n(X(s,0,X_0) - X(s,0,X_0)\|^2
\end{aligned}
$$

(with $\Omega_k^n := \{\omega : X(s,0,X_0,\omega) \in B_k^n\}$)

$$
\le c_t \sum_k E1_{\{\Omega_k^n\}} 3^{-n} = c_t 3^{-n} \to 0, \text{ as } n\to\infty.
$$

Further, by the same argument

$$
E\|\tilde{X}(t,s,g_n(X(s,0,X_0)) - \tilde{X}(t,s,X(s,0,X_0\|^2 \to 0
$$

as $n\to\infty$. This proves (4.2). ■

5. Spatially homogeneous random fields as solutions of bilinear stochastic partial differential equations.

We now specialize to the bilinear SPDE

$$
(5.1) \qquad \begin{cases} dZ = & D\triangle Z dt - \sqrt{2D}\,\nabla\cdot(Z\int\Gamma(\cdot - p)w(dp,dt)) \\ & Z_0 \in \boldsymbol{H}, \end{cases}
$$

where $\boldsymbol{H} := \boldsymbol{H}_{0,\Phi}$, equipped with scalar product $\langle\cdot,\cdot\rangle := \langle\cdot,\cdot\rangle_{0,\Phi}$ and norm $\|\cdot\| := \|\cdot\|_{0,\Phi}$, respectively. $D > 0$ is a diffusion constant and Γ is given by (2.10). The following (a priori) estimate (5.2) has been derived in Kotelenez [16], which follows from a generalization of Itô formula (2.24) to $\|Z(t)\|^2$, where now $\|\cdot\| = \|\cdot\|_{0,\Phi}$.

Lemma 5.1. *Suppose $Z_0 \in \boldsymbol{H}$ s.t. $E\|Z_0\|^2 < \infty$ and that (5.1) has a weak solution $Z(t,Z_0)$ with $Z(0,Z_0) = Z_0$. Then for any $T > 0$ there is a finite constant c_T such that*

$$
((5.2)) \qquad \sup_{0\le t\le T} E\|Z(t,Z_0)\|^2 \le c_T E\|Z_0\|^2.
$$

Let $L_{2,\mathcal{F}}([0,T]\times\Omega;\boldsymbol{H})$ be the space of adapted $dP\otimes dt$-measurable $\boldsymbol{H}$-valued square integrable processes.

Theorem 5.2. *To each $Z_0 \in \boldsymbol{H}$ such that $E\|Z_0\|^2 < \infty$ there is a unique weak solution $Z(t,Z_0)$. Moreover, $Z(t,Z_0) \geq 0$, if $Z_0 \geq 0$.*

Proof . (i) Since $\cup_{n>0}\boldsymbol{M}_n \cap \boldsymbol{H}_0$ is dense in $\boldsymbol{H}$ we may choose a sequence $Z_{0,n} \in \boldsymbol{M}_n \cap \boldsymbol{H}_0$ with $E\|Z_{0,n} - Z_{0,m}\|^2 \to 0$, as $n,m \to \infty$. However, for total finite mass we have unique solutions $Z(t,Z_{0,n})$, $Z(t,Z_{0,m})$ and $Z(t,Z_{0,n}) - Z(t,Z_{0,m}) = Z(t,Z_{0,n} - Z_{0,m})$ by Theorem 2.9. (5.2) implies for any $T>0$ $\int_0^T E\|Z(t,Z_{0,n}) - Z(t,Z_{0,m})\|^2 dt \to 0$, as $n,m\to\infty$. By the completeness of $L_{2,\mathcal{F}}([0,T]\times\Omega;\boldsymbol{H})$ there is a unique process $Z(t,Z_0)$ such that for any $T>0$ $n\to\infty$ implies

$$
(5.3) \qquad \begin{cases} \sup_{0\leq t\leq T} E\|Z(t,Z_{0,n}) - Z(t,Z_0)\|^2 & \to 0 \\ \int_0^T E\|Z(t,Z_{0,n}) - Z(t,Z_0\|^2 dt & \to 0. \end{cases}
$$

Moreover, (5.2) holds for $Z(t,Z_0)$.
(ii) Let $C_c^m(\Re^d,\Re)$ be the subspace of $C_0^m(\Re^d,\Re)$ whose elements have compact support. Suppose $\psi \in C_c^0(\Re^d,\Re)$. Then $\Phi^{-1}\psi \in \boldsymbol{H}_0$, and we have:

$$
\begin{aligned}
(5.4) \qquad & \sup_{0\leq t\leq T} E|\langle Z(t,Z_0) - Z(t,Z_{0,n}),\psi\rangle_0| \\
& \leq \sup_{0\leq t\leq T} E\|(Z(t,Z_0) - Z(t,Z_{0,n})\|^2 \cdot \|\Phi^{-1}\psi\|_0 \\
& \to 0, \text{ as } n\to\infty
\end{aligned}
$$

by (5.3). Now let $\varphi \in C_c^2(\Re^d,\Re)$. (5.4) implies

$$
\sup_{0\leq t\leq T}\int_0^t E|\langle Z(s,Z_0) - Z(s,Z_{0,n}), D\Delta\varphi\rangle_0| ds \to \infty,
$$

as $n\to\infty$. Moreover,

$$
\begin{aligned}
& E\int_0^t \langle Z(s,Z_0) - Z(s,Z_{0,n}), \int \Gamma(\cdot - p) w(dp,ds)\cdot\nabla\varphi\rangle_0^2 ds \\
= & \sum_{\ell=1}^d E\int_0^t \int\int\int (Z(s,Z_0,r) - Z(s,Z_{0,n},r)(Z(s,Z_0,q) - Z(s,Z_{0,n},q))\cdot \\
& \cdot \tilde{\Gamma}_\varepsilon(r-p)\tilde{\Gamma}_\varepsilon(q-p)\partial_\ell\varphi(r)\partial_\ell\varphi(q) dr dq dp ds \\
\leq & \sum_{\ell=1}^d \int_0^t E\langle |Z(s,Z_0) - Z(s,Z_{0,n}|, |\partial_\ell\varphi|\rangle_0^2 ds \to 0, \text{ as } n\to\infty
\end{aligned}
$$

(by (5.4) - using $\int \tilde{\Gamma}_\varepsilon(r-p)\tilde{\Gamma}_\varepsilon(q-p)dp = \exp(\frac{-|r-q|^2}{8\varepsilon}) \leq 1$).
(iii) Since $Z(t, Z_{0,n})$ is a weak solution of (5.1), step (ii) implies the same for $Z(t, Z_0)$.
(iv) The uniqueness follows from (5.2).
(v) If $Z_0 \geq 0$ we choose $Z_{0,n} \geq 0$ approximating Z_0. By construction $Z(t, Z_{0,n}) \geq 0$. Hence by choosing a subsequence n_k for which $Z(t, Z_{0,n_k}, r, \omega) \to Z(t, Z_0, r, \omega)$ $dr \otimes dP$ almost everywhere, we obtain $Z(t, Z_0) \geq 0$. ■

We define the shift operator U_h, $h \in \Re^d$, of $\boldsymbol{H}$ by

$$(U_h f)(r) := f_h(r) := f(r+h).$$

Further, w_h is the shifted Brownian sheet as mentioned in Section 2. If ξ and η are random variables with values in some measurable space we will write

$$\xi \sim \eta,$$

if ξ and η have the same distributions. Clearly, by (2.13) we have the following:

If η is an $\mathcal{F}_t$-measurable $\boldsymbol{H}_0$-valued random variable with $E\|\eta\|_0^2 < \infty$, then

$$\int \eta(r) w_h(dr,t) \sim \int \eta(r) w(dr,t) \tag{5.5}$$

Theorem 5.3. *Suppose $E\|Z_0\|^2 < \infty$ in addition to*

$$Z_{0,h} \sim Z_0. \tag{5.6}$$

Then for any $t \geq 0$

$$(U_h Z)(t, Z_0) \sim Z(t, Z_0). \tag{5.7}$$

Proof. (i) Now we consider (5.1) driven both by w_0 (the unshifted sheet) and w_h. The solutions of (5.1) with initial conditions Z_0 and $Z_{0,h}$ and driving noise w_0 will be denoted by $Z(t, Z_0, w_0)$ and $Z(t, Z_{0,h}, w_0)$ respectively. $Z(t, Z_{0,h}, w_h)$ is the solution of (5.1) with initial condition $Z_{0,h}$ and driving noise w_h.
(ii) Theorem 4.2 implies

$$Z(t, Z_0, w_0) \sim Z(t, Z_{0,h}, w_0). \tag{5.8}$$

(iii) Let $\varphi \in C_c^2(\Re^d, \Re)$, since differentiation and shift commute on $C_c^2(\Re^d, \Re)$ we obtain:

$$\begin{aligned}\langle U_h Z(t, Z_0, w_0), \varphi\rangle_0 &= \langle Z(t, Z_0, w_0), U_{-h}\varphi\rangle_0 \\ &= \langle Z_0, U_{-h}\varphi\rangle_0 + \int_0^t \langle Z(s, Z_0, w_0), U_{-h} D\Delta\varphi\rangle_0 ds \\ &\quad + \int_0^t \langle Z(s, Z_0, w_0), \int \Gamma(\cdot - p) w_0(dp, ds) \cdot U_{-h} \nabla \varphi\rangle_0 \\ &= \langle Z_{0,h}, \varphi\rangle_0 + \int_0^t \langle U_h Z(s, Z_0, w_0), D\Delta\varphi\rangle_0 \\ &\quad + \int_0^t \langle U_h Z(s, Z_0, w_0), \int \Gamma(\cdot - p) w_h(dp, ds) \cdot \nabla\varphi\rangle_0\end{aligned}$$

(by change of variables in the stochastic integral). Hence,

$$U_h Z(t, Z_0, w_0) = Z(t, Z_{0,h}, w_h). \tag{5.9}$$

(iv) Next, denote the solutions of the microscopic SODE's (2.12) on $\Re^{d\cdot N}$ driven by w_0 and w_h and with initial conditions $r_{0,N,h}$ by $r_N(t, r_{0,N,h}, w_0)$ and $r_N(t, r_{0,N,h}, w_h)$, respectively, where we assume that for $\mathcal{Z}_{0,N,h} := \sum a_i \delta_{r^i_{0,h}} \in \boldsymbol{M}_n$ $((r^1_{0,h}, \ldots, r^N_{0,h}) = r_{0,N,h})$, $E\gamma^2_{2,n}(\mathcal{Z}_{0,N,h}, Z_{0,n,h}) \to 0$ as $N \to \infty$, and $E\|Z_{0,n,h} - Z_{0,h}\|^2 \to 0$ as $n \to \infty$. By (2.13) both $r_N(t, r_{0,N,h} w_0)$ and $r_N(t, r_{0,N,h}, w_h)$ can be approximated by SODE's driven by $\Re^{d\cdot M}$-valued standard Brownian motions $\beta_{d\cdot M}$, and $\beta_{d\cdot M,h}$ $(M \to \infty)$ whose solutions are weakly unique (as a consequence of strong uniqueness - cf. Ikeda and Watanabe [11]). Hence

$$r_N(t, r_{0,N,h}, w_0) \sim r_N(t, r_{N,0,h}, w_h). \tag{5.10}$$

Hence,

$$Z(t, \mathcal{Z}_{0,N,h}, w_0) \sim Z(t, \mathcal{Z}_{0,N,h}, w_h). \tag{5.11}$$

The extension (cf. Theorem 2.5) implies

$$Z(t, Z_{0,n,h}, w_0) \sim Z(t, Z_{0,n,h}, w_h), \tag{5.12}$$

whence by (5.3) (both for (5.1) driven by w_0 and w_h)

$$Z(t, Z_{0,h}, w_0) \sim Z(t, Z_{0,h}, w_h). \tag{5.13}$$

Now (5.13), (5.9) and (5.8) together imply (5.7). ∎

References

[1] Adler, Robert J. (1981) *The Geometry of Random Fields.* John Wiley & Sons, Chichester.

[2] Arnold, L., Curtain, R.F., Kotelenez, P. (1980) Nonlinear stochastic evolution equations in Hilbert space. Universität Bremen, FDS, Report No. 17.

[3] Chorin, A.J. (1973) Numerical study of slightly viscous flow. *J. Fluid Mech.* **57**, 785-796.

[4] Chorin, A.J. (1991) Statistical mechanics and vortex motion. *Lectures in Applied Mathematics*, **28**.

[5] Dawson, D.A. (1975) Stochastic evolution equations and related measure processes, *J. Multivariate Analysis*, **5**, 1-52.

[6] Dawson, D.A. Vaillancourt, J. (1994) Stochastic McKean-Vlasov Equations. (Preprint - Technical Report No. 242, Carleton University Lab. Stat. Probab.)

[7] De Acosta, A. (1982) Invariance principles in probability for triangular arrays of B-valued random vectors and some applications, *Ann. Probability*, **2**, 346-373.

[8] Dudley, R.M. (1989) *Real Analysis and Probability*, Wadsworth and Brooks, Belmont, California.

[9] Dynkin, E.B. (1965) *Markov Processes*, **I**, Springer Verlag, Berlin.

[10] Fife, P. (1992) Models for phase separation and their mathematics. To appear in: *Nonlinear Partial Differential Equations and Applications*, M. Mimura and T. Nishida (eds.) Kinokuniya Pubs., Tokyo.

[11] Ikeda, V., Watanabe, S.(1981) *Stochastic Differential Equations and Diffusion Processes.* North Holland, Amsterdam.

[12] Il'in, A.M., Khasminskii, R.Z. (1964) On equations of Brownian motion. *Probab. Th. Appl.* **9** No.**3** (in Russian).

[13] Kotelenez, P. (1992) Existence, uniqueness and smoothness for a class of function valued stochastic partial differential equations, *Stochastics and Stochastic Reports*, **41**, 177-199.

[14] Kotelenez, P. (1995a) A stochastic Navier-Stokes equation for the vorticity of a two-dimensional fluid, *Ann. Appl. Probability.*

[15] Kotelenez, P. (1995b) A class of quasilinear stochastic partial differential equations of McKean-Vlasov type with mass conservation, *Prob. Th. Rel. Fields*, **102**, 159-188.

[16] Kotelenez, P. (1995c) Smooth and homogeneous solutions of bilinear stochastic partial differential equations arising in systems of particles and vortices, In preparation.

[17] Kotelenez, P. (1995d) Macroscopic limit theorems for the vorticity distribution in an incompressible $2D$-fluid, In preparation.

[18] Kotelenez, P., Wang, K. (1994) Newtonian particle mechanics and stochastic partial differential equations. In D.A. Dawson (eds.) *Measure Valued Processes, Stochastic Partial Differential Equations and Interacting Systems*, Centre de Recherche Mathématiques, CRM Proceedings and Lecture Notes, Volume 5, 139-149.

[19] Krylov, N.V., Rozovskii, B.L. (1979) On stochastic evolution equations, *Itogi Nauki i Tehniki*, VINITI, 71-146.

[20] Lebowitz, J.L., Rubin, E. (1963) Dynamical study of Brownian motion. *Phys. Rev.* **131**, 2381-2396.

[21] Marchioro, C., Pulvirenti, M. (1982) Hydrodynamics in two dimensions and vortex theory, *Comm. Math. Phys.* **84**, 483-503.

[22] Nelson, E. (1972) *Dynamical Theories of Brownian Motion*, Princeton University Press, Princeton, N.J.

[23] Pardoux, E. (1975) *Equations aux derivees partielles stochastique non lineaires monotones. Etude de solutions fortes de type Itô.* These.

[24] Pitt, L.D. (1978) Scaling limits of Gaussian rector fields, *J. Multivariate Anal.* **8**, 45-54.

[25] Puckett, E.G. (199x) Vortex methods: An introduction and survey of selected research topics. *Centre for Mathematics and its Applications, Australian National University, Canberra* (Preprint CMA-MR22-91).

[26] Walsh, J.B. (1986) An introduction to stochastic partial differential equations, In P.L. Hennequin (ed.) *Ecole d'Ete de Probabilites de Saint-Flour XIV-1984*, **LNM 1180**, Springer.

Department of Mathematics
Case Western Reserve University
Cleveland, Ohio 44106

email: pxk4@po.cwru.edu

Nongaussian autoregressive sequences and random fields

Keh-Shin Lii and Murray Rosenblatt

Introduction

In this paper we discuss estimation procedures for the parameters of autoregressive schemes. There is a large literature concerned with estimation in the one-dimensional Gaussian case. Much of our discussion will however be dedicated to the nonGaussian context, some aspects of which have been considered only in recent years. Results have also at times been obtained in the broader context of autoregressive moving average schemes. We restrict ourselves to the case of autoregressive schemes for the sake of simplicity. Also they are the discrete analogue of simple versions of stochastic differential equations with constant coefficients. It is also apparent that nonGaussian autoregressive stationary sequences have a richer and more complicated structure than that of the Gaussian autoregressive stationary sequences.

Discussion

Consider the system of equations

(1)
$$\sum_k \varphi_k x_{t-k} = \xi_t$$

with $k = (k_1, \ldots, k_d)$, $t = (t_1, \ldots, t_d)$ lattice points in $\mathbb{Z}^d$, only a finite

Research supported by ONR Grant N00014-92-J-1086 and ONR Grant N00014-90-J-1371 number of the real φ_k nonzero, and the ξ_t independent and identically distributed random variables with $E\xi_t \equiv 0$, $E\xi_t^2 = \sigma^2 > 0$ for $t \in \mathbb{Z}^d$. We are concerned with stationary solutions of such a system. Let

$$\varphi(z) = \sum_k \varphi_k z^k \tag{2}$$

be the filter polynomial of the system with $z^k = z_1^{k_1} \dots z^{k_d}$. One can easily show that if the polynomial $\varphi(z)$ has no zero on the d-torus $|z_1| = \cdots = |z_d| = 1$ there is a unique stationary solution that is an infinite moving average of the ξ random field

$$x_t = \sum_k \alpha_{t-k}\xi_k, \quad \sum |\alpha_k|^2 < \infty. \tag{3}$$

In the one-dimensional case ($d = 1$) it is conventional to write

$$\sum_{k=0}^{p} \varphi_k x_{t-k} = \xi_t, \quad \varphi_0 = 1. \tag{4}$$

In this case there is a solution of (4) if and only if $\varphi(z)$ has no zeros on $|z| = 1$. However, in the case of higher dimension, e.g. $d = 5$, one can give examples of polynomials $\varphi(z)$ with zeros on the d-torus yet such that there exists a solution of (1) which is stationary and is of the form (3). Such a polynomial is $\varphi(z) = 1 - \frac{1}{5}(z_1 + z_2 + z_3 + z_4 + z_5)$.

Notice that the grid mesh of the set of lattice points is one as we have taken it. However, in the case of a stochastic differential equation the approximating autoregressive scheme would be one for which the grid mesh would be h with h small.

Let us note that a stationary solution of the system (1) of the form (3) has an absolutely continuous second order spectrum with spectral density

$$f_2(\lambda) = (2\pi)^{-d} \left| \sum \varphi_k e^{-ik\cdot\lambda} \right|^{-2} \sigma^2. \tag{5}$$

where $\lambda = (\lambda_1, \dots, \lambda_d)$. In the case of a Gaussian solution (here the ξ_t's are Gaussian) the full probability structure is determined by the spectral density or equivalently the second order moments. If the solution is non-Gaussian (the ξ_t's are nonGaussian) the probability structure is not fully determined by the spectral density. The knowledge of the probability structure in the Gaussian case is equivalent to knowledge of the modulus of the filter function

$$|\varphi(e^{-i\lambda})|.$$

In the nonGaussian case some knowledge of the phase of the filter function

$$arg\{\varphi(e^{-i\lambda})\}$$

is also required. If there are p distinct real roots of the polynomial $\varphi(z)$ in the one-dimensional nonGaussian case, there are 2^p different probability schemes with the same second order spectral density (5). Given any specific real root z_v, there are the two alternatives of its absolute value being less than one or greater than one—of leaving z_v as it is or replacing it by z_v^{-1}. In the case of complex roots, they appear in conjugate pairs because of the assumption of real coefficients φ_k but the appropriate modified remarks are obvious. In the Gaussian case the random variables ξ_t for the 2^p alternate schemes have the same probability distribution except for rescaling of the variance σ^2 so as to ensure the same spectral density.

If all moments of the ξ random variable are finite and ξ is nonGaussian, there is some cumulant $\mu_k \neq 0$ for $k > 2$. The Fourier transform of kth order cumulants

$$c_{j_1,\dots,j_{k-1}} = cum(x_t, x_{t+j_1}, \dots, x_{t+j_{k-1}}) \tag{6}$$

exists and is given by

$$\begin{aligned} f_k(\lambda^{(1)}, \dots, \lambda^{(k-1)}) &= \mu_k \{2\pi\}^{(k-1)(-d)} \varphi(e^{-i\lambda^{(1)}})^{-1} \dots \\ &\qquad \varphi(e^{-i\lambda^{(k-1)}})^{-1} \varphi(e^{i[\lambda^{(1)}+\dots+\lambda^{(k-1)}]})^{-1} \\ &= \mu_k \{2\pi\}^{(k-1)(-d)} \sum c_{j_1,\dots,j_{k-1}} \\ &\qquad \exp\{-i[j_1 \cdot \lambda^{(1)} + \dots + j_{k-1} \cdot \lambda^{(k-1)}]\}. \end{aligned} \tag{7}$$

A representation for kth order cumulant spectra like (5) holds also for general infinite moving averages (3). If

$$\alpha(e^{-i\lambda}) = \sum_k \alpha_k \, e^{-ik\cdot\lambda}, \quad \sum |\alpha_k|^2 < \infty, \tag{8}$$

then the corresponding kth order cumulant spectral density has the form

$$\begin{aligned} f_k(\lambda^{(1)}, \dots, \lambda^{(k-1)}) =& \mu_k \{2\pi\}^{(k-1)(-d)} \\ & \alpha(e^{-i\lambda^{(1)}}) \dots \alpha(e^{-i\lambda^{(k-1)}}) \alpha(e^{i[\lambda^{(1)}+\dots+\lambda^{(k-1)}]}) \end{aligned} \tag{9}$$

if kth order moments are finite. Nonparametric methods for the estimation of the transfer function (8) based on estimation of second and higher order cumulant spectra are given in Lii and Rosenblatt [7].

Finite Parameter Estimates

Let us first consider the classical Gaussian one-dimensional case. Due to lack of identifiability it is conventional here to assume that all the roots of the polynomial $\varphi(z)$ are outside the unit disc in the complex plane. This convention restores the identifiability of the coefficients φ_j. For convenience the representation of the process

$$x_t = \varphi(B)^{-1}\xi_t = (1 - \varphi_1 B - \cdots - \varphi_p B^p)^{-1}\xi_t \tag{10}$$

is considered with B the backward shift operator. The parameters to be estimated are $\varphi_1, \ldots, \varphi_p$ and we list σ as the $(p+1)$st φ_{p+1}. Let us introduce the matrix $\sum$ with elements

$$\sigma_{ij} = \begin{cases} \sigma^{-2}\gamma_x(i-j) & \text{if } 1 \le i, j \le p \\ 2\sigma^{-2} & \text{if } i = j = p+1 \\ 0 & \text{otherwise} \end{cases} \tag{11}$$

where $\gamma_x(\cdot)$ is the covariance function of the sequence x_t. One can then show that there is a sequence of estimates of $\varphi_1, \ldots, \varphi_{p+1}$ that is a solution of the likelihood equations and is unbiased and asymptotically normal with covariance matrix $n^{-1}\sum^{-1}$. A discussion of different variants of maximum likelihood estimates and solutions of the likelihood equations can be found in Brockwell and Davis [2].

The condition that all the roots of $\varphi(z)$ lie outside the unit disc in the complex plane in the one-dimensional case is often referred to as the minimum phase condition and we shall use that designation. In the case of the minimum phase condition the same estimates as those used in the Gaussian case can be used to estimate $\varphi_1, \ldots, \varphi_p$ consistently even if the ξ random variables have a nonGaussian distribution. However, it is clear that if one knew the distribution of the ξ's one could use that knowledge to get better estimates than the quasiGaussian estimates we have just referred to. Kreiss [6] has analyzed this nonGaussian minimum phase context under the assumption of a smooth positive density function for the ξ random variables. He has also proposed an adaptive procedure which does not require knowledge of the density function of the ξ random variables. Heuristically Kreiss' procedure can be described as follows. Given the minimum phase assumption, use a $\sqrt{n}$ consistent procedure for initial estimation of the φ_j coefficients. Of course, the estimates used in the Gaussian case would be such a $\sqrt{n}$ consistent procedure. Let the preliminary set of estimates of $\varphi_1, \ldots, \varphi_p$ be $\tilde{\varphi}_1, \ldots, \tilde{\varphi}_p$. An approximate deconvolution of the x_t sequence can be effected by

$$\hat{\xi}_t = x_t - \tilde{\varphi}_1 x_{t-1} - \cdots - \tilde{\varphi}_p x_{t-p}. \tag{12}$$

In terms of the $\hat{\xi}_t$'s one can obtain estimates of the probability density of the ξ's and appropriate functionals of the density. The preliminary estimates of $\varphi_1, \ldots, \varphi_p$, namely $\tilde{\varphi}_1, \ldots, \tilde{\varphi}_p$ can then be upgraded using the information on the density to obtain estimates $\hat{\varphi}_1, \ldots, \hat{\varphi}_p$ of the φ's that are asymptotically equivalent to the maximum likelihood estimates of the φ's.

In the nonminimum phase case estimates based on the Gaussian likelihood are not consistent. If one knows the scaled density function of the ξ's

$$f_\sigma(\xi) = \frac{1}{\sigma} f\left(\frac{\xi}{\sigma}\right) \tag{13}$$

with σ an unknown scaling, and f a positive sufficiently smooth function one can determine the asymptotic behavior of the maximum likelihood estimates. If the system is not minimum phase we factor the polynomial

$$\begin{aligned} \varphi(z) &= (1 - \varphi_1 z - \cdots - \varphi_p z^p) = \varphi^+(z)\varphi^*(z) \\ &= (1 - \psi_1 z - \cdots - \psi_r z^r)(1 - \psi_{r+1} z - \cdots - \psi_p z^s) \end{aligned} \tag{14}$$

where $\varphi^+(z)$ has no roots on the closed unit disc in the complex plane and $\varphi^*(z)$ has all its roots in the interior of the unit disc. Here $r, s \geq 0$ are nonnegative integers with $r + s = p > 0$. Let

$$\varphi^+(z)^{-1} = \sum_{j=0}^{\infty} \alpha_j z^j, \quad \varphi^*(z)^{-1} = \sum_{j=s}^{\infty} \beta_j z^{-j} \tag{15}$$

and

$$c1 = E\left(\frac{f'_\sigma}{f_\sigma}(z)\right)^2 \sigma^2, \quad c2 = E\left(z\frac{f'_\sigma}{f_\sigma}(z)\right)^2. \tag{16}$$

Set

$$\psi_{p+1} = \sigma.$$

We introduce the positive definite matrix $\sum = (\sigma_{u,v}; u, v = 1, \ldots, p+1)$

with

$$
(17) \qquad \sigma_{u,v} = \begin{cases} c1 \sum_{j=0}^{\infty} \alpha_j \alpha_{j+|u-v|} & \text{if } u, v = 1, \ldots, r \\ c1 \sum_{j=s}^{\infty} \beta_j \beta_{j+|u-v|} & \text{if } u, v = r+1, \ldots, p \\ & \quad (u,v) \neq (p,p) \\ \beta_s^2(c2-1) + c1 \sum_{j=s}^{\infty} \beta_{j+1}^2 & \text{if } u = v = p \\ \sum_j \alpha_{j-u} \beta_{j+v-\sigma} & \text{if } u = 1, \ldots, r \\ & \quad v = r+1, \ldots, p \\ \sigma^{-1} \beta_s (c2-1) & \text{if } u = p, v = p+1 \\ \sigma^{-2}(c2-1) & \text{if } u = v = p+1 \\ 0 & \text{otherwise.} \end{cases}
$$

In the nonminimum phase case one can show there is an approximate maximum likelihood sequence of estimates of $\varphi_1, \ldots, \varphi_{p+1}$ that is unbiased and asymptotically normal with covariance matrix $n^{-1} \sum^{-1}$ (see Breidt et al. [1]).

A discussion of the estimation of parameters in the case of a continuous time parameter Gaussian ARMA process can be found in the paper of Pham-Dinh-Tuan [8]. It is an interesting but generally unresolved question as to when a discrete time parameter Gaussian ARMA scheme can be obtained from a continuous time parameter Gaussian ARMA scheme by discrete sampling.

R. Wiggins [9] introduced a method he termed minimum entropy deconvolution. Let $c_m(x)$ be the mth cumulant ($m \geq 2$) of the random variable x assuming that mth absolute moments of x exist. The standardized mth cumulant of x is

$$
(18) \qquad k_m(x) = c_m(x) / (c_2(x))^{m/2} \, .
$$

It is clear that if x_t is a stationary process satisfying (1) then

$$
(19) \qquad k_m(x_t) = k_m(\xi_t) \frac{\sum_t \alpha_t^m}{(\sum \alpha_t^2)^{m/2}}, \quad m > 2 \, .
$$

In the case of the autoregressive scheme (1) we can normalize the problem by setting $\varphi_0 = 1$. Wiggins idea was to estimate the coefficients φ_j by passing the sequence x_t through a finite filter with weights c_j to obtain

$$\sum c_j x_{t-j} = (c * x)_t = \xi(c)_t \tag{20}$$

and then maximize the absolute value of the sample standardized fourth cumulant as a function of the c_j's given say in the one-dimensional case the observations $x_1, x_2, \ldots, x_n$. Of course the corresponding procedure could be carried for the mth cumulant as long as m is greater than two. Aspects of Wiggins original idea and its parallel for an mth cumulant have been discussed by Donoho [4], Rosenblatt and Lii [7], Gassiat [5] and Cheng [3]. Let

$$E_n\phi(\xi(c)) = \frac{1}{n^d} \sum_{t \in T_n} \phi(\xi(c)_t) \tag{21}$$

with $T_n = \{t = (t_1, \ldots, t_d) : 1 \le t_j \le n,\ j = 1, \ldots, d\}$ for a function or vector $\phi = (\phi_1, \ldots, \phi_q)$. The statistic to be maximized is of the form

$$J_n(c) = |k_m(\xi(c)_t;\ t \in T_n)|^2 = L(E_n\phi_1(\xi(c)), \ldots, E_n\phi_q(\xi(c))) \tag{22}$$

for some integer $m > 2$ and for appropriate L and $(\phi_1, \ldots, \phi_q)$ determined by k_m. Let

$$J(c) = L(E\phi_1(\xi(c)_0, \ldots, E\phi_q(\xi(c)_0)). \tag{23}$$

Under appropriate smoothness and moment conditions one can show that if $\hat{\varphi}_n$ is the estimate of autoregressive parameters φ maximizing (22) as a function of c then

$$n^{d/2}\,(\hat{\varphi}_n - \varphi) \tag{24}$$

is asymptotically normal with mean zero and covariance matrix $\sum$ with $\sum = A'^{-1}\Lambda \cdot A^{-1}$. The matrices A and Λ are given by

$$\begin{aligned} A_{ij} &= \sum_{k=1}^{q} E(\phi_k''(\xi_0)x_{-i}x_{-j})D_k L(E\phi(\xi_0)) \\ &\quad + \sum_{k,\ell=1}^{q} E(\phi_k'(\xi_0)x_{-i})E(\phi_\ell'(\xi_0)x_{-j})D_k D_\ell L(E\phi(\xi_0)) \end{aligned} \tag{25}$$

and

$$
\begin{aligned}
\Lambda_{ij} &= \sum_s E\{\psi(\xi_0)\psi(\xi_s)x_{-i}x_{s-j}\} \\
&+ \sum_\ell \sum_s a_{\ell,i} E\{\phi_\ell(\xi_0)\psi(\xi_s)x_{s-j}\} \\
&+ \sum_\ell \sum_s a_{\ell,j} E\{\phi_\ell(\xi_0)\psi(\xi_s)x_{s-i}\} \\
&+ \sum_{k,\ell} \sum_s a_{k,i} a_{\ell,j} \operatorname{cov}(\phi_k(\xi_0), \phi_\ell(\xi_s))
\end{aligned} \tag{26}
$$

with

$$
\psi(u) = \sum_{k=1}^{q} \phi_k'(u) D_k L(E\phi(\xi(c)_0)) \tag{27}
$$

and

$$
a_{k,i} = \sum_{\ell=1}^{q} D_k D_\ell L(E\phi(\xi_0)) E(\phi_\ell'(\xi_0)x_{-i}). \tag{28}
$$

Here D_k is $\partial/\partial\phi_k$, $k = 1, \ldots, q$. Closely related results have been given by Gassiat [5].

Random Fields

We shall make a few remarks on two-dimensional autoregressive random fields and then consider a simple but interesting example. The comments though not generally characteristic of the multidimensional case give substantial insight into that case.

A two-dimensional autoregressive random field $x_{t,\tau}$ satisfies a system of equations

$$
\varphi(B_1, B_2)x_{t,\tau} = \varepsilon_{t,\tau} \tag{29}
$$

with the B_i the backshift operators with

$$
B_1^j B_2^k x_{t,\tau} = x_{t-j,\tau-k} \tag{30}
$$

for all integers j, k. Also

$$
\varphi(B_1, B_2) = 1 - \sum_{(j,k)\in S} a_{j,k} B_1^j B_2^k \tag{31}
$$

with S a finite subset of $Z^2 - \{(0,0)\}$ (Z is the set of integers) and the random variables $\varepsilon_{t,\tau}$ are assumed to be independent and identically distributed nonGaussian with finite second moment. If $\varphi(z_1, z_2) \neq 0$ on the polydisc $D^2 = \{(z_1, z_2) : |z_1| = |z_2| = 1\}$ there is as was already noted a unique stationary solution of (29). There is then a Laurent series expansion for $\varphi^{-1}(z_1, z_2)$ in an "annulus"

$$\{(z_1, z_2) : r < |z_1|, |z_2| < r^{-1}\} \tag{32}$$

for some r, $0 < r < 1$. The expansion

$$\varphi^{-1}(z_1, z_2) = \sum_{j,k=-\infty}^{\infty} b_{j,k} z_1^j z_2^k \tag{33}$$

converges uniformly on D^2. The $b_{j,k}$'s are the Fourier coefficients

$$b_{j,k} = \frac{1}{(2\pi)^2} \iint_{-\pi}^{\pi} \varphi^{-1}(e^{-i\lambda_1}, e^{-i\lambda_2}) e^{ij\lambda_1 + ik\lambda_2} d\lambda_1 d\lambda_2. \tag{34}$$

The minimum entropy procedure is a method that can be used to estimate the coefficients $a_{j,k}$ and the variance of the $\varepsilon_{t,\tau}$'s.

A simple example of a random field is given by

$$x_{t,\tau} + a x_{t-1,\tau} + b x_{t,\tau-1} = \varepsilon_{t,\tau}. \tag{35}$$

If $|a| + |b| < 1$ the stationary solution is given by

$$x_{t,\tau} = (1 + aB_1 + bB_2)^{-1} \varepsilon_{t,\tau} = \sum_{k=0}^{\infty} (-1)^k (aB_1 + bB_2)^k \varepsilon_{t,\tau}. \tag{36}$$

There are two other obvious cases in which there is a stationary solution. If $1 + |a| < |b|$

$$x_{t,\tau-1} + \frac{a}{b} x_{t-1,\tau} + \frac{1}{b} x_{t,\tau} = \frac{1}{b} \varepsilon_{t,\tau} \tag{37}$$

and the stationary solution is given by

$$\begin{aligned} x_{t,\tau-1} &= \left(1 + \frac{a}{b} B_1 B_2^{-1} + \frac{1}{b} B_2^{-1}\right)^{-1} \frac{1}{b} \varepsilon_{t,\tau} \\ &= \sum_{k=0}^{\infty} (-1)^k \left(\frac{a}{b} B_1 B_2^{-1} + \frac{1}{b} B_2^{-1}\right)^k \frac{1}{b} \varepsilon_{t,\tau}. \end{aligned}$$

In the last case $1 + |b| < |a|$

$$x_{t-1,\tau} = \left(1 + \frac{1}{a}B_1^{-1} + \frac{b}{a}B_1^{-1}B_2\right)^{-1} \frac{1}{a}\varepsilon_{t,\tau}$$

with an infinite expansion that is the counterpart of (36) and (37).

It is clear that in the case of the system (35) there is a stationary solution if and only if the trigonometric polynomial

$$1 + ae^{-i\lambda} + be^{-i\mu} \tag{38}$$

has no zero for real λ and μ. A zero for the polynomial (38) is equivalent to a solution for the pair of equations

$$\begin{aligned} 1+a\cos\lambda + b\cos\mu &= 0 \\ a\sin\lambda + b\sin\mu &= 0\ . \end{aligned} \tag{39}$$

Then

$$\begin{aligned}(1 + a\cos\lambda)^2 &= b^2\cos^2\mu = b^2(1 - \sin^2\mu) \\ &= b^2\left(1 - \frac{a^2}{b^2}\sin^2\lambda\right) = b^2 - a^2 + a^2\cos^2\lambda\end{aligned}$$

implying that

$$\cos\lambda = \frac{b^2 - a^2 - 1}{2a}$$

if $a \neq 0$. Similarly

$$\cos\mu = \frac{a^2 - b^2 - 1}{2b}$$

if $b \neq 0$. It is clear that if there is a solution we must have

$$\left|\frac{b^2 - a^2 - 1}{2a}\right| \le 1, \quad \left|\frac{a^2 - b^2 - 1}{2b}\right| \le 1. \tag{40}$$

The conditions could be written in the following equivalent form

$$|a - 1| \le |b| \le |a + 1|, \quad |b - 1| \le |a| \le |b + 1|. \tag{41}$$

In the accompanying graph the shaded range denotes the points (a, b) for which the polynomial (38) has a zero and the unshaded range the points (a, b) for which (38) has no zero.

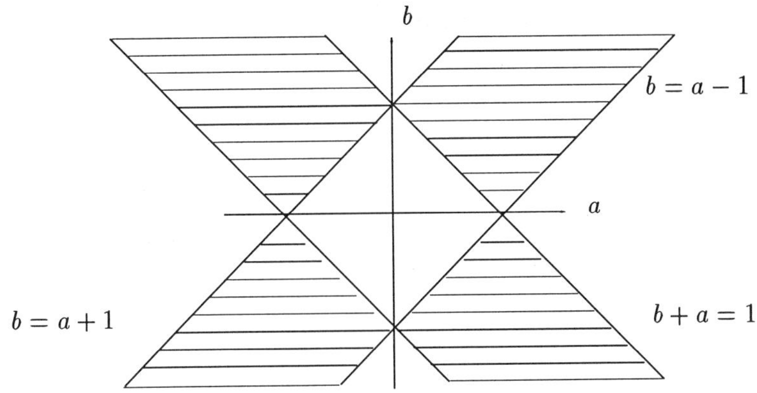

Computations

In the first simulation we consider an AR(2) scheme

$$(1 + \varphi_1 B + \varphi_2 B^2)x_t = \varepsilon_t$$

with the ε_t's a sequence of independent, identically distributed double exponential random variables with scale parameter 1. An approximate maximum likelihood estimate is obtained by maximizing

$$\frac{1}{n-q}\sum_{t=q}^{n} \log f_\sigma \left(\widehat{\varepsilon}_t(\underline{\varphi}, \sigma)\right) + \frac{1}{2\pi}\int_{-\pi}^{\pi} \log |\varphi(e^{-i\lambda})| d\lambda$$

with

$$f_\sigma(x) = \frac{1}{\sigma} f\left(\frac{x}{\sigma}\right)$$

where f is the double exponential density and

$$\widehat{\varepsilon}_t(\underline{\varphi}, \sigma) = \varphi(B)x_t$$

with

$$\varphi(B) = 1 - \sum_{j=1}^{p} \varphi_j B^j, \quad p = 2.$$

The maximization is carried out with respect to the parameters φ_j, σ. The approximation

$$\frac{1}{2\pi}\int_{-\pi}^{\pi} \log|\varphi(e^{-1\lambda})|d\lambda \cong \frac{1}{m}\sum_{j=0}^{m-1} \log|\varphi(e^{-2\pi ij/m})|$$

is used with $m = 64$. Details concerning the computing scheme are given in Lii and Rosenblatt [7]. The maximum likelihood estimates are compared with the minimum entropy estimates.

The three specific models are given by $(\varphi_1, \varphi_2) = (-0.05, -1.5)$, $(0.02, -0.67)$ and $(0.4, -0.96)$. The corresponding roots (r_1, r_2) are $(0.8, -0.8333)$, $(-1.206, 1.2367)$ and $(1.25, -0.833)$. Each of these models are simulated 200 times with a sample size n (of x_t) equal to 50, 100 and 200. Results are summarized in Tables 1–3. In these tables the maximum likelihood method ($m\ell e$) and the maximum entropy method (me) are compared. Generally the me method has a larger standard deviation than that of $m\ell e$ method. In each method the accuracy improves as the sample size increases as would be expected. Out of the 200 independent simulations for each case the number of solutions which are in the correct region (in terms of location of roots inside or outside of the unit circle) is denoted by n_1. It is seen that the $m\ell e$ method has a better accuracy than the me method and the accuracy increases as the sample size increases.

The computation for the $m\ell e$ method is described in Breidt et al. [1]. The computation for the me method amounts to maximizing

$$\left\{\frac{c_4 - 3c_2^2}{c_2^2}\right\}^2$$

as a function of the parameters where c_i is the ith order sample central moment. In all these computations, the search for the maximum is carried out by a search on a grid of step size 0.05 with $\varphi_1 \in (-1, 1)$, $\varphi_2 \in (-2, 0)$. We note that in this range all possible regions are covered.

Table 1

$\varphi_1 = -0.05, \quad \varphi_2 = -1.5, \quad \sigma = 1.0$
$r_1 = 0.8, \quad r_2 = -0.8333$

n	50		100		200	
method	mle	me	mle	me	mle	me
n_1	114	74	179	140	199	177
$\hat{\varphi}_1$	−0.0658	−0.0027	−0.0440	−0.0343	−0.0471	−0.0592
$\widehat{sd}$	0.2183	0.2788	0.1132	0.2192	0.0770	0.1666
$\hat{\varphi}_2$	−1.6311	−1.6818	−1.5638	−1.5746	−1.5330	−1.5634
$\widehat{sd}$	0.2338	0.2754	0.1741	0.2440	0.1079	0.2259
$\hat{\sigma}$	1.0377	1.1421	1.0182	1.0584	1.0127	1.0629
$\widehat{sd}$	0.2684	0.3003	0.1720	0.2167	0.1130	0.1876

Table 2

$\varphi_1 = 0.02, \quad \varphi_2 = -0.67, \quad \sigma = 1.0$
$r_1 = 1.206, \quad r_2 = 1.2367$

n	50		100		200	
method	mle	me	mle	me	mle	me
n_1	132	111	168	153	193	173
$\hat{\varphi}_1$	0.0155	0.1479	0.0173	0.0158	0.0206	0.0314
$\widehat{sd}$	0.1152	0.3473	0.0709	0.2605	0.0505	0.1409
$\hat{\varphi}_2$	−0.6401	−0.4677	−0.6479	−0.5753	−0.6595	−0.6333
$\widehat{sd}$	0.1074	0.2832	0.0628	0.2196	0.0450	0.1569
$\hat{\sigma}$	0.9447	1.1795	0.9764	1.0896	0.9928	1.0392
$\widehat{sd}$	0.1924	0.4247	0.1064	0.3107	0.0815	0.1444

Table 3

$\varphi_1 = 0.4, \quad \varphi_2 = -0.96, \quad \sigma = 1.0$
$r_1 = 1.25, \quad r_2 = -0.833$

n	50		100		200	
method	mle	me	mle	me	mle	me
n_1	143	123	181	162	199	180
$\hat{\varphi}_1$	0.3502	0.2376	0.4133	0.3682	0.4185	0.4350
$\widehat{sd}$	0.3734	0.5746	0.1710	0.3846	0.0673	0.1624
$\hat{\varphi}_2$	−0.9844	−0.9185	−0.9656	−0.9475	−0.9624	−0.9681
$\widehat{sd}$	0.1599	0.3489	0.0854	0.2176	0.0552	0.1333
$\hat{\sigma}$	1.0007	1.1755	1.0017	1.0740	1.0016	1.0449
$\widehat{sd}$	0.2294	0.3952	0.1247	0.2119	0.0853	0.1543

The simulation examples for the two-dimensional case are for the model

$$x_{t,\tau} - ax_{t-1,\tau} - bx_{t,\tau-1} = \varepsilon_{t,\tau} \tag{42}$$

with $\varepsilon_{t,\tau} = \eta_{t,\tau} - 1$, and $\{\eta_{t,\tau}\}$ independent exponential with mean one. When $|a| + |b| < 1$ in (42) we have

$$\begin{aligned} x_{t,\tau} &= \frac{1}{1 - aB_1 - bB_2} \varepsilon_{t,\tau} \\ &= \sum_{j=0}^{\infty} (aB_1 + bB_2)^j \varepsilon_{t,\tau}. \end{aligned} \tag{43}$$

When $1 + |a| < |b|$ in (1) then

$$x_{t,\tau} = -(bB_2)^{-1} \sum_{j=0}^{\infty} \left((bB_2)^{-1} - \frac{a}{b} B_1 B_2^{-1}\right)^j \varepsilon_{t,\tau} \ . \tag{44}$$

In the simulations the infinite sums in (43) and (44) are truncated at $j = 15$.

For each simulation $\{x_{t,\tau}\}_{t,\tau=-1}^{20}$ are generated and for each pair (a, b)

$$\varepsilon_{t,\tau} = (1 - aB_1 - bB_2)x_{t,\tau}$$

are computed for $t, \tau = 1, \ldots, 20$. Then $(\hat{a}, \hat{b})$ attains the maximum of

$$\left\{ \left(\frac{c_4 - 3c_2^2}{c_2^2} \right)^2 \right\}$$

(where c_i is the ith central sample moment of $\varepsilon_{t,\tau}$) and is searched for on a grid of step size 0.025.

Four models are considered with

$$(a, b) = (-0.4, 0.5),\ (0.4, 0.5),\ (-0.4, 1.7),\ (-0.4, -1.7).$$

For the first two models the search is in the range $a, b \in (-1, 1)$ while the range of search for the last two models is $a, b \in (-2, 2)$. Each model is independently simulated 100 times. Results are summarized in the following Table 4. The estimation procedure is seen to be effective in this simple example.

Table 4. Estimates of parameters for 100 simulations

(a,b)	$-0.4,\quad 0.5$	$0.4,\quad 0.5$	$-0.4,\quad 1.7$	$-0.4,\quad -1.7$
$(\bar{\tilde{a}},\bar{\tilde{b}})$	$-0.4114, 0.4939$	$0.3848, 0.5050$	$-0.4070, 1.7000$	$-0.3970, -1.7100$
s.d.	$0.0836, 0.0778$	$0.0716, 0.0665$	$0.0900, 0.1353$	$-0.0842,\quad 0.1137$

References

[1] F. Breidt, R. Davis, K. S. Lii, and M. Rosenblatt, 1991: Maximum likelihood estimation for noncausal autoregressive processes. *J. Mult. Anal.*, **36**, 175–198.

[2] P. Brockwell and R. Davis, 1991 *Time Series: Theory and Methods* ., Springer Verlag.

[3] Q. Cheng, 1990: Maximum standardized cumulant deconvolution of non-Gaussian linear processes. *Ann. Statist.*, **18**, 1774–1783.

[4] D. Donoho, 1981: On minimum entropy deconvolution, in *Applied Time Series Analysis II* (ed. D. F. Findley) 565–608.

[5] E. Gassiat, 1990: Estimation semi-paramétrique d'un modèle autorégressif stationnaire multiindice non nécessairement causal. *Ann. Inst. Henri Poincaré.*, **26**, 181–205.

[6] J.-P. Kreiss 1987: On adaptive estimation in autoregressive models when there are nuisance functions. *Statist. Decisions.*, **5**, 59–75.

[7] K. S. Lii and M. Rosenblatt 1982: Deconvolution and estimation of transfer function phase and coefficients for non-Gaussian linear processes. *Ann. Statist.*, **10**, 1195–1208.

[8] Pham-Dinh-Tuan 1977: Estimation of parameters of a continuous time Gaussian stationary process with rational spectral density. *Biometrica.*, **64**, 385–399.

[9] R. A. Wiggins 1978: Minimum entropy deconvolution. *Geoexploration.*, **17**.

Mathematics Department
University of California, San Diego
La Jolla, CA 92093-0112

mrosenblatt@ucsd.edu

Feature and contour based data analysis and assimilation in physical oceanography

Arthur J. Mariano and Toshio M. Chin

1. Introduction

The goal of this contribution is to summarize current research in physical oceanography on identifying, analyzing, modeling, or assimilating (into numerical ocean models) coherent structures seen in data. The structures considered here are dynamical features such as vortices and fronts, sets of contour positions, and the location of property extrema. The key analysis assumption allows the structure present in the data to determine the coordinate system for analyzing the data. Recent developments in the last decade are emphasized here rather than attempting a thorough historical review. Also because of space limitations, only the aspects central to our theme from the quoted studies will be noted. Because of the authors' research interests, examples will focus on meso- and large-scale phenomena in the Northwest Atlantic Ocean.

For an example of this, see the colour plate at the beginning of this volume, which we shall refer to as Fig. 1 for this paper.

An example of an ocean field, namely sea surface temperature in an area encompassing the Gulf Stream, is shown in Fig. 1. Parts of the field look random but the field is clearly heterogeneous with well-defined features. Evident in this figure is a broad range of spatial scales from Gulf Stream streamers/shingles, with cross-feature scale of a few *km* and along-feature scale of tens of *km*, to the horizontal scale of the subtropical gyre, whose northern boundary is demarked by the eastward penetration of the Gulf Stream, of a few thousand *km*. The diameter of coherent vortices such as warm-core rings, visible north of the stream in Fig. 1, and cold-core rings, visible south of the stream, is about couple hundred *km* (Olson, 1992). The width of the Gulf Stream is about 100 *km* with a dominant meander wavelength of 300-400 km (Watts, 1983). These features are a product of heterogeneous nonlinear dynamics in space and time over a broad range of scales and stochastic forcing. Coupled nonlinear dynamics of the state variables and sparse noisy observations of both the primary state variables and the stochastic forcing functions make inverse problems in physical oceanography extremely challenging (Bennett, 1992). Sparse noisy observations of the ocean should be assimilated into imperfect and simplified dynamical models to maximize the information content of both (Ghil and Malanotte-Rizzoli, 1991).

Oceanographic and atmospheric sampling strategies, analysis methods, and assimilation techniques are mostly formulated in an Eulerian framework, *i.e.* fixed in space such as the common Cartesian rectangular coordinate system. In general, every analysis variable, $Q(x, y, z, t)$, is assumed to be a function of 2 horizontal coordinates (x, y) such as (longitude, latitude), (along-, cross-stream), or (radial, azimuthal), a vertical coordinate z usually parallel to gravity, and time t. In most oceanographic studies, an Eulerian viewpoint is adopted and the bulk of analysis techniques concentrate on measuring and analyzing fields or time series at one or a small number of observing locations. It is even common to estimate the mean Eulerian velocity and eddy kinetic energy levels from Lagrangian observations of ocean flow (*e.g.* Rossby *et al.*, 1983).

A non-Eulerian framework, on the other hand, frequently leads to a more natural representation for oceanic phenomena. In particular, usefulness of feature-based sampling in field experiments, for certain studies, has been well-known for a long time. Iselin and Fuglister (1948) stated that a ship-based investigation of Gulf Stream paths using planned sections or a fixed grid would be an inherent waste of time. Feature-based sampling would yield the most useful information for studying Gulf Stream paths (Fuglister, 1963). In the early 1980s feature-based sampling strategy was employed in the Warm Core Ring (WCR) experiment, to study the closed currents found just north of the Gulf Stream that retain a circular structure with diameter of a few 100 km for several months. Satellite images, like the one shown in Figure 1, were consulted to make repeated surveys of the same warm-core ring, the infamous "82B", over a span of 2 years for the purpose of testing dynamical theories of ocean vortices and for quantifying the contribution of warm-core rings in exchanging heat, salt, biological communities, etc. between the Sargasso Sea and the Slope Water. Satellite data and seeding rings and other ocean features with drifters in combination with dedicated ship-based surveys has yielded a wealth of information on ocean dynamics and its influence on biology. (see volume 90 (C5) of the Journal of Geophysical Research, 1985 for a special issue dedicated to the

WCR experiment; The Ring Group, 1981; Rossby *et al.*, 1983; Brown *et al.*, 1986; Richardson *et al.*, 1989; Olson *et al.*, 1994).

Lagrangian, or particle following, coordinate system is a commonly used non-Eulerian framework in studies of fluids, and it is inherently feature-based. A Lagrangian approach for estimating terms in the conservation of potential vorticity equation was adopted by Kennelly (1984) for warm core ring "82B" and by Arhran and Colin de Verdiere (1985) for a submesoscale coherent vortex observed in the Eastern Atlantic Ocean. Mariano and Rossby (1989) have shown that estimating the terms in the potential vorticity equation in a local space/time coordinate system defined by the centroid of SOFAR float clusters, i.e., a Lagrangian viewpoint, and a feature-based temporal differencing scheme yielded more accurate estimates than traditional Eulerian field estimates and fixed interval differencing schemes.

The following example from Halkin and Rossby (1985) illustrates another data analysis situation where Eulerian-based averaging yielded poor estimates. In their study of Gulf Stream dynamics, temperature and velocity data as a function of depth and cross-stream location from 16 transects, some partial, centered at 73° W were collected on a quasi-bi-monthly schedule for two and one-half years. The Gulf Stream velocity and temperature transects averaged in a "stream" coordinate system, defined by placing the origin of each transect at the location of maximum transport for that transect, produced a remarkably coherent Gulf Stream structure relative to Eulerian estimates smeared by spatial meandering (Fig. 2). The maximum amplitude of the Eulerian-based average (Fig. 2a) down-stream velocity is only 100 cm/s, almost a factor of two smaller than measured down-stream surface velocities in the Gulf Stream core (Fig. 2c). The maximum amplitude of a feature-based average (Fig. 2b) is over 170 cm/s. The stream-based average transect (Fig. 2b) also has stronger gradients and exhibits the well-known (Halkin and Rossby, 1985) offshore deepening of the high velocity core to a greater degree than the Eulerian-based average transect (Fig. 2a). This example clearly shows the importance of choosing the right coordinate system for averaging data. The Eulerian average (Fig. 2a) does not resemble any of the data realizations (Fig. 2c) used in the averaging. The stream-based average (Fig. 2b) is more realistic since it more closely resembles each of the data realizations that are used in the averaging.

Kwok *et al.*, (1990) used a feature-based approach for estimating the motion of ice floes from Synthetic Aperture Radar (SAR) images of the ocean surface. The boundaries of ice floes were identified from a sequence of SAR images and tracked in time for the purpose of estimating trans-

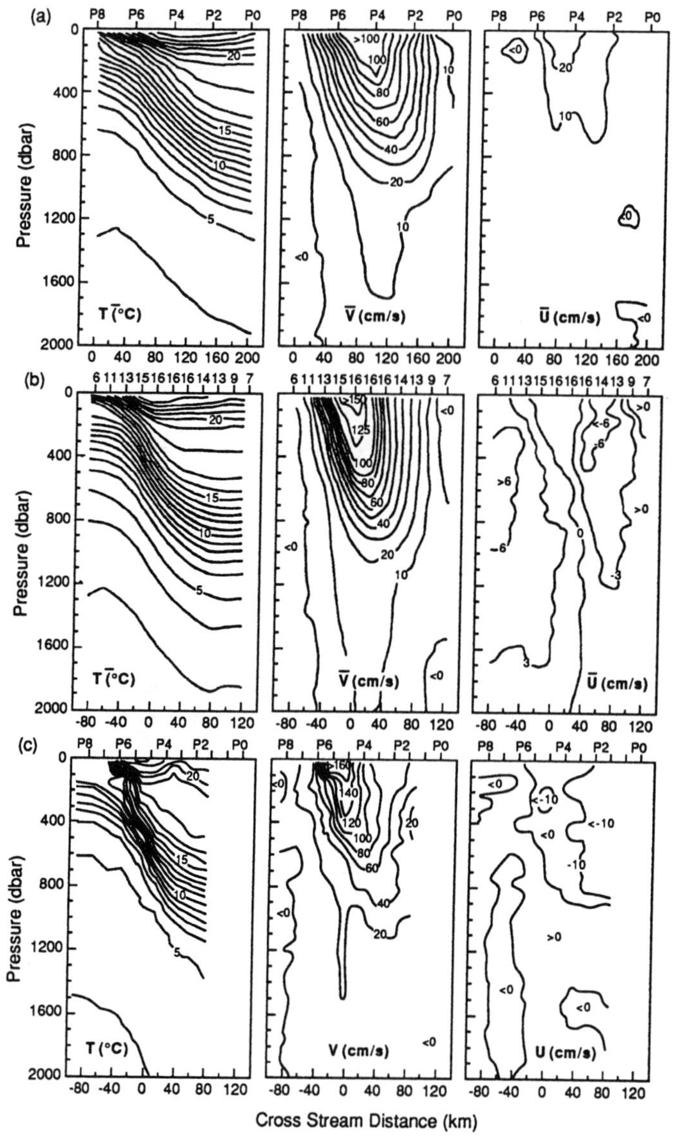

Figure. 2 (a) Eulerian average temperature and velocity transects from repeated surveys of the Gulf Stream near 73° W calculated by averaging the data at each of the nine stations, $P0$,$P1$, ...,$P8$. (b) A stream-based average (see text) for the same data that was used to generate (a). Note that the high velocity core for the stream-averaged data is 50% greater than the Eulerian-based average. The number of Pegasus profiles used in the average is given above the lower set of plots. (c) Transect from March, 1982 showing a typical realization. Note the seasonal outcropping of the isotherms due to winter cooling. (from Halkin and Rossby, 1985)

lational and rotational velocities of the ice field. Their approach involves transforming two-dimensional boundary points (*longitude* x, *latitude* y) into one-dimensional curves (ψ, s) whose first coordinate ψ is defined by the angle between the local tangent at a boundary point and a reference line.

The second coordinate is the arclength s calculated along the two-dimensional boundaries. Ice floes are matched between a pair of images by finding the maximum cross-correlation between the one-dimensional curves from each image. This approach matches ice floes based on most similar shapes. For a unit time between a pair of images, translational velocity is estimated as the difference of the position of the ice floe's centroid $(\hat{x}, \hat{y})$ from each images while the rotational motion of each ice floe is calculated by differencing the average angle $\hat{\psi}$ from each image. One conclusion of this study is that this feature-based approach is far more economical for large satellite data sets than traditional two-dimensional cross-correlation methods between the set of boundary points.

The main premise of this introduction is that non-Eulerian approaches for sampling the ocean and analyzing ocean data may be more fruitful than traditional Eulerian-based approaches in many applications. It will be argued below that non-Eulerian data assimilation methods should also be developed and compared to traditional Eulerian-based methods. In the next two sections 2 and 3, a general technique that allows features in the data to define the coordinate system for further analysis and assimilation of the data into numerical models is applied to a diverse set of oceanographic data. This technique, named contour analysis by Mariano (1990), is a methodology for analyzing and assimilating data that is in the spirit of the computational fluid dynamical technique of contour dynamics (Deem and Zabusky, 1978; Zabusky, 1981; Zabusky *et al.*, 1979; 1983; Pratt and Stern, 1986; Stern and Pratt, 1985; Pullin, 1992). In sections 2 and 3, our contour-based technique for combining information about contour position and for quantifying contour variability, respectively, is detailed. In section 4, the use of contours, a Lagrangian viewpoint and/or feature-models for assimilating data is discussed. A few concluding remarks and a contour-based approach to nonlinear parameter estimation are given in section 5.

2. Contour Analysis

As mentioned in the introduction, the combining (or averaging, blending, melding) of different analyses of geophysical fields is an important component of a data analysis system. The previous Gulf Stream example (Fig. 2) and the next example (Fig. 3) show that it may be necessary to devise alternate methods of melding the data often by transforming the data into a *local* coordinate system defined by features in the data. Mariano (1990) presents one such method, contour analysis, and this section concentrates on this approach by illustrating it with a number of examples. The oceanographic examples presented below were chosen to indicate the variety of data sources to which the contour analysis method can be applied. The criteria for the success of the method is that the resulting field resembled that of the initial fields, in terms of spatial gradients and feature shape, to a greater degree than the results of more traditional estimators.

Suppose that a dynamical model produces a forecast of the streamfunction field at 100 m depth and a statistical analysis of *in situ* data produces another estimate of the 100 m streamfunction field. What is the "optimal" method for melding the two 100 m streamfunction field estimates and producing one 100 m streamfunction field estimate that is more accurate than both initial estimates? Let $\psi_1(x,y)$ and $\psi_2(x,y)$ denote the two fields, which have corresponding estimation variance fields $e_1^2(x,y)$ and $e_2^2(x,y)$, respectively. The classic "optimal" estimator (*e.g.* Gelb; 1986, Ghil *et al.*, 1981) for melding these fields is,

$$\hat{\psi} = \frac{(e_2^2\psi_1 + e_1^2\psi_2)}{(e_1^2 + e_2^2)}. \tag{1}$$

If ψ_1 and ψ_2 are independent estimates, the variance field of (1) is

$$\hat{e}^2 = \frac{e_1^2 e_2^2}{(e_1^2 + e_2^2)}. \tag{2}$$

If ψ_1 and ψ_2 are unbiased estimates, $\hat{\psi}$ is also unbiased and has minimum variance with respect to all linear combinations of ψ_1 and ψ_2. The estimation error of our melded field, $\hat{e}$, is less than the estimation error of both initial fields, e_1 and e_2.

To understand the difficulties in this approach, consider the following example which illustrates some undesirable and well-known side effects of the classic estimator, $\hat{\psi}$ (1). Suppose, as it is shown in Figs. 3a and 3b, that each 100 m streamfunction field consists of a circular ring and a front. The frontal location differences between the two fields may be due to inadequate data or model resolution, incorrect model physics, or simplifying dynamical/statistical assumptions in the analyses. Assume that the estimation variance of each field is the same, $e_1^2 = e_2^2 =$ constant. Hence the melded field (Fig. 3c) is an arithmetic average of the two fields at each point in the domain. The melded front is weakened and the melded ring is no longer circular and is also of low amplitude (Fig. 3c). Estimate (1) smears the signatures of important dynamical features that are present in each initial analysis. This smearing is an unavoidable problem with minimizing the L_2 norm, since minimizing the L_2 norm is a smoothing operation. This inherent smearing will be enhanced when melding data from heterogeneous regions during periods of explosive instabilities.

The contour analysis method was developed in order to obtain a formal procedure for determining the coordinate transformation that will eliminate the smearing associated with the classic "optimal" estimator (e.g., Fig. 3d shows the resulting melded field for the example above using the contour analysis estimator introduced next). In contrast to the studies mentioned in the introduction which pick a single point for a coordinate origin, contour analysis considers the whole field for the coordinate transformation.

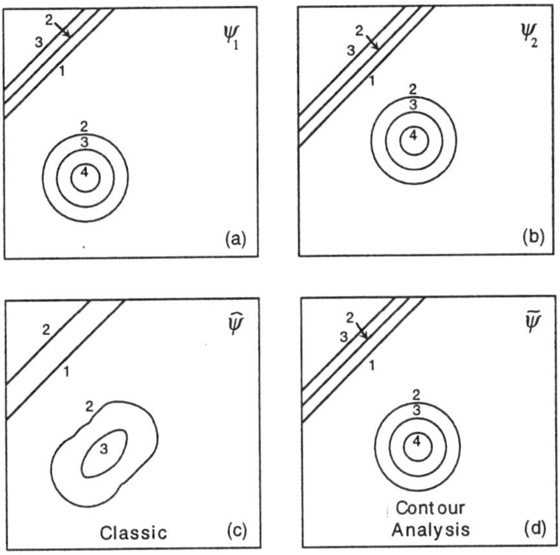

Figure. 3 (a) Streamfunction field one in nondimensional units. (b) Streamfunction field two. (c) The classic estimator's (1) streamfunction field from melding fields one and two. (d) Contour analysis' (3) streamfunction field from melding fields one and two.

The fundamental difference between classic estimators such as (1) and the contour analysis approach is that contour analysis averages the position of the field values, rather than the field values at each point. The first step is to apply a contouring program to each of the estimates of the geophysical field to convert each data field into a series of contours. A pattern recognition program must then be used to determine which contours from the individual field estimates should be paired together: the position data from the representations of the contour are then melded to find the average position of the contour. Consider the melding of the two fields shown in Figs. 3a and 3b. Each contour from the first field, $c_1(n)$, is represented by an ordered set of $\boldsymbol{x}_1$ and $\boldsymbol{y}_1$ positions,

$$\{x_{1k}(n), \quad y_{1k}(n): \quad k = 1, 2, ..., n_1(n)\},$$

where $c_1(n)$ is read as the nth contour of field 1 and $n_1(n)$ is the number of position data points determined by the resolution of the first field estimate and the contouring program. Each of the n contours from the first field should be combined with the corresponding contours from the second field, $c_2(n)$, represented by an ordered set of $\boldsymbol{x}_2$ and $\boldsymbol{y}_2$ positions,

$$\{x_{2k}(n), \quad y_{2k}(n): \quad k = 1, 2, ..., n_2(n)\}.$$

The $n_i(n)$ points must be connected, so the values of $c_l(n)$ are not necessarily distinct. For example, for the fields in Fig. 3, $c_l(1) = 3$, $c_l(2) = 2$, $c_l(3) = 1$, $c_l(4) = 2$, $c_l(5) = 3$ and $c_l(6) = 4$ for $l = 1, 2$. The ideal case is when $c_1(n) = c_2(n)$ for all n. The case of left-over unmatched contours is discussed below.

In general, $n_1(n) \neq n_2(n)$ for all n. Also, in order to find the average location of the nth contour it is necessary to find points on each contour which correspond: they must be at a similar position in the geophysical feature. A practical solution is to interpolate the contour position data so that the Kth point for both contours is located at the same fraction of the total respective arc length along each contour. This practical solution is not a unique solution and other approaches should be studied. The arc length is measured relative to the same major feature that is evident in both contours, *e.g.* an inflection point, an extrema, or a common starting point on both features such as the eastern most point on a circular contour or the southern most position of a straight contour. The arc length $\boldsymbol{S}$ for each contour n is calculated by

$$S_l(n) = \sum_{k=2}^{n_l(n)} (\Delta x_{lk}^2 + \Delta y_{lk}^2)^{\frac{1}{2}};$$

$$\Delta x_{lk} = (x_{lk}(n) - x_{l(k-1)}(n)), \quad \Delta y_{lk} = (y_{lk}(n) - y_{l(k-1)}(n)),$$

where $l = 1, 2$ for the field estimates 1 and 2.

Corresponding contours $c_1(n)$ and $c_2(n)$ are both divided into $N(n)$ segments and the position of the segment nodes:

$$\{X_{lK}(n), Y_{lK}(n) : K = 0, 1, ..., N(n)\},$$

are found by using cubic splines to interpolate both the latitude and the longitude to the same fraction of the arclength along both contours. For each pair of contours, four cubic spline interpolations with arclength as the independent variable are necessary: one cubic spline interpolation for each of the two latitude data sets and one cubic spline interpolation for each of the two longitude data sets. The choice of arc length as the independent variable permits melding of multi-valued features. The number $N(n)$, which determines the resolution $\Delta S_l(n) = \frac{S_l(n)}{N(n)}$ of the interpolated contours is picked empirically and depends on $n_1(n)$, $n_2(n)$, resolution requirements, and computational speed. For most applications $N(n) = N$, *i.e.* all contours are divided into the same number of segments.

The melded contour field, denoted by $\tilde{\psi}$, can then be calculated for m contour pairs by

$$(3) \qquad \tilde{\psi}(X_K(n), Y_K(n)) = \tilde{\psi}(\sum_{l=1}^{2} w_{lK}(n) X_{lK}(n), \quad \sum_{l=1}^{2} w_{lK}(n) Y_{lK}(n)),$$

$$n = 1, 2, \ldots, m \quad \textit{and} \quad K = 0, 1, 2, \ldots, N(n).$$

Note that the contour-based estimator can not be written in terms of ψ since only contour positions, $(X_{lk}(n), Y_{lk}(n))$, from the lth realization of the kth segment node of the nth contour are averaged and no operations are performed on the ψ values after the initial selection of m ψ values for analysis. The averaging of the nth contour positions yields field estimates at points, $(X_K(n), Y_K(n))$, that are irregularly distributed in space. For many applications, such as initialization of a numerical model, the values of the analysis variable are needed on a regular grid. A practical solution is to use a large value for both m, the number of contour values, and for $N(n)$, the number of segment nodes, insuring that there are contours present all over the analysis domain for interpolation. A two-dimensional bicubic spline, a generalization of the one-dimensional splines described earlier, have been used for interpolating the $\tilde{\psi}$ values onto a regular grid (Mariano, 1990).

Although the contour analysis approach has been described for melding two field estimates, the procedure is not limited: if more than two fields are to be melded, the index l used above would simply range from 1 to the total number of estimates.

The weights, $w_{lK}(n)$, are picked for the specific problem to be solved and their sum must equal one to prevent biases. *Optimal* weights analogous to (1) for melding two fields are

$$(4) \qquad w_{1K}(n) = \frac{e_2^2(X_{2K}(n), Y_{2K}(n))}{(e_1^2 + e_2^2)}, \quad w_{2K}(n) = \frac{e_1^2(X_{1K}(n), Y_{1k}(n))}{(e_1^2 + e_2^2)}.$$

The estimation variance field of (3) with weights given by (4) is given by (2) evaluated at the position of the melded contours. Fig. 3d is a plot of the analytic solution for the first example using equal weights, $w_{1K}(n) = w_{2K}(n) = 0.5$. Other weights are given in the examples presented below and the estimation variance field associated with these weights are most easily calculated at the melded segment nodes using the estimation variance of the segment nodes of the initial contours.

Figs. 4a and 4b show two realizations of the Harvard Quasigeostrophic Open Ocean Model (Robinson and Walstad, 1987) output streamfunction fields (31 $\times$ 31) at 100 m depth that differ due to differing initial conditions. The melded field which results from using the classic estimator (1) yields a very smeared representation of reality (Fig. 4c). It would be difficult to claim that the field shown in Fig. 4c is an "optimal" estimate of the 100 m streamfunction field in the Gulf Stream. Contour analysis (Fig. 4d) produces an output more similar to the data from which it was calculated; the Gulf Stream maintains a uniform cross-stream structure and the rings are not smeared in strength. Also, the ring positions in 4d are the averaged positions of the ring positions in 4a and 4b and the shape of the Gulf stream front is the average of the frontal shapes shown in 4a and 4b. Generalization of this example to melding numerical model output from various initial conditions implies that contour analysis is ideal for averaging the results of Monte Carlo simulations or for producing a mean field from a multi-year numerical simulation of the ocean circulation. However, the formulation of a robust algorithm for these types of problems will be difficult due to the complex feature-matching that is needed for realistic basin-wide fields that contain features that change their topology.

Under certain circumstances, pre-processing steps will be necessary if there is not a one-to-one correspondence between contours of the initial fields. See Mariano (1990) for details of possible pre-processing steps, contour-matching algorithm, and computational tricks. However at times, there will still be a mismatch. For example for a given contour value, realization A contains one occurrence of that contour while realization B contains two occurrences of the contour. Regardless of what metric or set of rules is used to define similar contours, there will be an unmatched contour from realization B. A decision has to be made-discard the extra contour if it has large error associated with it, include the contour in your final estimate, or include it with a footnote detailing frequency of occurrence.

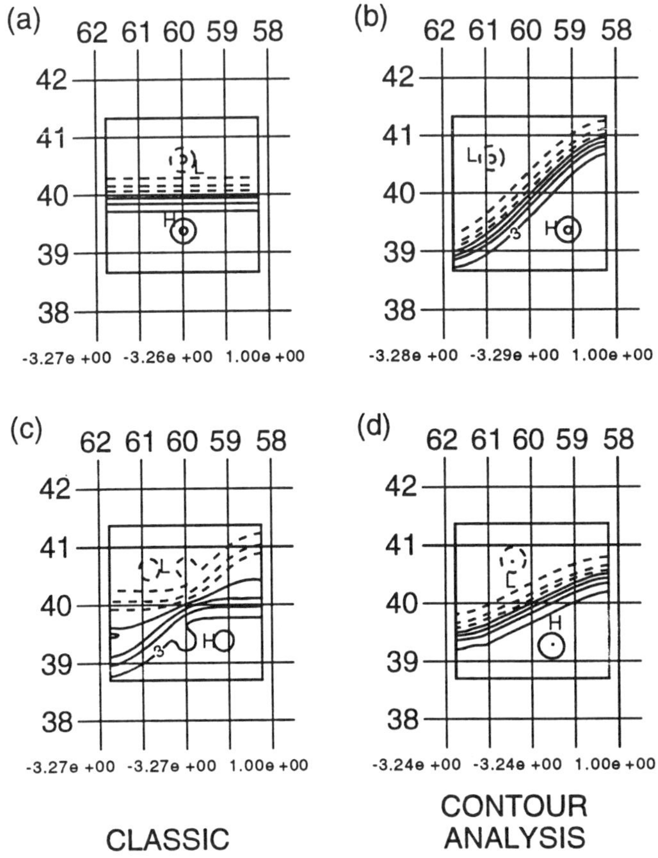

Figure 4. (a) Harvard open ocean quasigeostrophic model output streamfunction field one. The three numbers below each plot are minimum streamfunction value, maximum streamfunction value and contour interval. (b) Streamfunction field two. (c) The melded field using the classic estimator (1). (d) The melded field using contour analysis (2).

The latter approach was adopted by the lead author for the analysis of temperature sections when the position of the 29° C isotherm was only measured in three out of the eight transects. 29° C contour was included in the final contour-based average but with a footnote pointing out that this contour only appeared in three out of the eight realizations. As this example shows, the formulation of a set of expert rules for decision making is a nontrivial and important component of the contour analysis technique.

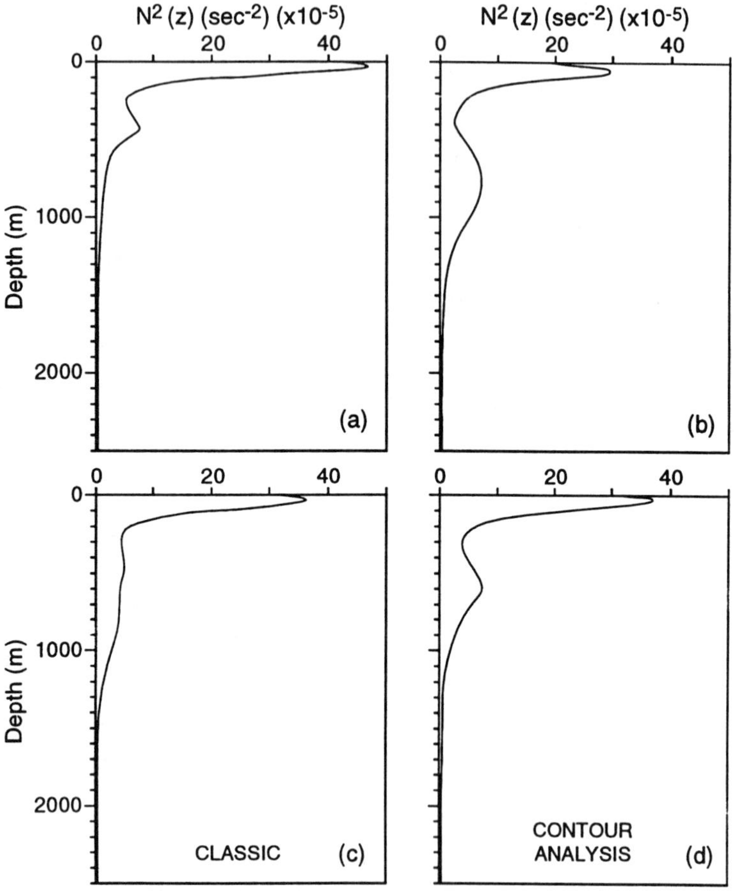

Figure 5. (a) Brunt-Väisälä (B-V) frequency versus depth curve from averaged slope water data. (b) B-V frequency versus depth curve from averaged Sargasso Sea data. (c) The results of averaging (a) and (b) together using the standard approach. (d) The results of averaging (a) and (b) together using contour analysis.

The next example shows how the contour analysis approach can be used for combining different estimates of a curve. Shown in Fig. 5a and 5b are two Brunt-Väisälä (B-V) frequency versus depth curves. The important differences between the curves are the strength and vertical location of the local maxima. Fig. 5c shows the smearing that results from the usual oceanographic practice of averaging the B-V frequencies at each depth. In order to preserve features, each curve is broken into segments; with break points located at local extrema in the curve and divided into $N+1$ points. For the independent depth variable, a linear interpolation is sufficient while for the dependent variable, a one-dimensional spline interpolation is necessary. These are averaged as before, but now since the contour values are variable, the contour value must be averaged too.

For the example shown in Fig. 5, the B-V curves (5a and 5b) were broken into four segments; the first data point to the first local maximum, the first data point after the first local maximum to the local minimum, the first data point after the local minimum to the second local maximum and the first data point after the second local maximum to the last point. For each segment, $N+1$ evenly spaced depth values as well as B-V frequency values are averaged. The resulting contour analysis curve is shown in Fig. 5d. Again, the contour analysis approach produces a profile whose shape is more similar to the initial analyses than the standard averaging procedure. Moreover, local maxima and vertical gradients, which are of extreme dynamical importance for determining necessary conditions for stability problems and for providing the vertical restoring force, are not smeared.

The next example extends the basic contour analysis algorithm for gappy and time-varying data such as the advanced very high resolution radiometers (AVHRR) remotely sensed sea surface temperature (SST) data. Fig. 1 is one of the best composite images of such data. In general, this data is gappy since the sensors can not "see" through the clouds and sense the much colder cloud-top temperatures instead of the ocean surface. By digitizing the location of maximum thermal contrast between the Gulf Stream and slope waters a time series of Gulf Stream Northern Edge Positions (GSNEP) can be generated. Nearby (in both space and time) positions are used to construct continuous paths of the Gulf Stream Northern Edge. These paths are important for navigation, fisheries research (*e.g.* Olson *et al.*, 1994), and for studying geophysical fluid dynamics. Processing and digitizing of two-day satellite data composites for studying Gulf Stream paths is described by Cornillon (1986). The goal is to fill in the space-time gaps in the data to produce a regularly-gridded path for further analysis and for input into dynamical models.

The space-time interpolation of the GSNEP is not a trivial interpolation problem since longitude versus latitude is often multi-valued. Contour analysis deals with the multi-valuedness by forming two single-valued space series for a given composite; latitude versus arc length and longitude versus arc length. The process of filling a spatial gap consists of first searching into the past (before the interpolation time) and then into the future (after the interpolation time) to find the nearest time where data exists in the region of the data gap. Fig. 6 shows an example of this procedure. To interpolate the gappy region in Fig. 6b between 66^oW and 58^oW, data from a longitudinal range, determined by the data gap and extended by the phase speed estimates, from Fig. 6a and 6c are used. Latitudes from the past and future space series are averaged together, as well as the longitudes.

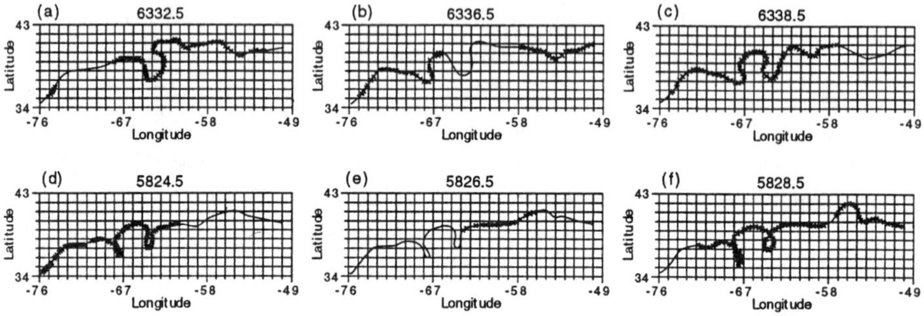

Figure 6. The Xs are Gulf Stream Northern Edge Positions (GSNEPs) from two day composites of satellite IR data digitized at URI. The contour analysis fit is given by the solid line. Truncated Julian day is atop of the longitude/latitude grid. (a)-(c) Data from two day composites 6332.5 and 6338.5 are used for interpolating the data gap on day 6336.5. (d)-(f) Another example of interpolating multi-valued features using contour analysis.

The basic contour analysis algorithm is used except for one major modification: contours must be propagated to a common analysis time and this requires estimating phase speeds from the data. Past data are propagated forward while future data are propagated backward using local (in space and time) phase speed estimates. Chin and Mariano (1993, 1995) have extended the basic algorithm of Mariano (1988, 1990) by incorporating two-dimensional phase speeds estimated from a better feature matching algorithm and a three-step Kalman smoother (forward Kalman filter, backward Kalman filter, inverse covariance weighting of both sweeps) for more optimal averaging. The feature matching algorithm relies on analyzing local extrema of contour position deviations from a reference curve. Put simply, phase speeds are first estimated by tracking the movement of these extrema calculated directly from the data input augmented by initial estimates of contour positions in the gappy regions. The initial phase speed and position estimates are improved by the Kalman smoother. The governing equations in the Kalman smoother for the position and phase speed estimates are coupled and this requires solving the resulting system in an iterative fashion. As expected, the GSNEPs were better estimated using the newer algorithm relative to the earlier algorithms in data-sparse regions.

Monte Carlo simulations were performed by removing all data from a set of 25 two-day composites that were randomly picked, using the contour analysis technique to estimate the path positions, and comparing the path estimates with data not used in the interpolation. These simulations showed that the position estimation error was on the order of 10 km for data-rich periods and for fairly weak Gulf Stream meandering. The error is on the order of 100 km for data-sparse periods and vigorous Gulf Stream meandering. Apparently poor interpolations (an example is shown in Fig. 7) generally occurred at times immediately following ring births during strong ring-stream interactions, or when the sparsity of data resulted in a poor estimate of the phase speed. The success rate of our latest algorithm

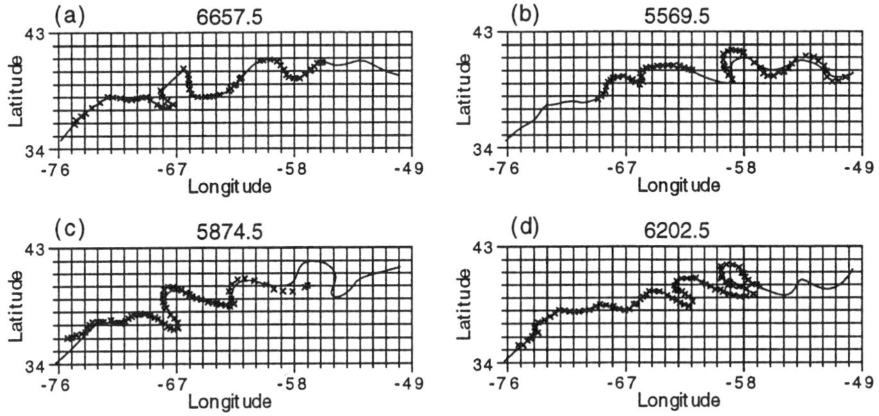

Figure 7. Same as (6) but showing poor interpolations. (a) Contour analysis produces an unrealistic interpolation of a meander which is about to form a warm core ring. Due to clouds, the closest input data are from 6 days before and 8 day after the interpolation time and do not contain a signature of the large meander. (b)-(d) Three of the worst interpolations from the simulation described in the text. None of the input data shown here was used for the interpolation.

is over 90% and it is being applied to path data from the Kuroshio current (the Pacific "equivalent" of the Gulf Stream). Poor estimates were improved by judicial addition of data points for the final data product (Chin and Mariano, 1995). It is evident in Fig. 6 and in careful inspection of over two thousand (2053) interpolated Gulf Stream paths, for the years 1982-1993, that the contour analysis approach is well suited for space-time interpolation of frontal paths containing multi-valued features.

The success rate of algorithms for repeated sampling of vortices in numerical simulations of two-dimensional and (quasi-) geostrophic turbulence (Weiss, 1981; Babiano *et al.*, 1987; Benzi *et al.*, 1988; McWilliams, 1990; McWilliams *et al.*, 1994) has improved to over 90% correct selections (determined by comparing the automatic algorithm to results from careful manual inspection of the fields). The largest failure rates occur during the first stage of eddy merger events when the algorithm only identifies one of the merging eddy. In our Gulf Stream northern edge example, the algorithms also have the most problems during events where the topology of the features of interest change. Reliable, robust, and fully automated pattern recognition software is of paramount importance for feature and contour-based analysis methods and is usually the most difficult component to develop.

3.Empirical Orthogonal Contours

Contour analysis preserves the shape of features by averaging in a data adaptive curvilinear coordinate system defined by contour position. A natural next step is the analysis of a system's variability in the same curvilinear coordinate system. Traditional oceanographic variability analysis (e.g. Weare et al., 1976, Halliwell and Mooers, 1983 and Preisendorfer, 1988) have relied on empirical orthogonal functions (EOFs), a technique popularized by Lorenz (1956). The proposed technique *melds* contour analysis with EOF analysis and hence its name, empirical orthogonal contour, hereafter EOC.

One may argue that EOCs are nothing more than complex EOFs of contour data. However, this argument is false because of the following reasoning. The most important components of contour-based techniques are contour matching, feature correspondence and contour segmenting, which are used for constructing the coordinate system, for analyzing the data, from the data themselves. The procedure for implementing these components are an integral part of the EOC technique and thus the technique detailed next deserves its own name. EOCs are derived by solving the eigensystem of the complex covariance matrix in contour space.

The efficient representation of two-dimensional contour variability by EOCs requires solving the eigensystem of the covariance matrix of contour segment nodes. Define $i = \sqrt{-1}$ and form the set of complex position data for each field realization, $l = 1, \ldots, R$, from the contour segment nodes,

$$(5)\quad \boldsymbol{S} = [S_{lK}(n) = X_{lK}(n) + iY_{lK}(n) : K = 1, 2, \ldots, N(n); n = 1, \ldots, m(l)]$$

For each contour value indexed by n and for each field realization l, K indexes the segment node positions. $m(l)$ is the number of contours in each realization l. In this formulation, it is not necessary for each realization to contain all the contours being analyzed. The corresponding set of mean positions, determined by averaging over the R realizations in contour space (eqn. 2), is

$$(6)\quad \bar{\boldsymbol{S}} = \left[\bar{S}_K(n) = \bar{X}_K(n) + i\bar{Y}_K(n) : K = 1, 2, \ldots, N(n); n = 1, \ldots, m\right],$$

where m is the maximum number of analysis contours contained in any field realization, $m = max(m(1), m(2), \ldots, m(R))$. It should be noted that this averaging is performed at the segment nodes, which are evenly spaced along the contour and have been measured relative to features in the contour. This allows the averaging of contours of different lengths. For each contour value, one is allowed to choose the number of segment nodes, $N(n)$. However, when averaging the segment nodes from all realizations of the nth contour value, the number of segment nodes must be equal.

The covariance matrix for the set of contour segments $\boldsymbol{S}$ is given by

$$(7)\qquad \boldsymbol{H} = <(\boldsymbol{S} - \bar{S})(\boldsymbol{S} - \bar{S})^{\dagger}> = <\boldsymbol{S}'S'^{\dagger}>,$$

where the expected value operator, $<>$, is calculated over the R realizations and $\dagger$ denotes complex conjugate transpose. The order of the Hermitian matrix $\boldsymbol{H}$ is $M = \sum_{n=1}^{m} N(n)$.

The eigensystem with eigenvalues, λ_e, is

$$(8)\qquad \boldsymbol{H}E_e = \lambda_e \boldsymbol{E}_e,$$

The position data can be expanded in terms of the mean contour and *nsig* significant orthogonal eigenvectors, $\boldsymbol{E}_e$,

$$S = \bar{S} + \sum_{e=1}^{nsig} \boldsymbol{p}_e E_e,$$

where the set of R principal components for each eigenvector $\boldsymbol{E}_e$ are

$$\boldsymbol{p}_e = p_{1e}, \ldots, p_{Re}.$$

p_{le} modulates the $\boldsymbol{e}^{th}$ spatial eigenvectors for the lth field realization. Each complex eigenvector explains some fraction of the total observed movement of the set of analyzed contours. They are ordered such that the first mode explains the largest fraction of contour position variance, *i.e.* the largest fraction of contour movement. The next mode explains the largest fraction of the remaining contour position variance and so on for the higher modes.

Not all of the calculated eigenvectors are used in decomposing the data for analysis, only *nsig* significant eigenvectors are used. The choice of *nsig* is difficult in practice (Preisendorfer, 1988). The eigenvectors are arbitrary to within multiplication by a complex number. It is customary to normalize each eigenvector to unit magnitude and form an ortho-normal basis, viz.

$$\boldsymbol{E}_\xi{}^{\dagger}\boldsymbol{E}_e = \begin{cases} 1, & \text{if } \boldsymbol{\xi} = \boldsymbol{e} \\ 0, & \text{if } \boldsymbol{\xi} \neq \boldsymbol{e} \end{cases}$$

However, an arbitrary phase factor still remains. Horel (1984) states that phase ambiguity has limited the usefulness of complex EOF analysis. Also, one of the main conclusions of Davis *et al.* (1991) is that different rules should be researched to eliminate the arbitrary phase problem. A new rule for setting the arbitrary phase factor is introduced next. The arbitrary phase factor may be chosen i) to force an agreement with independent information or a theoretical model, ii) for graphical or visual reasons, or ii) by a traditional rule. Hardy and Walton (1978) suggested that the arbitrary phase factor should be chosen so that the mean value of the phase of the complex principal components is approximately zero when the data is expanded in terms of all the eigenvectors. This is a very reasonable choice since the expected value of all the detrended data, $\boldsymbol{S}'$, is $\overline{X'} = \overline{Y'} = 0$, which is defined to be zero phase (Preisendorfer, 1988). The arbitrary phase factor can also be chosen so that the mean value is exactly zero when the data is expanded in terms of only the significant eigenvectors or just the first eigenvector. This phase selection rule has been used in most of the studies cited above. It is shown below that Hardy and Walton's rule, termed HW, may lead to an undesirable orientation of complex statistical modes. Our working definition of a desirable modal orientation is that the principal component phases corresponding to the first statistical mode, eigenvector $\boldsymbol{E}_1$ of (8), cluster about 0 and 180 degrees.

Suppose that a complex EOF analysis is performed on a data set consisting of an infinite number of realizations (Fig. 8) and all the variance is explained by one eigenvector or equivalently one statistical mode with an arbitrary phase factor (Fig. 8b). The corresponding series of the principal component phases are either $+90°$ or $-90°$. The average phase of this principal components is $0°$ and the calculated mode requires no rotation to insure Hardy and Walton's criterion of zero average principal component phase. However, this mode (Fig. 8c) is not the primary variability pattern, as defined above, since it must be rotated by $\pm 90°$ for each realization (Fig. 8d). Also, it is obvious that the calculated mode (Fig. 8b) does not have the same variability as the initial data (Fig. 8a). This example illustrates the need for an alternate rule.

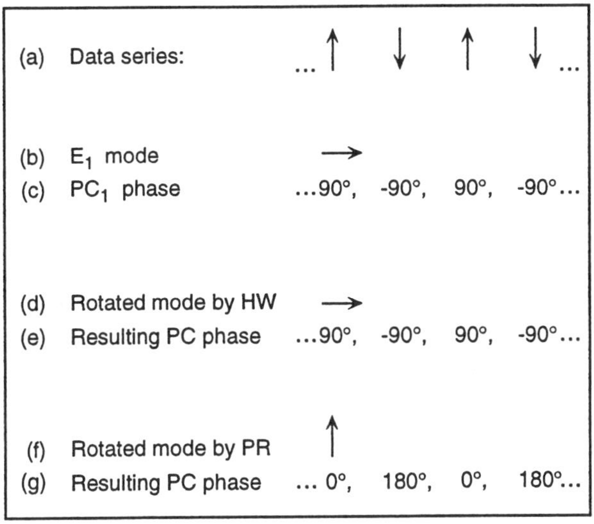

Figure 8. (a) A time series of some vector quantity. (b) The eigenvector associated with the covariance matrix of this time series and (c) its corresponding principal component phase time series. (d) The rotated eigenvector according to the HW phase rule and (e) its corresponding principal component phase time series. Note the undesirable phase distribution of ± 90 degrees. (f) The rotated eigenvector according to the PR phase rule and (g) its corresponding principal component phase time series.

The new criterion, termed the Phase Rule (PR), is to average the principal components phases in the following manner and rotate the eigenvectors by this average;

- the phases are defined from $\pm 180°$.
- if the phase is positive, average in as is.
- if the phase is negative, average in the phase + $180°$.

The average phase using PR for our example is now $90°$ and the calculated mode needs to be rotated by $90°$ (Fig. 8f). This rule insures us that the principal component phases will cluster near $0°$ or $180°$ on the average (Fig. 8g), which is the desirable distribution of phases for a primary mode. The reader can easily verify that a desirable pattern using PR is obtained for other data scenarios analogous to Fig. 8.

Fig. 9 shows one application of EOCs to the GSNEP data set described earlier. For this application, there is only $n = 1$ contour value for analysis; each data realization has only one contour, so $m = 1$; each realization of this contour value is divided into the same number of segment nodes, $N(1) = 400$; and the number of realizations R is 2053. Recall that EOCs are the dominant complex modes of contour position variability so they are represented as vectors at each of the average segment node positions. The vectors are scaled so that the length of the vector in Fig. 9 corresponds to the average distance moved by a point on the contour that is explained by each given mode. By inspecting the principal component time series, periods of vigorous Gulf Stream meandering, corresponding to complex principal components with large magnitudes, can be identified. Likewise, when the magnitude of the principal components is near zero corresponds to time periods of weak Gulf Stream meandering. Principal component spectra can be used for identifying the dominant frequencies in the data and a spatial spectra of the EOCs yields information on the dominant wavenumbers observed by the data. A thorough analysis of the principal component time series (not shown) is in progress and shows a clear annual signal in the first two modes. On the average, the Gulf Stream is further south in the spring and further north in the fall. During the course of the year the Gulf Stream moves, on the average, approximately 90 km. However, there is a great deal of interannual variability to this signal.

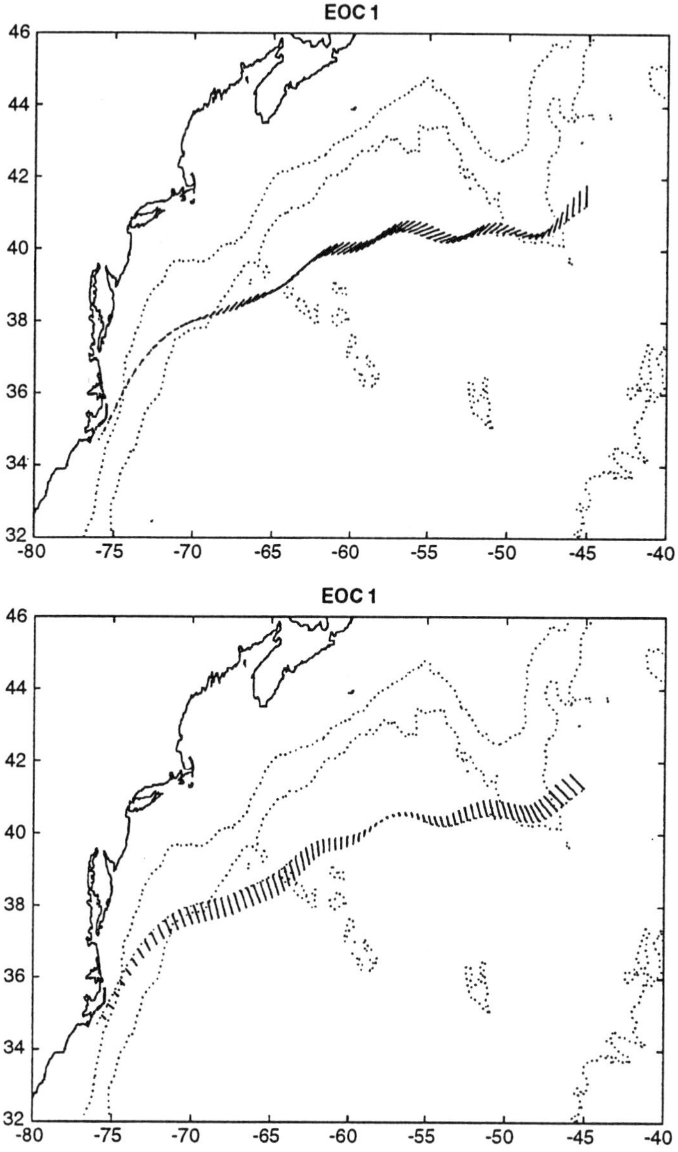

Figure 9. The two dominant modes of Gulf Stream path variability about the mean location of the GSNE. (a) The first EOC mode explains 41% of path variability and consists of fairly coherent SW-NE movements of the stream between 70° W and 50° W. (b) The second EOC mode explains 20% of path variability and corresponds to the Gulf Stream path becoming either predominantly zonal (east-west) or a Gulf Stream that flows more northeasterly.

4. Feature Models and Data Assimilation

In some cases, it is not possible to get the model analysis variables directly from the data. For instance, in the Gulf Stream, there is a large quantity of satellite-derived SST that can not be directly converted to shallow streamfunction or velocity values because of the lack of a tight T/S relationship in the upper few hundred meters of the water column in the Gulf Stream ring and meander region. Since information is precious and AVHRR images are information rich, it is advantageous to formulate procedures to extract as much information as possible. AVHRR images do a great job of tracking fronts and features. Feature models can be used to convert the AVHRR observations into streamfunction values. This approach was developed by Spall, Pinardi and Robinson at Harvard University in the mid 1980s (Robinson *et al.*, 1989; Spall and Robinson, 1990).

Weekly forecasts of the Gulf Stream ring and meander region using the multi-level Harvard open ocean quasigeostrophic (QG) model have been routinely calculated during the Gulfcasting program (Robinson *et al.*, 1989). Recent calculations have used primitive equation models; adjoint-based assimilation techniques; blended regional features such as the Gulf Stream recirculation, deep western boundary current and slope water current models. The available input data consists of NOAA AVHRR maps, *in situ* (A)XBTs, floats and/or satellite altimetric data and are not sufficient for direct calculation of the streamfunction fields at each model level needed for initializing the QG model nor for the calculation of initial and boundary values for a primitive equation model. The procedure for assimilating streamfunction features is discussed next. The feature model approach is quite general and allows for defining other fields, such as velocity, usually consistent with the prescribed dynamics.

These data are used in an initialization procedure based on a feature model approach: analytical models of warm core rings, cold core rings, and the Gulf Stream front are used to represent the full three-dimensional streamfunction fields given only surface frontal locations. The feature models used for Gulf Stream studies (Spall and Robinson, 1990) are shown in Fig. 10. A Gaussian thin jet model is based on four parameters; maximum surface velocity, v_{top}, depth of the thermocline, TC, bottom velocity, v_{bottom}, and the half width cross-stream distance where the velocity is $1/2 v_{top}$. Each ring model has four adjustable parameters. By analyzing all the data, the Gulf Stream front is located as well as the center and horizontal extent of Gulf Stream rings. At each ring location, an analytical three-dimensional stream function is centered at the location determined from the data. The along-stream coordinate of the Gulf Stream is digitized from a composite of data sources and the Gaussian thin jet model is centered about it (Fig. 10d). This approach has been successful for predicting strong mesoscale dynamical events in the Gulf Stream (Robinson *et al.*, 1989; Spall and Robinson, 1990).

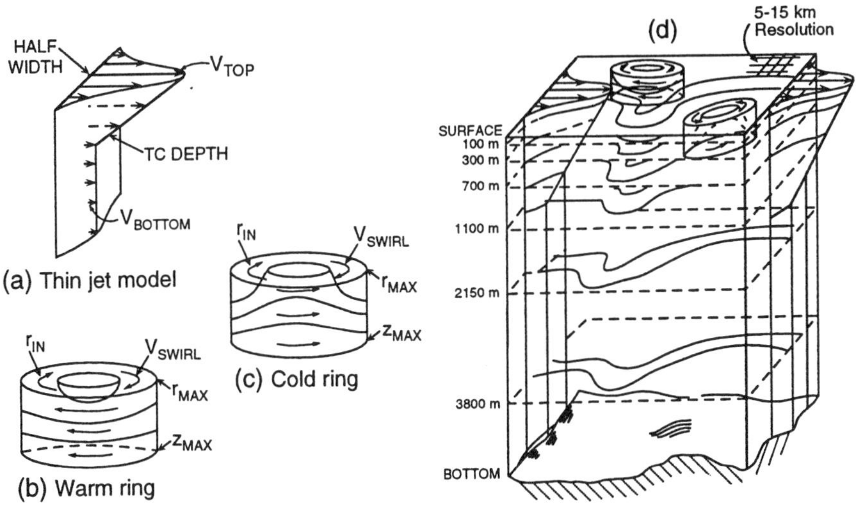

Figure 10. Three feature models for the (a) Gulf Stream, (b) warm and (c) cold core rings. (d) The initial streamfunction field based on feature model assimilation in the Harvard QG model.

In the first studies using feature models in ocean models, the feature models are put into a motionless ocean and a new reinitialization is performed when sufficient new data becomes available. Information about the fields outside the domain of the initialization features is lost from one Gulfcast prediction to the next. Consequently, the model must spend part of its integration time in spinning up the initial field to a balanced state. The first attempt to solve this problem, a generalization of the contour analysis procedure was developed with Spall and Robinson to incorporate both the model generated background field from the previous forecast and new frontal location information (Mariano, 1990).

For each field, a curvilinear coordinate system (τ, η) is defined by each frontal feature. For the Gulf Stream, the τ coordinate is defined by the geographical location of the Gulf Stream axis or north wall: the τ coordinate is given by the arc length along the current. The η coordinate is perpendicular to the τ coordinate at every point: it measures the distance between the background field and the stream axis. A polar coordinate system is used for each ring, where τ, η are the azimuthal and radial coordinates, respectively.

Using the notation of section 2, let the model generated field be the first field and the new satellite frontal segment information be the second field. Frontal information from field two, e.g. the center position and radius of each ring and the location of the Gulf Stream, is assimilated into field one. Assimilation consists of moving the frontal features from field one to the new locations of the frontal features from field two by averaging the τ coordinate using (3) with two sets of weights. If new information is available, $w_{1K}(j)$ is zero and $w_{2K}(j)$ is one. If there is no new information or the front did not move, $w_{1K}(j)$ is one and $w_{2K}(j)$ is zero. In essence, the points where $\eta = 0$ in field one are mapped to a new position determined by the position of the points where $\eta = 0$ in field two. The rest of the points, $\eta \neq 0$, are mapped from field one to field two by the mapping determined by the $\eta = 0$ points. In other words, the τ coordinate axis for each frontal segment is mapped from field one to field two and the points perpendicular to each τ coordinate axis are mapped by exactly the same transformation. Of course, variable weighting can also be used and this experimental technique has been dubbed rubber sheeting.

The use of feature-based models has been adopted by the Fleet Numerical Meteorology and Oceanography Center for operational forecasts of the ocean temperature and salinity fields (Clancy *et al.*, 1990). Their scheme is named OTIS, the Optimum Thermal Interpolation System, and utilizes both direct observations of the ocean and a "bogus" ocean data base. The "bogus" ocean data base uses frontal information obtained from satellites and water mass climatologies consisting of typical temperature profiles for over 30 ocean regions in the North Atlantic, W. Pacific and Polar seas. The temperature profiles are represented by three vertical Empirical Orthogonal Functions (EOFs) for each region and a canonical correlation analysis is used to adjust model variables given a set of EOF amplitudes, longitude, latitude, and time (Cummings, 1994). To determine what region or water mass, *e.g.* Slope Water, Sargasso Sea, or Gulf Stream, a model grid point is in, the location of the model grid point relative to frontal paths, determined from features in satellite AVHRR data, is found. Three-dimensional feature models of fronts and eddies, similar to the models shown in Fig. 10, are also blended into the analysis using an objective analysis scheme.

Similar feature-based data assimilation procedures have been used by hurricane forecasters (Mathur, 1991; DeMaria *et al.*, 1992). Because hurricanes form over data-sparse open oceans and have spatial scales smaller than the most frequently observed scales in the atmosphere, forecasters have relied on airplane flights into hurricanes to gather the necessary information for initializing smaller-scale nested models that are embedded in larger-scale weather models. However, the data from these reconnaissance flights may not be sufficient for mapping the dynamical fields associated with hurricanes. For instance, the spatial scales associated with the most extreme winds is on the order of tens of *km* in a vortex whose radius is on the order of a four to six hundred *km*. Bogus wind observations, based on a hurricane climatology archived at the National Hurricane Center and a feature model, are added into the part of the analysis domain occupied by the vortex.

Carter (1989) assimilated velocity and height (using a feature model based on Fig. 2b and temperature, pressure observations) data from isopycnal RAFOS floats sampling the main thermocline of the Gulf Stream. Carter developed an efficient Kalman filtering algorithm, based on pointers and exploiting the Lagrangian nature of the weekly observations of the model's state variables, for a one-layer shallow water equation model. The floats sample the position of the Gulf Stream over large spatial scales and a simple but realistic Gulf Stream with meanders is produced in his numerical simulations.

Ghil and Ide (1994) and Ide and Ghil (1995) used an Extended Kalman Filter to assimilate both Eulerian velocity measurements (available from current meters) and observations of the position of the vortex center (available from floats or satellite images) into dynamical systems defined by a sum of point vortices. Two key results of this study are that assimilation of vortex positions is preferable to the assimilation of Eulerian velocity observations and that feature models can reduce the dimension of the state space. Todling and Ghil (1994) showed that well-placed observations, based on predicted features, can efficiently track atmospheric instabilities. Strong instabilities lead to dominant spatial patterns and measurements should be placed in the regions of maximum amplitude.

Using a reduced-gravity shallow-water channel model to simulate an unstable oceanic front, Smeed (1995) showed that data assimilation based on averaging feature model parameters produces accurate forecasts and that feature models can do a better job of interpolating sparse data than standard objective analysis techniques. Smeed's reason for the success of his feature model approach is that *a priori* information-the number and shape of features-is added to the analysis. The main disadvantage of his approach is that information about the number and type of features is needed and it may not be readily available. In summary, these recent data assimilation studies for simple dynamical systems have indicated that feature-based assimilation is a viable technique that is probably superior to traditional methods.

5. Discussion

A contour-based approach to nonlinear parameter estimation is presented next. Mariano and Brown (1992) developed the parameter matrix algorithm for efficient objective analysis of large satellite data sets. This algorithm requires estimating nine parameters, in different time and space bins of the analysis domain, for the following anisotropic and time-dependent simplification of some true ocean correlation function,

$$C(dx, dy, dt) = C(1)[1. - (DX/C(4))^2 - (DY/C(5))^2] \times \exp -[(DX/C(6))^2 + (DY/C(7))^2 + (dt/C(8))^2], \tag{9}$$

$$DX = dx - C(2) \times dt \text{ and } DY = dy - C(3) \times dt,$$

where

- dx is the east-west lag; dy is the north-south lag; and dt is the time lag.
- $C(1)$ is the correlation at zero lag and equals one minus the normalized (by the field variance) measurement variance (which includes a subgrid scale component).
- $C(2)$ and $C(3)$ are the mean phase speeds in the east-west and north-south direction, respectively.
- $C(4)$ and $C(5)$ are the zero-crossing scales in the east-west and north-south direction, respectively.
- $C(6)$ and $C(7)$ are the spatial decay scales (or the e-folding scales) in the east-west and north-south direction, respectively.
- $C(8)$ is the temporal decay scale.

This correlation function can be rotated in space by $C(9)$, an arbitrary angle.

The estimation of the nine correlation parameters, $C(1), C(2), \ldots, C(9)$, from discrete three-dimensional (east-west, north-south, time) correlation functions is a difficult nonlinear fitting problem. All of the common nonlinear least-squares fitting algorithms do poorly. A relatively easier fitting problem results from fitting ellipses to the position of the zero-contour and e-folding contour to find the orientation ($C(9)$), length of the major and minor axis in lag space ($C(4)-C(7)$). Phase speeds ($C(2), C(3)$) are simply calculated from the average movement of the centroid of these contours. $C(1)$ is the value of the raw correlation at zero lag. $C(8)$ is calculated by fitting a temporal exponential function to the maximum values of the main group of contours, the "bullet", from the different temporal lags. Preliminary Monte Carlo experiments show that the contour analysis approach is superior to typical nonlinear least-square fitting methods. An example of the results from using this algorithm on real SST data is shown in Fig. 11. The algorithm is still not perfect! In operational use, a median filter is used to remove outliers.

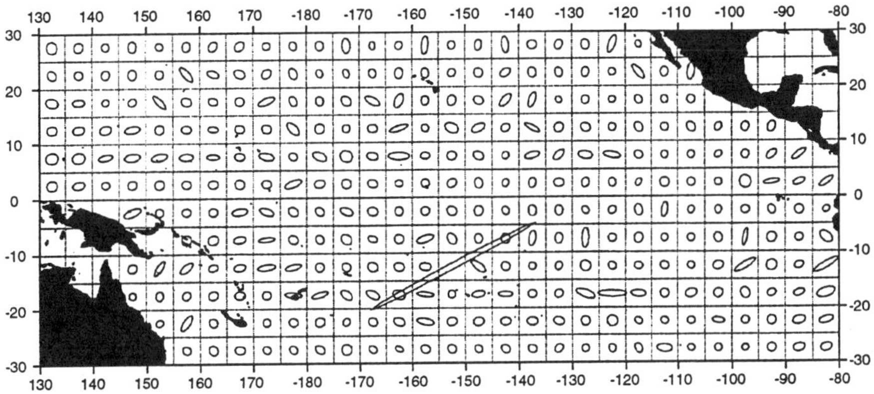

Figure 11. The spatial distribution of the east-west, $C(6)$,, north-south, $C(7)$, e-folding scales and rotation angle $C(9)$ calculated from SST correlation functions for the tropical Pacific Ocean. The ellipses are scaled so that the one-half of the length of the major and minor axis are the e-folding scales. The ellipses are rotated by $C(9)$ which is measured relative to due east in the usual sense of positive angle being counter-clockwise.

As the GSNEP example and the last example illustrate, it is difficult to obtain a perfect contour analysis algorithm. Reliable implementation of the contour analysis approach is constrained by pattern recognition and matching limitations. Pattern recognition is used for a) matching similar contours (e.g., the 2-D field example), identifying features for phase speed calculations (e.g., the GSNEP example), selecting local significant extremum (e.g., the 1-D curve example) and for selecting and inserting data patches when melding gappy data (e.g., the GSNEP example). The extent of practical applications of contour analysis in physical oceanography approach will ultimately depend on advances in pattern recognition algorithms.

Pattern recognition algorithms are used in other fields of studies including artificial intelligence (AI), manufacturing and robotics, audio (voice recognition) and video processing (image morphing), medical research, people identification and weapons research. For example, a quick reference check on medical abstracts published in the last three years reveal well over one hundred studies per year using some form of contour analysis. The reason for the wide-spread use of such methods in medicine is that the size and shape of body parts is a critical component of medical diagnosis. Is the organ swollen? Is the tumor growing or shrinking? Is the diameter of your critical arteries large enough for good blood flow? In all these fields, the use of contour and feature-based analysis is exploding mainly due to new technological advances especially in computing power and smaller-cheaper-smarter sensors.

Unfortunately it is not easy to port techniques from one field to another. Research into AI and pattern recognition has shown that the best algorithms are highly problem dependent. Oceanographers can learn, but not borrow directly, from research in other fields. Oceanographers are blessed with well-defined coherent features in our data set but are cursed with basic flow statistics that are highly nonstationary over the scales of the observing system due to dynamical instabilities. Current research has shown that it is extremely difficult to achieve a 100% success rate. Researchers, such as ourselves, are satisfied with an algorithm if its success rates is over 90%. Consequently, post-processing is usually necessary with most pattern recognition software. Based on these findings, it will be a long time before machines will totally replace skilled specialists in oceanography. On the other hand, mechanization of analysis and assimilation procedures would certainly enhance objectivity and repeatability. The authors thus believe that the approaches described here will become more commonplace in physical oceanography.

References

[1] Arhan, M., and A. Colin de Verdiere 1985: Dynamics of eddy motions in the eastern North Atlantic. *J. Phys. Oceanogr.*, **15**, 153-170.

[2] Babiano, A., Basdevant, C., Legras, B. and R. Sadourny 1987: Vorticity and passive-scalar dynamics in two-dimensional turbulence. *J. Fluid Mech.*, **183**, 379-397.

[3] Bennett, A.F. 1992: *Inverse methods in physical oceanography.* Cambridge University Press.

[4] Benzi, R., Patarnello, S. and P. Santangelo 1988: Self-similar coherent structures in two- dimensional decaying turbulence. *J. Phys. A: Math. Gen.*, **21**, 1221-1237.

[5] Brown, O.B., P.C. Cornillon, S.R. Emmerson, and H.M. Carle 1986: Gulf Stream warm rings: a statistical study of their behavior. *Deep-Sea Research*, **33** 11,12, 1459-1473.

[6] Carter, E.F. 1989: Assimilation of Lagrangian data into a numerical model. *Dyn. Atmos. Oceans*, **13**, 355-348.

[7] Chin, T.M. and A.J. Mariano. Optimal space-time interpolation of gappy frontal position data, 1993: In: Proceedings of the Aha Hulikona Hawaiin Winter Workshop. Honolulu, Hawaii, January 1993, 265-289.

[8] Chin, T.M. and A.J. Mariano 1995: Space-time Interpolation of Oceanic Fronts. *IEEE J. of Geoscience and Remote Sensing* (Submitted)

[9] Clancy, R.M., P.A. Phoebus, and K.D. Pollak 1990: An operational global-scale thermal analysis system. *J. Atmos. Oceanic. Tech.*, **7**, 233-254.

[10] Cornillon, P. 1986: The effect of the New England Seamounts on Gulf Stream Meandering as Observed from Satellite IR Imagery. *J. of Physical Oceanogr.*, **16**, 386-389.

[11] Cummings, J.A. 1994: Global and Regional Ocean Thermal Analysis Systems at Fleet Numerical Meteorology and Ocean Center. *Proceedings MTS 94 Conference*, 433-439.

[12] Davis, J.M., F.L. Estis, P. Bloomfield and J.F. Monahan 1991: Complex Principal Component Analysis of Sea-level Pressure over the Eastern USA. *Int. J. Clim.*, **11**, 27-54.

[13] Deem, G. S. and N.J. busky 1978: Vortex waves: Stationary "V states," interactions, recurrence and breaking. *Phys. Rev. Lett.*, **40**, 859-62.

[14] De Maria, M., S.D. Aberson, K.V. Ooyama, and S.J. Lord 1992: A nested spectral model for hurricane track forecasting. *Mon Wea. Rev.*, **120**, 1628-1643.

[15] Fuglister, F.C. 1963: Gulf Stream 60. In. M Sears [ed.], Progress in Oceanography, Pergamon Press, **1**, 265-373.

[16] Ghil, M. and P. Malanotte-Rizzoli 1991: Data assimilation in meteorology and oceanography. *Adv. Geophys.*, **33**, 141-266.

[17] Ghil, M. and K. Ide 1994: Extended Kalman Filtering for Vortex Systems: An Example of Observing-System Design. *Data Assimilation: NATO ASI Series*, **19** Series I, 167-191.

[18] Halkin, D., and H. T. Rossby 1985: The structure and transport of the Gulf Stream at 78° W. *J. Phys. Oceanogr.*, **15**, 1439-14562.

[19] Halliwell, G.R. and C.N.K. Mooers 1983: Meanders of the Gulf Stream downstream from cape Hatteras, 1975-1978. *J. Phys. Oceanogr.*, **13** (7), 1275-1292.

[20] Hardy, D.M. and J.J. Walton 1978: Principal Components Analysis of Vector Wind Measurements. *J. Appl. Meteorol.*, **17**, 1153-1162.

[21] Ide, K. and M. Ghil 1995: Extended Kalman Filtering for Vortex Systems: Part I: Point Vortices. *In press, Dyn. Atmos. Oceans*

[22] Iselin, C.O'D., and F.C. Fuglister 1948: Some recent developments in the study of the Gulf Stream. *J. Marine Res.*, **7**, 317-329.

[23] Kennelly, M.A. 1984: The velocity structure of Warm Core Ring 82B and associated cyclonic features. *Tech. Rep.* 84-10 Woods Hole Oceanographic Institute, 100pp

[24] Kwok, R., Curlander, J.C., McConnell, R., and S.S. Pang 1990: An Ice-Motion Tracking System at the Alaska SAR Facility. *J. of Oceanic Engineer.*, **15**, 44-54.

[25] Lorenz, E.N. 1956. Empirical Orthogonal Functions and Statistical Weather Prediction. Rep. 1, *Statist. Forecasting Project*, MIT.

[26] Mariano, A.J. and H.T. Rossby 1989: The Lagrangian Potential Vorticity Balance during POLYMODE. *J. Phys. Oceanogr.*, **19**(7), 927-939.

[27] Mariano and O. Brown 1992: Efficient objective analysis of dynamically heterogeneous and nonstationary fields via the parameter matrix. *Deep-Sea-Res.*, **39** (7/8), 1255-1271.

[28] Mariano, A.J. 1990: Contour Analysis: A New Approach for Melding Geophysical Fields. *J. Atmos. Oceanogr. Tech.*, **7** (2), 285-295.

[29] Mariano, A.J. 1988: Space-time interpolation of Gulf Stream north wall positions. Harvard Open Ocean Model Reports, (29), *Reports in Meteorology and Oceanography*, **29**, Harvard University.

[30] Mathur, M.B. 1991: The National Meteorological Center's quasi-Lagrangian model for hurricane prediction. *Mon. Wea. Rev.*, **119**, 1419-1447.

[31] McWilliams, J.C. 1990: The vortices of geostrophic turbulence. *J. Fluid Mech.*, **219**, 387-404.

[32] McWilliams, J.C., Weiss, J.B, and I. Yavneh 1994: Anisotropy and Coherent Vortex Structures in Planetary Turbulence. *Science*, **264**, 410-413.

[33] Olson, D. B. 1991: Rings in the ocean. *Annu. Rev. Earth Planet. Sci.*, **19**, 283-311.

[34] Olson, D. B., G. Hitchcock, A. Mariano, C. Ashjian, G. Peng, R. W. Nero, and G. Podestá 1994: Life on the edge: Marine life and fronts. *Oceanogr.*, **7** (2), 52-60.

[35] Pratt, L.J. and M.E. Stern 1986: Dynamics of potential vorticity fronts and eddy detachment. *J. Phys. Oceangr.*, **16**, 1101-20.

[36] Preisendorfer, R. W. 1988: *Principal Component Analysis in Meteorology and Oceanography*, Elsevier, NY.

[37] Pullin, D.I. 1992: Contour Dynamics Methods. *Annu. Rev. Fluid Mech.*, **24**, 89-115.

[38] Richardson, P.L., D. Walsh. L. Armi, M. Schroder, and J.F. Price 1989: Tracking Three Meddies with SOFAR Floats. *Journal of Physical Oceanography*, **19** No. 3, 371- 383.

[39] Ring Group 1981: Gulf Stream Cold-Core Rings: Their Physics, Chemistry, and Biology. *Science*, **212**, 1091-1100.

[40] Robinson, A.R., M.A. Spall, L.J. Walstad, and W.G. Leslie 1989: Data assimilation and dynamical interpolation in gulfcast experiments. *Dyn. Atmos Oceans* **13**, 269- 300.

[41] Robinson, A. R., A. Hecht, N. Pinardi, J. Bishop, W.G. Leslie, Rosentroub, A.J. Mariano, and S. Brenner 1987: Small synoptic/mesoscale eddies and energetic variability of the eastern Levantine basin. *Nature*, **327**, 131.

[42] Robinson, A.R. and L.J. Walstad 1987: The Harvard Open Ocean Model: Calibration and Application to Dynamical Process, Forecasting and Data Assimilation. *J. Appl. Numer. Math.*, **3** (1-2), 89-131.

[43] Rossby, H.T., S.C. Riser, and A.J. Mariano 1983: The Western North Atlantic - A Lagrangian Viewpoint *Eddies in Marine Science* Springer-Verlag, New York, 66-88.

[44] Smeed, D. 1995: Feature models and data assimilation. *J. Atmos. Oceanogr. Tech.* (submitted)

[45] Spall, M.A. and A. R. Robinson 1990: Regional primitive equation studies of the Gulf Stream meander and ring formation region. *J. Phys. Oceanogr.*, **20**, 985-1016.

[46] Stern, M.E. 1985: Large scale wave breaking and shingle formation. *J. Phys. Oceanogr.*, **15**, 1274-83.

[47] Stern, M.E. and L.J. Pratt 1985: Dynamics of vorticity fronts. *J. Fluid Mech.*, **161**, 513-32.

[48] Todling, R. and M. Ghil 1994: Tracking atmospheric instabilities with the Kalman filter. Part I: Methodology and one-layer results. *Mon. Weather Rev.*, **122** 183-204.

[49] Watts, D.R. 1983: Gulf Stream Variability *Eddies in Marine Science* Springer-Verlag, New York, 114-143.

[50] Weare, B. C., A. R. Navato, and R. E. Newell 1976: Empirical orthogonal analysis of Pacific sea-surface temperatures. *J. Phys. Oceanogr.*, **6**, 671-678.

[51] Weiss, J. 1981: The dynamics of enstrophy transfer in two-dimensional hydrodynamics. *La Jolla Inst. Tech. Rep.*, **LJI-TN-81-121**

[52] Zabusky, N.J. 1981: Computational synergetics and mathematical innovation. *J. Comput. Phys.*, **43**, 195-249.

[53] Zabusky, N.J., Hughes, M.H., and K.V. Roberts 1979: Contour dynamics for Euler equations in two dimensions. *J. Comput. Phys.*, **30**, 96-106.

[54] Zabusky, N. J. and E.A. Overman II 1983: Regularization of contour dynamical algorithms. I. Tangential regularizations *J. Comput. Phys.*, **52**, 351-73.

[55] Robinson, A.R., M. Spall, and N. Pinardi 1988: Gulf Stream Simulations and the Dynamics of Ring and Meander Processes. *J. Phys. Oceanogr.*, **18**, 1811–1853.

RSMAS/MPO
University of Miami
4600 Rickenbacker Causeway
Miami, FL 33149-1098

mariano@sgrmag.r.rmas.miami.edu

Topics in Statistical Oceanography

S. Molchanov

"The trail that is always new" (R. Kipling)

0. Introduction

The aim of this paper is to review some recent results in Lagrangian analysis. These results concentrate on the very old and classical problem of the transition from Lagrangian statistics, drifter trajectories, to Eulerian data. In this area the number of physics publications is incredibly large, and it is difficult to find formulas which are not already known in the fluid dynamics literature. Orientation in such literature is a problem: when, how and under what additional assumptions can one use a given formula in a specific practical situation?

Pure mathematical results in Lagrangian analysis are poor and often based on non-physical models such as δ -correlated (in time) random velocity fields, and the splitting of different correlations.

Recent mathematical progress in turbulent transport, an aspect of Lagrangian analysis, can be found in the work of the Princeton group (M. Avellaneda, A. Majda) [1] - [4]. Analogous methods produce further new results. These and similar methods will be discussed in this paper. This paper is based on the lecture notes "Statistical problems of oain not only a deterministic part, which can be seen on geographical maps asceanography" of lectures given during the Spring of 94 at UNCC and to be repeated in the Spring of 95, in a different format.

The author thanks R.F. Anderson and Z. Zhang (UNCC), R. Carmona (UCI) and L. Piterbarg (USC) for very interesting discussions and help in preparing this article.

1. Models for turbulence and the Kolmogorov spectrum

It is well known that ocean currents contain not only a deterministic part, which can be seen on geographical maps as a system of arrows, but also a well-developed multiscale, random, turbulent, component. Separation of these components and spectral analysis of the random part is one of the fundamental problems of oceanography. Knowledge of the energy spectrum of the turbulent component of the ocean flows yields knowledge of the distribution of the kinetic energy in the upper layer of the ocean. Information about this spectrum is a key element for understanding many theoretical and practical problems, including interaction in the ocean-atmosphere system, meteorology, temperature and salinity fields and their prediction. A popular method to obtain information about ocean currents is based on the analysis of drifter trajectories. The drifter method generates several very old and complicated problems about the relationship between the Eulerian and Lagrangian approaches in hydrodynamics. These problems have probabilistic, statistical, ODE and SPDE elements and must be the subject of an interdisciplinary investigation. The aim of this section is to give a mathematical introduction to the subject, formulate precise mathematical statements and to describe the boundaries of applicability of the different approximations. This paper is to be considered as a mathematical text rather than a physical or oceanographical dicussion. Such an approach is typical in modern applied mathematics.

The following model will be central to our considerations. Let $\overrightarrow{V}(t,x)$: $t \geq 0, x \in R^2$, be a two-dimensional vector (velocity) field, homogeneous in space and time. Suppose that,

1) $<\overrightarrow{V}(t,x)>=\overrightarrow{a} \in R^2$

2) $\overrightarrow{v}(t,x) = \overrightarrow{V}(t,x) - \overrightarrow{a}$ is Gaussian and isotropic. (That is, the corresponding probability law is invariant with respect to the full group of Euclidean transformations of R^2 - shifts, rotations and reflections.)

3) $div \overrightarrow{V} = div \overrightarrow{v} = 0$. (Incompressibility)

4) $\overrightarrow{v}(t,x)$ is a Markovian homogeneous field in time. (Note that together with the Gaussian hypothesis and homogeneity this yields, for all $s < t$ and $x \in R^2$,

$$(1.1) \qquad \langle \overrightarrow{v}(t,x) \mid \sigma\left(\overrightarrow{v}(s,\cdot) : s \leq t\right)\rangle$$
$$= \int_{R^2} k(t-s, x-z)\overrightarrow{v}(s,z)dz, -s, x-z)\overrightarrow{v}(s,z)dz,$$

where k is the kernel of an L^2 bounded integral operator.

5) Let

$$B_{i,j}(\tau, z) = < v_i(t_1, x_1) v_j(t_2, x_2) >: i = 1, 2, \quad t_1, t_2 > 0,$$
$$z = x_2 - x_1, \quad \tau = t_2 - t_1,$$

be the correlation tensor of the field $\overrightarrow{v} = (v_1, v_2)$ in a fixed coordinate system. Suppose that $B(\tau, z) = \{B_{i,j}(\tau, z)\}$ is a sufficiently smooth (at least of class $C^{2+\delta}$) matrix function and for $| \tau |, | z | \longrightarrow \infty$ its norm decreases rapidly. For example,

$$||B(\tau, z)|| \leq \frac{c}{| \tau |^{\alpha} | z |^{\beta}} : \alpha > 1, \beta > 2.$$

In this case, there exists a matrix spectral density $E(\omega, k)$ such that

$$B(\tau, z) = \int_{R^1 \times R^2} \exp(i(k, z) + i\omega\tau) E(\omega, k) d\omega dk. \tag{1.2}$$

Here ω is time frequency and k is space frequency. Of course, $E(\omega, k) = E^*(\omega, k)$, and $E(\omega, k)$ is positive definite.

Properties 3, (incompressibility) 4, (Markov structure) and isotropy imply a very special form of the correlation $B(\omega, z)$ and the spectral density $E(\omega, k)$. To construct this representation assume, for simplicity, that

$$\overrightarrow{v}(t, x) = curl\psi(t, x) = (-\frac{\partial}{\partial x_2}, \frac{\partial}{\partial x_1})\psi(t, x), \tag{1.3}$$

where $\psi(t, x) : (t, x) \in R^1 \times R^2$ is a scalar **stationary** and isotropic in space and homogeneous in time Gaussian field, (potential or stream function), and Markovian in time, in the sense of (1.1). Formula (1.3) will guarantee us incompressibility and the Markovian property of the initial velocity field $\overrightarrow{V}(t, x)$. A priori the stream function $\psi(t, x)$ can only be a field with **stationary increments** in space. Never the less the final result does not depend on the hypothesis that $\psi(t, x)$ is stationary and isotropic in space and applies in a general situation.

Suppose that

$$< \psi(t, x) >= 0$$

and that

$$\beta(\tau, |z|) = < \psi(t_1, x_1)\psi(t_2, x_2) >, \quad |z| = |x_2 - x_1|, \quad \tau = |t_2 - t_1|$$

is the (scalar) correlation function of the stream function. (Note the isotropy in space and homogeneity in time.) Then

$$\{B_{i,j}(\tau, z)\} = \begin{pmatrix} \frac{\partial^2\beta}{\partial x_2^2} & -\frac{\partial^2\beta}{\partial x_1 \partial x_2} \\ -\frac{\partial^2\beta}{\partial x_1 \partial x_2} & \frac{\partial^2\beta}{\partial x_1^2} \end{pmatrix} \tag{1.4}$$

and if

$$\widehat{\beta}(\tau, k) = \int_{R^2} e^{i(k,x)}\beta(\tau, x)dx$$

then we have

$$\begin{aligned} \{E_{i,j}(\tau, k)\} &= \{\textstyle\int B_{i,j}(\tau, x)e^{i(k,x)}dx\} \\ &= \widehat{\beta}(\tau,k)\begin{pmatrix} k_2^2 & -k_1k_2 \\ -k_1k_2 & k_1^2 \end{pmatrix} \end{aligned} \tag{1.5}$$

The time structure of the correlator $\beta(\tau, |z|)$ is also very simple. If $s < t$ we have

$$E\left[\psi(t, x) \mid \sigma(\psi(s, \cdot) : s \leq t)\right] = \int_{R^2} k(t - s, x - z)\psi(s, z)dz, \tag{1.6}$$

and then the relation for $(t_0 \leq t_1 \leq t_2)$

$$\begin{aligned} &E\left[\psi(t_2, x) \mid \sigma(\psi(s, \cdot) : s \leq t_0)\right] \\ =&E\left[E\left[\psi(t_2, x) \mid \sigma(\psi(s, \cdot) : s \leq t_1)\right] \mid \sigma(\psi(s, \cdot) : s \leq t_0)\right] \end{aligned}$$

gives

$$\int_{R^2} k(t_2 - t_1, x - z)k(t_1 - t_0, z)dz = k(t_2 - t_0, x).$$

A Fourier transformation in space shows that

$$\widehat{k}(t_2 - t_1, k)\widehat{k}\,(t_1 - t_0, k) = \widehat{k}\,(t_2 - t_0, k)\,.$$

Using in addition the isotropy

$$\widehat{k}(t, k) = \exp(-|t|\Omega(|\,k\,|))$$

where $\Omega(| k |) > 0$, we can obtain, from (1.6), that

$$\widehat{\beta}(\tau, k) = \widehat{k}(\tau, k)\widehat{\beta}(0, k),$$

so that

$$\widehat{\beta}(\tau, k) = \exp(-|t|\Omega(| k |))\widehat{\beta}(0, k).$$

An additional Fourier transformation in time shows that the energy spectrum of $\overrightarrow{v}$ is given by the formula

$$\begin{aligned}\widehat{\widehat{\beta}}(\omega, \tau) &= \textstyle\int \exp(i\omega\tau)\,\widehat{\beta}(\tau, k)d\tau \\ &= \frac{1}{2\pi}\frac{\Omega(| k |)}{\omega^2 + \Omega^2(| k |)}\widehat{\beta}(0, k)\begin{pmatrix} k_2^2 & -k_1k_2 \\ -k_1k_2 & k_1^2 \end{pmatrix}.\end{aligned}$$

or in different notations,

$$E(\Omega, k) = \frac{1}{2\pi}\frac{\Omega(| k |)}{\omega^2 + \Omega^2(| k |)}\frac{E(|k|)}{|k|}\left(\delta_{i,j} - \frac{k_ik_j}{|k|^2}\right) \tag{1.7}$$

This is the canonical representation, known in statistical hydrodynamics [6],[7], of the energy spectrum when conditions 1)-5) are assumed.

In the multidimensional case with an of isotropic Gaussian flow, Markovian in time, we have essentially the same result, and now the spectral density has the form

(1.8)

$$E_{i,j}(\omega, k) = \frac{1}{2\pi}\frac{\Omega(| k |)}{\omega^2 + \Omega^2(| k |)}\frac{E(|k|)}{|k|^{d-1}}\left(\delta_{i,j} - \frac{k_ik_j}{|k|^2}\right) \quad i, j = 1, 2, \cdots d.$$

The functions $E(|k|)$, $\Omega(|k|)$ have a very elegant physical meaning. According to the classical Kolmogorov picture we have to understand the turbulence as a system of eddies of different sizes $\ell = \frac{1}{|k|}$ in some inertial interval of frequencies $\kappa_0 \leq |k| \leq \kappa_i$. Every eddy may include subeddies of frequencies $\geq \frac{1}{|k|}$, and has a finite, and random, life-time. After its death it generates several smaller eddies. The largest eddies have characteristic frequency κ_0. $\ell_0 = \frac{1}{\kappa_0}$ is the so-called basic scale. In the case of the ocean, $\ell_0 \simeq$ 500-1000 km and the corresponding initial eddies are generated by synoptic processes in the atmosphere.

The function $E(|k|)$ gives the density of the kinetic energy corresponding to modes with frequency $|k|$, (i.e. the kinetic energy per unit volume for frequencies of magnitude $|k|$). The total density of energy for the eddy with scale $\ell = \frac{1}{|k|}$ is of order

$$\int_{|k|}^{\infty} E(\rho)d\rho$$

and the corresponding typical velocity $v(|k|)$ is of order

$$v(|k|) = \left(\int_{|k|}^{\infty} E(\rho)d\rho\right)^{\frac{1}{2}} \tag{1.9}$$

The turn-time for a $|k|$-eddy, is, consequently

$$\tau(|k|) = \frac{\ell}{v(|k|)} = \frac{1}{kv(|k|)} = \frac{1}{k\left(\int\limits_{|k|}^{\infty} E(\rho)d\rho\right)^{\frac{1}{2}}}.$$

The second basic function in the canonical representation $\Omega(|k|)$ is related with the life-time of an eddy with frequency $|k|$. Namely, the inverse value $t(|k|) = \Omega^{-1}(|k|)$ is the mean of a typical life-time.

There exist three different possibilities:

1. $|t(|k|)| \gg \tau(|k|), \quad \kappa_0 \leq |k| \leq \kappa_i$. Every eddy makes many turns during its life-time. The extremal case of this type are the so-called steady flows, where $\overrightarrow{a} = 0$, $\overrightarrow{v}(t, x) = \overrightarrow{v}(x)$. (See also the end of Section 1.)

2. $|t(k)| \sim \tau(|k|)$. This possibility corresponds to the classical Kolmogorov turbulence (see below).

3. $|t(k)| \ll \tau(|k|)$. In this case, which is probably realistic for ocean turbulence, at least for low frequencies there is an approximation by δ-correlations in time, and the methods of SPDE theory apply (see below). A better approximation for case 3 may be based on a multiscale analysis.

Consider the especially interesting case of self-similar flows of generalized Kolmogorov turbulence. By definition, the random flow $\overrightarrow{V}(t, x) = \overrightarrow{a} + \overrightarrow{v}(t, x)$ has generalized Kolmogorov energy spectrum if for some inertial interval

$$|k| \in [\kappa_0, \kappa_i]$$

the functions $\Omega(|k|)$, $E(|k|)$ in representation (1.7) (or (1.8) have the form

$$\begin{aligned} E(|k|) =& c_1|k|^{1-\varepsilon}, \qquad \varepsilon > 2, \\ \Omega(|k|) =& \ \ c_2|k|^{\zeta}, \qquad \zeta > 0. \end{aligned}$$

Outside the inertial interval, these functions must be smooth and $E(|k|)$ must be rapidly decreasing

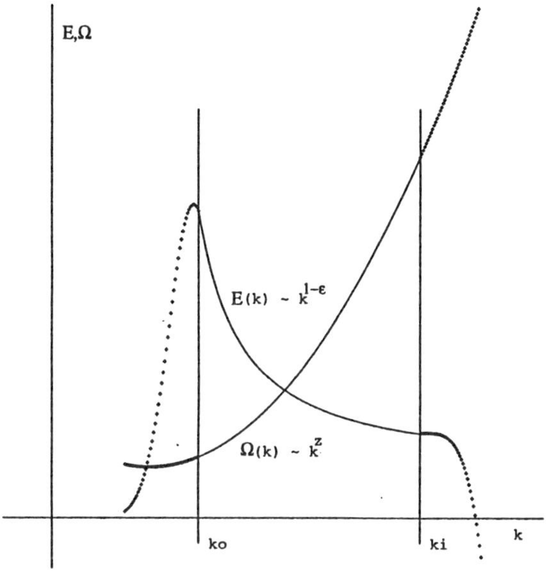

Figure 1

Classical Kolmogorov turbulence corresponds to the case

$$\varepsilon = \frac{8}{3}, \zeta = \frac{2}{3}$$

The constants $\kappa_0, \kappa_i, c_1, c_2, \varepsilon, \zeta$, and $\overrightarrow{a}$ give the principal description of generalized turbulence. The fundamental problems in oceanography for a given, region of homogeneity, in the ocean, are as follows: to check the hypothesis of Kolmogorov turbulence, to find the corresponding parameters ε and $\zeta$and to understand the meaning of these parameters in the context of global ocean circulation. An additional problem is to explain the relationship between energy spectra for the potential and velocity fields of the ocean. There is strong evidence that the Kolmogorov spectrum matches that of the potential field of the ocean.

Finally we note that the following two cases of turbulent flows are mathematically interesting, but are not very realistic.

a) $\overrightarrow{V}(t,x) = \overrightarrow{V}(x)$ (steady flows) In this case the correlations are also time independent and

$$E_{i,j}(\omega, k) = \delta_0(\omega) \frac{E(k)}{|k|} \left(\delta_{i,j} - \frac{k_i k_j}{|k|^2} \right). \tag{1.10}$$

This case corresponds to the situation when $\Omega(|k|) \gg 1$, uniformly the spectral interval $\kappa_0 \le |k| \le \kappa_i$.

If $\kappa_0 \ge 1$, then steady flows are related with the limit $\zeta \to \infty$ and the other Kolmogorov parameters $\varepsilon, \kappa_0, \kappa_i$ are fixed! (However, this is really only an issue of the proper choice of units of measurements.)

b) δ-correlated in time flows. In this situation

$$B_{i,j}(\tau, z) = \delta_0(\tau) B_{i,j}(z).$$

i.e. $\vec{V}(t,x)$ is a white noise in time. In the language of spectral analysis this means that $\Omega(|k|) \ll 1$ for $\kappa_0 > 1$ when $\zeta \to -\infty$. The spectrum of $E_{i,j}(\omega, k)$ does not depend on the time frequency ω.

The physical theory of isotropic turbulence can be found in the Landau-Lifshitz course (vol. 6), see [5]. For the mathematical aspects see also Batchelor [6], Monin, Jaglom [7]. For the theory of turbulent transport for Kolmogorov turbulence see the renormalization approach in [1] and [4].

2. Planar wave representation, simulation of velocity, and Lagrangian trajectories (LT)

Let $\vec{V}(t,x)$ be a random Gaussian incompressible flow with the canonical representation (1.7) for the energy spectrum. The Lagrangian trajectory (LT) $x(t)$, by definition, is a solution of the random ODE

$$\frac{dx}{dt} = \vec{V}(t, x_t).$$

That is

$$\begin{aligned} x_t &= x_0 + \int_0^t \vec{V}(s, x_s)ds \\ &= x_0 + \vec{a}t + \int_0^t \vec{v}(s, x_s)ds \end{aligned} \tag{2.1}$$

A more general form of LT includes molecular diffusivity. In fact, the molecular diffusivity in the ocean is related more with the short wave currents than heat conductivity, which is extremely small in the case of water. In different terms, we can include some part of the energy spectrum the high frequency region into the diffusivity.

This form of LT is given by the SPDE

$$dx_t = \sigma dw_t + \overrightarrow{V}(t, x_t)dt,$$

or

$$(2.2) \qquad x_t = x_0 + \sigma w_t + \overrightarrow{a}t + \int_0^t \overrightarrow{v}(s, x_s)ds.$$

For a smooth Gaussian field $\overrightarrow{V}(t,x)$, which cannot increase faster than a linear function in space and time, a unique solution to (2.2) exists. This solution, $x_t = x_t(x_0, \omega_m, \omega)$, depends on two different random parameters: ω_m is an index for the realization $\overrightarrow{V}(,,\omega_m)$ of the drift, and, ω is an index for the Brownian trajectory $w_t(\omega)$.

We will prove in Section 3 that the process x_t has stationary increments and the Lagrangian velocity

$$V_t^L = \overrightarrow{V}(t, x_t)$$

is a stationary process.The condition $div\overrightarrow{v} = 0$ and the ergodicity of $\overrightarrow{v}$ are the key elements of this proof.

Unfortunately, the Lagrangian velocity is an extremely complicated non-linear function of the Eulerian field $\overrightarrow{V}(t,x)$, and the process V^L is non-Gaussian. As a result, theoretical methods for the investigation of $\overrightarrow{V^L}$ are not well-developed. Furthermore experimental data for drifter (Lagrangian) trajectories in the ocean are relatively poor.

In such a situation numerical simulation of the Eulerian field $\overrightarrow{V}$, the Lagrangian trajectories x_t and the Lagrangian velocities $\overrightarrow{V^L}$ can be very useful, especially from the point of view of development of new statistical methods. To construct $\overrightarrow{V}(t,x)$ we can use the formula

$$(2.3) \qquad \overrightarrow{V}(t,x) = \overrightarrow{a} + curl\psi(t,x)$$

and the spectral representation of the stream function, potential ψ :

$$(2.4) \qquad \psi(t,x) = \int_{R^1 \times R^2} \exp(i\omega t + i(k,x))Z(d\omega, dk)$$

where $Z(d\omega, dk)$ is a complex scalar Gaussian measure with non-correlated increments:

$$\begin{aligned} &\angle Z(k' + dk, \omega' + d\omega)\bar{Z}(k + dk, \omega + d\omega)\rangle = 0 \\ (2.5) \qquad &\quad \text{if}(k' - k)^2 + (\omega' - \omega)^2 \neq 0 \\ &\angle Z(k + dk, \omega + d\omega)\bar{Z}(k + dk, \omega + d\omega)\rangle = \tilde{E}(k, \omega) dk d\omega \\ &\quad \text{if}(k' - k)^2 + (\omega' - \omega)^2 = 0 \end{aligned}$$

According to Section 1, in this situation we have Markovianess in time and isotropy, and so

$$(2.6) \qquad \tilde{E}(\omega, k) = \frac{E(|k|)}{|k|^3} \frac{\Omega(|k|)}{\omega^2 + \Omega^2(|k|)}$$

In the language of $\overrightarrow{v}(t, x) = \overrightarrow{V}(t, x) - \overrightarrow{a}$, formula (2.6) gives the matrix spectral density $E_{i,j}$:

$$(2.7) \qquad E_{i,j}(\omega, k) = \frac{1}{2\pi} \frac{\Omega(|\ k\ |))}{\omega^2 + \Omega^2(|\ k\ |))} \frac{E(|k|)}{|k|} \left(\delta_{i,j} - \frac{k_i k_j}{|k|^2} \right) : i = 1, 2$$

This is the same as (1.7). For the numerical simulation of $\overrightarrow{V}$ we can use a discretization $\tilde{Z}$ of the random spectral measure Z in representation (2.4) and then formula (2.3) for $\overrightarrow{\tilde{V}}(t, x)$.

We will carry out this discretization in a few distinct steps. First of all, consider the support of $E(\omega, k)$ in space, the inertial interval $R = [\kappa_0 \leq |k| \leq \kappa_i]$, a finite number of space modes $k^{(i)}, i = 1, 2, ...N$ in the ring, see ($pic2$), and a corresponding partition $R^{(i)}$, $\bigcup_{i=1}^{N} R^{(i)} = R$. Then put

$$\begin{aligned} (2.8) \qquad \psi(t, x) \simeq \tilde{\psi}(t, x) &= \sum_{i=1}^{N} \exp(i\omega t) \int_{R^1 \times R^{(i)}} Z(dk, d\omega) \\ &= \sum_{i=1}^{N} e^{i(k^{(i)}, x)} A^{(i)}(t) \end{aligned}$$

This is the complex form of the planar wave representation for the stream function $\tilde{\psi}(t, x)$.

If the system $k^{(i)}$, and partition $R^{(i)}$, are symmetric with respect to the k_1, k_2 axis, (see ($Figure2$)) we can use the following real planar wave representation.

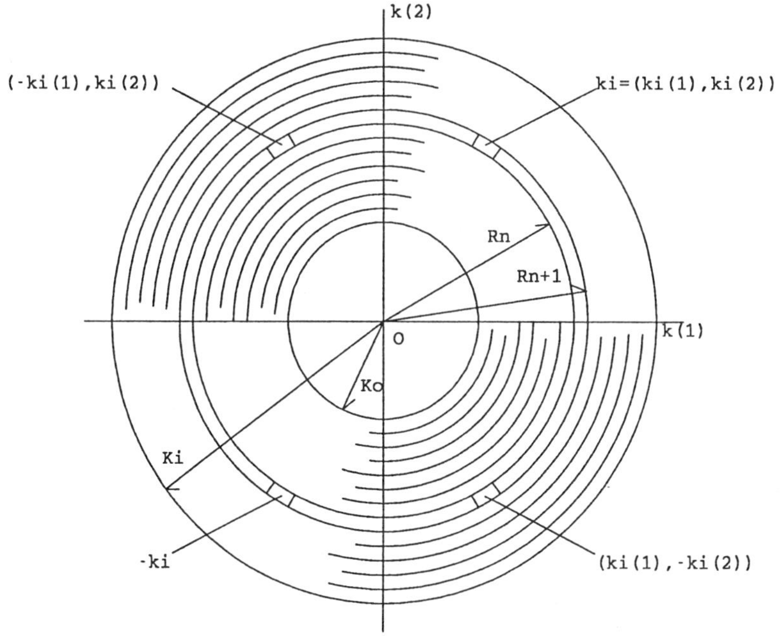

Figure 2

(2.9)

$$\widetilde{\psi}(t,x) = \sum_{i=1}^{N}(A_1^{(i)}(t)\cos(k_1^{(i)}x_1)\cos(k_2^{(i)}x_2) + A_2^{(i)}(t)\cos(k_1^{(i)}x_1)\sin(k_2^{(i)}x_2)$$
$$+ A_3^{(i)}(t)\sin(k_1^{(i)}x_1)\cos(k_2^{(i)}x_2) + A_4^{(i)}(t)\sin(k_1^{(i)}x_1)\sin(k_2^{(i)}x_2)$$

The total number of modes is now equal $4N$. The formula

$$\overrightarrow{V}(t,x) = \overrightarrow{a} + curl\widetilde{\psi}(t,x)$$

will give the planar wave representation for the velocity field. In this case $A_j^{(i)}(t), i = 1,2,...N;\quad j = 1,2,3,4$ are independent stationary processes. Now change $\Omega(|k|)$ in $R^{(i)}$ to $\Omega\left(\left|k^{(i)}\right|\right)$. Elementary calculations, for fixed i, j, give,

(2.10)

$$\beta^{(i)}(t-s) = < A_j^{(i)}(t)A_j^{(i)}(s) > = \sigma_i^2\exp(-\mid t-s\mid \Omega(\mid k^{(i)}\mid),$$
$$\sigma_i^2 = \int_{R^{(i)}} \frac{E(|k|)}{|k|^3}dk \simeq |R^{(i)}|\frac{E(|k^{(i)}|)}{|k^{(i)}|^3}, \qquad |R^{(i)}| = \text{Area}R^{(i)}$$

This implies that the $A_j^{(i)}(t)$ are Ornstein-Uhlenbeck processes with different correlation scales $\alpha^{(i)} = \Omega(|k^{(i)}|)$ and variances σ_i^2. Of course, this special structure of the planar wave decomposition (2.9) with random O-U amplitudes $A_j^{(i)}$ reflects the initial Markovian property on the velocity field $\vec{V}(t,x)$.

The simulation of a single O-U process with correlation

$$\beta(\tau) = \sigma^2 \exp(-\alpha|\tau|)$$

is a simple problem. If Δ is a time step, and $a_n = A(n\Delta)$, then

$$a_{n+1} = a_n \exp(-\alpha\Delta) + \sigma\sqrt{1-\exp(-2\alpha\Delta)}\xi_n \tag{2.11}$$

where ξ_n are i.i.d N(0, 1) random variables, and a_n is independent of noise. So for $\alpha\Delta \gg 1$, the auto-regressive sequence a_n becomes practically pure white noise. This explains our previous statement, that high frequency eddies can be included into diffusivity.

Now we can describe the algorithm for generating the velocity field $\vec{V}(\cdot,\cdot)$ and Lagrangian trajectories for given set of parameters:

I. Drift $\vec{a}$, inertial interval $[\kappa_0, \kappa_i]$, density $E(|k|)$, $|k| \in [\kappa_0, \kappa_i]$, intensity of the eddy life $\Omega(|k|)$, $|k| \in [\kappa_0, \kappa_i]$.

II. Additional parameters: Number of modes N and time step Δ in the discrete model. Partition $R^{(i)}, R = \{|k| \in [\kappa_0, \kappa_i]\} = \bigcup_{i=1}^{4N} R^{(i)}$ with central modes $\{k^{(i)}\} : i = 1, 2, ..., 4N$, amplitude factors

$$\sigma_i = \left(\int_{R^{(i)}} \frac{E(|k|)}{|k|^3} dk \right)^{\frac{1}{2}}$$

and time scales

$$\alpha_i = \Omega\left(\left|k^{(i)}\right|\right), \qquad i = 1, 2, ..., 4N.$$

choose initial points $X^{(1)}, ..., X^{(K)}$ for the future Lagrangian trajectories $X_{n\Delta}^{(i)}$, $n = 0, 1, ...,$ $X_0^{(i)} = X^{(i)}$ $i = 1, 2, ..., K$ (K is the number of trajectories).

III. From the point of view of generalized Kolmogorov turbulence the best possible partition of the ring R has the following structure. Fix a small number δ and consider the system of circles of radii

$$r_0 = \kappa_0, \quad r_1 = \kappa_0(1+\delta), \cdots, r_m = \kappa_0(1+\delta)^m = \kappa_i.$$

Then $\Delta r_i = r_{i+1} - r_i = \kappa_0 (1+\delta)^i \delta$. Divide the circle of radius r_i into equal arcs of the length $\simeq \Delta r_i$. There are

$$M \simeq \frac{2\pi r_i}{\Delta r_i} = \frac{2\pi\kappa_0 (1+\delta)^i}{\kappa_0 (1+\delta)^i \delta} = \frac{2\pi}{\delta}$$

such arcs if $M_1 = \left[\frac{\pi}{2\delta}\right]$, we can represent the ring $r_i \le |k| \le r_{i+1}$ as a union of $4M$ almost equal squares $Q_{i,j}$ with side length $\Delta r_i = \kappa_0 (1+\delta)^i$. (cf. Figure 2.) The fluctuations of frequency k inside $Q_{i,j}$ are of order Δr_i. The fluctuation of the corresponding wave length is of order

$$\Delta \ell = \frac{1}{r_i} - \frac{1}{r_i + \Delta r_i} \simeq \frac{\Delta r_i}{r_i^2}$$

and

$$\frac{\Delta \ell_i}{\ell_i} = \frac{\Delta r_i}{r_i} = \frac{\kappa_0 (1+\delta)^i \delta}{\kappa_0 (1+\delta)^i} = \delta,$$

where δ is the relative error in the representation of all modes by the discrete system $k^{(i,j)}$of the centers of the squares $Q_{i,j}$. The total number of the modes, and independent O-U processes in the formula (2.9) are equal to:

$$N = 4M_1 m = 4\left[\frac{\pi}{2\delta}\right], \qquad (1+\delta)^m = \frac{\kappa_i}{\kappa_0}.$$

Even for $\delta = \frac{1}{2}, m = 4, r_i \simeq 5\kappa_0, 4M_1 = 12$, a very good almost isotropic model is obtained for 48 modes.

IV. The construction of the Lagrangian trajectories has a recurrent structure. At the moment $t_n = n\Delta$ we have to know $X_{tn}^{(1)}, X_{tn}^{(2)}, \cdots X_{tn}^{(K)}$ the present positions of the drifters and the amplitude factors $a_{(n-1)\Delta}^{(i)}$ along with the parameters $k^{(i)}, \overrightarrow{a}, \alpha^{(i)}, \sigma^{(i)}$, (cf I, II).

Generating N new $N(0,1)$ random variables $\xi_n^{(i)} : i = 1, 2. \cdots, N$ and we can recalculate the amplitudes according to formula (2.11) via

$$a_{n\Delta}^{(i)} = a_{(n-1)\Delta}^{(i)} \exp\left(-\alpha^{(i)}\Delta\right) + \sigma^{(i)}\sqrt{1 - \exp\left(-2\alpha^{(i)}\Delta\right)}\xi_n^{(i)},$$

The velocities $\overrightarrow{V}(n\Delta, X_{n\Delta}^{(i)})$ are now given by the expression

$$\overrightarrow{\tilde{V}}(n\Delta, X_{n\Delta}^{(j)}) = curl\tilde{\psi}\left(n\Delta, X_{n\Delta}^{(j)}\right) + \overrightarrow{a} : j = 1, 2, \cdots, K.$$

If we use the expression (2.9) for $\widetilde{\psi}$, then

$$\overrightarrow{V}(n\Delta, X_{n\Delta}^{(j)})$$
$$= (\sum_{i=1}^{N} -(-k_2^{(i)} A_1^{(i)}(t)\cos(k_1^{(i)}x_1)\sin(k_2^{(i)}x_2) + k_2^{(i)} A_2^{(i)}(t)\cos(k_1^{(i)}x_1)\cos(k_2^{(i)}x_2)$$

$$(2.12)\quad - k_2^{(i)} A_3^{(i)}(t)\sin(k_1^{(i)}x_1)\sin(k_2^{(i)}x_2) + k_2^{(i)} A_4^{(i)}(t)\sin(k_1^{(i)}x_1)\cos(k_2^{(i)}x_2)),$$
$$\sum_{i=1}^{N}(-k_1^{(i)} A_1^{(i)}(t)\sin(k_1^{(i)}x_1)\cos(k_2^{(i)}x_2) - k_1^{(i)} A_2^{(i)}(t)\sin(k_1^{(i)}x_1)\sin(k_2^{(i)}x_2)$$
$$+ k_1^{(i)} A_3^{(i)}(t)\cos(k_1^{(i)}x_1)\cos(k_2^{(i)}x_2) + k_1^{(i)} A_4^{(i)}(t)\cos(k_1^{(i)}x_1)\sin(k_2^{(i)}x_2)))$$

and

$$(2.13)\quad X_{(n+1)\Delta}^{(i)} = X_{n\Delta}^{(i)} + \overrightarrow{V}\left(n\Delta, X_{n\Delta}^{(i)}\right)\Delta + \sqrt{\Delta}\sigma\overrightarrow{\xi_n^{(i)}} : i = 1, 2, \cdots, K$$

The last term includes the diffusivity of the Lagrangian trajectory given in the most general form (2.2). The $\overrightarrow{\xi_n^{(i)}}$ are two additional independent N(0, 1) random variables. (The total number to be generated at moment $t_n = n\Delta$ is $N + 2K$). Formulas (2.12) and (2.13) involve the calculation of a trigonometrical polynomial of degree N.

In terms of implimenting this algorithm, it is clear that it is not too demanding in terms of computer memory and it seems that for a number of modes of order $N \simeq 500$ the algorithm can be realized on a PC or Sun station. The only problem is the generator of the random numbers must be fast enough. Additional randomization for the N(0,1) between different modes can take care of possible correlation in the generators of the random number.

Realization of this algorithm was carried out by Z. Zhang (UNCC) (cf. Figure 3) and R. Carmona (UCI). Carmona used a parallel processor and as a result, constructed the Lagrangian mapping :

$$x \xrightarrow{L_t} X_t^{(x)},$$

where $X_t^{(x)}$ is a Lagrangian trajectory for a given turbulent velocity field and molecular diffusivity $\frac{\sigma^2}{2}$ (maybe $\sigma^2 = 0!$). Estimation of the Lyapunov exponent for this mapping showed, at least empirically, its fast stabilization

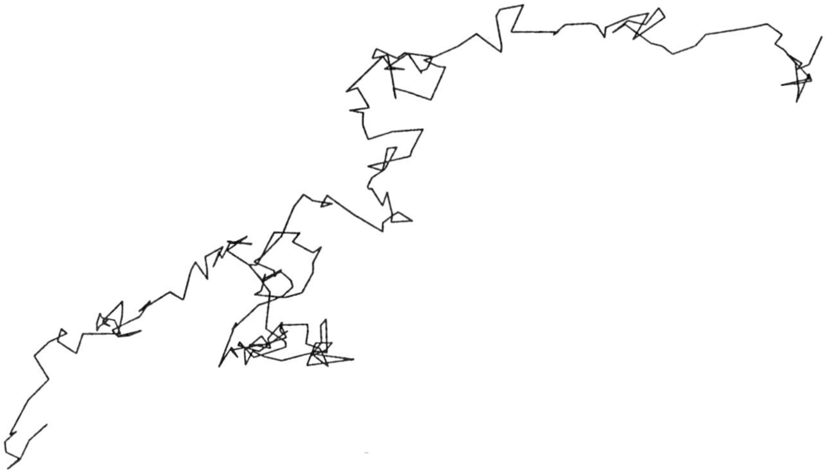

Figure 3

and positivity. This last fact reflects the very good mixing properties of the Lagrangian mapping and is important in the analysis of, for example, the behavior of oil spills on the surface of the ocean. To conclude this section we give without proof the following result.

Theorem 2.1. *Let $\psi_\varepsilon(t,x)$ be the stream function generated by a partition P_ε as described above and $\overrightarrow{V}_\varepsilon(t,x) = \overrightarrow{a} + curl\psi_\varepsilon(t,x)$ its corresponding velocity field. One can construct these stream functions on one probability space, in fact the probability space of 3-dimensional white noise $Z(d\omega, dk)$: $(\omega, k) \in R^1 \times R^2$), in such a way that for arbitrary $n > 0$*

$$\overrightarrow{V}_\varepsilon(t,x) \underset{\varepsilon \to 0}{\longrightarrow} \overrightarrow{V}(t,x) \quad in \quad C_{loc}(R^1) \times C_{loc}^{(n)}(R^2)$$

i.e. the convergence is uniform over all compact subsets in space and time together with all spacial derivative of order $\leq n$.

This relatively simple theorem gives the possibility to obtain information about Lagrangian motion for turbulence with a Kolmogorov spectrum from systems with discrete spectrum, such as systems periodic in space. We know the theorem generates uniform convergence of the Lagrangian trajectories $X^{\varepsilon,x}(t)$, related with $\overrightarrow{V}_\varepsilon(t,x)$, to the initial Lagrangian trajectories $X^x(t)$, introduced at the beginning of this section.

More details about planar wave decomposition, numerical simulation and Lagrangian mapping can be found in [8].

3. General properties of Lagrangian trajectories and homogenization

Recall that for a given velocity field $\overrightarrow{V}(t,x)$ (see Section 1, 2) the Lagrangian trajectory (LT) is a solution of the random ODE

$$dx_t = \overrightarrow{V}(t, x_t)\, dt \tag{3.1}$$

or

$$x_t = x_0 + \int_0^t \overrightarrow{V}(s, x_s)\, ds = x_0 + \overrightarrow{a}t + \int_0^t \overrightarrow{v}(s, x_s)\, ds \tag{3.2}$$

The Lagrangian velocity is a random process given by the right hand side of (3.1):

$$\overrightarrow{V^L}(t) = \overrightarrow{V}(t, x_t) = \dot{x}_t \tag{3.3}$$

If, in our hydrodynamic system the ocean we have a molecular diffusivity σ^2, then instead of (3.1),(3.2), and (3.3) we have to consider the SDE :

$$dx_t = \sigma dw_t + \overrightarrow{V}(t, x_t)\, dt, \tag{3.4}$$

$$x_t = x_0 + \sigma w_t + \overrightarrow{a}t + \int_0^t \overrightarrow{v}(s, x_s)\, ds, \tag{3.5}$$

$$\overrightarrow{V^L}(t) = \overrightarrow{V}(t, x_t)\, . \tag{3.6}$$

Suppose that the assumptions of Sections 1 and 2 hold, so the field $\overrightarrow{V}(t,x)$ is homogeneous in space and time, incompressible, Gaussian, Markovian in time, and the space correlations are smooth and fast decreasing. Then $< \overrightarrow{V}(t,x) >= \overrightarrow{a}$, and $\overrightarrow{v}(t,x) = \overrightarrow{V}(t,x) - \overrightarrow{a}$ is isotropic in space.

Theorem 3.1. *For any fixed, non-random, initial position, x_0 we have the following:*
a) The trajectory $x_t : t \geq 0$ is a process with stationary ergodic increments for both the ODE and SDE models: ie for any fixed $h > 0$ the process

$$\overrightarrow{V_h^L}(t) = \frac{x_{t+h} - x_t}{h}$$

is ergodic and stationary in time. It is essential that this fact is not asymptotic ($t \to \infty$) but is true for all t. In particular, it is important that the time series

$$\overrightarrow{V_h^L}(n) = \frac{x_{(n+1)h} - x_n}{h}, \qquad n = 0, 1, \cdots$$

is stationary and homogeneous. Such series are typical for practical oceanography where drifter measurements are discrete in time.
b) The Lagrangian velocities (3.3),(3.5) are stationary and ergodic. As a result, the one-dimensional distributions for $\overrightarrow{V^L}(t) : t \geq 0$ are Gaussian and

$$< \overrightarrow{V^L}(t) >=< \overrightarrow{V}(0,0) >= \overrightarrow{a}.$$

That is, the Stokes drift is equal to the mean Eulerian drift.
c) For large t, the trajectory $x_t : t \geq 0$ after a rescaling and normalization has the structure of Brownian motion with diffusion parameter D (Taylor's diffusivity). More precisely, for $T \longrightarrow \infty$ on every bounded time interval $[0, \tau]$,for $s \in [0, \tau]$

$$x_s^* = \frac{x_{sT} - \overrightarrow{a} sT}{\sqrt{T}} \overset{dist}{\longrightarrow} \sqrt{D} w_s \tag{3.7}$$

where $w_s : 0 \leq s \leq \tau$ is a standard Wiener process.

It thus follows that the processes $\overrightarrow{V^L}(t) : t \geq 0$ and $x_t : t \geq 0$ are "approximately" Gaussian or "quasi-Gaussian" (see the previous statement b).

The statements (a), (b), (c) for the theory of equation (3.1) and (3.4) are all well-known in the physics literature, but I have no knowledge of any mathematical reference containing these results. The proof in [7] is not correct, because it does not use the incompressibility condition, which is crucial in this context.

To understand the idea of the proof, which is not trivial and contains additional important information, so-called homogenization result.

We start from the following example, which is interesting by itself. I will follow one old paper of M. Freidlin [9] and my recent lectures [10]. In the beginning we will study the case of periodic steady flows.

3.1 Homogenization of periodic steady flows. Consider a time-independent steady, periodic in space, deterministic incompressible flow $\vec{a}(x) : x \in R^d$. Suppose that the lattice of periods is a standard cubic lattice Z^d, i.e. $\vec{a}(x+T) \equiv \vec{a}(x) : T \in Z^d$. In addition assume $div\,\vec{a}(x) = 0$.

The velocity field $\vec{a}(x)$ is not random, but we can include it into the probabilistic ensemble (Ω_m, μ) according to formula

$$\vec{a}(x,\omega) = \vec{a}(x+\omega) : \omega \in S^d = R^d/Z^d.$$

Take the probability measure μ to be normalized Lebesque measure on the unit torus S^d. Assume that the initial point $x_0 = 0$ of our Lagrangian trajectory is uniformly distributed on S^d. Of course, we can use the alternate interpretation: x_0, but the cell of periodicity S^d is randomly shifted with uniform distribution.

We will prove Theorem 1 of this section under the condition $\sigma > 0$, i.e. in the situation of a SDE. We have

$$x_t = \omega + \sigma w_t + \int_0^t \vec{a}(x_s)\, ds \tag{3.8}$$

Let

$$x_t^* = x_t\,(mod 1)$$

be the projection of LT x_t on our "probabilistic space" S^d.

The generator of the diffusion process $x_t : t \geq 0$ is given by the formula

$$L = \frac{\sigma^2}{2}\Delta + \left(\vec{a}(x), \nabla\right) \tag{3.9}$$

Note that L commutes with the shifts $x \to x + T, T \in Z^d$. This means that the space of Z^d-periodic smooth functions is invariant with respect to L, so that L can be considered as the generator of a diffusion process on $S^d = R^d/Z^d$. This last process is exactly $x_t^* : t \geq 0$. For each $t > 0$ the transition probability of x_t^* has density $p^*(t,x,y) : x, y \in S^d$and

$$\begin{aligned} \frac{\partial p^*}{\partial t} &= L_x p^* = L_y^T p^* \\ p^*(0,x,y) &= \delta_y(x) \end{aligned} \tag{3.10}$$

Of course $p^* > 0$ for $t > 0$ since the operator L is non-degenerated. Furthermore, since the phase space S^d is compact, the Markov process $x_t^* : t \geq 0$ has the best possible ergodic properties, satisfying for example Döblin's condition.Thus

$$p^*(t, x, y) \underset{t \to \infty}{\rightarrow} \pi(y), \qquad L^T \pi = 0. \tag{3.11}$$

But

$$\begin{aligned} L^T &= \frac{\sigma^2}{2}\Delta - div\left(\overrightarrow{a}\cdot\right) \\ &= \frac{\sigma^2}{2}\Delta - \left(\overrightarrow{a}, \nabla\right) + div\,\overrightarrow{a} \\ &= \frac{\sigma^2}{2}\Delta - \left(\overrightarrow{a}, \nabla\right) \end{aligned}$$

Only at this point do we use the condition $div\,\overrightarrow{a} = 0$! As a result

$$L^T \pi = \frac{\sigma^2}{2}\Delta\pi - \left(\overrightarrow{a}, \nabla\pi\right) = 0 \tag{3.12}$$

and

$$\pi = 1.$$

(Note that the solution of the above problem in the class of probability densities is unique).

As a result, the process $x_t^* : t \geq 0$ on S^d with uniformly distributed initial point $x_0^* = \omega$ is stationary Markov with the best possible mixing coefficient. Thus

$$\overrightarrow{V^L}(t) = \overrightarrow{a}(x_t) = \overrightarrow{a}(x_t^*), \quad t > 0,$$

is also stationary. But

$$x_t = \omega + \sigma w_t + \int_0^t \overrightarrow{V^L}(s)\, ds, \tag{3.13}$$

so $x_t : t \geq 0$ is a process with stationary increments and we have proved parts a) and b) of Theorem 1. In addition

$$\left\langle \overrightarrow{V^L}(t) \right\rangle = \left\langle \overrightarrow{a}(x_t^*) \right\rangle = \int\limits_{S^d} \overrightarrow{a}(x)\, dx = \overrightarrow{a},$$

So that the means of the Lagrangian and Eulerian velocities coincide.

Next we can prove next the CLT for $x_t : t \geq 0$. The best method which will give us the expression for the Taylor's diffusivity D, is based on martingale theory. We construct the almost linear harmonic function for the operator L, which is the solution of the heat equation

$$\frac{\partial \overrightarrow{H}}{\partial t} + L\overrightarrow{H} = 0, \tag{3.14}$$

where

$$\overrightarrow{H} = \overrightarrow{x} - \overrightarrow{\beta} t + \overrightarrow{h}(x), \tag{3.15}$$

with periodic third term $\overrightarrow{h}$.

It follows from (3.14) and (3.15) that on S^d

$$L h_k(x) = -a_k(x) + \beta_k, \qquad k = 1, 2, \cdots d. \tag{3.16}$$

According to the Fredholm alternative equation, (3.16) is solvable if

$$\int\limits_{S^d} (-a_k(x) + \beta_k)\, \pi(x)\, dx = 0,$$

i.e. if

$$\beta_k = \int\limits_{S^d} a_k(x)\, dx, \quad \overrightarrow{\beta} = \overrightarrow{a} = \int\limits_{S^d} \overrightarrow{a}(x)\, dx.$$

We next introduce the process $\vec{\xi}_t = \vec{H}(t, x_t) : t \geq 0$. Ito's formula gives

$$\vec{\xi}_t = \vec{\xi}_0 + \int_0^t \nabla_x \vec{H}(s, x_s)\, dw_s + \int_0^t \left(\frac{\partial}{\partial s} + L\right) \vec{H}(s, x_s)\, ds$$

(3.17)

$$= \vec{H}(0, \omega) + \int_0^t \nabla_x \vec{H}(s, x_s)\, dw_s$$

$$= \vec{H}(0, \omega) + \int_0^t \left(I + \nabla_x \vec{h}(x_s^*)\right) dw_s$$

$$= \sigma w_t + \int_0^t \nabla_x \vec{h}(x_s^*)\, dw_s + \omega + \vec{h}(\omega) \quad (\omega = x_0 = x_0^*)$$

The process $\vec{\xi}_t : t \geq 0$ is now represented as a stationary stochastic integral with ergodic increments. According to the classical Billingsley theorem, we can apply the functional form of the CLT for $\vec{\xi}_t$. Since

(3.18) $$\vec{\xi}_t = x_t - \vec{a}t + O(1),$$

or more precisely for $s \in [0, \tau]$

$$\frac{x_{Ts} - \vec{a}Ts}{\sqrt{T}} = \frac{\sigma w_{Ts} + \int_0^{Ts} \nabla \vec{h}(x_u^*)\, dw_u}{\sqrt{T}} + O\left(\frac{1}{\sqrt{T}}\right),$$

standard calculations shows that

$$\left\langle \frac{x_{Ts_1} - \vec{a}Ts_1}{\sqrt{T}} \times \frac{x_{Ts_2} - \vec{a}Ts_2}{\sqrt{T}} \right\rangle \underset{T \to \infty}{\longrightarrow} D|s_1 - s_2|,$$

where the Taylor tensor of the turbulent diffusion is equal

(3.19) $$D = I\sigma^2 + \int_{S^d} \left(\nabla \vec{h} \times \nabla \vec{h}\right) dx.$$

Equivalently in coordinates

(3.19') $$D_{i,j} = \sigma^2 \delta_{i,j} + \int_{S^d} (\nabla h_i \times \nabla h_j)\, dx.$$

Billingsley's theorem gives

$$\frac{x_{Ts} - \overrightarrow{\alpha} Ts}{\sqrt{T}} \underset{T\to\infty}{\overset{\text{dist}}{\to}} \sqrt{D} w_s.$$

Unfortunately, the formula (3.19) is not effective because it contains a solution of the PDE (3.16). In the case when the molecular diffusivity $\frac{\sigma^2}{2}$ is small we can try to use asymptotic methods. Two different asymptotic approaches are known. One is based on variational method (A. Fannjiang, G. Papanicolaou (11). The second uses the projection of the diffusion process onto the graph of the level lines of the steam function $A(x)$: $\overrightarrow{a}(x) = curl\ A(x)$ (see the recent very interesting paper (12).

Now we will discuss the time-dependent, turbulent, random flows. The idea is the following one. As a first step we analyse the turbulence with a finite number $4N$ of modes, periodic in space. Then we send $N \to \infty$ and using the uniformity of our estimates get the final homogenization result and CLT.

3.2 Two mode model. We will prove our result even in the case when the molecular diffusivity σ vanishes. To minimize technical difficulties let us start from the following model of LT $\left(\overrightarrow{a} = 0\right)$:

$$\begin{aligned} dx_t &= \overrightarrow{V}(t, x_t)\, dt \\ \overrightarrow{V}(t,x) &= curl \psi_2(t,x) \\ \psi_2(t,x) &= \xi_1(t)\psi_1(x) + \xi_2(t)\psi_2(x) \end{aligned} \tag{3.20}$$

Here $\psi_i(x) = \psi_i(x+T) : T \in Z^2 : i = 1,2$, are periodic stream functions on R^2 and the amplitudes $\xi_i(t) : i = 1,2$ are Ornstein-Uhlenbeck processes with generators

$$L_i = \alpha_i \frac{\sigma_i^2}{2}\frac{\partial^2}{\partial a_i^2} - \alpha_i a_i \frac{\partial}{\partial a_i} \qquad i = 1,2. \tag{3.21}$$

This is, in some sense a two-mode model only two O-U amplitudes. We stress that the molecular diffusivity $\frac{\sigma^2}{2}$ in (3.20) is equal to 0. As such, it provides a case which is most difficult and interesting. The triplet $(\xi_1(t), \xi_2(t), x_t) \in R^1 \times R^1 \times R^2$ is a Markov diffusion process with generator

$$\begin{aligned} L = {} & \frac{\sigma_1^2\alpha_1}{2}\frac{\partial^2}{\partial a_1^2} - \alpha_1 a_1 \frac{\partial}{\partial a_1} + \frac{\sigma_2^2\alpha_2}{2}\frac{\partial^2}{\partial a_2^2} - \alpha_2 a_2 \frac{\partial}{\partial a_2} \\ & + \left(-\left(a_1\frac{\partial\psi_1}{\partial x_2} + a_2\frac{\partial\psi_2}{\partial x_2}\right)\frac{\partial}{\partial x_1}, \left(a_1\frac{\partial\psi_1}{\partial x_1} + a_2\frac{\partial\psi_2}{\partial x_1}\right)\frac{\partial}{\partial x_2}\right). \end{aligned}$$

The operator L is degenerate. The rank of the diffusion tensor is equal to 2 in R^4 and it has the form

$$L = X_1^2 + X_2^2 + Y,$$

where X_1, X_2, Y are first order operators (vector fields). According to classical Hörmander hypoellipticity considerations we have to construct the system of commutators (Lie algebra)

$$[X_1, Y] = -\alpha_1 \frac{\partial}{\partial a_1} - \frac{\partial \psi_1}{\partial x_2}\frac{\partial}{\partial x_1} - \frac{\partial \psi_1}{\partial x_1}\frac{\partial}{\partial x_2}, [X_2, Y], [Y, [X_2, Y]], \qquad etc.$$

If the functions are generic, the rank of this family of the vector fields is equal to 4 at every point of the phase state and, as a result, the transition density

$$p(t, (a_1, a_2, x), (a_1', a_2', x')) : \quad (a_1, a_2, x), (a_1', a_2', x') \in R^1 \times R^1 \times R^2$$

exists and are continuous.

The process $(\xi_1(t), \xi_2(t), x_t(modT))$ is now ergodic and it is easy to see that the corresponding invariant density π, which is the solution of the problem.

$$L^T \pi = 0,$$

is identically equal to $const \exp(-\frac{a_1^2}{2\sigma_1^2} - \frac{a_2^2}{2\sigma_2^2})$. Again the condition

$$\overrightarrow{V} = curl\psi(t, x), \quad div\overrightarrow{V} = 0$$

is crucial!

Now we can repeat the previous considerations, as in the steady case, to prove that the process $x_t : t \geq 0$ has stationary ergodic increments, assuming that the initial point x_0 is uniformly distributed on the torus $S^2 = [0, T]^2$. T is the period of $\psi_i : i = 1, 2$.

To prove a CLT for x_t we can construct an almost linear harmonic coordinate. Consider the function

$$\overrightarrow{H_2}(a_1, a_2, x) = x + \overrightarrow{h_2}(a_1, a_2, x)$$

where $\overrightarrow{h_2}$ is T-periodic in x and bounded. The equation of L-harmonicity for H_1 gives

$$L\overrightarrow{h_2} = \left(\left(a_1\frac{\partial \psi_1}{\partial x_2} + a_2\frac{\partial \psi_2}{\partial x_2}\right), -\left(a_1\frac{\partial \psi_1}{\partial x_1} + a_2\frac{\partial \psi_2}{\partial x_1}\right)\right) \tag{3.22}$$

The right part is orthogonal to π and this fact, together with the fast decreasing of $\pi(a_1, a_2)$, which will compensate the unboundedness of a_1, a_2, will guarantee the solvability of this equation.

The classical martingale form of CLT will give us the asymptotic Wiener structure of LT x_t.

We can extend this idea for the special class of turbulent models with a discrete spectrum.

3.3 Modified finite number of modes model. Let n be an integer and let $Z^2_{(n)} = \frac{1}{n}Z^2$, a lattice with step size $h_n = \frac{1}{n}$. Suppose that $\widetilde{Z}^2_{(n)}$ is a partition of R^2 into squares of the size h_n, centered at $Z^2_{(n)}$. The modes

$$k^{(i)} = \frac{(i_1, i_2)}{n} \in R = I_{\{\kappa_0 \leq |k| \leq \kappa_{ii}\}}$$

will be the basic frequencies of our model. As in Section 2, we can construct the stream function

(3.23)

$$\widetilde{\psi}_n(t,x) = \sum_{i=1}^{N} (A_1^{(i)}(t)\cos(k_1^{(i)}x_1)\cos(k_2^{(i)}x_2) + A_2^{(i)}(t)\cos(k_1^{(i)}x_1)\sin(k_2^{(i)}x_2)$$
$$+ A_3^{(i)}(t)\sin(k_1^{(i)}x_1)\cos(k_2^{(i)}x_2) + A_4^{(i)}(t)\sin(k_1^{(i)}x_1)\sin(k_2^{(i)}x_2)$$

The number N is equal to the number of squares $R^{(i)} \subset \widetilde{Z}^2_{(n)}$ which have a nontrivial intersection with sup $E(|k|) = R$, the support of the space spectrum for the field $\overrightarrow{V}(t,x)$. The amplitudes $A_j^{(i)}(t) : i = 1, 2, \cdots N \quad j = 1, 2, 3, 4$ are independent stationary Gaussian Markovian O-U processes with parameters

$$\sigma^2_{(i)} = \int_{R^{(i)}} \frac{E(|k|)}{|k|^3} dk \simeq \frac{E(|k^{(i)}|)}{|k^{(i)}|^3}\frac{1}{n^2}, \tag{3.24}$$
$$\alpha^{(i)} = \Omega\left(\left|k^{(i)}\right|\right),$$

and correlation functions

$$\beta^{(i)}(t_2 - t_1) = \sigma^2_{(i)} \exp\left(-\alpha^{(i)}(t_2 - t_1)\right)$$

(compare with Section 2).

The only, but essential, difference between this model and the model of Section 2 is the nature of partition $R = \bigcup_{i=1}^{4N} R^{(i)}$. Again take $\overrightarrow{a} = 0$. In our case the stream function $\widetilde{\psi}_n(t,x)$ and the corresponding velocity field

$$\overrightarrow{\widetilde{V}}_n(t,x) = curl\widetilde{\psi}_n(t,x) = \left(-\frac{\partial}{\partial x_2}, \frac{\partial}{\partial x_1}\right)\widetilde{\psi}_n(t,x)$$

are periodic functions with the lattice of periods $\left(\widetilde{Z}^2_{(n)}\right)^* = 2\pi n Z^2$. Let

$$S_n^2 = 2\pi n S^2 = 2\pi n\left([0,1]^2\right)$$

be the corresponding torus.

The O-U processes $A_j^{(i)}(t)$ are stationary 1-dimensional diffusion processes with generators

$$L_j^{(i)} = \alpha^{(i)}\frac{\sigma^2_{(i)}}{2}\frac{\partial}{\partial\left(a_j^{(i)}\right)^2} - \alpha^{(i)}a_j^{(i)}\frac{\partial}{\partial a_j^{(i)}} \qquad i = 1,2,\cdots N \quad j = 1,2,3,4$$

For a given molecular diffusivity σ^2 we can consider the $(4N+2)$ dimensional homogeneous in time Markov process with generator

$$L = \sum_{i=1}^{N}\sum_{j=1}^{4}\alpha^{(i)}\frac{\sigma^2_{(i)}}{2}\frac{\partial}{\partial\left(a_j^{(i)}\right)^2} - \alpha^{(i)}a_j^{(i)}\frac{\partial}{\partial a_j^{(i)}} + \frac{\sigma^2}{2}\Delta_y + \left(\overrightarrow{\widetilde{V}}(\overrightarrow{a},y), \nabla_y\right), \tag{3.25}$$

where

$$\Delta_y = \frac{\partial^2}{\partial y_1^2} + \frac{\partial^2}{\partial y_2^2} \qquad \nabla = \left(\frac{\partial}{\partial y_1}, \frac{\partial}{\partial y_2}\right)$$

and

(3.26)

$$\overrightarrow{\widetilde{V}}\left(\overrightarrow{a},y\right) = curl_y\widetilde{\psi}_n(\overrightarrow{a},y)$$

$$\widetilde{\psi}_n(\overrightarrow{a},y) = \sum_{i=1}^{N}(a_1^{(i)}\cos(k_1^{(i)}y_1)\cos(k_2^{(i)}y_2) + a_2^{(i)}\cos(k_1^{(i)}y_1)\sin(k_2^{(i)}y_2) + a_3^{(i)}\sin(k_1^{(i)}y_1)\cos(k_2^{(i)}y_2) + a_4^{(i)}\sin(k_1^{(i)}y_1)\sin(k_2^{(i)}y_2)$$

Here $\overrightarrow{a} = \left\{a^{(i)}, i = 1,2,\cdots N\right\} = \left\{a_j^{(i)}, i = 1,2,\cdots N \quad j = 1,2,3,4\right\}$.

The first$4N$ terms in the generator (3.25) represent $4N$ independent O-U amplitudes $A_j^{(i)}(t)$ and the last term corresponds to the diffusion process y_t such that

$$dy_t = \sigma dw_t + \left(curl\widetilde{\psi}_n(left\overrightarrow{a}(t), y_t)\right)$$

where $\widetilde{\psi}_n(\overrightarrow{a}, y)$ is given by formula (3.26).

The sense of this construction is very simple. To use the ergodic theory of homogeneous Markov processes and additive functional of such processes we include the time-dependence of the drift $\overrightarrow{V}_n(t, \cdot)$ into a system of coefficients $A_j^{(i)}(t)$ which are Markov process by themselves. After this procedure the amplitudes of the plane waves in the "potential" (3.26) became constant in time.

As earlier we can consider the projection y_t^* of y_t on the torus S_n^2 . i.e.

$$y_t^* = y_t \, (mod 2\pi n) \, .$$

The process

$$\left(\overrightarrow{a}(t), \, y_t^*\right) = \left(A_j^{(i)}(t), i = 1, 2, \cdots m \quad j = 1, 2, 3, 4, \quad y_t^*\right)$$

again has hypoelliptic generator L, but now on the functional space $C^2\left(R^{4N} \times S_n^2\right)$ and it is easy to see that it is ergodic. The corresponding invariant density $\pi\left(\overrightarrow{a}, y\right)$; i.e. the non-negative integrable solution of the equation

$$L^T \pi = 0,$$

is equal to

$$\pi\left(\overrightarrow{a}, y\right) = \prod_{i=1}^{N} \prod_{j=1}^{4} \left(\frac{\exp - \frac{\left(a_j^{(i)}\right)^2}{2\sigma_{(i)}^2}}{\sqrt{2\pi}\sigma_{(i)}} \right), \tag{3.27}$$

which is independent on y.

Of course, as before, the key moment of the calculations is related to the incompressibility:

$$div_y \overrightarrow{V}\left(\overrightarrow{a}, y\right) = 0.$$

As a result, the operator L^T does not dependence on y in the potential part and (3.27) is a solution of the basic equation (3.25).

Note that the stream function (3.23) can be written down in the form

$$\psi_n(t,y) = \sum_{i=1}^{N} \widetilde{A}_1^{(i)}(t)\cos\left(k_1^{(i)}y_1 + k_2^{(i)}y_2 + \theta_1(t)\right) + \widetilde{A}_2^{(i)}(t)\cos\left(k_1^{(i)}y_1 + k_2^{(i)}y_2 + \theta_2(t)\right)$$

with uniformly distributed phases $\theta_1(t), \theta_2(t)$ and new, again independent, O-U amplitudes $\widetilde{A}_2^{(i)}(t), j = 1,2 \quad i = 1,2,\cdots N$. It means that any initial point, say $x_0 = 0$ can be considered as uniformly distributed on the torus S_n^2 since the processes $\overrightarrow{a}(t)$ has for arbitrary t the stationary distribution (3.27), we have proved the two first statements of Theorem 1 of this section.

The proof for the last statement, the CLT follows the same strategy as in the previous two-modes special case.

To finish the proof for continuous spectrum we have to make limiting transition $n \to \infty$ from discrete to continuous spectrum. This transition is based on the following important lemma:

Lemma 1. *Consider the sequence of n-dimensional O-U process $A_n(t)$ with generators*

$$L_n = \sum_{i=1}^{4N}\left(\alpha_i \frac{\sigma_i^2}{2}\frac{\partial^2}{\partial a_i^2} - \alpha_i a_i \frac{\partial}{\partial a_i}\right)$$

and suppose that for all n,

$$a) \quad \sum_i \sigma_i^2 \le B_0 < \infty,$$

$$b) \quad 0 < c_0 \le \alpha_i \le c_i < \infty.$$

Then the distributions P_n of the energy processes

$$\xi_n(t) = |A_n(t)|^2 = \sum_{i=1}^{4N} a_i^2(t)$$

are tight.

This lemma together with non-trivial hypoelliptical estimations gives, in the limit $n \to \infty$, the final statement of Theorem 1 of this section. The same approach gives essential information about the Lagrangian mapping compare [8], but we will not discuss this subject here.

4. Relating and estimating Eulerian and Langrangian spectra

We have proven already in (Section 3) that the Lagrangian velocity $\overrightarrow{V^L}(t)$ is a stationary, vector valued, process which has the same 1-dimensional Gaussian distributions as the initial Eulerian field $\overrightarrow{V}(t,x)$. As a result, all moments of $\overrightarrow{V^L}(t)$ and its corresponding correlations functions are finite. Of course, $\left\langle \overrightarrow{V^L}(t) \right\rangle = \overrightarrow{a} = \left\langle \overrightarrow{V} \right\rangle$ and the next (second order) Lagrangian correlation tensor

$$B(\tau) = \langle (\overrightarrow{V^L}(t+\tau) - \overrightarrow{a}) \otimes (\overrightarrow{V^L}(t) - \overrightarrow{a}) \rangle \tag{4.1}$$

describes the energy properties of the Lagrangian velocity.

The Fourier transform

$$\widehat{B}(\omega) = \int \exp(i\omega\tau)\, B(\tau)\, d\tau \tag{4.2}$$

which exists as an analytic function of ω gives the Lagrangian energy spectrum. This spectrum, in principle, can be evaluated in terms of drifter's trajectories and, as it was mentioned above, Section 1, the goal of our theoretical analysis is to relate the Eulerian spectrum $E(\omega,k)$ and the Lagrangian spectrum $\widehat{B}(\omega)$.

The two following results are central to our approach. (For simplicity again take $\sigma^2 = 0$ and $\overrightarrow{a} = 0$)

Theorem 4.1. *If*

$$X(t) = \int_0^t \overrightarrow{V^L}(s) ds$$

is the Lagrangian trajectory starting from the point $x_0 = 0$, then

$$B(\tau) = \int d\omega dk E(\omega,k) \langle \exp(i(k, x_\tau)) \rangle . \tag{4.3}$$

Theorem 4.2. *The Lagrangian spectral density is compound Cauchy, and representable as*

$$\widehat{B}(\omega) = \int \frac{h}{(\omega-\omega_1)^2 + h^2} \rho(\omega_1, h)\, d\omega_1 dh, \tag{4.4}$$

where $\rho(\omega, h)$ is a positive exponentially decreasing function of both variables.

We start with the physical proof of Theorem 1 of this section which is a key element in the proof of Theorem 2, of this section, followed by a discussion on underwater stones. The Eulerian field has spectral representation

$$\vec{V}(t,x) = \int \exp\left(i\omega t + i(k,x)\right) Z(d\omega, dk) \tag{4.5}$$

where the Gaussian random measure $Z(d\omega, dk)$ has non-correlated, complex, increment. Then, by the definition,

$$\overrightarrow{V^L}(t) = \vec{V}(t, x_t),$$

so that

$$\begin{aligned}
&\langle \overrightarrow{V^L}(t+\tau) \otimes \overrightarrow{V^L}(t)\rangle \\
&= B(\tau) \\
&= \langle \iint e^{(i(\omega(t+\tau)-\omega_1 t)+(i(k,x_{t+\tau})-(k_1,x_t)))} Z(d\omega, dk) \otimes Z(d\omega_1, dk_1)\rangle \\
&= \iint e^{(i\omega(t+\tau)-i\omega_1 t)} \langle e^{(i(k,x_{t+\tau})-(k_1,x_t))}\rangle \delta_{(\omega-\omega_1)}\delta_{(k-k_1)}\langle |Z(d\omega, dk)|^2\rangle \\
&= \int \exp(i\omega\tau) \langle \exp(i(k, x_{t+\tau} - x_t))\rangle E(\omega, k)\, d\omega dk \\
&= \int \exp(i\omega\tau) \langle \exp(i(k, x_\tau))\rangle E(\omega, k)\, d\omega dk
\end{aligned} \tag{4.6}$$

Similar calculations are correct for Ito integrals. If $a(s,\omega)$ is non-anticipating with respect to the Wiener filtration then

$$\left\langle \left(\int_0^T a(s,\omega) dw_s \right)^2 \right\rangle = \int_0^T \left\langle a^2(s,\omega) \right\rangle ds$$

The main problem, which will be discussed below, is that the Lagrangian trajectory x_t is a global functional of our noise $Z(d\omega, dk)$ and we cannot use directly the Ito formulation. Moreover, the multidimensional case, the notions of future and past, which are essential in the Ito calculus are not obvious. I will return to this problem later, but for now let us prove Theorem 2 of this section assuming Theorem 1.

Lemma 2. *For fixed $k \in R^d$ and arbitrary stationary process $\overrightarrow{\xi}_s \in R^d, \left\langle \overrightarrow{\xi}_s \right\rangle = 0$ the following function*

$$\Psi(t) = \left\langle \exp\left(i\left(k, \int_0^t \overrightarrow{\xi}_s ds\right)\right)\right\rangle \tag{4.7}$$

is a characteristic function (i.e. the Fourier transform of a probability measure).

Proof. According to Bochners theorem it is suffcient to check that for any integer $n \geq 2$, and points $t_1, t_2, \cdots t_n$

$$det(\Psi(t_i - t_j)) \geq 0.$$

(The function $\Psi(t)$ is obviously continuous). But

$$\begin{aligned}\Psi(t_i - t_j) &= \left\langle \exp\left(i\left(k, \int_0^{t_i - t_j} \overrightarrow{\xi}_s ds\right)\right)\right\rangle \\ &= \left\langle \exp\left(i\left(k, \int_0^{t_i} \overrightarrow{\xi}_s ds\right)\right) \exp\left(-i\left(k, \int_0^{t_j} \overrightarrow{\xi}_s ds\right)\right)\right\rangle \\ &= \left\langle \eta_i \overline{\eta}_j \right\rangle,\end{aligned}$$

where

$$\eta_i = \exp\left(i\left(k, \int_0^{t_i} \overrightarrow{\xi}_s ds\right)\right)$$

In other words $det(\Psi(t_i - t_j))$ is the Graham determinant for the system of random variables $\eta_i : i = 1, 2, \cdots n$, and it is non-negative.

It follows from Lemma 2 that

$$\begin{aligned}\Psi(t) &= \left\langle \exp\left(i\left(k, \int_0^t \overrightarrow{\xi}_s ds\right)\right)\right\rangle \\ &= \int_{R^1} \exp(i\omega_1 t)\, \mu_k(d\omega_1)\end{aligned}$$

for some probability measure $\mu_k(d\omega_1)$, depending on the parameter k. The family $\mu_k(d\omega_1)$ is continuous with respect to k in the weak topology. Now in the context of Theorem 1,

$$\begin{aligned}
B(t) \\
&= \iint \frac{E(|k|)}{|k|}\left(\delta_{i,j}-\frac{k_ik_j}{|k|^2}\right)\frac{\Omega(|k|)}{\omega^2+\Omega^2(|k|)}\exp(i\omega t)\exp(i\omega_1 t)\,\mu_k(d\omega_1)\,d\omega dk \\
&= \iint \frac{E(|k|)}{|k|}\left(\delta_{i,j}-\frac{k_ik_j}{|k|^2}\right)\frac{\Omega(|k|)}{(\omega-\omega_1)^2+\Omega^2(|k|)}\exp(i\omega t)\,\mu_k(d\omega_1)\,d\omega dk.
\end{aligned}$$

That is, for a suitable positive matrix measure $\mu(dk, d\omega_1)$

$$\begin{aligned}
\widehat{B}(\omega) &= \int \frac{E(|k|)}{|k|}\left(\delta_{i,j}-\frac{k_ik_j}{|k|}\right)\frac{\Omega(|k|)}{(\omega-\omega_1)^2+\Omega^2(|k|)}\mu_k(d\omega_1)\,dk \\
&= \frac{1}{\pi}\int \frac{k}{(\omega-\omega_1)^2+k^2}\mu(dk, d\omega_1),
\end{aligned}$$

and so the spectral density $\widehat{B}(\omega)$ is compound Cauchy. This completes the proof. The following result can be considered as a specialization of Lemma 2. It will be useful in the future.

Lemma 3.. *If the process $\overrightarrow{\xi}_s$ in the statement of Lemma 2 is Gaussian, then the characteristic function $\Psi(t)$ is infinitely divisible.*

Proof. If $D(\tau) = Cov\left(\overrightarrow{\xi}_{t+\tau} \times \overrightarrow{\xi}_t\right)$ is a correlator of the (Gaussian) process $\overrightarrow{\xi}_s$ then

$$\begin{aligned}
\Psi(t) &= \exp\left(-\frac{1}{2}\left(\iint_0^t D(s_2-s_1)\,ds_1ds_2k, k\right)\right) \\
&= \exp\left(-\frac{1}{2}\iint_0^t \int_{R^1} e^{i\omega_1(s_2-s_1)}ds_1ds_2\,(d\mu(\omega_1)\,k, k)\right),
\end{aligned}$$

where $d\mu(\omega_1)$ is the matrix spectral density of $\overrightarrow{\xi}$, i.e. $(d\mu(\omega_1)\,k, k) = dv(\omega_1)$ is a positive scalar measure. But

$$\begin{aligned}
\iint_0^t \exp i\omega_1(s_2-s_1)\,ds_1ds_2 &= \left|\int_0^t \exp(i\omega s)\,ds\right|^2 \\
&= \frac{\sin^2\frac{\omega t}{2}}{(\frac{\omega}{2})^2}
\end{aligned}$$

As result, the function

$$\Psi(t) = \exp\left(-\frac{1}{2}\left(\iint_0^t D(s_2 - s_1)\, ds_1 ds_2 k, k\right)\right)$$
$$= \exp\left(-\frac{1}{2}\int \frac{1-\cos t\omega_1}{\omega_1^2} dv(\omega_1)\right)$$

According to the classical Kolmogrorov theorem the last expression is the representation of an infinitely divisible distribution with finite second moment.

As a result, the spectral measure of the Lagrangian velocity $\overrightarrow{V^L}(t)$ is a convolution of two densities, a Cauchy, and a density with characteristic function $\Psi(t)$) with an additional integration with respect to parameter k. Thus in some sense, the spectrum is doubly infinitely divisible.

We return to the proof of the basic Theorem 1 of this section. The following trivial example shows, that, in the case of an anticipating functional, Ito formula, which we have used in the physical proof of the Theorem 1, generally speaking, does not apply.

Let $w_s : s \in [0,1]$ be a Wiener process and $a(s,\omega) \equiv w_1$. Then

$$\eta_t = \int_0^t a(s,\omega)\, dw_s = w_1 w_t$$

is an anticipating Ito's integral. But $\left\langle \int_0^t a(s,\omega)\, dw_s \right\rangle = \langle w_1 w_t \rangle = t$, i.e. η_t is not a martingale. Moreover

$$\langle \eta_1^2 \rangle = \left\langle \left(\int_0^t a(s,\omega)\, dw_s\right)^2 \right\rangle = \langle w_1^4 \rangle = 3$$
$$\neq \int_0^t \langle a^2(s,\omega) \rangle\, ds = 1.$$

To prove Theorem 1 we will use the following approach. Consider first instead of the velocity field $\vec{V}(t,x)$, its approximation $\vec{V}_n(t,x)$, as in Section 3. Thus

$$\vec{V}_n(t,x) = \sum_{i=1}^{4N} A_i(t)\, \psi_i(x)$$

where $\psi_i(x)$ are the planar waves with wave vectors $k^{(i)} \in \frac{1}{n}Z^2$. Let us fix two modes $k^{(i_0)}, k^{(i_1)}$ and consider SDE equation for LT:

$$\begin{aligned} dx_t &= \overrightarrow{V}_n(t, x_t) \\ &= A_{i_0}(t)\, dt\psi_{i_0}(x) + A_{i_1}(t)\, dt\psi_{i_1}(x) + \sum_{i \neq i_0, i_1} A_i(t)\, dt\psi_i(x) \\ &= \sigma dw_t + A_{i_0}(t)\, dt\psi_{i_0}(x) + A_{i_1}(t)\, dt\psi_{i_1}(x) + \overrightarrow{\widetilde{V}_n}(t, x_t) \end{aligned}$$

For large n the amplitudes $A_{i_0}(t), A_{i_1}(t)$ are small in probability and using standard perturbation techniques we can represent LT in the form

$$x_t = \widetilde{x}_t + l_1(\widetilde{x}_t, A_{i_0}(t), A_{i_1}(t)) + remainder$$

here $\widetilde{x}_t$ is independent of $A_{i_0}(t), A_{i_1}(t)$: l_1 is the linear functional of the amplitudes A_{i_0}, A_{i_1} and the remainder is a smaller order term.

After this we can repeat the standard Ito calculations and represent the correlator $B_n(\tau)$ as an integral sum and let $n \to \infty$. Such an approach works for all anticipating functional which are smooth with respect to white noise. The required calculations are relatively long and we will not describe the details here. The fundamental equation (4.3) in Theorems 1 of this section is known in the physics literature. It is one of the, weak, consequences of the Kraichman DIA method [14]. General DIA method is only an approximation in the theory of transport processes (L. Piterbarg recently constructed a corresponding counterexample for heat flow), but our consequence (4.3) is correct.

Formula (4.3) represents in some sense the integral equation for the unknown Lagrangian velocity correlation tensor $B(\tau)$. But this equation is not closed, because

$$\langle \exp(i(k, x_t)) \rangle = \left\langle \exp\left(i \left(k, \int_0^t \overrightarrow{V^L}(s) ds \right) \right) \right\rangle$$

must be expressed in terms of $B(\tau)$!

We can use the following Gaussian approximation. Suppose the Lagrangian velocity has a Gaussian distribution. It is not correct exactly, but looks to be a reasonable approximation, the one-dimensional distributions of $\overrightarrow{V^L}(t)$ are Gaussian, the process $x_t : t \to \infty$ is asymptotically Gaussian. We must stress, at this point, that the real Eulerian velocity field (see Section 1) is also Gaussian only approximately. For more information about Gaussian approximation see also the well-known review [16].

In the framework of Gaussian approximation we can rewrite equation (4.3) in the following form:

$$(4.8) \qquad = \int\limits_{R^1 \times R^2} d\omega dk E(\omega, k) \exp\left(i\omega t - \frac{1}{2} \iint\limits_0^t (B(s_2 - s_1) k, k) \, ds_1 ds_2 \right)$$

This is a closed nonlinear integral equation for $B(\tau)$.

Equation (4.8) and its further transformations and simplifications is the basis of the practical spectral analysis of LT and the transition from Lagrangian to Eulerian data.

There are two different opportunities in the study of (4.8). One is based on a method of sequential approximation which produces with some precision the theoretical formula for $B(\tau)$ in terms of $E(\omega, k)$. In this way, one can get the formula for the turbulent diffusivity

$$D = \int\limits_{R^1} B(\tau) \, d\tau = 2\pi \widehat{B}(0)$$

in terms of the Eulerian spectrum E. This formula, using an absolutely different argument was deduced in our recent paper [15] with L. Piterbarg. I will not discuss this opportunity here.

The second approach is based on the empirical approximation of the LT in the expectation $\langle \exp(i(k, x_\tau)) \rangle$ by a simpler Gaussian process of the given parametric type.

In the simplest version of this method we can approximate $x_t : t \geq 0$ by the 2-dimensional isotropic Wiener process with empirical turbulent diffusivity $\widetilde{D}$, i.e.

$$x_t = \left(x_t^1, x_t^2\right) = \sqrt{\widetilde{D}} \left(w_t^1, w_t^2\right)$$

where $\overrightarrow{w}_t$ is a standard Wiener process in R^2. Such approximation is natural if we expect apriori, that the time correlation of the velocity field $\overrightarrow{V}$ is short, (i.e the parameter ζ is large).

In the framework of this approximation we have

$$(4.9) \qquad B(\tau) = \int\limits_{R^1 \times R^2} d\omega dk E(\omega, k) \exp\left(i\omega\tau - \frac{\widetilde{D} |k|^2 \tau}{2} \right)$$

Using the formula

$$(4.10) \qquad \int_{R^1} \exp(i\omega k)\frac{a}{\pi(\omega^2+a^2)}d\omega = \exp(-a|\tau|)$$

and the canonical representation (1.7) for the Kolmogorov's spectrum we have, after integration over ω and angular integration in polar coordinates on R^2 the following expression:

$$(4.11) \qquad B_{,}(\tau) = \int_{\kappa_0}^{\kappa_i} dr E(r)\exp\left(-\left(\Omega(r)+\frac{\widetilde{D}r^2}{2}\right)|\tau|\right)$$

where

$$E(r) = c_1 r^{1-\varepsilon}, \qquad \Omega(r) = c_2 r^{\zeta}$$

are the density of energy and reciprocal time scale in the canonical representation.

To find the parameters $\kappa_0, \kappa_1, \varepsilon, \zeta, c_1, c_2$ we can use the method moments. Let us consider the correlation moment

$$m_l = \int_0^{\infty} B(\tau)\tau^l d\tau$$

Integration over τ in (4.11) gives us immediately

$$(4.12) \qquad m_l = l!\int_{\kappa_0}^{\kappa_i}\frac{c_1 r^{1-\varepsilon}dr}{\left(c_2 r^{\zeta}+\frac{\widetilde{D}}{2}r^2\right)^{l+1}}, \qquad l = 0, 1, \cdots$$

The integral in the right part of (4.12) can be simplified, in any case it is an analytic of all parameters $\kappa_0, \kappa_1, \varepsilon, \zeta, c_1, c_2$.

If we use instead of m_l the empirical correlation moments,

$$\widetilde{m}_l = \int_0^{\infty} \widetilde{B}(\tau)\tau^l d\tau,$$

for the empirical Lagrangian correlation function we will get a closed system of transcendental equations for Kolmogorov parameters. After simplification of (4.12) as mentioned above, this system will be elementary from the point of view of numerical computer solution. I will not discuss further details.

An essentially better approach which is valid for all values of the parameter ζ is based on the approximation of x_t by physical Brownian motion, the integral of an O-U process. Such an approximation was studied in detail, at the level of real data and simulations, in the recent paper by K. Owen, L. Piterbarg and B. Rozovskii [17]. Using empirical estimates for the parameters σ^2, α of the O-U processes we have

$$\begin{aligned}\langle \exp\left(i\left(k, x_\tau\right)\right)\rangle &= \exp\left(-\frac{|k|^2}{2}\iint_0^t \sigma^2 \exp\left(-\alpha\left|s_2 - s_1\right|\right) ds_1 ds_2\right) \\ &= \exp\left(-\frac{|k|^2\sigma^2}{\alpha}\left(t - \frac{1-\exp\left(-\alpha t\right)}{2}\right)\right)\end{aligned}$$

Instead of (4.11) we now get the equation
(4.13)

$$B\left(\tau\right) = \int_{\kappa_0}^{\kappa_i} dr c_1 r^{1-\varepsilon} \exp\left(-\tau c_2 r^\zeta - \frac{|k|^2 r^2}{\alpha}\left(\tau - \frac{1-\exp\left(-\alpha\tau\right)}{\alpha}\right)\right)$$

The evaluation of the correlation moments m_l, and integration over τ is also possible. It gives a more complicated form in comparison with (4.12), but transparent equations for the parameters $\kappa_0, \kappa_1, \varepsilon, \zeta, c_1, c_2$.

To conclude this section I will provide a brief description of the possible algorithms for the reconstruction of Kolmogorov parameters for the kinetic energy turbulent spectrum in the ocean.

1. For a given region of homogeneity in the ocean, the Lagrangian data

$$x_{t_1}, x_{t_2}, \cdots, x_{t_n}, \quad \left(t_n = T\right), \text{at moment} \quad t_i : i = 1, 2, \cdots, n$$

are not usually equidistant . Using spline interpolation construct a continuous and smooth trajectory $x_t : t \in [0, T]$.

2. To check homogeneity of Lagrangian velocity, probably the best method is based on the Wilcoxon non-parametric test for the Lagrangian velocities

$$v_{t_1} = \frac{x_{t_2} - x_{t_1}}{t_2 - t_1}, \; v_{t_2} = \frac{x_{t_3} - x_{t_2}}{t_3 - t_2}, \; etc.$$

of two different parts of the trajectory say, first half and second one.

3. Estimate Stokes drift $\overrightarrow{a} = \frac{x_t}{t}$ and represent the Lagrangian velocities in the coordinate system $\left(\overrightarrow{a}, \overrightarrow{a}^\perp\right)$.

4. Evaluate the Lagrangian correlation function $B_{1,2}(\tau)$ along $\overrightarrow{a}$ and along $\overrightarrow{a}^{\perp}$ using standard moving average method. Evaluate the correlation moments $\widetilde{m}_l$ for $B_{1,2}$.

5. Approximate x_t either by Brownian motion or by physical Brownian motion (as was discussed above) and find numerically the parameters $\kappa_0, \kappa_1, \varepsilon, \zeta, c_1, c_2$.

6. Substitute these parameters in the canonical expression for $E(\omega, k)$ and compare with the right part in the equation (4.11) or (4.13) with the empirical correlator $B_{1,2}(\tau)$. In the good cases we have to find some smoothing of $B_{1,2}(\tau)$. If the agreement is not very precise we can repeat our calculation with different smooth version of $B_{1,2}(\tau)$.

We stress that in this approach we do not need the Lagrangian spectrum $\widetilde{E}(\omega)$. If we need this spectrum one can use the canonical representation (4.9). This last fact is important, because as is usual in time-series theory, the quality of the spectral estimate is essentially lower than the quality of the correlator's estimate. Of course, the practical realization of this algorithm will probably require some modifications, but the principal ideas will survive.

References

[1] M. Avellaneda, A. Majda, 1992: Approximate and exact renormalization theories for a model for turbulent transport. *Phys. Fluids, A.*, **4(1)**, 45.

[2] M. Avellaneda, A. Majda, 1990: Mathematical model for exact renormalization for turbulent transport. *Comm. Math. Ph.*, **131**, 181-229.

[3] M. Avellaneda, A. Majda, 1991: An integral representation and the effective diffusivity in passive advection by laminar and turbulent flows. *Comm. Math. Ph.*, **138**, 339-354.

[4] M. Avellaneda, A. Majda, 1992: Renormalization theory for eddy diffusivity in turbulent transport. *Phys. Rew. Lett.*, **68, 20**, 3026-3031.

[5] L. Landau, E. Lifshitz, 1986: *Fluid mechanics.* Pergamon Press, Oxford, New York, Paris.

[6] G. K. Batchelor, 1982: *The theory of homogeneous turbulence.* Cambridge Univers. Press, Cambridge.

[7] A. Monin, A. Jaglom, 1975: *Statistical fluid mechanics*; *Mechanics of turbulence.* Cambridge, MA.

[8] R. Carmona, S. Grishin, S. Molchanov, (see present volume).

[9] M. Freidlin, 1964: Dirichlet problem for an equations with periodic coefficients. *Prob. Theory and Appl.*, **9**, 133-139.

[10] D. Bakry, R. Gill, S. Molchanov, 1992: Lectures in probability theory. Sant-Flour Summer School. Springer-Verlag. 1994: Lecture Notes in

Math, 1581.
[11] A. Fannjiang, G. Papanicolaou, 1994: Connection enhanced diffusion for periodic flows. *SIAM Journ. Appl. Math.*, **54**, 333-408.
[12] M. Freidlin, A. Wentzell, 1993: Diffusion processes on graphs and the averaging principle. Ann. Prob.
[13] R. Kraichnan, 1959: The structure of isotropic turbulence at very high Reynolds numbers. *J. Flui Mech.*, **5**, 497-543.
[14] S. Molchanov, L. Piterbarg, 1992: Heat propagation in random flow. *Russian J. Math. Phys.*, **1.1**, 1-22.
[15] R. Davis, 1982: On relating Eulerian and Lagrangian velocity statistics: single particle in homogeneous flow. *J. Fluid. Mech.*, **114**, 1-26.
[16] K. Owen, L. Piterberg, B. Rozowskii (see present volume).

Department of Mathematics
University of North Carolina
Charlotte
Charlotte, NC 28270

smolchan@unccsun.uncc.edu

Stochastic forcing of quasi-geostrophic eddies

Peter Müller

Abstract

The basic theoretical framework is presented for the study of the stochastic atmospheric forcing of barotropic quasi-geostrophic oceanic eddies. The governing equation is the linearized potential vorticity equation with a stationary stochastic forcing function representing random fluctuations in the atmospheric windstress curl. The frequency autospectrum of the response and the frequency cross-spectrum between forcing and response are spatial convolutions of the frequency spectrum of the forcing with the Green's function of the potential vorticity equation. Coherence maps have been calculated for simple geometries (meridional channel or infinite ocean) and simple forcing spectra (Gaussian or white in wavenumber space). They reproduce basic features of observed coherence maps. The major open question is whether the more complex structure of the observed maps is due to the more complex structure of the forcing spectrum or due to the processes neglected in the potential vorticity equation.

1. Introduction

Oceanic motions exhibit fluctuations over space and time scales ranging from centimeters to thousands of kilometers and from seconds to decades and beyond. Some of these fluctuations are thought to be caused by internal instability processes, others by external, mainly atmospheric, forcing. Often the atmospheric forcing fields must be regarded as random and the forcing problem becomes a stochastic forcing problem. Examples include the stochastic forcing of surface gravity waves by atmospheric pressure fluctuations (e.g., Phillips, 1957), of internal gravity waves by fluctuations in the atmospheric windstress (e.g., Rubenstein, 1994), of sea surface temperature anomalies by heat flux fluctuations (e.g., Frankignoul and Hasselmann, 1977), and of anomalies in the global thermohaline circulation by fluctuations in precipitation and evaporation (e.g., Mikolajewicz and Maier-Reimer, 1990). In this paper we consider the stochastic forcing of barotropic quasi-geostrophic eddies by fluctuations in the atmospheric windstress curl.

Barotropic quasi-geostrophic eddies are oceanic fluctuations for which the velocity and pressure fields are independent of the vertical coordinate and in geostrophic balance. Such fluctuations have been observed in the ocean at time scales from a few days to a few months and at length scales $O(500\text{ km})$ during the BEMPEX (Barotropic Electromagnetic and Pressure Experiment). The observed barotropic quasi-geostrophic fluctuations have also been found to be correlated with the atmospheric windstress curl as inferred from FNOC (Fleet Numerical Oceanographic Center) winds (Luther et al. 1990, Chave et al. 1992). These observed correlations provide direct empirical evidence of the stochastic forcing of these motions.

Theoretical research has focussed on whether or not the observed fluctuations and correlations can be "explained" by a stochastic forcing model. These theoretical attempts are reviewed here. First we discuss the potential vorticity equation which is believed to govern the dynamical evolution of barotropic quasi-geostrophic motions. The free wave solutions are barotropic planetary Rossby waves. Next we introduce the basic statistical definitions and assumptions that are employed in solving this equation and then give the formal solution under these assumptions. The final two sections discuss the two simplified cases that have been solved explicitly.

This review emphasizes the basic theoretical framework. Many details are omitted that can be found in the cited literature.

2. Potential vorticity equation

The dynamical evolution of barotropic quasi-geostrophic fluctuations or eddies is described by the barotropic quasi-geostrophic potential vorticity equation (e.g., Pedlosky, 1987). When nonlinear advection is neglected and a rigid surface and flat bottom are assumed, then this equation takes the form

$$\partial_t \Delta \psi(\vec{x},t) + \beta_0 \partial_x \psi(\vec{x},t) = F(\vec{x},t) + D(\vec{x},t) \tag{1}$$

Here $\vec{x} = (x, y)$ is the horizontal position vector, t time, $\Delta = \partial_x \partial_x + \partial_y \partial_y$ the Laplacian, ψ the streamfunction, and β_0 a constant reference value of the beta parameter that describes the meridional gradient of the Coriolis parameter. The terms on the right hand side describe the forcing and dissipation. For windstress forcing

$$F = \frac{1}{\rho_0 H_0} \vec{k} \cdot (\nabla \times \vec{\tau}) \tag{2}$$

where ρ_0 is a constant reference density, H_0 the constant ocean depth, $\vec{\tau}$ the atmospheric windstress, and $\vec{k}$ the vertical unit vector. It is the curl of the windstress that forces quasi-geostrophic motions. The exact nature of the processes that dissipate quasi-geostrophic eddies is not known. A simple parameterization of these unknown processes that is often employed is called Rayleigh damping or bottom friction and takes the form

$$D = -r\Delta\psi \tag{3}$$

with a constant damping coefficient r. The streamfunction ψ determines the zonal velocity component u, the meridional velocity v, and the pressure p via the relations

$$u = -\partial_y \psi \tag{4}$$

$$v = \partial_x \psi \tag{5}$$

$$p = \rho_0 f_0 \psi \tag{6}$$

where f_0 is a constant reference value of the Coriolis parameter.

The potential vorticity equation (1) together with (2) and (3) is the starting point for most analyses of the stochastic atmospheric forcing of barotropic quasi-geostrophic eddies.

On an infinite domain the unforced ($F = 0$) inviscid ($D = 0$) potential vorticity equation has wave solutions of the form

$$\psi(\vec{x}, t) = e^{i(\vec{k}\cdot\vec{x} - \omega_0 t)} \tag{7}$$

where $\vec{k} = (k, l)$ is the wavenumber vector and where the frequency ω_0 is determined by the dispersion relation

$$\omega_0 = -\frac{\beta_0 k}{k^2 + l^2} \tag{8}$$

These waves are planetary Rossby waves. In the $\vec{k}$-plane lines of constant frequency are circles with center at $(-\frac{\beta_0}{2\omega_0}, 0)$ and radius $\frac{\beta_0}{2\omega_0}$ (See Fig. 1).

When bottom friction is included, ω_0 is replaced by

$$\tilde{\omega}_0 = \omega_0 - ir \tag{9}$$

and the waves decay exponentially at rate r.

The solution of the initial value problem with forcing and friction can be obtained by introducing the Green's function $G(\vec{x}, t/\vec{x'}, t')$ which satisfies

$$(\partial_t \Delta' + r\Delta' + \beta_0 \partial_{x'}) G(\vec{x}, t/\vec{x'}, t') = -\delta(t - t')\delta(\vec{x} - \vec{x'}) \tag{10}$$

where Δ' is the Laplacian operator acting on $\vec{x'}$. The solution is then given by

$$\psi(\vec{x}, t) = \int_0^{t+} dt' \iint d^2x' G(\vec{x}, t/\vec{x'}, t') F(\vec{x'}, t') + \iint d^2x' G(\vec{x}, t/\vec{x'}, 0) \psi_0(\vec{x'}) \tag{11}$$

where ψ_0 is the initial streamfunction. An integral representation of the Green's function is given in Wood and Willmott (1988).

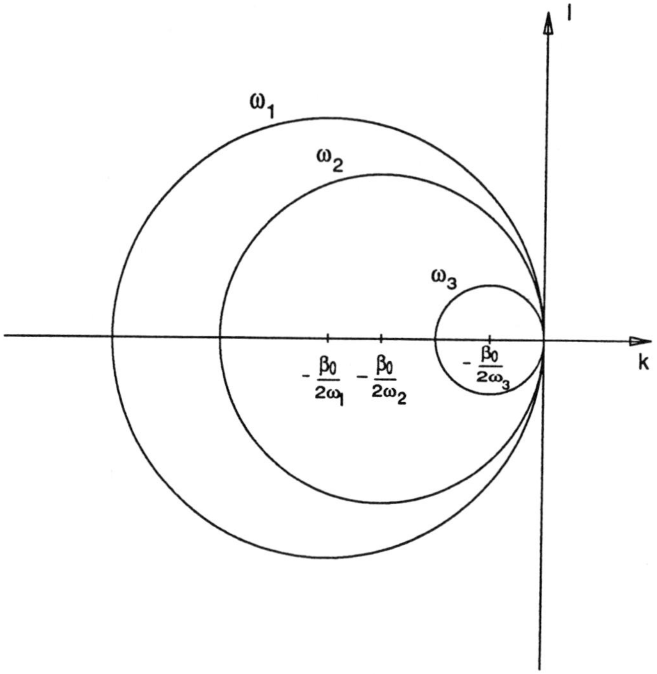

Figure 1: Dispersion relation of barotropic planetary Rossby waves ($\omega_1 < \omega_2 < \omega_3$).

3. Stochastic forcing

The forcing function $F(\vec{x}, t)$ is assumed to be a stationary random process with zero mean

(1) $$< F(\vec{x}, t) >= 0$$

and autocovariance function

(2) $$R_{FF}(\vec{x}, \vec{x'}, \tau) =< F(\vec{x}, t)F(\vec{x'}, t + \tau) >$$

Here cornered brackets denote ensemble means and τ is the time lag. We consider the response that is also a zero mean stationary random process with autocovariance function

(3) $$R_{\psi\psi}(\vec{x}, \vec{x'}, \tau) =< \psi(\vec{x}, t)\psi(\vec{x'}, t + \tau) >$$

and cross-covariance function

(4) $$R_{F\psi}(\vec{x}, \vec{x'}, \tau) =< F(\vec{x}, t)\psi(\vec{x'}, t + \tau) >$$

Stationary processes can be analyzed by Fourier transforming from the time to the frequency domain. Define the Fourier-transform pair by

(5) $$g(t) = \int_{-\infty}^{+\infty} d\omega g(\omega) e^{i\omega t}$$

(6) $$g(\omega) = \frac{1}{2\pi} \int_{-\infty}^{+\infty} dt g(t) e^{-i\omega t}$$

The Fourier transform of the covariance function $R(\tau)$ defines the frequency spectrum $S(\omega)$. Thus

(7) $$S_{FF}(\vec{x}, \vec{x'}, \omega) = \frac{1}{2\pi} \int_{-\infty}^{+\infty} d\tau R_{FF}(\vec{x}, \vec{x'}, \tau) e^{-i\omega\tau}$$

with similar formulae for $S_{\psi\psi}$ and $S_{F\psi}$. In general, the spectrum

(8) $$S(\omega) = \frac{1}{2\pi} \int_{-\infty}^{+\infty} d\tau < a(t) b(t+\tau) > e^{-i\omega\tau}$$

is related to the Fourier components $a(\omega)$ and $b(\omega)$ by

(9) $$< a^*(\omega) b(\omega') >= S(\omega) \delta(\omega - \omega')$$

which reflects the fact that the Fourier components are uncorrelated for stationary processes. Thus

(10) $$< F^*(\vec{x}, \omega) F(\vec{x'}, \omega') >= S_{FF}(\vec{x}, \vec{x'}, \omega) \delta(\omega - \omega')$$

with similar formulae for $S_{\psi\psi}$ and $S_{F\psi}$.

The frequency spectrum $S_{FF}(\vec{x}, \vec{x'}, \omega)$ has been estimated by Chave et al. (1991) from FNOC winds. An example is shown in Figure 2. It shows a map of the squared coherence

(11) $$\gamma_{FF}^2(\vec{x}, \vec{x'}, \omega) = \frac{|S_{FF}(\vec{x}, \vec{x'}, \omega)|^2}{S_{FF}(\vec{x}, \vec{x}, \omega) S_{FF}(\vec{x'}, \vec{x'}, \omega)}$$

between F at a fixed point $\vec{x}$ and $\vec{x'}$ over the North Pacific Ocean for a period of 25 days. The barotropic velocity components u and v and the pressure p are usually measured at a particular mooring site $\vec{x}_0$ in the ocean. One can thus estimate $S_{pp}(\vec{x}_0, \vec{x}_0, \omega)$ and the squared coherences

(12) $$\gamma_{Fp}^2(\vec{x}, \vec{x}_0, \omega) = \frac{|S_{Fp}(\vec{x}, \vec{x}_0, \omega)|^2}{S_{FF}(\vec{x}, \vec{x}, \omega) S_{pp}(\vec{x}_0, \vec{x}_0, \omega)}$$

and γ_{Fu}^2 and γ_{Fv}^2. An example is given in Figure 3, which shows maps of the squared coherence between the FNOC windstress curl over the North Pacific Ocean and the pressure (top), zonal velocity (middle), and meridional velocity (bottom) at the BEMPEX site (marked by the solid square).

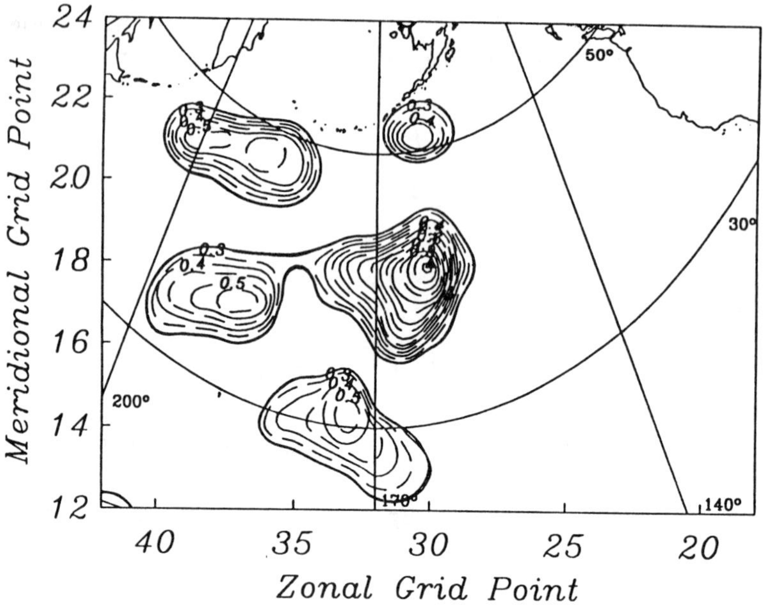

Figure 2: Maps of squared coherence between windstress curl at FNOC grid point (30, 18) and windstress curl over North Pacific Ocean at a period of 25 days for the period 8/86—6/87. The contour interval is 0.05. Only values larger than 0.3 are contoured. These values are statistically larger than zero at the 88% confidence level (from Chave et al., 1991).

4. Formal solution

The barotropic quasi-geostrophic response to stochastic forcing can be calculated by Fourier transforming the potential vorticity equation (1) which then becomes

$$(i\tilde{\omega}\Delta + \beta_0 \partial_x)\psi(\vec{x},\omega) = F(\vec{x},\omega) \tag{1}$$

where $\tilde{\omega} = \omega - ir$. The solution to this equation can be written in the form

$$\psi(\vec{x}) = e^{ikx}\Phi(\vec{x}) \tag{2}$$

where

$$k = \frac{\beta_0}{2\tilde{\omega}} \tag{3}$$

and where $\Phi(\vec{x})$ satisfies

$$\Delta\Phi(\vec{x}) + k^2\Phi(\vec{x}) = -\frac{iF(\vec{x})}{\tilde{\omega}}e^{-ikx} \tag{4}$$

The parametric dependence on the frequency ω has been suppressed.The Green's function that satisfies

$$\Delta' G(\vec{x}/\vec{x'}) + k^2 G(\vec{x}/\vec{x'}) = -\delta(\vec{x} - \vec{x'}) \tag{5}$$

is given by (Morse and Feshbach, 1953)

$$G(\vec{x}/\vec{x'}) = \frac{i}{4} H_0^{(1)}(kr) \tag{6}$$

where $r = \sqrt{(x - x')^2 + (y - y')^2}$ and $H_0^{(1)}$ is the Hankel function of the first kind of zeroth order. This particular Hankel function is chosen so that G decays towards infinity (for complex k) or so that the group velocity vector points outward for large radii (for the frictionless case which implies k real). With the help of this Green's function the solution of (1) becomes

$$\psi(\vec{x}, \omega) = -\frac{i}{\tilde{\omega}} \int\!\!\int d^2x' \tilde{G}(\vec{x}/\vec{x'}) F(\vec{x'}, \omega) \tag{7}$$

where

$$\tilde{G}(\vec{x}/\vec{x'}) = G(\vec{x}/\vec{x'}) e^{ik(x - x')} \tag{8}$$

To emulate an ocean basin of domain $\mathcal{A}$ the integration in (7) is limited to that domain. The implied boundary condition is that energy propagates out of the domain with no reflection. Physically, the energy is assumed to be absorbed in boundary layers. With this boundary condition no friction is needed to limit the response. The response is fetch limited. Reflecting boundaries could be handled by the method of images (Morse and Feshbach, 1953). In this case friction must be invoked to keep the response finite.

Having constructed the formal solution of the potential vorticity equation one can calculate the autospectrum

$$S_{\psi\psi}(\vec{x}_0, \vec{x}_0, \omega) = \frac{1}{\tilde{\omega}\tilde{\omega}^*} \int\!\!\int_{\mathcal{A}} d^2x' \int\!\!\int d^2x'' \tilde{G}^*(\vec{x}_0/\vec{x'}) \tilde{G}(\vec{x}_0/\vec{x''}) S_{FF}(\vec{x'}, \vec{x''}, \omega) \tag{9}$$

and the cross-spectrum

$$S_{F\psi}(\vec{x}, \vec{x}_0, \omega) = -\frac{i}{\tilde{\omega}} \int\!\!\int d^2x' \tilde{G}(\vec{x}_0/\vec{x'}) S_{FF}(\vec{x}, \vec{x'}, \omega) \tag{10}$$

The spectra for u, v, and p can be constructed by using the relations (4).

The expressions (9) and (10) represent the formal solution to the stochastic forcing problem, up to second order moments. Using observed windstress curl spectra $S_{FF}(\vec{x}, \vec{x'}, \omega)$, as, e.g., given in Figure 2, one could calculate maps of the squared coherences γ^2_{Fu}, γ^2_{Fv}, and γ^2_{Fp} and compare them to the observed maps, as given in Figure 3. This has not been done yet. Instead only simple geometries and simple models of the windstress curl have been considered.

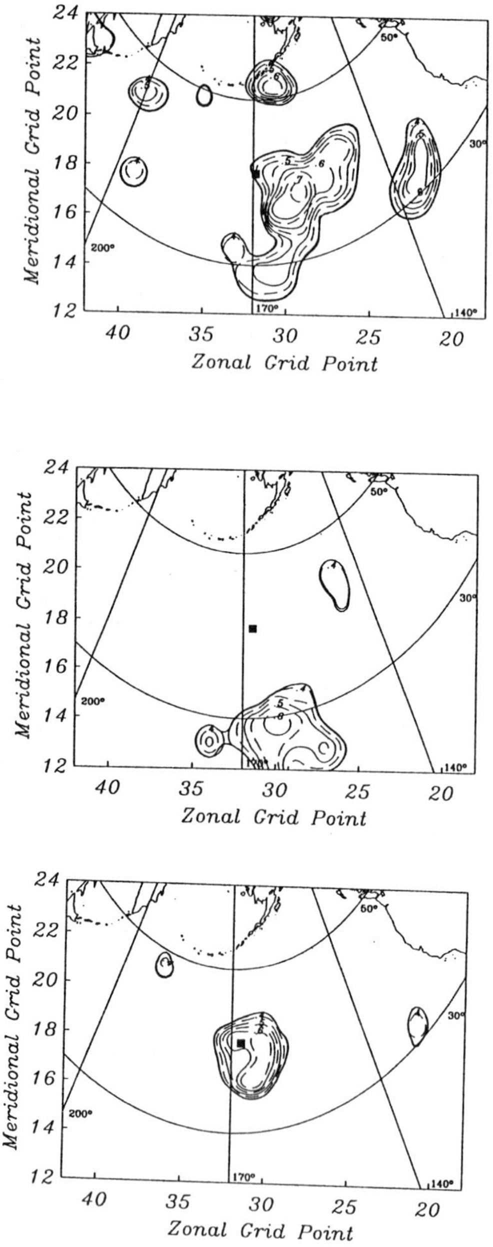

Figure 3: Maps of squared coherence between the FNOC windstress curl over the North Pacific Ocean and the bottom pressure (top), zonal velocity component (middle), and meridional velocity component (bottom) at the BEMPEX site (marked by the solid square). The map for the pressure is at a period of 28 days; the maps for the velocity components are at a period of 38 days. The contour interval is 0.05. Only values larger than 0.3 are contoured. These values are statistically larger than zero at the 88% confidence level (from Luther et al., 1990, and Chave et al., 1992).

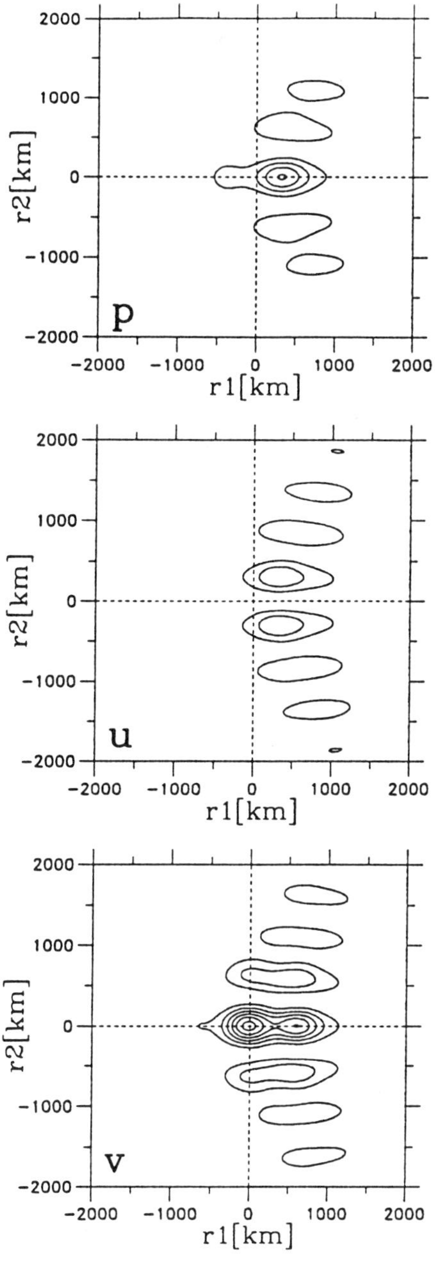

Figure 4: Maps of squared coherence between windstress curl and p (top), u (middle), and v (bottom) at $\omega = \frac{2\pi}{50d}$ in the resonant range. The contour interval is 0.02. (from Lippert and Müller, 1995).

5. Meridional channel

The specific case of a meridional channel with forcing that is statistically homogeneous in the y-direction has been considered by Brink (1989) and later by Samelson (1990). The channel is bounded by "absorbing" walls at $x = 0$ and $x = L$. Fourier transformation in the meridional variable leads to

$$(1) \qquad [i\tilde{\omega}(\partial_x \partial_x - l^2) + \beta_0 \partial_x]\psi(x, l, \omega) = F(x, l, \omega)$$

where l is the meridional wavenumber. As before the solution can be written in the form $\psi(x) = e^{ikx}\Phi(x)$, where $\Phi(x)$ satisfies

$$(2) \qquad \partial_x \partial_x \Phi(x) + (k^2 - l^2)\Phi(x) = -\frac{i}{\tilde{\omega}} F(x) e^{-ikx}$$

The Green's function for this equation is (Morse and Feshbach, 1953)

$$(3) \qquad G(x/x') = \frac{i}{2m_+} e^{im_+|x-x'|}$$

where m_+ is that root of $k^2 - l^2$ that has a positive imaginary part and leads to a decay towards infinity.

In the inviscid case (k real) the Green's function and the response becomes singular for $|l| = |k|$. In this case l, k, and ω satisfy the dispersion relation (8) of free planetary Rossby waves. The excited wave has a group velocity vector pointing in the y-direction (see Figure 1). It thus never encounters an absorbing wall and its amplitude therefore grows without any bound.

The solution for the streamfunction is

$$(4) \qquad \psi(x, l, \omega) = \frac{1}{2m_+\tilde{\omega}} \int_0^L dx' e^{im_+|x-x'|} e^{ik(x-x')} F(x', l, \omega)$$

from which we can construct the wavenumber-frequency spectra

$$(5) \qquad S_{\psi\psi}(x_0, x_0, l, \omega) = \frac{1}{4m_+ m_+^* \tilde{\omega}\tilde{\omega}^*} \int_0^L dx' \int_0^L dx'' e^{-im_+^*|x_0-x'|} e^{-ik^*(x_0-x')} \cdot e^{im_+|x_0-x''|} e^{ik(x_0-x'')} S_{FF}(x', x'', l, \omega)$$

$$(6) \qquad S_{F\psi}(x, x_0, l, \omega) = \frac{1}{2m_+\tilde{\omega}} \int_0^L dx' e^{im_+|x_0-x'|} e^{ik(x_0-x')} S_{FF}(x, x', l, \omega)$$

and the frequency spectra

$$(7) \qquad S_{\psi\psi}(x_0, x_0, y_0, y_0, \omega) = \int_{-\infty}^{+\infty} dl S_{\psi\psi}(x_0, x_0, l, \omega)$$

$$(8) \qquad S_{F\psi}(x, x_0, y_0 - \Delta y, y_0, \omega) = \int_{-\infty}^{+\infty} dl S_{F\psi}(x, x_0, l, \omega) e^{il\Delta y}$$

In performing the l integration one must note that $m_+ = m_+(l)$.

The spectra (7) have been numerically evaluated by Brink (1989) for a forcing spectrum of the form

$$(9) \qquad S_{FF}(x, x+\Delta x, y, y+\Delta y, \omega) = A(\omega) e^{-2\sigma \Delta x^2 - 2s\Delta y^2}$$

which has the wavenumber-frequency spectrum

$$(10) \qquad S_{FF}(x, x+\Delta x, l, \omega) = \frac{1}{2\pi}\sqrt{\frac{\pi}{2s}} A(\omega) e^{-2\sigma\Delta x^2} e^{-l^2/8s}$$

and have been compared with observations at a mooring site in the western North Atlantic. Samelson (1990) used the same approach and forcing spectrum to compare calculated coherence maps with observed maps based on a mooring site in the eastern North Atlantic.

Note that the theoretical model is statistically homogeneous in the y direction and that the response is limited by friction. The calculated maps thus depend on the poorly established parameterization of frictional processes.

6. Infinite ocean

The case of an infinite ocean was analyzed by Müller and Frankignoul (1981) and Lippert and Müller (1995). The forcing is now assumed to be statistically homogeneous in the x and y directions. Fourier transformation yields

$$(1) \qquad [-i\tilde{\omega}(k^2+l^2) + i\beta_0 k]\psi(k, l, \omega) = F(k, l, \omega)$$

or

$$(2) \qquad -i(\tilde{\omega} + \omega_0)\psi(k, l, \omega) = \frac{F(k, l, \omega)}{k^2 + l^2}$$

where $\omega_0(\vec{k})$ is the dispersion relation (8) of planetary Rossby waves. The solution is

$$(3) \qquad \psi(k, l, \omega) = H(k, l, \omega) F(k, l, \omega)$$

with transfer function

$$(4) \qquad H(k, l, \omega) = \frac{i}{(k^2 + l^2)(\tilde{\omega} + \omega_0)}$$

For an inviscid ocean the response becomes singular when ω, k, and l satisfy the dispersion relation for Rossby waves. These waves are excited resonantly. The wavenumber-frequency spectra of the response are given by

$$(5) \qquad S_{\psi\psi}(k, l, \omega) = |H(k, l, \omega)|^2 S_{FF}(k, l, \omega)$$

$$(6) \qquad S_{F\psi}(k, l, \omega) = H(k, l, \omega) S_{FF}(k, l, \omega)$$

and are directly proportional to the wavenumber-frequency spectrum of the forcing. The frequency spectra are given by

$$(7) \qquad S_{\psi\psi}(\vec{x}_0, \vec{x}_0, \omega) = \int_{-\infty}^{+\infty} dk \int_{-\infty}^{+\infty} dl S_{\psi\psi}(k, l, \omega)$$

$$(8) \qquad S_{F\psi}(\vec{x}, \vec{x}_0, \omega) = \int_{-\infty}^{+\infty} dk \int_{-\infty}^{+\infty} dl S_{F\psi}(k, l, \omega) e^{i\vec{k}\cdot(\vec{x}_0 - \vec{x})}$$

They have been numerically calculated by Lippert and Müller (1995) for the specific windstress curl spectrum

$$(9) \qquad S_{curlT}(\vec{k}, \omega) = \frac{1}{2} S_{curlT}(0) \frac{S_{curlT}(k)}{2\pi k}$$

where

$$(10) \qquad S_{curlT}(k) = \frac{1}{k_c - k_b} \begin{cases} 1 & \text{for } k_b < k < k_c \\ 0 & \text{otherwise} \end{cases}$$

Here k_b and k_c are low and high cut-off wavenumbers and k denotes the magnitude of wavenumber vector $\vec{k}$ (and not the x component as in the other parts of this paper). The model spectrum (9) is white in frequency space and white and isotropic in wavenumber space. $S_{curlT}(0)$ is the white noise level. These features are also basic to observed spectra (e.g., Chave et al., 1991).

Figure 4 shows a set of squared coherence maps for a period of 50 days. Explicitly, the coherence squared between F and p is given by

$$(11) \qquad \gamma^2_{Fp}(\vec{x}, 0, \omega) = \frac{|\int d\vec{k} e^{-i\vec{k}\cdot\vec{x}} H_p(\vec{k}, \omega) S_{FF}(\vec{k}, \omega)|^2}{\int d\vec{k} |H_p(\vec{k}, \omega)|^2 S_{FF}(\vec{k}, \omega) \int d\vec{k} S_{FF}(\vec{k}, \omega)}$$

where $H_p = \rho_0 f_0 H$. The squared coherence maps for u and v are obtained by substituting $H_u = -ilH$ and $H_v = ikH$, which are a consequence of $u = -\partial_y \psi$ and $v = \partial_x \psi$.

The calculated coherence maps show decaying periodic structures with local and non-local primary maxima, not unlike the observed maps shown in Figure 3. The patterns depend on the variable for which the coherence is calculated. This point is trivial since the relations $u = -\partial_y \psi$ and $v = \partial_x \psi$ imply that the map for u is the meridional and the map for v the zonal derivative of the map for ψ (or p). The simplicity of the model allows to explicitly relate features in the coherence maps to features in the forcing spectrum and transfer function. The location of the primary maximum depends on the symmetry of the integrand. The periodicity and decay scale depend on the distance at which the windstress curl autocorrelation function first crosses zero.

Like observed coherence maps, the calculated maps change with frequency. These changes reflect the frequency dependence of the transfer function. For small damping rates ($r \ll \omega$) the transfer function is given by

$$(12) \quad H(\vec{k},\omega) = \begin{cases} -\frac{i}{\beta_0 k} & \text{for } \omega \ll \omega_0 \\ \frac{i}{k^2+l^2}[i\pi\delta(\omega+\omega_0) + P(\frac{1}{\omega+\omega_0})] & \text{for } \omega \approx \omega_0 \\ \frac{i}{(k^2+l^2)\omega} & \text{for } \omega \gg \omega_0 \end{cases}$$

where P denotes the principal value. These different regimes are called Sverdrup, resonant, and off-resonant, respectively. The maps shown in Figures 3 and 4 are for a frequency in the resonant regime.

7. Summary and perspective

The stochastic atmospheric forcing of barotropic quasi-geostrophic eddies has been analyzed by using a simple theoretical framework: a linear evolution equation and a statistically stationary forcing function. Stationarity implies that all second order properties are completely determined by frequency spectra. Linearity implies that the autospectrum of the response and the cross-spectrum between forcing and response are determined by the spectrum of the forcing. The actual expressions consist of a convolution in physical space of the forcing spectrum with the Green's function of the evolution equation. The Green's function is a Hankel function of the first kind of zeroth order.

Actual calculations have considered simple geometries and simple forcing spectra: a meridional channel with forcing that is statistically homogeneous in the along channel coordinate and has a Gaussian spatial correlation function and an infinite ocean with forcing that is statistically homogeneous and has a white wavenumber spectrum. In both cases spatial Fourier transformation was employed rather than calculating the convolution integral. Also friction must be invoked in these cases to keep the response finite.

Comparison of calculated with observed coherence maps (although not presented in detail) shows that the simple models can reproduce some of the basic features of the observed maps. The open question is whether or not the more complex structure seen in observed coherence maps is due to the more complex structure of the forcing spectrum or due to processes neglected in the evolution equation. Some of the neglected processes have been studied: nonlinear interactions by Treguier and Hua (1987), meridional reflecting boundaries by Lippert and Käse (1985), vertical shear and topography by Samelson (1989) and meridionally inhomogeneous winds by Samelson and Shrayer (1991). Comprehensive data sets, such as BEMPEX, need to be further analyzed to find an answer to this question.

Acknowledgment: This work was supported by the Office of Naval Research.

7. References

Brink, K. H., 1989: Evidence for wind-driven current fluctuations in the Western North Atlantic. *J. Geophys. Res.*, **94**, 2029-2044.

Chave, A. D., D. S. Luther and J. H. Filloux, 1991: Variability of the wind stress curl over the North Pacific: Implications for the oceanic response. *J. Geophys. Res.*, **96**, 18361-18379.

Chave, A. D., D. S. Luther and J. H. Filloux, 1992: The barotropic electromagnetic and pressure experiment: 1. Barotropic current response to atmospheric forcing. *J. Geophys. Res.*, **97**, 9565-9593.

Frankignoul, C. and K. Hasselmann, 1977: Stochastic climate models, Part II Application to sea-surface temperature anomalies and thermocline variability. *Tellus*, **29**, 289-305.

Lippert, A. and R. F. Käse, 1985: Stochastic wind forcing of baroclinic Rossby waves in the presence of a meridional boundary. *J. Phys. Oceanogr.*, **15**, 184-194.

Lippert, A. and P. Müller, 1995: Direct atmospheric forcing of geostrophic eddies, Part II: coherence maps. *J. Phys. Oceanogr.*, (in press).

Luther, D. S., A. D. Chave, J. H. Filloux, and P. F. Spain, 1990: Evidence for local and nonlocal barotropic responses to atmospheric forcing during BEMPEX. *Geophys. Res. Lett.*, **17**, 949-952.

Mikolajewicz, U. and E. Maier-Reimer, 1990: Internal secular variability in an OGCM. *Climate Dyn.*, **4**, 145-156.

Morse P. M. and H. Feshbach, 1953: Methods of Theoretical Physics. McGraw-Hill.

Müller P. and C. Frankignoul, 1981: Direct atmospheric forcing of geostrophic eddies. *J. Phys. Oceanogr.*, **11**, 287-308.

Pedlosky, J., 1987: Geophysical Fluid Dynamics. Springer Verlag.

Phillips, O. M., 1957: On the generation of waves by turbulent winds. *J. Fluid Mech.*, **2**, 417-445.

Rubenstein, D., 1994: A spectral model of wind-forced internal waves. *J. Phys. Oceanogr.*, **24**, 819-831.

Samelson, R. M., 1989: Stochastically forced current fluctuations in vertical shear and over topography. *J. Geophys. Res.*, **94**, 8207-8215.

Samelson, R. M., 1990: Evidence for wind-driven current fluctuations in the eastern North Atlantic. *J. Geophys. Res.*, **95**, 11359-11368.

Samelson, R. M., and B. Shrayer, 1991: Currents forced by stochastic winds with meridionally varying amplitude. *J. Geophys. Res.*, **96**, 18425-18429.

Treguier, A. M., and B. L. Hua, 1987: Oceanic quasi-geostrophic turbulence forced by stochastic wind fluctuations. *J. Phys. Oceanogr.*, **17**, 397-411.

Wood, R. G., and A. J. Willmott, 1988: The generation of baroclinic Rossby waves by stationary and translating currents. Part 1: Stationary current forcing. *Geophys. Astrophys. Fluid Dyn.*, **41**, 287-311.

Department of Oceanography
University of Hawaii
1000 Pope Road
Honolulu, HI 96822

pmuller@iniki.soest.hawaii.edu

Maximum likelihood estimators in the equations of physical oceanography

L. Piterbarg and B. Rozovskii[1]

Abstract

We study parametric models of the upper ocean variability described by a certain class of stochastic partial differential equations. The maximum likelihood (ML) method is used to estimate the unknown model parameters for continuous or discrete (in time) observations. Necessary and sufficient conditions are given for the consistency, asymptotic normality and efficiency of the ML estimators when the number of observed spatial modes increases.

We demonstrate that the diffusivity estimate in the advection-diffusion equation is always consistent, while the consistency of the feedback parameter estimate depends on the spatial dimension. Further, the ML estimate of the velocity is consistent if and only if the diffusivity is zero. Estimation of the horizontal diffusivity and the bottom friction coefficient in the linearized barotropic vorticity equation will be discussed to illustrate our results.

0. Introduction

The problem discussed in this paper is an important part of a more comprehensive problem, commonly referred to as "data assimilation" in the oceanographical and meteorological literature [M. Ghil and P. Malanotte-Rizzoli, 1991]. Generally speaking the main purpose of data assimiliation is to fit a model to observations in some (preferably optimal) way and estimate the unknown model characteristics. These unknown characteristics might include initial and boundary conditions, observation errors, parameters such as velocities, diffusivity coefficients, etc. Here we focus on

[1]This work was supported by ONR Grant No. N00014-91-J-1526.

estimating unknown parameters in some commonly used equations of physical oceanography, such as the heat transport equation and the linearized equation for quasi-geostrophic motion in the upper ocean. In general, both equations can be written in the form

$$\partial u/\partial t + A_{\boldsymbol{\theta}} u = S(t, \boldsymbol{x}). \tag{1}$$

Here $u = u(t, \boldsymbol{x})$ is the observed quantity (temperature, height of the sea level, etc.) at point $\boldsymbol{x}$ at time t. $A_{\boldsymbol{\theta}}$ is a linear operator, which governs advection, diffusion, and/or other dissipation terms. It depends on an unknown vector parameter $\boldsymbol{\theta} = (\theta_1, \ldots \theta_K)$ with entries such as velocity, diffusion coefficients and the friction coefficients. Finally, $S(t, \boldsymbol{x})$ is a random field representing distributed sources (e.g. heat fluxes and wind stress), and/or model noise. The structure of $A_{\boldsymbol{\theta}}$ and $S(t, \boldsymbol{x})$ for some important applications will be discussed below in more detail.

To explain the structure of the observations which is assumed throughout this article, let us fix an arbitrary orthogonal basis $\{\varphi_m(\boldsymbol{x})\}_1^\infty$. It will be assumed that we observe (measure) only the first M amplitudes

$$u_1(t), \ldots, u_M(t), \quad t \in [0, T] \tag{2}$$

in the spatial Fourier expansion

$$u(t, \boldsymbol{x}) = \sum_m u_m(t) \varphi_m(\boldsymbol{x}) \tag{3}$$

Both types of observations, continuous and discrete in time, are considered.

In this work we discuss the methodology of parameter estimation based on the classical maximum likelihood approach, and demonstrate that this approach can be successfully extended to infinite-dimensional systems like the stochastic PDE (1).

Before proceeding with the detailed explanation of our methodology of parameter estimation, we have to (or at least attempt to) answer these preliminary questions:

- Why can/should we view the equations of physical oceanography as stochastic PDE's?
- What is the utility of the spectral representation (3) of oceanographical data?
- What is a "good" estimator, and why are the maximal likelihood estimators "good"?

We will try to answer these questions (at least partially) in the remaining part of the Introduction.

The equations of physical oceanography are phenomenological equations representing an approximate macroscopic description of ocean dynamics. To justify these equations on a deeper level would require a much more refined and therefore more difficult analysis involving microscopic considerations. In most cases, such an analysis is practically impossible. Therefore, to account for the neglected microscopic nature of the system, measurement error and other fine structure, one often considers appropriate perturbations of the macroscopic equations. An important class of such models is represented by random perturbations of forcing terms, initial and boundary conditions etc., which naturally lead to stochastic PDE models.

The idea of stochastically perturbed distributed parameter models of type (1) for ocean dynamics has been explored in the oceanographic literature for quite some time. Below we give some related examples.

One of the first studies in this area was done by Frankingnoul and Reynolds [1983]. They considered the following equation governing the large-scale long-term variability of sea surface temperature (SST) anomalies

$$\frac{\partial T}{\partial t} + \boldsymbol{u} \cdot \nabla T + \lambda T = S(t, \boldsymbol{x}), \tag{4}$$

where λ is an atmosphere and ocean coupling parameter, and $S(t, \boldsymbol{x})$ is a source term including heat fluxes at the air-water interface, vertical heat fluxes due to turbulent entertainment and anomalous advection of the mean temperature gradient. The random field $S(t, \boldsymbol{x})$ is assumed to be a "white noise" process in time. The reason for this assumption is due to the difference between the synoptic time scale in atmosphere (3-5 days) and the typical variability time of the sea surface temperature (2-5 months). In the discussed paper it was supposed that the velocity field $\boldsymbol{u}(x)$ is known (from the observations of drifting trade ships) and the problem stated was to estimate λ and variance S of $S(t, \boldsymbol{x})$.

For this purpose the following functional was minimized

$$J = \sum_i (F_T(\omega_i; \lambda, S) - \hat{F_T}(\omega_i))^2 \tag{5}$$

at each space point on a prescribed grid. Here $F_T(\omega; \lambda, S)$ is the frequency spectrum of the temperature for a fixed grid point derived from model (4), $\hat{F_T}(\omega)$ is the frequency spectrum for the same point obtained from observations of temperature, $\{\omega_i\}$ is the collection of frequencies due to a given time step and duration of observations. The choice of the functional (5) is reasonable because it is a measure of the discrepancy between model generated data and true observations. In a later paper [Frankingnoul at el., 1993] a more complicated model for the sea surface temperature was considered which allows the assimilation of atmospheric parameters.

One of the most important inverse problems in oceanography is the estimation of the velocity from passive scalar observations (see e.g. Wunsh [1978], [1985], Fiadeiro and Veronis [1984], Kelly [1989], Ghill and Malanotte-Rizzoli [1991] and references therein).

In [Ostrovskii, Piterbarg, 1985] and [Herterich, Hasselman, 1987] efforts were undertaken to estimate velocities, dissipation and generation parameters simultaneously. Both papers dealt with the same model

$$\frac{\partial T}{\partial t} + \boldsymbol{u} \cdot \nabla T + \lambda T - D\nabla^2 T = S(t, \boldsymbol{x}) \tag{6}$$

which differs from (4) due to the diffusion term. As before $S(t, \boldsymbol{x})$ is white noise in t. Additionally in [Herterich, Hasselman, 1987], it was assumed that in the neighborhood of each point of the grid, $S(t, \boldsymbol{x})$ is homogeneous with space covariance function of the form

$$R_S(x_1, x_2) = S \exp[-x_1^2/R_1^2 - x_2^2/R_2^2], \tag{7}$$

where S, R_1, R_2 are unknown parameters. The problem of interest was to estimate $\boldsymbol{u}, \lambda, D, S, R_1, R_2$ at each point of the grid. This was done by minimizing the functional

$$J = \Delta \boldsymbol{F}^t M \Delta \boldsymbol{F}, \tag{8}$$

where $\Delta \boldsymbol{F} = \boldsymbol{F} - \hat{\boldsymbol{F}}, \boldsymbol{F} = (\boldsymbol{F}(\omega_1), \ldots, \boldsymbol{F}(\omega_n))$, $\boldsymbol{F}(\omega)$ is a collection of temperature frequency auto and cospectra at a fixed grid point and 4 neighboring points, deduced from model (6), $\hat{\boldsymbol{F}}(\omega)$ is a similar collection of spectrum estimators obtained from observations and M is a suitably defined weighing matrix.

In [Ostrovskii, Piterbarg, 1985] to estimate $\boldsymbol{u}, D$ and λ the stochastic partial differential equation (6) was reduced to a multidimensional autoregressive model. Then some standard techniques for estimating unknown autoregression coefficients were applied and the obtained estimates were used to compute the physical parameters $\boldsymbol{u}, D, \lambda$.

Despite the difference in these approaches, the results are quite similar. In the Northeast Pacific, the analysis revealed circulations which resembled the overall structure of the surface current field inferred from ship-drift observations. By contrast, in the Northwest Pacific those studies did not exhibit advection patterns consistent with the circulation in the subtropical gyre and the Kurosio extension region. The reason for this is the poor spatial resolution of the temperature data. In both works [Ostrovskii,

Piterbarg, 1985] and [Herterich, Hasselman, 1987] the Namias array was used with a $5^0 \times 5^0$ grid, obtained from trade ship measurements.

In [Ostrovskii, Piterbarg, 1994] this approach was extended to the case when

$$\dot{S} + \beta S = \dot{W}(t, \boldsymbol{x}) \tag{9}$$

where $\dot{W}(t, \boldsymbol{x})$ is a white-noise process and β is a supplemental unknown parameter. This improvement is extremely important for studying heat processes with a time scale of less than 1 month.

Finally we consider one example related to the assimilation of the sea surface height data. Tsiperman and Thacker (1989) used the quasi-geostrophic barotropic vorticity equation with unknown parameters to compute steady-state circulation from simulated vorticity and streamfunction observations. In time-dependent form their model is

$$\frac{\partial}{\partial t}\nabla^2\psi + \frac{\partial \psi}{\partial x} + R_0 J(\psi, \nabla^2\psi) + D_h \nabla^2\psi - b\nabla^4\psi = \text{curl}\ \vec{\tau}(t, x, y) \tag{10}$$

where ψ is the barotropic streamfunction, R_0 the Rossby number, D_h the horizontal eddy diffusivity, b the bottom friction coefficient, $\vec{\tau}(t, x, y)$ the wind stress field and J the Jacobian. In both the linear ($R_0 = 0$) and nonlinear ($R_0 > 0$) cases, the unknown parameters D_h and b were estimated by the optimization of some misfit function involving the energy and the enstrophy of the system. A sinusoidal forcing was assumed and the initial field was considered as unknown.

Assimilation of altimeter data in the more complicated two-layer model of the primitive equations was discussed in the recent paper [Smedstad, Fox, 1994] where a review of the assimilation of sea surface height (SSH) observations is also given.

One can see that the heat balance equations (4), (6) as well as equation (10) for the stream function (in the case of $R_0 = 0$) can be treated in the general model (1).

For example, eq. (6) can be written in form (1) by setting

$$A_\theta = \theta_1 \frac{\partial}{\partial x} + \theta_2 \frac{\partial}{\partial y} - \theta_3 \nabla^2 + \theta_4 , \tag{11}$$

where $\theta_1 = u$, $\theta_2 = v$, are the unknown velocity components in the directions x and y, respectively $\theta_3 = D$, $\theta_4 = \lambda$.

To reduce the linear version of (10) to the general scheme we set

$$u = \Delta^{-1}\psi \tag{12}$$

where $\Delta = \nabla^2$, then

$$\frac{\partial u}{\partial t} + \Delta^{-1} u_x + D_h u - b\Delta u = = \text{curl } \Delta^{-1} \vec{\tau}(t, x, y) \; . \tag{13}$$

Therefore we have in this case

$$A_{\boldsymbol{\theta}} = \Delta^{-1} \frac{\partial}{\partial x} + \theta_1 u - \theta_2 \Delta u \tag{14}$$

where $\theta_1 = D_h, \theta_2 = b$ and we do not have any unknown parameters in the forcing.

So we see that (1) is general enough to cover a number of oceanographic models of the evolution of SST and SSH.

Now let us discuss briefly the utility of the spectal representation (3) for oceanographic data.

The most extensive of the new data sets are remotely sensed from space. Active and passive instruments operating in the microwave, infrared, and visible portions of the electromagnetic spectrum provide spatial and temporal coverage of the upper ocean variability. Among the most important and well-observed physical quantities from space are the sea surface temperature (SST) and the sea surface height (SSH) (altimetry data). SST images usually have a horizontal resolution of approximately 1.1 km, with temporal separations of 4 to 8 hours. While clouds often obscure much of the ocean, there are occasionally periods of 1 to 3 days with relatively few clouds during which 4 to 12 images can be collected [Statistics and Physical Oceanography, 1993].

The satellite altimeter takes measurements roughly every 7 km along a track with a distance of approximately 100 km between tracks. In contrast with the radiometer, the altimeter, employing active microwave radar, operates in both day and night-time hours, in the presence of clouds or in clear weather. For this reason, relatively long time series of SSH are available.

In summary, modern satellite facilities provide accurate measurements of SST and SSH with a high space resolution, but the time series of SST are often short. For this reason, (2,3) are suitable models for satellite observations, where M is large enough and by contrast the observation time T is not assumed to be large. Note that there are also a couple of other parameters which are measured well from space such as the sea surface wind velocity, color, etc. Since here we concentrate on the equation governing SST and SSH we will not discuss those satellite data.

Remote sensing is not the only method of measuring of SST and SSH. An alternative way of collecting SST data is by aggregating observations from trade ships as, for example, the Namias array [Frankignoul, Reynolds, 1983]. SSH is measured also from coastal stations. Such methods provide a long time series of observations but their space resolution is poor. A reasonable model for these observations is covered by (2) with small M and large T.

The general problem of parameter estimation can be formulated in a rather straightforward way: given the chosen model and the observed data, find the "optimal" values of the unknown parameters satisfying the imposed restrictions. In spite of the plentitude of possible approaches, this problem as a rule reduces to minimization of some misfit functional of the data and the parameters with respect to the latter. The misfit functional is a measure of discrepancy between the model and the data. It might be chosen in a number of reasonable ways. Some examples of misfit functionals popular in oceanographic literature were mentioned above (see (5), (8)), others can be found in Kelly [1989], Tsiperman and Thacker [1989], Malanotte-Rizzoli [1991], etc. Most of them lead to some form of least square procedure.

Although the task of minimization of a particular misfit functional might be technically very difficult, a far more difficult if at all solvable, problem is the comparison between different "optimal" estimates. In fact standard examples in statistics show that in many cases an optimal estimate simply does not exist (at least for a fixed sample size). However, the asymptotic approach to statistics gives a convenient framework for comparison and classification of estimators (see e.g., LeCam [1986], Ibragimov and Has'minskii [1981] and references therein). Of course the asymptotic approach to data analysis becomes practical if reasonably large samples are available (at least potentially). Fortunately, often this is not such a crucial limitation. As we mentioned above, the recent development of technical capabilities for collecting and processing of marine data has made possible the accumulation of very large data sets. For example, remote (satellite) sensing allows for simultaneous measurements of SST and SSH at thousands of gridpoints.

One of the most important concepts of the asympotic approach to statistical analysis is *asymptotic efficiency*.

A sequence of estimates $\hat{\theta}_n$ of the parameter θ is said to be asymptotically efficient with respect to the loss (risk) function w if these estimates mimimize the quantity

$$\overline{\lim_{n\to\infty}} Ew(\tilde{\theta}_n, \theta). \tag{15}$$

Here E stands for expectation and n is the asymptotic parameter. It characterizes the amount of data used to calculate the estimator $\tilde{\theta}_n$. For example, n might be the sample size or the observation time, or the signal to noise ratio, etc. The distance between $\tilde{\theta}_n$ and θ is a good example of the loss function $w(\tilde{\theta}_n, \theta)$.

Another very desirable asymptotic property of an estimator is its *consistency.* The estimator $\hat{\theta}_n$ is *consistent (asympototically unbiased)* if it converges (in a suitable sense) to the true value of the parameter as $n \to \infty$.

If $\hat{\theta}_n$ is a consistent estimate of parameter θ it is often possible to find a normalizing function $\sigma_n(\theta)$ such that the probability distribution $(\hat{\theta}_n - \theta)/\sigma_n(\theta)$ converges to one of the classical distributions (e.g. Gaussian, in which case the estimate $\hat{\theta}_n$ is referred to as asymptotically normal). This property is very important for the derivation of approximate confidence intervals.

Consistency, asymptotic efficiency and asymptotic normality (or existence of another limiting distribution) are the most important properties which must be addressed when judging "the quality" of estimates. Of course other important properties (e.g., computational complexity) should not be ignored either.

For the sake of concreteness in this paper we concentrate on the maximum likelihood (ML) approach to parameter estimation (see e.g. Cramer, 1946, Ibragimov and Khasminskii, 1981). The advantage of this approach as applied to the models of type (1) is twofold: firstly the ML estimates are usually simple and computationally effective, secondly they often provide an explicit criteria for consistency, asymptotic normality and asymptotic efficiency.

As one can see, the model of observation (2) includes two parameters related to the volume of the sample, the number of observed modes, M, and the time of observation, T. Here we completely focus on the case when M goes to infinity, while T remains fixed. *In other words the growth of the sample's volume in our sense means an indefinitly increasing space resolution.* So our study addresses first of all remotely sensed obervations.

As was mentioned before, the case when T goes to infinity and M is small and fixed, is also faced in oceanography, when climate time series are being considered. From the mathematical viewpoint, this asymptotic is comparatively straightforward. When T tends to infinity, under quite general conditions many popular estimators, such as ML estimators, least square estimators, etc., are consistent and asymptotically normal (see e.g. Loges, 1984). Meanwhile, in the case we are focusing on ($M \to \infty$, T fixed) the answer is much less trivial.

Roughly speaking, we prove that if $A_\theta = A_0 + \theta_1 A_1 + \ldots \theta_K A_K$, where A_k are partial differential operators and A_θ is an elliptic operator which satisfies some additional assumptions, then the MLE estimator for the parameter θ_k is consistent, asymptotically normal and efficient if and only if

(16)

$$order\, A_k \geq \begin{cases} \frac{1}{2[order(A_\theta)-d]} & \textit{in the case of continuous observations} \\ order(A_\theta) - \frac{1}{2}d & \textit{in the case of discrete observations.} \end{cases}$$

This result indicates that if the above condition holds the MLE can be made extremely accurate even on a very short time interval given that sufficiently dense spatial measurements are available.

1. Problem statement

In this section as well as in the following two we consider the problem of estimating a vector parameter $\boldsymbol{\theta} = (\theta_1, \ldots, \theta_K)$ in the equation

$$\frac{\partial u}{\partial t} + (A_0 + \theta_1 A_1 + \cdots + \theta_K A_K)u = S(t, \boldsymbol{x}), \tag{17}$$

where $u = u(t, \boldsymbol{x})$ is the observed random field, $\boldsymbol{x} \in G$, G is a bounded domain in the d-dimensional Euclidean space R^d, A_k, $k = 0, \ldots, K$ are linear self-adjoint operators defined on a convenient function space, and $S(t, \boldsymbol{x})$ is a zero mean Gaussian "white noise" process in t that does not contain unknown parameters.

We construct the maximum likelihood estimator for $\boldsymbol{\theta}$ using an expansion of the observed field in terms of an orthogonal basis and find necessary and sufficient conditions under which the estimator based on the first M expansion terms ("modes" in the physics language) is consistent, asymptotically Gaussian, and asymptotically efficient when M goes to infinity.

More precisely, let us assume that there exists a complete orthonormal system of eigenfunctions $\{\varphi_m(\boldsymbol{x}), m = 1, 2 \ldots\}$ common to the operators $A_0, \ldots, A_K$ such that

$$A_k \varphi_m = \lambda_{km} \varphi_m, \quad k = 0, 1, \ldots, K \tag{18}$$

and

$$\int_G \varphi_m(\boldsymbol{x}) \varphi_n(\boldsymbol{x}) d\boldsymbol{x} = \delta_{mn} \tag{19}$$

where λ_{km} are the corresponding real eigenvalues and δ_{mn} is the Kronecker delta. This implies of course that each φ_m is an eigenfunction for the operator

$$A_\theta = A_0 + \theta_1 A_1 + \cdots + \theta_K A_K \tag{20}$$

with eigenvalue given by

$$\lambda_m(\boldsymbol{\theta}) = \lambda_{0m} + \theta_1\lambda_{1m} + \cdots + \theta_K\lambda_{Km}. \tag{21}$$

Suppose that the unknown parameter $\boldsymbol{\theta}$ belongs to a bounded region Θ of R^K and denote

$$\overline{\lambda}_m = \sup_{\Theta} \lambda_m(\boldsymbol{\theta}) \tag{22}$$

Let us agree to number the functions φ_m, $m = 1, 2, \ldots$ in such a way that the sequence $\{\overline{\lambda}_m\}$ is non-decreasing

$$\overline{\lambda}_1 \le \overline{\lambda}_2 \le \ldots \le \overline{\lambda}_m \le \ldots \tag{23}$$

In so doing we refer to the first eigenfunctions $\varphi_1(\boldsymbol{x}), \ldots, \varphi_M(\boldsymbol{x})$ as "large-scale" or "higher" modes.

Since the system $\{\varphi_m(\boldsymbol{x})\}$ is assumed to be complete we can expand any solution of (17) (obeying the choosen boundary conditions) as

$$u(t, \boldsymbol{x}) = \sum_{m=1}^{\infty} u_m(t)\varphi_m(\boldsymbol{x}) \tag{24}$$

where the amplitudes are determined by

$$u_m(t) = \int_G u(t, \boldsymbol{x})\varphi_m(\boldsymbol{x})d\boldsymbol{x}\,. \tag{25}$$

Now we can give the exact statement of the problem:
To estimate the unknown multidimensional parameter $\boldsymbol{\theta}$ *from observations of the amplitudes*

$$u_1(t), \ldots, u_M(t), \tag{26}$$

assuming that θ *is confined to a (fixed) bounded region* Θ, *and* t *runs over the interval* $(0, T)$ *or a finite number of times* $t_1, \ldots t_N$.

In the first case we refer to continuous observations and in the second one to discrete observations.

In this paper we focus on the asymptotics of the proposed estimator only if M goes to infinity, while T or N are fixed. In other words, we address the asymptotics of "increasing space resolution" provided the time interval is finite.

In the case of discrete observations the set of observations $u_m(t_n)$, $m = 1, \dots, M$, $n = 1, \dots, N$ forms a Gaussian vector which we denote by $u^{\boldsymbol{\theta},M}$ stressing that the sample is chosen from the model involving θ as an unknown parameter. Let $P^{\boldsymbol{\theta},M}(u)$ be the probability distribution of the vector $u^{\boldsymbol{\theta},M}$ and

$$p^{\boldsymbol{\theta},M}(u) = \frac{dP^{\boldsymbol{\theta},M}(u)}{du} \tag{27}$$

the density function. We define the likelihood function by

$$L_M(\boldsymbol{\theta}) = L_M(\boldsymbol{\theta};\boldsymbol{\theta}_0) = \log p^{\boldsymbol{\theta},M}(u^{\boldsymbol{\theta}_0,M}) \tag{28}$$

where $\boldsymbol{\theta}_0 = (\theta_{01}, \dots, \theta_{0K})$ is the true value of the unknown parameter. Note that the likelihood function depends on unknown $\boldsymbol{\theta}_0$ only indirectly via the sample data. A vector $\hat{\boldsymbol{\theta}}_M$ which maximizes $L_M(\boldsymbol{\theta})$ is refered to as a *maximum likelihood estimator* (MLE). In many important cases and in particular in all the cases considred below this definition determines the MLE uniquely.

Obviously, if the observations are continuous, the sample is a continuous function in R^M rather than a finite dimensional vector as in the case of discrete observations. Thus, in the continuous case the density function cannot be defined in manner given by (27). This obstacle could be circumvented by replacing $p_M^{\boldsymbol{\theta}}$ in (28) by the the Radon-Nikodim density $dP^{\boldsymbol{\theta},M}/dP^{\boldsymbol{\theta}_0,M}$. Note that in the discrete case the latter definition of MLE is equivalent to the former one. Indeed in this case $dP^{\boldsymbol{\theta},M}/dP^{\boldsymbol{\theta}_0,M} = p^{\boldsymbol{\theta},M}/p^{\boldsymbol{\theta}_0,M}$, so $\log(dP^{\boldsymbol{\theta},M}/dP^{\boldsymbol{\theta}_0,M})(u^{\boldsymbol{\theta}_0,M})$ achieves its maximum at the same points that maximize $\log p^{\boldsymbol{\theta},M}(u^{\boldsymbol{\theta}_0,M})$

For the sake of uniformity everywhere below we assume that

$$L_M(\boldsymbol{\theta};\boldsymbol{\theta}_0) = \log \frac{dP^{\boldsymbol{\theta},M}}{dP^{\boldsymbol{\theta}_0,M}}(u^{\boldsymbol{\theta}_0,M}) \tag{29}$$

In the next section we give explicit formulas for the likelihood function in some simple cases.

To ensure effective computation of MLEs we need additional conditions on the noise $S(t,\boldsymbol{x})$. Namely, we assume that the functions $\varphi_m(\boldsymbol{x})$ are also eigenfunctions for the spatial covariance operator R of $S(t,\boldsymbol{x})$. From this it follows that

$$S(t,\boldsymbol{x}) = \sum_{m=1}^{\infty} \sigma_m s_m(t)\varphi_m(\boldsymbol{x}) \tag{30}$$

where σ_m^2 is the eigenvalue of R corresponding to φ_m and $\{s_m(t)\}$ is a sequence of independent Gaussian "white noise" processes with zero mean. To ensure finiteness of the noise variance and the variance of $u(t,\boldsymbol{x})$ we suppose

$$\sum_1^\infty \sigma_m^2 < \infty, \quad \sum_1^\infty \sigma_m^2/2\lambda_m(\boldsymbol{\theta}) < \infty \tag{31}$$

If condition (30) holds, the amplitudes $u_m(t)$ from (25) are independent and satisfy the following ordinary stochastic diferential equations

$$\dot{u}_m = \lambda_m(\boldsymbol{\theta})u_m + \sigma_m s_m(t), \ \ m = 1, 2, \ldots \tag{32}$$

As we will see later on, the independence property and the simple structure of equation (32) enable us to derive explicit expressions for the likelihood function and to study the MLE's asymptotics effectively. Before proceeding to the formulation of our main results, let us remark that we consider only the statistically stationary in time solution $u(t,\boldsymbol{x})$ of (17). This solution can be obtained by solving the Cauchy problem with an initial condition corresponding to the invariant distribution. In terms of amplitudes this invariant distribution is given by

$$u_m \Big|_{t=0} \sim N\left(0, \frac{\sigma_m^2}{2\lambda_m(\theta)}\right) \tag{33}$$

In doing so, we automatically assume that the operator $A_{\boldsymbol{\theta}}$ is strictly positive, $\lambda_m(\boldsymbol{\theta}) > 0$ for all m and $\boldsymbol{\theta}$. This assumption is made in order to simplify computations and resulting formulas. The main results remain true under arbitrary reasonably smooth initial conditions and without the assumption of strict positivity.

2. The case of a single unknown parameter and a self-adjoint operator.

For the sake of clarity, in this section we restrict ourselves to only one unknown parameter $\theta \equiv \theta_1$, i.e. we consider the equation

$$\frac{\partial u}{\partial t} + (A_0 + \theta A_1)u = S(t, \boldsymbol{x}). \tag{34}$$

In the case of continuous observations, the likelihood function for the first M components of (32) is given by

(35)

$$L_M(\theta,\theta_0) = (\theta-\theta_0)\sum_{m=1}^{M}\int_0^T\left(\frac{\lambda_{1m}}{\sigma_m^2}u_m(t)du_m(t) - \frac{\lambda_{1m}\lambda_{0m}}{\sigma_m^2}u_m^2(t)dt\right) - \\ -\frac{\theta^2-\theta_0^2}{2}\sum_{m=1}^{M}\int_0^T\frac{\lambda_{1m}^2}{\sigma_m^2}u_m^2(t)dt$$

(see [Huebner, Rozovskii, 1993]) and

$$L_M(\theta,\theta_0) = \sum_{m=1}^{M}\left\{N\ln\frac{\lambda_m(\theta)}{\lambda m\theta_0} - \sum_{n=1}^{N-1}\ln\frac{1-e^{-2\lambda_m(\theta)\Delta t_n}}{1-e^{-2\lambda_m(\theta_0)\Delta t_n}} - \right.$$

$$(36)\qquad -(\theta-\theta_0)\frac{\lambda_{1m}}{\sigma_m^2}u_m(t_1) - \frac{1}{\sigma_m^2}\sum_{n=1}^{N-1}\left[\frac{\lambda(\theta)}{1-e^{-2\lambda_m(\theta)\Delta t_n}}(u_m(t_{n+1}) \right. \\ -e^{-\lambda_m(\theta)\Delta t_n}u_m(t_n)^2 -$$

$$\left.\left. -\frac{\lambda_m(\theta_0)}{1-e^{-2\lambda_m(\theta_0)\Delta t_n}}(u_m(t_{n+1}) - e^{-\lambda_m(\theta_0)\Delta t_n}u_m(t_n))^2\right]\right\}$$

for the discrete observations [Piterbarg, Rozovskii, 1995], where $\Delta t_n = t_{n+1} - t_n$, $n = 1,\dots,N-1$ and $t_1 = 0$.

One can see that in the first case the functional to be maximized is quadratic in θ and hence the calculation of $\hat\theta_M$ is straightforward.

$$(37)\qquad \hat\theta_M = \frac{\sum_{m=1}}{M}\int_0^T\left(\frac{\lambda_{1m}}{\sigma_m^2}u_m(t)du_m(t) - \right. \\ \left.\frac{\lambda_{1m}\lambda_{0m}}{\sigma_m^2}u_m^2(t)\right)dt\frac{\sum_{m=1}^{M}\int_0^T\lambda_{1m}^2}{\sigma_m^2}u_m^2(t)dt$$

In the second case the situation is more difficult, because $\hat\theta_M$ can not be expressed in an explicit form. Nevertheless, it can be computed in a simple way by solving numerically the equation

$$\frac{\partial L_M(\theta,\theta_0)}{\partial\theta} = 0,$$

and asymptotic analysis of the behavior of $\hat\theta_M$ is feasible.

Proposition 1. *(i)*[Huebner, Rozovskii, 1993] *For continuous observations in model (34), the MLE of the unknown parameter θ is consistent,* *i.e.*

(38) $$\lim_{M\to\infty} \hat{\theta}_M = \theta_0$$

if and only if

(39) $$\sum_1^\infty \frac{\lambda_{1m}^2}{\lambda_m(\theta_0)} = \infty.$$

(ii)[Piterbarg, Rozovskii, 1995] *In the case of discrete observations, it is necessary and sufficient for consistency that*

(40) $$\sum_1^\infty \frac{\lambda_{1m}^2}{\lambda_m(\theta)^2} = \infty, \text{ for all } \theta \text{ in some neighborhood of } \theta_0 .$$

Let A_1, A_θ be elliptic differential operators of orders p_1 and p respectively, complemented by appropriate boundary conditions. It is well known that under condition (23) for large m

(41) $$\lambda_{1m} \sim m^{p_1/d}, \quad \lambda_m(\theta) \sim m^{p/d}$$

where symbol $\sim$ henceforth means that the ratio of the expressions on either side of this symbol is bounded above and below. Under (41) the conditions of consistency can be rewritten in terms of the order of the operators A_θ and A_1.

The consistency condition for continuous observations is given by

(42) $$d \geq p - 2p_1$$

and for the discrete observations by

(43) $$d \geq 2p - 2p_1$$

One can see that the minimal dimension, for which the unknown parameter can be estimated consistently, in the discrete case is larger. Equivalently (42), (43) imply that for the same dimension from the same order of A_θ in the discrete case the order of operator A_1 has to be higher to insure consistent estimation. Intuitively it is clear that the larger is $A_1 u$, the easier it is to estimate θ. Due to irregularity of the random forcing $S(t,x)$ one could expect the field $u(t,x)$ also to be quite irregular. So the term $A_1 u$ should be larger if the order of the operator A_1 is higher. On the other hand the likelihood function in the discrete case is less informative than in the continuous one. Higher order derivatives in the operator A_1 are needed to compensate for the aforementioned "built in" disadvantage of discrete time estimates.

Of course this explanation is very loose. For more rigorous analysis see [Huebner, Rozovskii, 1993]

As an example apply (42), (43) to the heat balance equation without advection

$$\frac{\partial u}{\partial t} + \lambda u - D\nabla^2 u = S(t, \boldsymbol{x}) \tag{44}$$

subject to some boundary (e.g. Dirichlet or periodic) conditions. First assume that $\lambda > 0$ is given and D is to be estimated. In this case $p = 2$ and $p_1 = 2$. Consequently conditions (42), (43) are satisfied for any D and we arrive at the following.

In the purely dissipative model (44), the ML estimator of diffusivity is consistent for all dimensions and both kinds of observations, continuous and discrete.

Now assume that $D > 0$ is given and we are interested in estimating λ. In this case $p = 2$, $p_1 = 0$. Then

The ML estimator of λ *is consistent iff*

$$d \geq 2 \tag{45}$$

for continous observations, and iff

$$d \geq 4 \tag{46}$$

for discrete observations.

So, by passing to discrete observations we lose two units of dimension. Thus, (46) means that in reality a consistent estimator of λ is impossible. It will be shown later that the above conditions hold for the full heat balance equation (9) where the velocity is given.

If conditions (39), (40) or equivalently (42), (43) hold, then the MLE of θ is asymptotically normal and efficient. More precisely the following statement holds true.

Proposition 2. [Huebner, Rozovskii, 1993] *Assume (39). Then in the case of continuous observations*

$$\lim_{M\to\infty} P\{c(M)(\hat{\theta}_M - \theta_0) < x\} = \Phi_{\sigma_c}(x) \tag{47}$$

and

(48)

$$\sigma_c^2 := \lim_{M\to\infty} E\left[(\hat{\theta}_M - \theta_0)^2 c(M)^2\right] = \lim_{M\to\infty} \inf_{\tau_M \in U} E\left[(\tau_M - \theta_0)^2 c(M)^2\right]$$

(ii) [Piterbarg, Rozovskii, 1995] *If (40) holds, then for discrete observations,*

$$\lim_{M\to\infty} P\{d(M)(\hat{\theta}_M - \theta_0) < x\} = \Phi_{\sigma_d}(x) \tag{49}$$

where

(50)

$$\sigma_d^2 := \lim_{M\to\infty} E\left[(\hat{\theta}_M - \theta_0)^2 d(M)^2\right] = \lim_{M\to\infty} \inf_{\tau_M \in U} E\left[(\tau_M - \theta_0)^2 d(M)^2\right]$$

and

$$\Phi_\sigma(x) = 1\sqrt{2\pi}\sigma \int_{-\infty}^{x} e^{-\frac{y^2}{2\sigma^2}} dy.$$

The normalization factors are given by

$$c(M) = \begin{cases} M^{\frac{2p_1 - p + d}{2d}} & d > p - 2p_1 \\ (\log M)^{1/2} & d = p - 2p_1 \end{cases} \tag{51}$$

$$d(M) = \begin{cases} M^{\frac{2p_1 - 2p + d}{2d}} & d > 2(p - p_1) \\ (\log M)^{1/2} & d = 2(p - p_1) \end{cases} \tag{52}$$

and U is the class of all admissible estimators of θ.

For the sake of brevity here as well as in Proposition 4 below we omit the detailed description of the class U and only remark that all bounded properly measurable functions of observations belong to this class. Note also that the limiting variances can be calculated explicitly (see e.g. Huebner, Rozovskii, 1993).

Relations (48), (50) show that the proposed estimators are asymptotically optimal in the mean square sense among all admissible estimators. One can see from (51), (52) that the convergence rate for the estimator of diffusivity is equal to $M^{-1/2-1/d}$ in the continuous case and $M^{-1/2}$ in the discrete case. The convergence rate for the estimate of λ in the continuous case for $d = 2$ is $(\log M)^{-1/2}$

3. Multidimensional parameter.

Again, we focus on equation (17) including $K > 1$ unknown parameters $\theta_1, \ldots, \theta_K$. As before we suppose that the operators $A_0, A_1, \ldots, A_K$ are self-adjoint, i.e. their eigenvalues $\lambda_{0m}, \lambda_{1m}, \ldots, \lambda_{Km}$, $m = 1, 2, \ldots$ are real.

For the multi-parameter case, we need an additional assumption of identifiability. In other words, we must assume the solvability of the system of equations for the maximum likelihood estimators of unknown parameters.

For continuous observations, this condition can be formulated in the following form [Hubner, 1993, Duncan and Pasik-Duncan, 1990]

$$\lim_{M\to\infty} \inf \det \left(\frac{\sum_{m=1}^{M} \lambda_{im}\lambda_{jm}\lambda_m^{-1}}{\sqrt{\sum_1^M \lambda_{im}^2 \lambda_m^{-1}} \sqrt{\sum_1^M \lambda_{jm}^2 \lambda_m^{-1}}} \right) \geq \delta \tag{53}$$

where $\lambda_m = \lambda_m(\theta) := \lambda_{0m} + \lambda_{1m}\theta_1 + \ldots + \lambda_{Km}\theta_K$, $\det(a_{ij})$ means the determinant of matrix (a_{ij}), $i, j = 1, \ldots, K$, and δ is a positive number.

In the case of discrete observations, asymptotic properties can be derived under the following assumption [Piterbarg, Rozovskii, 1995].

$$\left(\sum_{m=1}^{M} \frac{\lambda_{im}\lambda_{jm}}{\lambda_m^2} \right) \sim \gamma(\boldsymbol{\theta}) B_M \tag{54}$$

where B_M is a $K \times K$ non-degenerate matrix for sufficiently large M, and the function $\gamma(\boldsymbol{\theta})$ satisfies the conditions

$$0 < c_0 \leq |\gamma(\boldsymbol{\theta})| \leq c_1 \leq \infty \tag{55}$$

for all $\boldsymbol{\theta} \in \Theta$, where c_0, c_1 are constants.

Again as in the case of one parameter, the first case leads us to a linear system of equations for the estimators of the unknown parameters. In contrast, the corresponding system for discrete observations is much more complicated. For this reason we use assumption (54) which is more restrictive than (53). Here we exhibit results related to differential operators only. Let us denote the order of operator A_k by p_k, $k = 1, 2, \ldots, K$ and the order of $A_{\boldsymbol{\theta}} = A_0 + \theta_1 A_1 + \cdots + \theta_K A_K$ by p. Of course, we assume that $A_{\boldsymbol{\theta}}$ has the same order for all $\boldsymbol{\theta}$.

Let us denote the MLE of $\boldsymbol{\theta}$ by

$$\hat{\boldsymbol{\theta}}_M = (\hat{\theta}_{1M}, \hat{\theta}_{2M}, \ldots, \hat{\theta}_{KM})$$

and the actual value of $\boldsymbol{\theta}$ by $\boldsymbol{\theta}_0 = (\theta_{10}, \theta_{20}, \dots, \theta_{K0})$.

Proposition 3.

(i) [Huebner, 1993] *If the identifiability condition (53) holds and the observations are continuous in time then* $\hat{\theta}_{kM}$ *is consistent i.e.*

$$\lim_{M\to\infty} \hat{\theta}_{kM} = \theta_{k0} \tag{56}$$

if and only if

$$d \geq p - 2p_k \tag{57}$$

(ii) [Piterbarg, Rozovskii, 1995] *If the identifiability condtition (54) holds and the observations are discrete in time, then the MLE of* θ_k *is consistent if and only if*

$$d \geq 2(p - p_k) \tag{58}$$

Thus, the conditions of consistency do not depend on whether all the parameters are estimated simultaneously or only one is estimated while the others are fixed. Only the identifiability condition must hold.

Let us go back to the diffusion equation (44). In this case the validity of identifiability conditions (53), (54) can be easily checked and we come to the same conclusion as before.

The MLE for D is consistent for all dimensions even if all parameters are estimated simultaneously. At the same time, the MLE of λ *is consistent only under conditions (45), (46).*

The multidimensional versions of asymptotic normality were established in [Huebner, 1993], [Piterbarg, Rozovskii, 1995].

Let $I \subset \{1, \dots, K\}$ be the set of indices for which (57) or (58) holds in the case of continuous or discrete observations respectively, $\boldsymbol{\theta}_I$ is the vector of the parameters $\boldsymbol{\theta}_k$ such that $k \in I$, $\hat{\boldsymbol{\theta}}_{I,M}$ is the MLE of $\boldsymbol{\theta}_I$ and $\boldsymbol{\theta}_{I,0}$ is the true value of $\boldsymbol{\theta}_I$.

Let us set

$$c_k(M) = \begin{cases} M^{\frac{2p_k - p + d}{2d}} & d > p - 2p_k \\ (\log M)^{1/2} & d = p - 2p_k \end{cases} \tag{59}$$

$$d_k(M) = \begin{cases} M^{\frac{2p_k - 2p + d}{2d}} & d > 2(p - p_k) \\ (\log M)^{1/2} & d = 2(p - p_k) \end{cases} \tag{60}$$

and let $C(M)$, $D(M)$ be diagonal matricies with entries $c_k(M)$ and $d_k(M)$ respectively $k \in I$. The first part of the following statement addresses continous observations and the second one discrete observations.

Proposition 4.

(i) [Huebner, 1993] *Under the conditions (53) and (57), the distribution of the vector* $C(M)(\hat{\boldsymbol{\theta}}_{I,M} - \boldsymbol{\theta}_{I,0})$ *converges to the Gaussian distribution* $N(0, \Sigma_c)$ *where* Σ_c *is the limiting covariance matrix.*

(ii) [Piterbarg, Rozovskii, 1995] *Under conditions (54) and (58) the distribution of the vector* $D(M)(\hat{\boldsymbol{\theta}}_M - \boldsymbol{\theta}_{I,0})$ *tends to the Gaussian distribution* $N(0, \Sigma_d)$ *In both cases* U *is the set of all admissible estimators.*

We also believe that the MLE in the both cases are efficient under the same conditions, i.e.

$$\begin{aligned} \sigma^2_{k,c} &:= \lim_{M\to\infty} E\left[(\hat{\theta}_{k,M} - \theta_{k,0})^2 c_k(M)^2\right] \\ &= \lim_{M\to\infty} \inf_{\tau_M \in U} E\left[(\tau_M - \theta_{k,0})^2 c_k(M)^2\right] \end{aligned} \tag{61}$$

in the case of continuous observations and

$$\begin{aligned} \sigma^2_{k,d} &:= \lim_{M\to\infty} E\left[(\hat{\theta}_{k,M} - \theta_{k,0})^2 d_k(M)^2\right] \\ &= \lim_{M\to\infty} \inf_{\tau_M \in U} E\left[(\tau_M - \theta_{k,0})^2 d_k(M)^2\right] \end{aligned} \tag{62}$$

in the case of discrete observatins, although we do not have a complete proof yet.

Thus, in the multidimensional case the method of maximum likelihood provides asymptotically Gaussian and efficient estimators under the consistency conditions.

4. Non-self-adjoint operators. Velocity estimations.

Finally, we discuss the case of not necessarily self-adjoint operators. Extension to non-self-adjoint operators enables us to study properties of MLE's for the velocity $\boldsymbol{u}$ in (6) and both the horizontal eddy diffusivity and the bottom friction coefficient in (13). However in this case we cannot assume anymore that the eigenvalues are real.

For the sake of brevity we restrict ourselves to a single unknown parameter and discrete observations. Thus, the problem is to estimate θ in the equation

$$\frac{\partial u}{\partial t} + (A_0 + \theta A_1)u = S(t, \boldsymbol{x}). \tag{63}$$

As before we assume a common basis $\{\varphi_m(\boldsymbol{x})\}$ of eigenfunctions for A_0, A_1 and the covariance operator of the noise. Complex eigenvalues are now allowed

$$\lambda_{0m} = \alpha_{0m} + i\beta_{0m}, \; \lambda_{1m} = \alpha_{1m} + i\beta_{1m} \tag{64}$$

We will number the functions φ_m, $m = 1, 2, \ldots$ in such a way that the sequence of the real part of the eigenvalues is non-decreasing

$$\overline{\alpha}_1 \leq \overline{\alpha}_2 \leq \ldots \leq \overline{\alpha}_m \leq \ldots \tag{65}$$

where

$$\overline{\alpha}_m = \sup_\theta (\alpha_{0m} + \alpha_{1m}\theta) \tag{66}$$

Assume that the observations are of the form

$$\{u_m^{(1)}(t_j), u_m^{(2)}(t_j)\}, \quad m = 1, \ldots, M; \; j = 1, \ldots, N, \tag{67}$$

where

$$u_m^{(1)}(t) = \mathrm{Re}u_m(t), \quad u_m^{(2)}(t) = \mathrm{Im}u_m(t) \tag{68}$$

Let us give an example. Consider the 2-dimensional advection-diffusion operator given by (11) in the rectangle $G = \{x, y : \; -a \leq x \leq a; \; -b \leq y \leq b\}$. Obviously the functions

$$\varphi_{p,q}(x, y) = \gamma e^{i\pi(\frac{px}{a} + \frac{qy}{b})} \tag{69}$$

where $\gamma = (4ab)^{-1}$ and p, q run over all integers, form the orthonormal basis of eigenfunctions for operator (14) complemented by the appropriate periodic boundary conditions. It is readily checked that

$$A_\theta \varphi_{p,q}(x, y) = \left(\frac{i\pi\theta_1 p}{a} + \frac{i\pi\theta_2 q}{b} + \pi^2\theta_3 \left(\frac{p^2}{a^2} + \frac{q^2}{b^2} \right) + \theta_4 \right) \varphi_{p,q}(x, y). \tag{70}$$

The double indexed set (69) can be arranged as follows

$$\varphi_{00}(x, y), \; \varphi_{01}(x, y), \; \varphi_{10}(x, y), \; \varphi_{02}(x, y), \; \varphi_{11}(x, y), \; \varphi_{20}(x, y), \ldots \tag{71}$$

One can easily see that condition (65) is satisfied if θ_3, θ_4 run over bounded sets of positive values. For example, the set of three highest modes ($M = 3$)

is

$$\left\{1, e^{\frac{i\pi x}{a}}, e^{\frac{i\pi y}{b}}\right\}, \tag{72}$$

the set of 6 highest modes, ($M = 6$) is

$$\left\{1, e^{\frac{i\pi x}{a}}, e^{\frac{i\pi y}{b}}, e^{\frac{2i\pi x}{a}}, e^{\frac{2i\pi y}{b}}, e^{\pi i(\frac{x}{a}+\frac{y}{b})}\right\}. \tag{73}$$

So, for $M = 3$, the observed quantities are

$$u_0^{(1)}(t_j) = \gamma \int_G u(t_j, r) dr$$

$$u_1^{(1)}(t_j) = \gamma \int_G u(t_j, r) \cos(\frac{\pi x}{a}) dr$$

$$u_1^{(2)}(t_j) = \gamma \int_G u(t_j, r) \sin(\frac{\pi x}{a}) dr$$

$$u_2^{(1)}(t_j) = \gamma \int_G u(t_j, r) \cos(\frac{\pi y}{b}) dr$$

$$u_2^{(2)}(t_j) = \gamma \int_G u(t_j, r) \sin(\frac{\pi y}{b}) dr, \tag{74}$$

where $r = (x, y), \quad j = 1, \dots, N$.

Returning to the general equation (63) note that $u_m^{(1)}(t)$ and $u_m^{(2)}(t)$ are not independent random processes, while the processes $u_m^{(i)}(t)$ and $u_n^{(j)}(t)$ for $m \neq n$ are independent for all possible combinations of i's and j's. Instead of the single equation (32) for the amplitude of the m-th mode in the self-adjoint case, now we have two coupled equations

$$\dot{u}_m^{(1)} + \alpha_{1m} u_m^{(1)} - \beta_{1m} u_m^{(2)} = \sigma_m s_m^{(1)}(t)$$

$$\dot{u}_m^{(2)} + \beta_{1m} u_m^{(2)} + \alpha_{1m} u_m^{(2)} = \sigma_m s_m^{(2)}(t) \tag{75}$$

Let us assume that the asymptotics of the real part of eigenvalue $\lambda_m(\theta)$ for the operator A_θ is of the form

$$\alpha_{0m} + \alpha_{1m}\theta \sim m^r, \quad r \geq 0, \tag{76}$$

which is typical for elliptic operators.

Again to compute the MLE $\hat{\theta}_M$ one has to minimize $L_M(\theta, \theta_0)$ given by (36) numerically. However the asymptotic properties the estimator still can be analyzed analytically.

Proposition 5. [Piterbarg, Rozovskii, 1995] *Assume that condition (76) holds. Then the MLE $\hat{\theta}_M$ is consistent if in some neighborhood of θ_0*

$$\sum_{m=1}^{\infty} \left[\frac{\alpha_{1m}^2}{(\alpha_{0m} + \alpha_{1m}\theta)^2} + h^2 \beta_{1m}^2 e^{-2h(\alpha_{0m}+\alpha_{1m}\theta)} \right] = \infty, \tag{77}$$

where $h = \min_n \Delta t_n$.

Under condition (77) the ML estimator is asymptotically normal and efficient.

Note that in the case of self-adjoint operators condition (77) becomes (40).

Further, let us assume that the sequence $\{\beta_{1m}\}$ is separated from zero, and its elements are at most of a polynomial growth, i.e.

$$\lim_{m\to\infty} \inf |\beta_{1m}| > 0, \quad \lim_{m\to\infty} \sup \left(|\beta_{1m}| m^{-s}\right) < C \tag{78}$$

for some $s > 0, C > 0$.

Then, if $r > 0$ then the second term in the right-hand side of (77) does not affect divergence or convergence of the series. If in turn $r = 0$ then the first term does not play any role since the series $\sum_m \beta_{1m}^2$ diverges. We can summarize this as follows.

Proposition 6. *Assume (76) and (78).*

(i) If $r > 0$, then $\hat{\theta}_M$ is consistent given that in some neighborhood of θ_0

$$\sum_{m=1}^{\infty} \frac{\alpha_{1m}^2}{(\alpha_{0m} + \alpha_{1m}\theta)^2} = \infty \tag{79}$$

(ii) If $r = 0$ then $\hat{\theta}_M$ is consistent.

First, let us apply Proposition 6 to the transport equation

$$\frac{\partial u}{\partial t} + (\boldsymbol{v} \cdot \nabla)u + \lambda u - D\nabla^2 u = S(t, \boldsymbol{x}), \ \boldsymbol{x} \in R^d \tag{80}$$

with the velocity vector $\boldsymbol{v} = (v_1, \dots, v_d)$.

Note that if we focus on estimating either λ or D then the results of section 2 remain true, because in both cases $\beta_{1m} = 0$ and everything is predetermined by condition (79). We stress that condition (77) does not contain β_{0m} at all. Therefore in equation (80) the MLE for D is always consistent while the estimator for λ is consistent only for $d \geq 4$ (discrete observations).

Now consider the problem of estimating one of the velocity components, say v_1. In doing so we have $\alpha_{1m} = 0$,

$$r = \begin{cases} 2/d & D > 0 \\ 0 & D = 0 \end{cases} \tag{81}$$

In addition to (65) let us assume that φ_m are numbered in such a way that for any large m_0 there is non-zero β_{1m} for $m > m_0$. For instance, if $d = 2$ the numbering (71) satisfies this condition. In this case relation (78) is readily checked. So from Proposition 6 it follows that *if we estimate one component of the velocity while other components as well as λ and D are given, the MLE is consistent if and only if*

$$D = 0. \tag{82}$$

Finally, consider the d-dimensional version of the barotropic vorticity equation (13)

$$\frac{\partial u}{\partial t} + \frac{\partial}{\partial x}\Delta^{-1} u + D_h u - b\nabla^2 u = S(t, \boldsymbol{r}), \tag{83}$$

where the wind forcing is supposed to be a white noise in time, $\boldsymbol{r} = (x, y)$

Since the eigenvalues of the operator $A_0 = \frac{\partial}{\partial x}\Delta^{-1}$ are imaginary, all the results for estimation of D_h and b are the same as in the case of equation (44) with $\lambda = D_h$ and $D = b$. Therefore:

The bottom friction coefficient b can be estimated consistently from observations of the vorticity in all dimensions. On the other hand, the MLE of D_h is consistent only in 4-dimensional space.

Apparently, in the 4-dimensional space, equation (83) makes no physical sense. However, one can show that under continuous observations, the MLE of b is always consistent and the MLE of D_h is consistent if $d \geq 2$.

Acknowledgements. The authors are greatful to M. Huebner and the Reviewer for helpful remarks.

References

[1] Cramer, H., 1946, *Mathematical Methods of Statistics*, Princeton, Princeton University Press.

[2] Duncan, T.E., Pasik-Duncan, B., 1990, Adaptive control of continuous-time linear stochastic systems,*Math. Cont. Sygnals Sys.,*,**3**,45-56.

[3] Emery, W.J., Thomas, A.C., Collins, M.J., Crawford W.R., and Mackas, D.L., 1986, An objective method for computing advective surface velocities from sequential infrared satellite images, *J. Geophy. Res.*, **91**, 12865–12878.

[4] Fiadeiro, M.E., and Veronis, G., 1984, Obtaining velocities from tracer distributions, *J. Phys. Oceanogr.*, **14**, 1734–1746.

[5] Frankignoul, C., and Reynolds, R.W., 1983, Testing a dynamical model for mid-latitude sea-surface temperature anomalies, *J. Phys. Oceanogr.*, **13**, 1131–1145.

[6] Frankignoul, C., Scoffier, N., and Cane, M.A., 1993, An adaptive inverse method for model tuning and testing, *Statistical Methods in Physical Oceanography, Proceedings Hawaiian Winter Workshop, University of Hawaii at Manoa, 1993*, 331–350.

[7] Ghil, M., and Malanotte-Rizzoli, P., 1991, Data assimilation in meteorology and oceanography, *Advances in Geophysics*, **33**, Academic Press, 141–266.

[8] Herterich, K., and Hasselmann, K., 1987, Extraction of mixed layer advection velocities, diffusion coefficients, feedback factors and atmospheric forcing parameters from the statistical analysis of North Pacific SST anomaly fields, *J. Phys. Oceangr.*, **17**, 2145–2156.

[9] Huebner, M. and Rozovskii, B.L., 1994, On asymptotic properties of MLE for pararabolic stochastic PDE's, (to appear in *Probability Theory and Related Topics*).

[10] Huebner, M., 1993, Parameter estimation for stochastic differential equations, Thesis, University of Southern California.

[11] Ibragimov, I.A., and Khasminskii, R.Z., 1981, *Statistical Estimation (Asymptotic Theory)*, Springer-Verlag, New York, Heidelberg, Berlin.

[12] Kelly, L.A., 1989, An inverse model for near-surface velocity from infrared images, *J. Phys. Oceanogr.*, **19**, 1845–1864.

[13] LeCam, L., 1986, *Asymptotic Methods in Statistical Decision Theory*, Springer.

[14] Loges, W., 1984, Girsanov's theorem in Hilbert space and an application to the statistics of Hilbert space valued stochastic differential

equations, *Stoch. Proc. Appl.*, **17**, 243-263

[15] Ostrovskii, A.G., and Piterbarg, L.I., 1985, Diagnosis of the seasonal variability of water surface temperature anomalies in the North Pacific, *Meteorology and Hydrology*, **12**, 51–58 (in Russian).

[16] Ostrovskii, A.G., and Piterbarg, L.I., 1994, Inversion for heat anomaly transport from SST time series, 1995 (in press).

[17] Piterbarg, L.I., and Rozovskii, B.L., 1995, Estimating unknown parameters in SPDE's under discrete observations in time (in preparation)

[18] Smedstat, O.M., and Fox, D.N., 1994, Assimilation of altimeter data in a two-layer primitive equation model of the Gulf Stream, *J. Phys. Oceanogr.*, **24**, 305–325, *Statistics and Physical Oceanography*, 1993, National Academy Press, Washington, D.C.

[19] Tziperman, E., and Thacker, W.C., 1989, An optimal control/adjoint equations approach to studying the oceanic general circulation, *J. Phys. Oceanogr.*, **19**, 1471–1485.

[20] Wunsch, C., 1978, The general circulation of the North Atlantic west of 50°W determined from inverse methods, *Rev. Geophys.*, **16**, 583–620.

[21] Wunsch, C., 1985, Can a tracer field be inverted for velocity?, *J. Phys. Oceanogr.*, **15**, 1521–1531.

Center for Applied Mathematical Sciences
University of Southern California
1042 W. 36th Place, DRB 155
Los Angeles, CA 90089-1113

piter@cams.usc.edu
rozovski@cams-00.usc.edu

Chaotic transport by mesoscale motions

R. M. Samelson

Abstract

A brief review is given of a new perspective on fluid transport by mesoscale motions in strongly inhomogeneous flows. This perspective draws on results from dynamical systems theory and has been stimulated by recent observations of the Gulf Stream and by laboratory experiments on meandering jets. It generally presumes that the flow is dominated by relatively simple coherent structures, and involves the detailed analysis of the kinematics of fluid exchange between flow regimes.

1. Introduction

The oceanic mesoscale, broadly characterized by horizontal scales of tens to hundreds of kilometers and time scales of tens to hundreds of days, contains the energetic oceanic motions that are analogous to the synoptic pressure systems of the atmospheric troposphere, which are familiar from the weather maps of our everyday experience. These motions are, in a sense, the ocean's subsurface storms.

It is generally accepted that the rectified meridional heat and momentum fluxes from synoptic mid-latitude atmospheric disturbances provide a critical link in the general circulation of the atmosphere. The degree to which the general circulation of the oceans depends on analogous fluxes from oceanic mesoscale motions is not well understood. Rectified fluxes from mesoscale motions appear to be small in the ocean interior, relative to advective fluxes associated with the mean circulation (Hall and Bryden, 1982), but they may nonetheless play an important role (Rhines and Young, 1982).

Many inferences about the structure of the ocean general circulation have been founded on the assumption that the distribution of properties, such as temperature and salinity, reflects the patterns of the mean circulation. The resulting picture has been supplemented over the last thirty years by a growing number of direct velocity measurements from fixed or drifting instruments, but the present data set is not sufficient to constrain the large-scale circulation (Schmitz and McCartney, 1993). Typically, over much of the ocean, the time-dependent part of the measured velocities has a standard deviation that is comparable to or larger than the amplitude of the mean velocities over the measurement period. Calculations with numerical models indicate that such time-dependent mesoscale motions can drive a significant rectified tracer flux (Spall, 1993), and so affect tracer distribution and the inferred structure of the general circulation.

For these and numerous other reasons, the question of transport by mesoscale motions is an important one, and the motivation to study it is broad. In the present chapter, a brief review is presented of a new perspective on transport by mesoscale motions in strongly inhomogeneous flows. The development of this perspective has been stimulated by recent observations of float trajectories in the Gulf Stream (Bower and Rossby, 1989), by laboratory experiments on quasi-geostrophic jets (Sommeria et al., 1989), and by recent progress in dynamical systems theory (Guckenheimer and Holmes, 1983), and involves the detailed analysis of the kinematics of fluid exchange between flow regimes in idealized models of meandering jets and traveling waves. It presumes that the flow is dominated by relatively simple coherent structures, and relies on the application to fluid mixing of mathematical techniques developed to study the geometric structure and qualitative behavior of solutions of ordinary differential equations in Hamiltonian and general nonlinear dynamical systems (Aref, 1984; Ottino, 1989; Wiggins, 1992). The associated calculations can be quite technical, and are not reproduced here, in order to provide an accessible review that outlines the general issues. The reader is referred to the original publications for more detail.

2. Transport and mixing in meandering jets

In the mid-1980's, Thomas Rossby and co-workers developed the RAFOS float, a freely-drifting subsurface oceanographic instrument that follows fluid motion on a constant density surface and reports its drift track by regularly recording acoustic signal arrivals from an array of fixed sound sources. They deployed about 40 of these instruments in the Gulf Stream off Cape Hatteras, and obtained a data set that for the first time described subsurface Lagrangian motion in the Gulf Stream in considerable detail (Bower and Rossby, 1989). A striking feature of this data set was the large fraction of floats that were rapidly (within 1 month or 1000 km) expelled from the Stream. In contrast, the "Eulerian" view provided by infrared satellite imagery suggests that the Stream maintains coherence over much longer space and time scales, although its path fluctuates. Bower (1991) argued that this contradiction could be resolved by considering the effect of coherent, steadily-propagating meanders on the float trajectories, and proposed a simple two-dimensional kinematic model of a meandering jet that illustrated this effect.

A shortcoming of this model is that, in a strict sense, no exchange between flow regimes occurs. This may be demonstrated by appealing to the analogy with Hamiltonian dynamics that has motivated some of the work to be discussed below. Since the model is two-dimensional (with the two components of velocity representing quasi-horizontal flow on an isopycnal surface) and incompressible, the velocity field may be written in terms of a streamfunction $\psi(x, y, t)$, so that

$$dx/dt = u = -\psi_y\,, \;\; dy/dt = v = \psi_x\,,$$

where x and y are the quasi-horizontal coordinates and t is time. These are just Hamilton's equations of motion for a one degree of freedom dynamical system with Hamiltonian function ψ and canonical coordinates x and y. For a streamfunction representing a jet with a steadily-propagating meander, such as the form chosen by Bower (1991), transformation to a coordinate frame propagating with the phase speed of the meander removes the explicit time dependence from ψ. Hamiltonian theory then implies that the flow is integrable in the moving frame: all fluid particles either follow periodic trajectories or remain at stagnation points, with the exception of particles following trajectories of separatrix type ("homoclinic" or "heteroclinic orbits," in the jargon of dynamical systems theory); these special trajectories asymptote to stagnation points and are isolated limiting curves that separate the xy-plane into regions made up of trajectories of qualitatively similar character.

For the model proposed by Bower (1991), there are three types of flow regime in the moving frame: downstream flow in the meandering core of the jet, recirculating flow along the flanks of the jet (with counter-rotating cells located alternately below meander crests and above meander troughs, forming "cat's eyes" patterns on each flank), and weak retrograde motion in the far field. When averaged over one recirculation period, the flow in the recirculating regions has zero mean velocity (mean velocity equal to the meander phase speed in the fixed frame), while the flow in the jet core is always downstream, and the flow in the far field is always upstream (with respect to the meander). Bower (1991) interpreted the flow in the recirculation regime as an indication of cross-jet exchange, since flow in it alternately resembles flow in the jet core and in the far field. Strictly speaking, however, there is no exchange, as fluid parcels in the recirculation regime oscillate regularly but do not escape.

In the moving frame, this flow has stagnation points above meander crests and below meander troughs, at the furthest upstream and downstream extents of the recirculation cells, and these stagnation points are connected by separatrix trajectories that form the boundaries between the flow regimes. It is well known by mathematicians that such separatrix connections between interior stagnation points (which, in effect, prevent exchange between flow regimes) in solutions of ordinary differential equations will be broken by arbitrarily small perturbations of almost any type. Poincare (1899) already recognized the complexity that the resulting set of trajectories would present in the case where the small perturbation was oscillatory, but it was not until relatively recently (Smale, 1967; Moser, 1973) that this complexity was given a detailed mathematical description by the demonstration that sets of solutions near the broken separatrix could be put into one-to-one correspondence with infinite sequences of binary symbols. This leads immediately to the conclusion that these sets of solutions are "chaotic," where in this case the word "chaotic" is precisely defined and means in part that they include an uncountably infinite set of nonperiodic trajectories, as well as a countably infinite set of periodic trajectories with periods equal to all the subharmonics of the perturbation frequency (Guckenheimer and Holmes, 1983).

The breaking of the separatrix connections implies, almost by definition, the presence of exchange between flow regimes. (One still has to show that a meaningful definition of the flow regimes can be made for the perturbed flow, but this can generally be done; see Wiggins (1992).) Melnikov (1963) developed a rigorous perturbation theory to test for the breaking of the separatrix and the existence of trajectories that pass between regimes by calculating (to first order in the perturbation amplitude) the size of the "gap" in the broken separatrix. Subsequently, the line of reasoning summarized in the preceding paragraph led to the recognition that Melnikov's method provided a (and to date the only general) rigorous technique for proving the existence of chaotic solutions of systems of ordinary differential equations (Guckenheimer and Holmes, 1983; Wiggins, 1992), but it is interesting that Melnikov's original motivation was to estimate the exchange between regimes (nonlinear stability), the problem of primary interest in the present case, rather than to prove the existence of chaos *per se*. Twenty-five years later, this motivation was rediscovered with the application of Melnikov's method to fluid transport problems by Knobloch and Weiss (1987) and Rom-Kedar et al. (1990).

In light of these ideas, Samelson (1992) re-examined the meandering-jet model of Bower (1991), with the object of crudely estimating the efficiency with which various types of disturbances would induce transport between the three flow regimes that arise in the model as described above. The results of those calculations indicated that the strength of the exchange induced by periodic fluctuations in the amplitude of the meander depended strongly on the frequency of the fluctuations, while the exchange due to propagating plane waves was generally largest when the plane-wave phase speed roughly matched flow velocities along the separatrices. Duan and Wiggins (submitted) have considered the more general case of quasi-periodic disturbances. Weiss (1991) has analyzed an area-preserving two-dimensional map that illustrates the characteristic behavior of particle trajectories that undergo exchange between flow regimes in general traveling wave fields.

Samelson (1992) argued that the model result of greatest physical relevance to the Gulf Stream problem was the estimate of exchange, but the presence of exchange induced by oscillatory disturbances also implies the existence of chaotic solutions. On time scales that are long compared to the disturbance period (and thus generally long compared to the residence times of fluid particles in the Gulf Stream region), the stretching and folding of fluid patches that is associated with the chaotic behavior results in "mixing" (or "stirring," since reversible in the absence of diffusion) of the fluid to exponentially small horizontal scales, which is related to the existence of a positive Lyapunov exponent, the asymptotic local exponential rate of stretching of fluid elements. A number of related analyses have appeared that focus on the mixing and transport properties of similar zonally-periodic jets (Behringer et al., 1991; Pierrehumbert, 1991; del-Castillo-Negrete and Morrison, 1993). These were motivated either by laboratory experiments in a rotating annulus (Sommeria et al., 1989) or by interest in large-scale atmospheric flows, for both of which the long-time behavior and periodic re-entrant geometry may be of more direct physical relevance than for the Gulf Stream. Studies motivated by other fluid problems have demonstrated that chaotic mixing can result in dispersion (mean square displacement) that asymptotically increases quadratically with time, rather than linearly as for stochastic diffusion (Pasmanter, 1988; Weiss and Knobloch, 1989; Ridderinkhof and Zimmerman, 1992). This behavior has been related to non-ergodicity of the flow, that is, to the presence of a mixture of regular and chaotic regions in the flow (Mezic and Wiggins, 1994a).

The remarkable experiments of Sommeria et al. (1989) demonstrated by dye injection that rapid mixing can occur along each flank of a meandering quasi-geostrophic jet while little or no mixing occurs across the jet axis. In an effort to explain these results, Del-Castillo-Negrete and Morrison (1993) used a combination of the *ad hoc* "resonance overlap" criterion developed by Chirikov (1979), kinematic arguments concerning the phase speeds of the meanders (neutral waves), and linearized dynamics. They argued that the tendency of the meandering jet to homogenize the fluid on each side of the jet while preserving gradients across the jet axis was due to the kinematic properties of the dynamical modes. In other words, the dynamics control critical properties of the neutral waves, such as their phase speeds relative to the maximum jet velocity. Once these parameters are determined, the mixing is controlled by the kinematic properties of the neutral waves. Meyers (1994) has applied these arguments to the Gulf Stream.

Bower (1991) showed that the cross-stream extent of the recirculation cells in the meandering jet increased as the meander phase speed approached the maximum jet velocity. Del-Castillo-Negrete and Morrison (1993) illustrated that a global bifurcation ("separatrix reconnection") of the velocity field occurs in this limit, as the central "jet core" region described above disappears and the recirculation cells come into direct contact. Small disturbances can then in principle cause mixing across the jet axis, but the latter authors argue from linearized dynamics that for the rotating annulus experiments the phase speeds of the neutral waves are generally smaller than the maximum jet velocity, so that this situation does not occur, consistent with the observed absence of mixing across the jet core. In a flow with vertical shear such as the Gulf Stream, however, a level often does exist at which a linear phase speed matches the maximum jet velocity, so this picture may well be relevant, and suggests that the induced transport across the Gulf Stream will depend on depth (Bower, 1991; Samelson, 1992; Pratt et al., 1994). In the reconnected limit, it should be possible to use Melnikov's method and "lobe dynamics" (Wiggins, 1992) to estimate the amplitude of exchange across the jet axis, but this has not been done.

In general, this type of model gives a picture of transport and mixing that is highly inhomogeneous. Here "transport" is defined with respect to flow regimes associated with the structure of coherent features in the velocity field (such as meandering jets), while "mixing" means the stretching and folding of fluid patches to exponentially small horizontal scales. Both transport and mixing are controlled by the interaction of the coherent features with time-dependent disturbances of smaller amplitude, and are primarily associated with stagnation points and separatrices (possibly in a translating frame of reference) in the undisturbed flow. The existence and strength of the transport and mixing depend on the details of the disturbances, and in simple models can be predicted by perturbation calculations. For finite amplitude disturbances, numerical solution of the full equations is necessary, but the broken separatrices still act as geometric "templates" for the fluid exchange (Beigie et al., 1994).

3. Potential vorticity dynamics: critical layers, instabilities, and homogenization

The approach summarized above has provided an innovative perspective on, and some quantitative predictions of, the manner in which time-dependent velocity fields cause fluid transport and mixing in strongly inhomogeneous flows. However, a basic objection may be raised against the kinematic models: the velocity fields are not derived from rationally motivated dynamical models, and so typically violate physical principles (such as potential vorticity conservation) that are fundamental to the theory of ocean dynamics. A similar objection can be raised with respect to the use of linear dynamics in these calculations, since finite amplitude particle displacements are in general incompatible with linearized dynamical equations. It is important to inquire whether the picture changes when the velocity fields are constrained to satisfy the appropriate nonlinear fluid dynamical equations.

Mesoscale ocean motions are believed often to conserve approximately (on quasi-horizontal trajectories) a variety of potential vorticity. In quasi-geostrophic theory (or in general if some balance condition holds), the instantaneous motion field is determined by the distribution of potential vorticity, along with appropriate boundary conditions, through the inverse of an elliptic operator (Pedlosky, 1987). The inviscid dynamics then reduces to the advection of the scalar potential vorticity field by the velocity field that is induced by the potential vorticity distribution. This is directly analogous to the situation for two-dimensional incompressible flow, when the dynamics reduce to advection of the scalar vorticity by a velocity that is related to the vorticity by a Poisson equation for the streamfunction.

It is not hard to see that the Lagrangian conservation of potential vorticity can in principle constrain particle motions in an important way, since the potential vorticity is diagnostically related to the streamfunction. For example, for models with time-periodic streamfunctions such as those discussed above, the diagnostic relation would require that the potential vorticity also be time-periodic. Since the potential vorticity is conserved along trajectories, this in turn evidently implies that the trajectories are periodic, not chaotic. The contradiction can be avoided, and chaotic trajectories can occur, if the potential vorticity is uniform in chaotic regions (Del-Castillo-Negrete and Morrison, 1993).

A related argument, which does not rely on the diagnostic relation between streamfunction and potential vorticity, leads to a similar conclusion. In a region of non-vanishing gradient of any scalar field (such as potential vorticity) that is conserved along particle trajectories in two-dimensional, incompressible flow, the trajectories can be obtained directly from the streamfunction and the scalar field–if these are known as a function of space and time–without solving any differential equations (Brown and Samelson, 1994). In other words, the flow is integrable by quadratures in the sense of classical Hamiltonian theory. Note that flows can have complicated advective behavior and still satisfy the hypotheses of this argument: see Warn and Gauthier (1989) for an example of a flow with non-vanishing potential vorticity gradient that (to leading order) is explicitly integrable but develops potential vorticity structure on arbitrarily small scales on sufficiently large time intervals.

In regions where the potential vorticity gradient vanishes, the Lagrangian constraint is trivial, and one might expect that the kinematic models should provide qualitatively accurate predictions of the character of the particle motions. Calculations for an elliptical vortex flow using a nonlinear dynamical model with piecewise-constant potential vorticity (Polvani and Wisdom, 1990), in which the chaotic trajectories were confined to regions of uniform potential vorticity, appear to support this expectation.

A more demanding test of these ideas is to examine the transport in nonlinear dynamical models with smooth potential vorticity fields. In general, the arguments of Del-Castillo-Negrete and Morrison (1993) and Brown and Samelson (1994) demonstrate that chaotic trajectories cannot arise in analytic solutions of the inviscid dynamical equations with non-uniform potential vorticity (since analytic functions are constant everywhere if they are constant in any region). This does not rule out the possibility of chaos in continuously-differentiable solutions with non-uniform potential vorticity (since these may have constant potential vorticity in finite regions) or in flows with momentum (vorticity) diffusion. It seems necessary to resort to numerical solution of the dynamical equations in order to examine the transport for general potential vorticity distributions. This implies that the rigid constraints of the exact Lagrangian conservation principle must be relaxed to some degree, if only as a result of discretization and round-off error, and the possibility must be considered that fundamental characteristics of particle motion in numerical solutions may be altered by numerical error of this type, or by explicit diffusion introduced to control smoothness or numerical stability.

Some numerical integrations of the nonlinear barotropic quasi-geostrophic equations have been carried out by Pierrehumbert (1991) with the chaotic transport picture in mind. He considered several initial value problems in which the initial states consisted of a superposition of two neutral Rossby waves of differing horizontal scale. The initial flow field resembled the meandering jet studied by Bower (1991), but with a broader and weaker jet core. Structures suggesting the presence of broken separatrices appear in his numerical solutions. He did not analyze these features in detail, but their appearance is of substantial significance, since it is in principle possible that the (near) conservation of potential vorticity could force the particle motion to be integrable, and prevent the breaking of separatrices. In that case, the qualitative picture suggested by the kinematic models would be fundamentally wrong.

Pierrehumbert (1991) characterized the mixing that occurred in these solutions by extracting empirical correlation and spectral scaling laws, and suggested that the "fine-grained" potential vorticity behaved as a passive scalar, while the "coarse-grained" potential vorticity homogenized rapidly in certain regions, in general agreement with predictions from a kinematic model. The nonlinear evolution did not result in fundamental changes to the large-scale structure of the flow. Pierrehumbert (1991) argued that this behavior was consistent with a scale separation arising from the mathematical "smoothing" effect of the inverse Laplacian operator that determines the streamfunction from the vorticity field, and additionally that the flow may divide itself into "small-scale" and "large-scale" components because of the proximity (in function space) of the large-scale flow component to a stable steady solution of the equations.

The division of the potential vorticity into "coarse-grained" and "fine-grained" fields is a natural description of the structure that is created by the complicated advective evolution. In general, however, it is not clear how the scale that divides these two regimes should be chosen *a priori*, nor even that this can be done at all, in the case where the energy spectrum of the (initial) flow is continuous. Holloway (1986) has argued that it is admissible to treat small-scale vorticity approximately as a passive tracer only for interactions with a wide scale separation in which the small-scale vorticity is advected directly by the large-scale flow. Pierrehumbert (1991) noted connections with the Rossby wave critical layer theory of Killworth and McIntyre (1985) and the weakly nonlinear baroclinic equilibration problem solved by Warn and Gauthier (1989), in both of which potential vorticity acts explicitly as a passive scalar in part of the flow. The analytic solution for inviscid, weakly nonlinear, baroclinic equilibration near minimum critical shear obtained by Warn and Gauthier (1989) provides a beautiful and explicit example of how a "coarse-grained" potential vorticity field may homogenize while the "fine-grained" potential vorticity retains all the structure associated with reversibility in the absence of diffusion. In general, since the homogenization process is evidently highly dependent on scale, a fundamental question must be, how does the flow determine what the "coarse-grain" and "fine-grain" scales are?

The streamfunction geometry encountered in the study of Rossby wave critical layers (Stewartson, 1978; Warn and Warn, 1978; Killworth and McIntyre, 1985; Haynes, 1985) is reminiscent of the kinematic meandering jet model discussed above. In critical layer theory, incident waves are forced linearly far from the jet, and the problem is to determine the structure of the resulting flow field near the critical line (where incident wave phase speed equals undisturbed jet flow velocity), as well as in the far field (where the transmitted and reflected waves are linear). The interaction of the incident forced wave with the jet leads to the formation of a "cat's eyes" streamline pattern along the critical line, with width proportional to the square root of the amplitude of the far-field waves. These features are similar to the recirculation regimes in the moving frame for kinematic models of the type considered by Bower (1991), which are also centered near a critical line, where the phase speed of the meander equals the velocity of the jet.

An important difference between the kinematics of the two cases is that for critical layer theory to apply, the waves must be of small amplitude outside the asymptotically thin region surrounding the critical line, whereas the meanders represent finite amplitude cross-stream displacements of the entire jet structure. In the critical layer theory, vortex roll-up takes place in the recirculations on the time scale of the background shear, while the modified potential vorticity field inside the recirculations can be unstable to smaller-scale disturbances with much shorter timescales (Killworth and McIntyre, 1985). These instabilities create small-scale, or "fine-grained," potential vorticity structure rapidly, while simultaneously homogenizing the locally-averaged, or "coarse-grained," potential vorticity field (Haynes, 1985).

The kinematic models would appear to suggest that for broad-band (rather than monochromatic) excitation, a secondary, chaotic, mixing layer surrounding the critical layer should exist, whose size is related to the amplitude of the additional wave perturbations. Ngan and Shepherd (submitted) have recently analyzed an asymptotic dynamical model of this interaction. This mixing should generally lead to the exchange of fluid between the critical layer and the exterior regions, and so violate at least one of the hypotheses of the reflection theorem of Killworth and McIntyre (1985).

There is little evidence for this sort of process in the numerical solutions displayed by Haynes (1985)–though the looping structure of the potential vorticity contour in his Figure 3d has a tantalizing resemblance to the familiar heteroclinic tangles–but these integrations have been carried out for only a short time, and with monochromatic forcing only. Can this process influence the evolution of the critical layer itself?

Arguments such as those of Del-Castillo-Negrete and Morrison (1993) and Brown and Samelson (1994) suggest that regions where the potential vorticity gradient vanishes have special kinematic properties. Such regions are also generally candidates for barotropic or baroclinic instabilities, since in a parallel flow geometry these instabilities generally require a reversal of the cross-stream potential vorticity gradient (Pedlosky, 1987). Linear theory suggests that it is the critical line and not the zero in the potential vorticity gradient that determines the location of the largest cross-stream displacements due to unstable waves (Lozier and Bercovici, 1992), but as noted above, linear theory does not properly account for finite particle displacements, and in general the connection between instabilities and the kinematic criteria is not clear. Are the mechanisms of chaotic transport relevant when the large-scale flow is unstable?

4. Summary

The geometric perspective of dynamical systems theory has provided new insights into the kinematics of fluid mixing and transport in strongly inhomogeneous time-dependent flows. This perspective is intrinsically appropriate for flows that are dominated by coherent features. The relevant transports are specifically defined with respect to flow regimes, not spatial coordinates. Similarly, the distribution of mixing regions is generally controlled in large part by the structure of the coherent features. The geometric perspective provides a conceptual contrast to the traditional statistical description of property transport in statistically homogeneous, or weakly inhomogeneous, flows (Davis, 1983; Haidvogel and Keffer, 1984).

Since the kinematics of the transport and mixing depend in detail on the instantaneous structure of the flow field, it is important that models of this process properly represent the dynamics. In dynamical models, the onset of chaotic particle motion is evidently associated with "coarse-grained" potential vorticity homogenization, while the "fine-grained" potential vorticity appears to be advected approximately as a passive scalar, but it is not clear what determines the corresponding scales. There appear to be connections between these processes, Rossby wave critical layer theory, and instabilities that have been only briefly explored.

An important practical question concerns the degree to which the geometric perspective can yield insight into transport and mixing associated with the complex velocity fields that are found in numerical ocean models and in the ocean itself, or whether the approach is only appropriate for the simple laminar flow fields of the type analyzed in the work discussed above. Only a few limited attempts to extend these ideas to more complex velocity fields have been made. Dutkiewicz et al. (1994) have calculated numerically the particle motion in a kinematic model with a laminar meandering jet similar to that of Bower (1991) and Samelson (1992) but in which the disturbances are generated by a stochastic-flight model rather than by simple harmonic fluctuations. It would be interesting to consider in detail how the induced exchange in this model varies through the progression from periodic to quasi-periodic and then to stochastic disturbances. Osborne and Caponio (1990) have analyzed a stochastic kinematic model in which the streamfunction has a power law wavenumber spectrum and random phase, and argue that the fractal character of individual trajectories depends on the degree of the power law. Some direct attempts have been made to evaluate fractal dimensions, Kolmogorov entropies, and Lyapunov exponents from ocean float trajectories (Osborne et al., 1986; Brown and Smith, 1990), but the data are somewhat limited for this application. Pierrehumbert and Yang (1994) have used approximate Lyapunov exponents to characterize the mixing due to advection by a velocity field obtained from a numerical atmospheric general circulation model. Some progress has been made on extending aspects of the mathematical theory to three dimensional velocity fields (Mezic and Wiggins, 1994b).

The mixing of fluid properties by mesoscale motions has traditionally been parameterized by gradient diffusion in numerical ocean models that do not explicitly resolve the mesocale variability. The geometric perspective gives a different view of transport and mixing due to mesoscale motions. Recently, an experiment has been carried out to determine mixing rates in the North Atlantic by repeatedly measuring an injected chemical tracer with an extremely low concentration detection limit (Ledwell and Watson, 1993). The observed horizontal tracer distribution did not fit simple advection-diffusion models, as localized regions of relatively large horizontal gradient of tracer concentration separated regions of roughly uniform tracer concentration. This pattern is suggestive of the inhomogeneous mixing that results from the processes discussed above, and may be a hint that some of these ideas are relevant to the ocean.

Acknowledgements. Preparation of this manuscript was supported by the Ocean Sciences Division of the National Science Foundation, Grant OCE-9114977, and by the Office of Naval Research, Grant N00014-92-J-1589, Code 322MM.

References

[1] Aref, H., 1984. Stirring by chaotic advection. *Journal of Fluid Mechanics*, **143**, 1-21.

[2] Beigie, D., A. Leonard, and S. Wiggins, 1994: Invariant manifold templates for chaotic transport. *Chaos, solitons, and fractals*, **4**(6), 749-868.

[3] Behringer, R., S. Meyers, and H. Swinney, 1991: Chaos and mixing in a geostrophic flow. *Physics of Fluids A*, **3**(5), 1243-1249.

[4] Bower, A., 1991: A simple kinematic mechanism for mixing fluid parcels across a meandering jet. *Journal of Physical Oceanography*, **21**(1), 173-180.

[5] Bower, A., and T. Rossby, 1989: Evidence of cross-frontal exchange processes in the Gulf Stream based on isopycnal RAFOS float data. *Journal of Physical Oceanography*, **19**(9), 1177-1190.

[6] Brown, M. G., and R. M. Samelson, 1994. Particle motion in vorticity-conserving, two-dimensional incompressible flows. *Physics of Fluids A*, **6**(9), 2875-2876.

[7] Brown, M. G., and K. B. Smith, 1990. Are SOFAR float trajectories chaotic? *Journal of Physical Oceanography*, **20**(1), 139-149.

[8] Chirikov, B., 1979. A universal instability of many-dimensional oscillator systems. *Physics Reports*, **52**, 263-379.

[9] Davis, R. E., 1983. Oceanic property transport, Lagrangian particle statistics, and their prediction. *Journal of Marine Research*, **41**, 163-194.

[10] Del-Castillo-Negrete, D., and P. Morrison, 1993. Chaotic transport by Rossby waves in shear flow. *Physics of Fluids A*, **5**(4), 948-965.

[11] Duan, J., and S. Wiggins, Fluid exchange across a meandering jet with quasi-periodic variability. *Journal of Physical Oceanography*, submitted.

[12] Dutkiewicz, S., A. Griffa, and D. Olson, 1994. Particle diffusion in a meandering jet. *Journal of Geophysical Research*, **98**(C9), 16,487-16,500; correction, *Journal of Geophysical Research*, **98** C(10), 18,313.

[13] Guckenheimer, J., and P. Holmes, 1983. *Nonlinear oscillations, dynamical systems, and bifurcations of vector fields.* Springer, New York.

[14] Haidvogel, D., and T. Keffer, 1984. Tracer dispersal by mid-ocean eddies, part 1: ensemble statistics. *Dynamics of Atmospheres and*

Oceans, **8**, 1-40.

[15] Hall, M., and H. Bryden, 1982. Direct estimates and mechanisms of ocean heat transport. *Deep-Sea Research*, **29**(3A), 339-359.

[16] Haynes, P., 1985. Nonlinear instability of a Rossby wave critical layer. *Journal of Fluid Mechanics*, **161**, 493-511.

[17] Holloway, G., 1986. Eddies, waves, circulation, and mixing: statistical geofluid mechanics. *Annual Reviews of Fluid Mechanics*, **18**, 91-147.

[18] Killworth, P., and M. McIntyre, 1985. Do Rossby wave critical layers absorb, reflect, or over-reflect? *Journal of Fluid Mechanics*, **161**, 449-492.

[19] Knobloch, E., and J. Weiss, 1987. Chaotic advection by modulated traveling waves. *Physical Review A*, **36**(3), 1522-1524.

[20] Ledwell, J., A. Watson, and C. Law, 1993. Evidence for slow mixing across the pycnocline from an open-ocean tracer-release experiment. *Nature*, **364**, 701-703.

[21] Lozier, M. S., and D. Bercovici, 1992. Particle exchange in an unstable jet. *Journal of Physical Oceanography*, **22**(12), 1506-1516.

[22] Melnikov, V., 1963. On the stability of the center for time-periodic perturbations. *Transactions of the Moscow Mathematical Society*, **12**, 1-57.

[23] Meyers, S., 1994. Cross-frontal mixing in a meandering jet. *Journal of Physical Oceanography*, **24**(7), 1641-1646.

[24] Mezic, I., and S. Wiggins, 1994a. On the dynamical origin of asymptotic t^2 dispersion of a nondiffusive tracer in incompressible laminar flows. *Physics of Fluids*, **6**(6), 2227-2229.

[25] Mezic, I., and S. Wiggins, 1994b. On the integrability and perturbation of three-dimensional fluid flows with symmetry. *Journal of Nonlinear Science*, **4**, 105.

[26] Ngan, K., and T. G. Shepherd. Chaotic mixing and transport in Rossby - wave critical layers. *Journal of Fluid Mechanics*, submitted.

[27] Moser, J., 1973. *Stable and random motions in dynamical systems.* Princeton University Press, Princeton, 198 pp.

[28] Osborne, A. R., and R. Caponio, 1990. Fractal trajectories and anomalous diffusion for chaotic particle motions in 2-D turbulence. *Physical Review Letters*, **64**, 1733-1739.

[29] Osborne, A. R., A. D. Kirwan, Jr., A. Provenzale, and L. Bergamasco, 1986. A search for chaotic behavior in large and mesoscale motion in the Pacific Ocean. *Physica D*, **23**, 75-83.

[30] Ottino, J. M., 1989. *The kinematics of mixing: stretching, chaos, and transport.* Cambridge University Press, Cambridge.

[31] Pasmanter, R., 1988. Deterministic diffusion, effective shear, and patchiness in shallow tidal flows. In *Physical processes in estuaries*, J. Dronkers and W. Van Leussen, eds., Springer, New York, 42-52.

[32] Pedlosky, J., 1987. *Geophysical fluid dynamics.* Springer, New York.

[33] Pierrehumbert, R., 1991. Chaotic mixing of tracer and vorticity by modulated traveling Rossby waves. *Geophysical and Astrophysical Fluid Dynamics*, **58**, 285-320.

[34] Poincare, H., 1899. Les methodes nouvelles sur la mecanique celeste. Gauthier-Villars, Paris.

[35] Polvani, L., and J. Wisdom, 1990. Chaotic Lagrangian trajectories around an elliptical vortex patch embedded in a constant and uniform background shear. *Physics of Fluids A*, **2**(2), 123-126.

[36] Pratt, L., M. S. Lozier, and N. Beliakova, Parcel trajectories in quasi-geostrophic jets, part 1: neutral modes. *Journal of Physical Oceanography*, submitted.

[37] Rhines, P., and W. Young, 1982. Homogenization of potential vorticity in planetary gyres. *Journal of Fluid Mechanics*, **122**, 347-367.

[38] Ridderinkhof, H., and J. T. F. Zimmerman, 1992. Chaotic stirring in a tidal system. *Science*, **258**, 1107.

[39] Rom-Kedar, V., A. Leonard, and S. Wiggins, 1990. An analytical study of transport, mixing, and chaos in an unsteady vortical flow. *Journal of Fluid Mechanics*, **214**, 347-394.

[40] Samelson, R. M., 1992. Fluid exchange across a meandering jet. *Journal of Physical Oceanography*, **22**(4), 431-440.

[41] Smale, S., 1967. Differentiable dynamical systems. *Bulletin of the American Mathematical Society*, **73**, 747-817.

[42] Sommeria, J., S. Meyers, and H. Swinney, 1989. Laboratory model of a planetary eastward jet. *Nature*, **337**, 58.

[43] Spall, M., 1993. Mechanism for low frequency variability and salt flux in the Mediterranean salt tongue. *Journal of Geophysical Research*, **99**(C5), 10,121-10,129.

[44] Stewartson, K., 1978. The evolution of the critical layer of a Rossby wave. *Geophysical and Astrophysical Fluid Dynamics*, **9**, 185-200.

[45] Warn, T., and P. Gauthier, 1989. Potential vorticity mixing by marginally unstable baroclinic disturbances. *Tellus*, **41A**, 115-131.

[46] Warn, T., and H. Warn, 1978. The evolution of a nonlinear critical level. *Studies in applied mathematics*, **59**, 37-71.

[47] Weiss, J., 1991. Transport and mixing in traveling waves. *Physics of Fluids A*, **3**(5), 1379-1384.

[48] Weiss, J., and E. Knobloch, 1989. Mass transport and mixing by modulated traveling waves. *Physics Reviews A*, **40**, 2579.

[49] Wiggins, S., 1992. *Chaotic transport in dynamical systems.* Springer, New York, 301 pp.

Woods Hole Oceanographic Institution
Woods Hole, MA 02543

rsamelson@whoi.edu

Chaotic Transport and Mixing by Ocean Gyre Circulation

Huijun Yang

Abstract

This paper reviews some recent development of the Lagrangian modeling of chaotic transport and mixing by the gyre-scale (basin-scale) ocean circulation, by use of some contemporary ideas and tools. Lagrangian trajectories in a simple oceanic flow can exhibit complex behaviors, called the chaotic advection. This chaotic advection can induce the chaotic transport and mixing in the fluid particles and finer structures in the tracer fields. The chaotic transport and mixing by gyre-scale circulation motion was proposed as one of fundamental mechanisms for the chaotic or fractal behaviors of the quasi-Lagrangian drifter trajectories observed in oceans, and water mass and water property exchanges on the gyre-scale or basin scale. It is this chaotic (or stochastic) nature of the Lagrangian trajectories that provides a possible linkage between the large-scale motion and the finer structure in the tracer fields. The results showed the fundamental importance of transients of the large scale oceanic motion in the Lagrangian transport and mixing processes. Some basic ideas and methodology were introduced in relation to the Lagrangian modeling of the water mass and water property exchanges. The discussion was focused on the chaotic transport and mixing by the gyre-scale (basin-scale) motion with some references to that by the meso-scale motions. The chief purpose of the paper is to introduce main ideas and methodology behind this Lagrangian modeling, rather than to give a complete literature review on the chaotic transport and mixing in physical oceanography or on the gyre-scale transport and mixing. A simple gyre scale ocean circulation model, mimicking the upper North Atlantic, was used to highlight the ideas and the results were found to be favorably comparable to observations.

1. Introduction

Understanding the transfer of fresh water, water masses, heat, and materials (or tracers) of biogeochemical importance by ocean circulation is one of oldest, most important problems still facing oceanographers today. The transport across the different gyres as well as from one hemisphere to the other hemisphere in the upper limb of world oceans, such as the Atlantic, is crucial in understanding ocean circulation and world ocean's role in regional and global climate variability on interannual, decadal or longer time scales. There is indication the North Atlantic plays a major role in establishing regional and global climate (Molinari et al., 1994). The ocean's main gyres in the Atlantic equatorward from the polar region consist of the subpolar gyre, subtropical gyre, tropical gyre and equatorial gyre. Observations show that the northward water mass transport of the Florida Current through the Straits of Florida off Miami is as high as 30 Sv (1 Sv $= 10^6 m^3 s^{-1}$) in the upper North Atlantic. The net northward transformation of warm water in the subtropical gyre into the cold water in the subpolar gyre in the North Atlantic is 13 Sv. It is belived, however, that only half of this net transport, i.e., 6.5 Sv, is accomplished by the gyre circulation mode. See Schmitz and McCartney (1993) for an excellent review on the water mass transport in the North Atlantic based on available observation data.

The mechanisms by which the water mass exchanges from one gyre to another are poorly understood. There have been some suggestions for the exchange such as the baroclinicity (Pedlosky, 1984), the titling of the zero wind stress curl line (Huang, 1990; Rhines and Schopf, 1991) and the strong meandering of the Gulf Stream off the western boundary of the North Atlantic (Bower, 1991). However, until recently little attention has been paid to the transients or temporal variation in the gyre-scale ocean circulation in the theoretic study. The reason is simple. Traditionally, the mean transport is based on the climatological mean Eulerian flow. The mean Eulerian meridional flow is exactly zero across the intergyre boundary line. Therefore it is unable to explain the mean intergyre water column or water mass exchange, which is observed and required. However, the picture will change dramatically when one examines the Lagrangian transport based on the Lagrangian modeling. It was discovered recently that there could exist the mean Lagrangian transport by temporally varying gyre scale or basin scale ocean circulation, called the *chaotic transport* (to be explained shortly).

Closely related to the water mass exchange are the water property exchange and modification. The water properties could include physical properties such as temperature and salinity, or biogeochemical properties such as sediment, nutrient and carbon. The lack of a clear understanding of water mass exchange prevents us from meaningfully interpreting the water properties and their variabilities, since without direct water mass exchange, the water property exchange is limited to diffusion process.

However the chaotic transport and mixing could provide a new mechanism for the large-scale transport and mixing processes in oceans. A simple ocean flow in Eulerian description, such as a two dimensional barotropic unsteady flow or steady (or unsteady) three dimensional baroclinic flow, can produce very complex Lagrangian trajectories. For instance, two nearby fluid particles will separate from each other exponentially, and eventually lose the information about their initial positions. This phenomenon is called the *chaotic advection* in fluid mechanics (Aref, 1984). Chaotic advection could produce a complicated transport in the flow and induce the so-called the chaotic transport. The chaotic transport, in turn, could mix fluid particles from different regions (or gyres) or particles with different properties and these cause the *chaotic mixing* (Chien et al., 1986; Khakhar et al., 1986). Conventional oceanographers may well call it chaotic stirring according to Eckart (1948). The (quasi) Lagrangian drifters in the oceanographic study now are increasingly deployed in the ocean field research project. These drifter trajectories in oceans were found to display many of the characteristics of fractal curves (Osborne et al., 1986; 1989; Sanderson and Booth, 1991) with implications in the ocean mixing. The chaotic mixing by ocean mesoscale flows such as a meandering jet like the Gulf Stream or tides has also been recognized (e.g., Brown and Smith, 1990, 1991; Ridderinkhof and Zimmerman, 1992; Samelson, 1992; Lozier and Bercovici, 1992; Cushman-Roising, 1993; Redderinkhof and Loder, 1994, Meyers, 1994; Pratt et al., 1995). (See, Samelson in this volume for a review on the transport by mesoscale motions.) Chaotic mixing renews interest in laboratory experiments of geophysical flows (Sommeria et al., 1989, 1991; Behringer et al., 1991), which reveal many fascinating features of chaotic mixing in jets. This concept has also been successfully applied to the atmosphere for both passive and active tracers (e.g., Pierrehumbert, 1991a,b; Yang, 1993a,b, 1995a; Pierrehumbert and Yang, 1993; Yang and Pierrehumbert 1994) in explaining potential vorticity mixing, dry air production, the cloud pattern formation, the moisture distributions and the polar vortex mixing. The chaotic transport and mixing has many potential applications in other sciences, technology and engineering. They range from mixing in the interior of stars in astrophysics, combustion in mechanical engineering, chemical reactors in chemical engineering, mixing in blood vessels in physiology, aeration in bioreactors in bioengineering, blending of additives in food engineering, polymer blending and compounding in polymer engineering, and mixing in the mantle of the earth. In fact, the chaotic transport and mixing has become a new subject in mathematics, sciences and engineering (Ottino, 1989; Wiggins, 1992). The reader is encouraged to consult Ottino's book for general introduction on the chaotic advection by various flows.

Recently the chaotic transport and mixing due to transients or temporal variability in the gyre circulation was proposed as one of fundamental mechanisms for general gyre (or basin) scale ocean transport and mixing (Yang and Liu, 1994; Liu and Yang, 1994; Yang, 1995b). Drifter trajectories in the Gulf Stream and Kuroshio extensions have been shown to have fractal properties (Osborne et al., 1986, 1989; Brown and Smith, 1990, 1991; Sanderson and Booth, 1991). The physical mechanism responsble for the fractal structure of the trajectories reamins unclear, but may be related to geostrophic turbulence (Osborne et al., 1989 and Provenzale et al. (1991). However, it also could be in part due to the chaotic transport and mixing by the gyre ocean circulation. It was demonstrated that the transients or temporal variations in the gyre circulation mode were of critical importance for the intergyre mass transport, and long term water property exchanges. The results are promising and provide some new tools for research in general oceanography when the knowledge of a long term transport is required.

The result provides a simple, yet fundamentally important linkage between the deterministic flow field on large spatial scale and the resulting chaotic (or stochastic) behaviors in the Lagrangian drifter trajectories and the tracer fields in world oceans. In general, the first step of this approach is to obtain a flow field from dynamic equations, such as potential vorticity equation. The flow field then is used to calculate the Lagrangian trajectories and to characterize the Lagrangian pathways. Some qualitative description and method from the Hamiltonian chaos theory are available for such characterization. The water mass transport can be directly discussed in relation to these Lagrangian trajectories and flow pathways. In addition, the water property change that occurs along the Lagrangian pathways can be calculated according to the equation governing the water property change rate. The Lagrangian water property evolution can be obtained. The statistic methods are then applied to further analyze the results.

The objectives of this paper were 1) to review some recent development on the application of the chaotic transport and mixing by the gyre-scale ocean circulation, 2) to introduce various ideas and methodology associated with the Lagrangian modeling approach. A simple double gyre-jet ocean model was used, which consists a subtropical gyre in the south and a subpolar gyre in the north and a meandering free jet in between, like the Gulf Stream east of Cape Hatteras in the North Atlantic. Several statistic methods were introduced, including correlation function, fractal dimension, dispersion diagram, probability distribution function (PDF), and the finite-time Lyapunov exponent. A contemporary perspective from the dynamical systems theory (i.e., Hamiltonian chaos theory) was presented for a deeper understanding of chaotic nature, along with other analytic methods in addressing the Lagrangian transport problem. The analyses and some results presented here are equally applicable to other intergyre boundary such as that between the subtropical gyre and tropical gyre, between the tropical and equatorial gyres and the inter hemisphere boundary. The approach can be used in the transport and mixing problems in oceans in general.

The paper was arranged as follows. A simple double gyre-jet ocean model was presented in the next section. The water mass exchange was addressed in section 3, followed by the related water property exchange in section 4. Section 5 was devoted to the discussion of chaotic nature of the transport from a view of Hamiltonian chaos and a computable method to detect the chaos and barrier to transport, called the finite-time Lyapunov exponent, with the results. Some concluding remarks were given in the final section.

2. A Simple Ocean Circulation Model

As first step, the dynamics of ocean gyre circulation was considered to obtain the flow field. The model used in this study was a simple double gyre-jet ocean circulation model, driven by surface wind and a meandering jet. This model was developed based on a simple double gyre ocean circulation model in Yang and Liu (1994) and Liu and Yang (1994), by including a meandering jet between two gyres. The model was fully developed in Yang (1995b) and was to represent the upper North Atlantic. The reader is referred to the original paper for detailed discussions and justifications of parameter values. The model ocean is homogeneous with a flat bottom. A double gyre circulation is forced by a wind field with a subpolar gyre in the north and a subtropical gyre in the south and a free meandering jet between the two gyres.

Under some approximations, the flow field in this circulation model could be derived from the dynamics consideration as follows

(2.1)
$$\psi = (1 - e^{-x/\delta})\ \mathrm{curl}_z\tau(x-2) + \Psi_0\big[1 - \tanh(\frac{Y}{\lambda})\big] + \epsilon_1 f(Y)\cos(k_1(x - c_1 t)) + \epsilon_2 f(Y)\cos(k_2(x - c_2 t))$$

where

(2.2)
$$f(y) = \Psi_0 \mathrm{sech}^2(\frac{y}{\lambda}).$$

ψ is the stream function of the flow field, and x, y and t denote the zonal direction, meridional direction and time, respectively. τ is the zonal wind stress. The variables have been nondimensionalized, the length scale by L, time by advection time t_{adv} of the interior Sverdrup circulation and the wind stress by τ_0. δ is the thickness of the Stommel boundary layer. The domain of the ocean is a square of length $4000km$ in horizontal with depth of 1 km. The thickness of the Stommel western boundary layer is about 100 km ($\delta = 0.05$), and the advection time scale is about 12.5 $years$. Ψ_0 is proportional to the strength of the meandering jet; λ represents the width of the jet; c_1 and c_2 are phase velocities and k_1 and k_2 are wave numbers of the meandering jet. ϵ_1 and ϵ_2 are amplitudes. The mean position of the intergyre boundary line and the jet is at $y = 0$. The values of the parameters are adapted after the Gulf Stream (See Yang, 1995b).

The zonal wind was allowed to slowly migrate north and south according to $\tau = \frac{1}{\pi}\cos(\pi(Y))$ and $Y = y - y_0(t) = y - a\sin(\omega t)$ where a is the amplitude and ω is the frequency of the migration. Observations have shown that over the extratropical oceans the wind field migrates about $1000km$ meridionally each year. Hence we chose $\omega = 50$, which corresponds to an annual frequency of the wind migration, and $a = 0.2$, which is about $800km$ from the northernmost to the southernmost in the current parameter setting.

The equations describing Lagrangian trajectory of a fluid particle in this slowly migrating double gyre-jet circulation system are

$$\frac{dx}{dt} = u = -\frac{\partial\psi}{\partial y}, \quad \frac{dy}{dt} = v = \frac{\partial\psi}{\partial x}. \tag{2.3}$$

The northern and southern boundary conditions are chosen to be at rest at $y = 1 \pm a$. The western and eastern boundary conditions were imposed as no normal flux. The equations were solved numerically by using fourth-order Runge-Kutta method. This double gyre-jet ocean circulation model reduces to the one used in Yang and Liu (1994) and Liu and Yang (1994) when the meandering jet is absent. When there is no wind forcing this ocean model reduces to the meandering jet model used by Bower (1991), del-Castillo-Negrete and Morrison (1993), Samelson (1992) and Meyers (1994) in addressing cross-frontal mass exchange (also see Lozier and Bervovici, 1992; Cushman-Roisin, 1993; Pratt et al., 1995). It should be pointed out that this model is an idealized one. Because of lack of dynamic interaction between the gyre circulation mode and the jet, it could be reviewed as a kinematic one even though the model is dynamically consistent under the reasonable approximation.

3. Water Mass Transport

3.1 Water Column Dispersion and Correlation Fractal Dimension. Using the fluid particle (or float) to represent the barotropic water column or water mass, we could examine the water mass transport in the ocean. Based on the observation the northward transport of the water of subtropical origin was considered. To examine the water mass dispersion process, we initially placed the 10,000 particles in the western boundary layer in the southern subtropical gyre centered at $(0.025, -0.5)$ with an area of 0.05×0.05. Instead plotting the trajectory of the particle, we plotted the distribution of the fluid particles at time t (Fig.1). It was shown that some subtropical water had been transported into the subpolar gyre at $t = 2$ and more at $t = 4$. The stretching and folding process was pronounced in the figure.

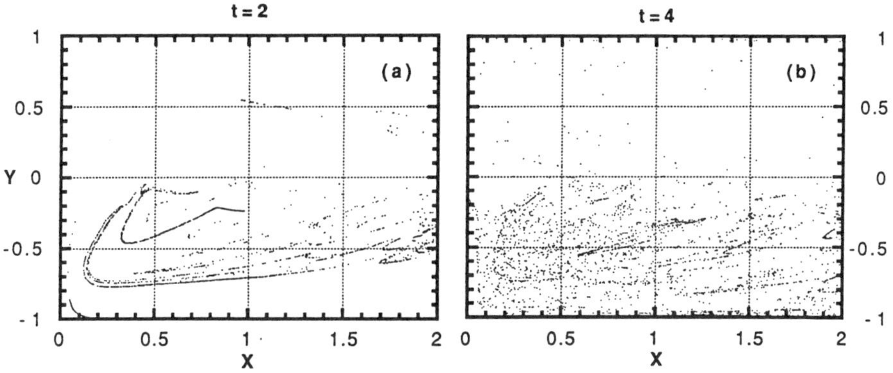

Figure 1. The water column dispersion. At (a) $t = 2$ and (b) $t = 4$. One unit time is about 12.5 *years*. The mean intergyre boundary is at $y = 0$. Initially all water particles were located near $(0.025, -0.5)$ with an area of 0.05×0.05 in the western boundary of the subtropical gyre.

In physical oceanography, the fractal dimension of fluid parcel trajectory (or float trajectory) has been used to quantify the chaotic

However, it is not good tool to quantify the dispersion process (see Fig.1). To better characterize the mixing and transport in the above experiments a two particle correlation function $H(r)$ is introduced. Given a cloud of N particles, one computes $H(r)$ in the following manner. First compute the $N(N-1)/2$ distances between all pairs of particles, then $H(r)$

is defined as the number of pairs with distance less than r, i.e.,

$$H(r) = \sum_{distance<r} \text{pair of particles} \tag{3.1}$$

$H(r)$ asymptotes to a constant at large r, and the distance at which $H(r)$ begins to flatten is indicative of the overall extent of the cloud. If $H(r)$ exhibits a self-similarity subrange in which

$$H(r) \sim r^{\alpha}, \tag{3.2}$$

then α is the *correlation dimension* characterizing the geometry at the corresponding length scales (Grassberger and Procaccia 1983). $\alpha = 0,\ 1$, and 2 correspond the particles clustered at one point, well-separated filaments, and an area-filling cloud, respectively. In a log-log plot, the slope corresponds to the correlation dimension. Obviously the correlation fractal dimension is not exactly the fractal of any fluid particle (float) trajectory. Nevertheless, they are related to each other (see references given in Introduction on related issue).

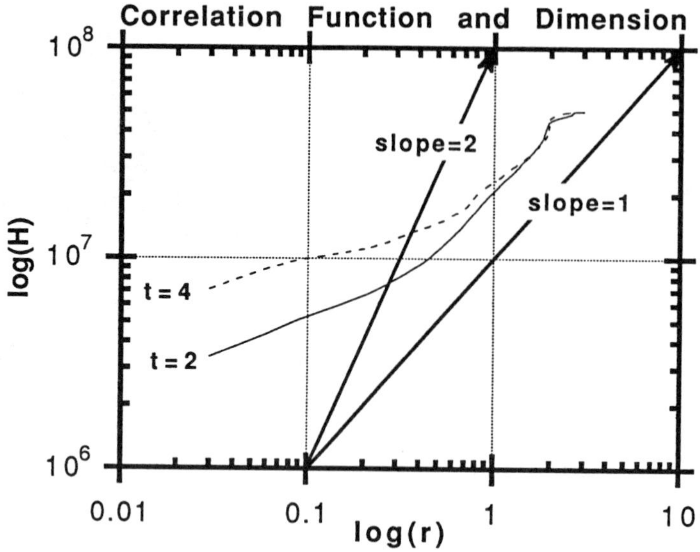

Figure 2. The correlation function and fractal dimension, corresponding to the experiment in Fig.1. The slope for the gyre scale was larger than that for the small scale, indicating that mixing occurred on the gyre scale first.

Fig. 2 showed the correlation function and corresponding fractal dimension for the present case. It was unlikely that the dispersion process could be modeled as the diffusion process. At beginning, the mixing took place on the gyre scale rather than small scale. As times went by, the fractal dimension approached 2, i.e., mixing over the whole domain. However even at time ($t = 4$) there was still some coherent structure in the distribution. Similar phenomena have been observed in the geophysical flows in general (e.g., Pierrehumbert and Yang, 1993; Yang, 1993a,b).

3.2 Zonal Water Column Probability Distribution. To examine how much and how fast the subtropical water is transferred into the subtropical gyre, the zonal water column probability distribution function could be introduced. First the 40,000 fluid particles (floats) were uniformly placed over the whole subtropical gyre. Then the Lagrangian trajectory and location for each water column were calculated at time t. Adding the water column in the zonal direction gave the zonal water column probability distribution at time t. Initially this was a step function with zero in the subpolar gyre and constant in the subtropical gyre. At later time, because of intergyre mass exchange, the subtropical water would be transferred into the subpolar gyre so that the number of the subtropical water column in the subtropical gyre would reduce whereas the number in the subpolar gyre would increase. This function describes the probability that the water of subtropical origin was found in a particular latitude (see, Liu and Yang, 1994).

3.3 Intergyre Exchange. To estimate how much the subtropical water has been transferred into the subpolar gyre during a lapse of time, the particle number in the subpolar gyre was computed. This provided some information about the Lagrangian transport of the subtropical water mass transport into the subpolar gyre. Fig.3 showed the evolution of the total particle number in the subpolar gyre in percentage in the same experiment as in last subsection. Initially there was no particle in the subpolar gyre. As time went by the particle number was increased dramatically at first and then slowly. For example, after $t = 1$ (about 12.5 years), about 6.7% subtropical water column had moved into the subpolar gyre. After 25 years, i.e., $t = 2$, about 10% subtropical water mass had been transferred into the subpolar gyre.

3.4 Mean Lagrangian Transport. How should the mean Lagrangian transport be quantified? Suppose that the volume in the subtropical gyre is V and each particle represents the unit water mass, or water column. Then the volume of the water that has been transported into the subpolar gyre is $V\gamma$, where γ is percentage of the particle number having arrived in the subpolar gyre at t. In general one has the Mean Lagrangian Transport (MLT)

$$MLT = \gamma \frac{V}{t} \tag{3.3}$$

where t is time, during which $\gamma \times 100\%$ particle number has been found in the subpolar gyre.

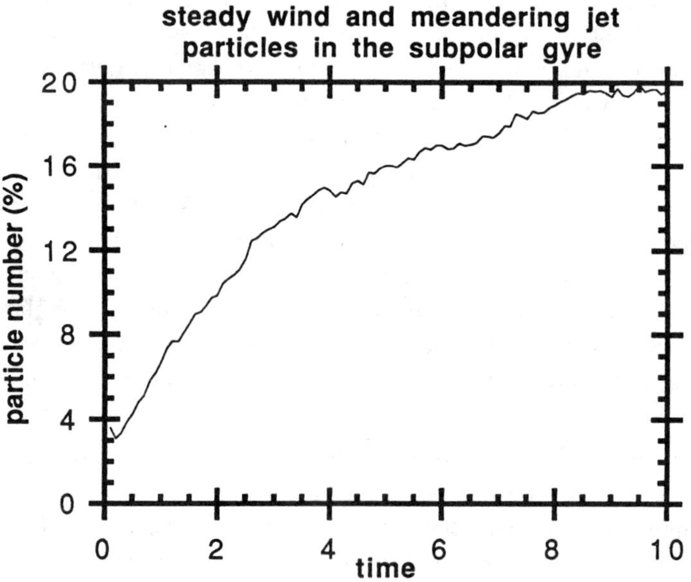

Figure 3. The northward transport of water mass of subtropical origin into the subpolar gyre when the wind was steady. At $t = 2$, 10% subtropical water mass has been transported into the subpolar gyre, which was equivalent to 2.6 Sv when applying to the North Atlantic. In annually migrating wind, it increased to 4.7 Sv. Maximum intergyre water mass transport occurred when the migration was in period of interannual to decadal time scales. From Yang (1995b).

Accordingly, in the present case one had $V = (4000 \times 10^3\ m) \times (2000 \times 10^3\ m) \times (10^3\ m) = 8 \times 10^{15}\ m^3$. The unit time is 12.5 years. Thus, one had

$$MLT = \frac{8 \times 10^{15}\ m^3}{12.5\ years} \frac{\gamma}{t} = 20 \frac{\gamma}{t}\ \text{Sv}, \tag{3.4}$$

where t is the nondimensional time; and the MLT is in unit of Sverdrup

(1 Sv $= 10^6 m^3 s^{-1}$). In the particle evolution diagram, e.g., Fig.3, (γ/t) represents the mean slope. Therefore, the mean Lagrangian transport was proportional to the mean slope in the particle number evolution diagram. For instance, at $t = 1$, about 6.7% subtropical water mass had been transferred into the subpolar gyre. So the mean Lagrangian transport $MLT = 20 \times 0.067 = 1.3$ Sv. That means that the 12.5 year mean Lagrangian transport in this case was about 1.3 Sv due to the meandering jet. The estimate at $t = 2$ was lower, with $MLT = 1.0$ Sv.

Similarly the Instantaneous Lagrangian Transport (ILT) can be defined by

$$ILT = \frac{dV}{dt} \tag{3.5}$$

where dV is the volume of water mass that has transferred into the subpolar gyre during dt period of time. The ILT is exactly the same as the instantaneous Eulerian transport.

3.5 Main Results and Applications to the North Atlantic. The results showed that the meandering jet caused strong intergyre mass transport and the exchange was greatly enhanced when the wind was allowed to migrate north and south. Extensive parameter sensitivity experiments showed that in the steady wind, the transport increases with the width of the jet, and the amplitude and wave number of the waves in the jet. The transport also increased with the amplitude of the waves in the migrating wind. Maximum transport occurred when the wind migrated interannually to decadally.

When applied to the North Atlantic, the transport number had to be multiplied by a factor 2.6 because of the volume difference in the subtropical gyre here and that in the North Atlantic. Hence the results showed that in the current case the intergyre water mass transport was 2.6 Sv when wind was steady whereas it increased to 4.7 Sv when the wind was migrating annually with a distance 800 km. The results further showed that the mean Lagrangian transport could be as high as 7.5 Sv in some other parameters. This amounts more than a half of the net intergyre transport observed in the North Atlantic, which is about 6.5 Sv (Schmitz and McCartney, 1993). The other half is likely accomplished by the meridional overturning mode due to the thermohaline circulation. The results hence were consistent with observations and current understanding of the intergyre mass transport. (See Yang, 1995b)

4. Water Property Exchange

4.1 The Equation. The equation for the tracer advection by this double gyre-jet ocean circulation can be written in general by

$$\frac{\partial \phi}{\partial t} + J(\psi, \phi) = Q + \kappa \nabla^2 \phi \tag{4.1}$$

where ϕ is the tracer field; ψ is the stream function derived in the last section; Q is the source or sink term of the tracer field and κ is the diffusivity of the tracer field. J is the Jacobean. The tracer can represent any water property such as temperature, salinity, carbon, and nutrients. By the method of characteristic, this equation can written as

$$\frac{d\phi}{dt} = Q + \kappa \nabla^2 \phi \tag{4.2}$$

along each characteristics path. The characteristics path is exactly the Lagrangian flow pathway determined earlier by (2.3). Therefore, the equations for the tracer Lagrangian modeling are (2.3) and (4.2) with proper boundary conditions. The equations were easily solved numerically.

4.2 Conserved Passive Tracer Mixing. When there was no source and sink, and the diffusivity was so small it could be negligible, the tracer was conserved. The evolution of this tracer was solely controlled by the advection process by the ocean circulation. Suppose that the initial distribution of the tracer concentration perturbation was a function of the gyre scale circulation's stream function, i.e., Ψ_G. The tracer isolines were the same as the time mean streamlines. Higher value was in the subtropical gyre centered at the anticyclonic gyre center and lower value was found in the subpolar gyre centered at the cyclonic gyre center. When there was no meandering jet and the wind was steady, then the tracer distribution was time independent. The time mean stream lines were the same as the tracer isolines, as shown by solid lines in Fig.4a,b. Thus there was no water property exchange between two gyres. However, if there was a meandering jet in the ocean circulation, even when the wind was steady, the meandering jet alone could cause the water property exchange between the two gyres and exchange across the time mean tracer isolines. The results showed that most exchange took place along the intergyre boundary in the interior and whole eastern basin. During each period of the meandering jet produced one tracer front from the western boundary layer near the intergyre line. The tracer fronts first moved eastward along the intergyre line, then

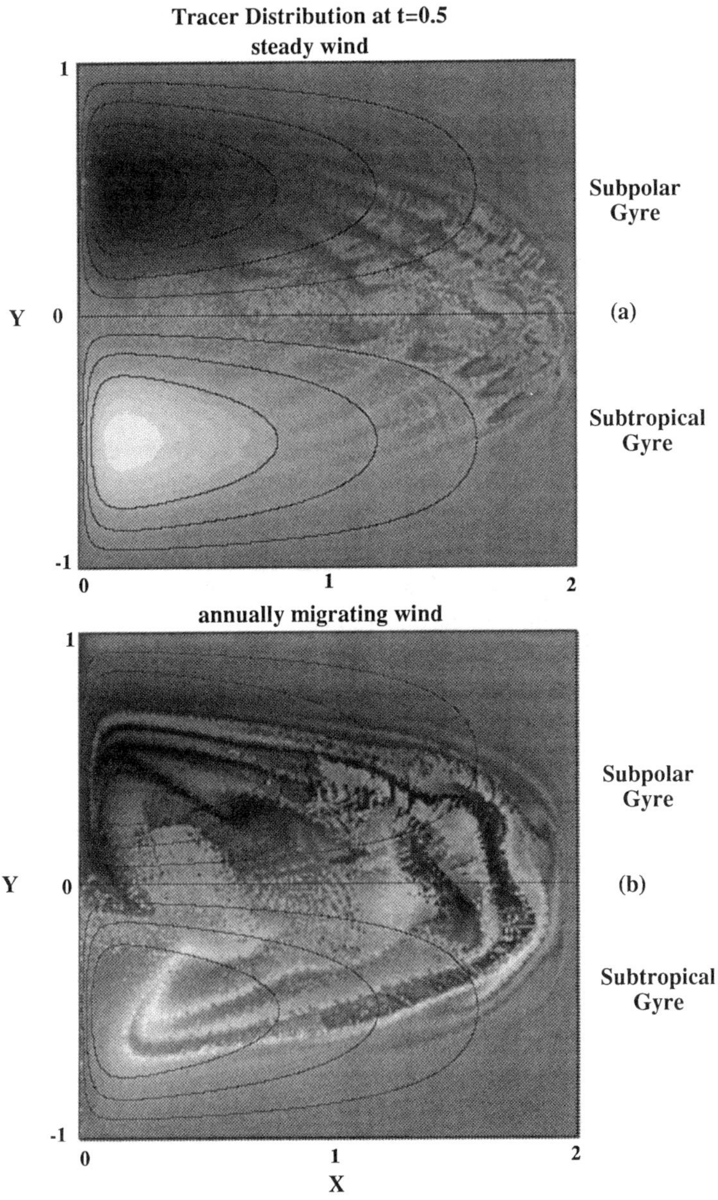

Figure 4. The passive tracer mixing (a) in steady wind and (b) annually migrating wind. The solid lines were the initial isolines of the tracer field and the mean streamlines. The overall intergyre transport and mixing was enhanced, but the meridional gradient of the tracer in the upstream increased when the wind was migrating.

pitched off to both sides beginning at middle western basin. The subtropical tracer front went to the subpolar gyre and the subpolar tracer went to the subtropical gyre. That resulted in the strong intergyre water property exchange (Fig.4a).

When the wind was migrating annually the results showed a much enhanced tracer exchange (Fig.4b). The results suggest that the meandering jet played two roles in the cross jet exchange. The first was to distort the tracer front produced by the wind migration. This meandering effect could result in an enhanced exchange in the eastern basin. The second was the zonal jet effect, which has a barrier effect on the intergyre exchange. Thus this factor could reduce the intergyre mixing, resulting in less cross jet mixing in the western basin.

4.3 Non-Conserved Passive Tracer: Water Temperature Experiments. The water temperature could be treated as a non-conserved passive tracer since there is diabatic heating or the air-sea interaction. In order to make the mechanism clear we use a simple parameterization to account for the diabatic heating. It was assumed that due to the diabatic heating the water was heating in the subtropical gyre and cooling in the subpolar gyre. In addition the heating and cooling were uniform in zonal direction and linear in meridional direction. Therefore, the source and sink term in the temperature equation took the form of $Q = -\alpha y$, where α is the rate of change in temperature. The diffusion effect was modeled by five-point averaging in space.

The temperature gradient should be small so that ocean circulation remains basically barotropic. Suppose that the water temperature change would not greatly alter the flow field in the ocean. Thus the leading order circulation was the same as before. The water temperature could be treated as a non-conserved passive tracer. Initially the water temperature was zonally uniform and linearly decreases with y, where the southern boundary is hottest with maximum temperature of 0.5^oC and the northern boundary is coldest with minimum temperature of -0.5^oC. The coefficient α is taken to be $0.1^oC/year/(2000\ km)$.

When the wind was steady the ocean circulation caused dramatically change in the water temperature pattern (Fig.5). Due to the mixing between two gyres and within its own gyre, and the diabatic heating, the zonal pattern became zonally asymmetric (Fig.5). At $t = 0.5$, in the western basin, there was no substantial exchange between the cold subpolar water and warm subtropical water. However, due to the recirculation the water temperature became uniform in the gyre center. Due to the efficient rapid transport through the western boundary, the warmest water in the subtropical gyre went into the intergyre boundary in the subtropical gyre side and the coldest water in the subpolar gyre went into the intergyre boundary in the subpolar gyre side. As a result, the transport produced high temperature gradient across the jet core in the western basin. Because of interaction between the meandering jet and the gyre circulation, when the warm water from the southern boundary passed the western basin, it separated from the subtropical gyre into the subpolar gyre. Likewise, the cold water from the northern boundary separated from the subpolar gyre in the middle and eastern basin into the subtropical gyre. Each period of the meandering, this process produced one warm ring in the subpolar gyre and one cold ring in the subtropical gyre. As these rings passed into the eastern basin, they detached from the jet. In the subtropical gyre there were about seven or eight cold rings and four warm rings in the subpolar gyre, as shown in Fig.5. Further north or south the ring disappeared due to the heating in the subtropical gyre and the cooling in the subpolar gyre. At later time, more rings were formed and detached from the jet. As time went by, the two recirculation regions became homogenized in the western basin, resulting in enhanced gradient across the jet core. On the other hand, in the eastern basin, because of the ring detachment and high transport, the meridional temperature gradient was much weaker near the intergyre boundary than in the western basin. Similar process took place in migrating wind.

4.4 Front and Exchange Across the Jet Core. The results capture many observed salient features associated with the water property front across the Gulf Stream in the upper North Atlantic. Most Lagrangian transport across the jet core occurred in the interior region and eastern basin whereas most Lagrangian exchange in the western basin took place between the Gulf Stream and its surrounding fluid, not across the jet core. The experiments on the conserved passive tracers and the water temperature showed clearly that there were strong, narrow, persistent tracer and water temperature fronts across the jet core in the western basin in upstream of the jet. The water front became weak as it went to downstream. These results agree with observations in the region (e.g., Robinsion, 1982 and references within; Bower et al., 1985; Bower, 1991; Bower and Lozier, 1994). Both the meandering Gulf and the migration of wind pattern enhanced the meridional gradient of the water property across the Gulf Stream.

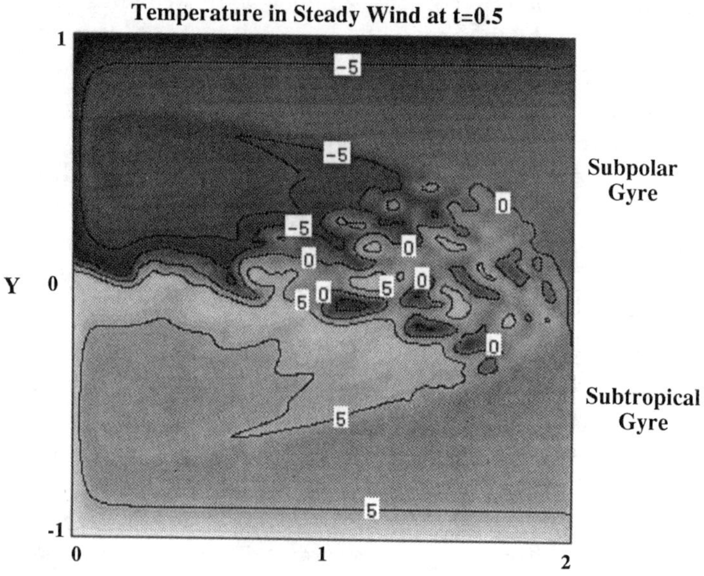

Figure 5. The heat exchange and water temperature distribution in steady wind. The temperature perturbation was in unit of tenth of degree Celsius. The initnial temperature distribution was linear in y and independent of x. As a result of gyre circulation and the meandering jet interaction, the warm anticyclonic rings in the subpolar gyre and cold cyclonic rings in the subtropical gyre were formed and separated from the jet.

5. Chaotic Zones and Barriers

5.1 A Perspective From Hamiltonian Chaos. This Lagrangian approach has a direct linkage to the contemporary dynamical systems theory because of special relationship between the system governing this two-dimensional, incompressible flow, i.e. equation (2.3) and the Hamiltonian system. The system (equation (2.3)) is exactly a Hamiltonian system with the stream function acting as the Hamiltonian, and two spatial variables acting as action and the momenta (e.g., Aref, 1984; or Yang, 1993a,b). Therefore, the kinematics of transport and mixing by a two-dimensional, incompressible flow such as (2.1) is equivalent to the dynamics of a Hamiltonian system with one degree of freedom if the flow is steady, one and a half degrees of freedom if the flow is time periodic, and two degrees of freedom if the flow is unsteady and nonperiodic in time. A Hamiltonian system of one degree of freedom such as a steady two-dimensional flow, is integrable and most Hamiltonian systems with two degrees of freedom are nonintegrable (e.g., Arnold, 1989). There has been a wealth of information about such system (e.g., Guckenheimer and Holmes, 1983; Lichtenberg and Lieberman, 1983).

If a Hamiltonian system is integrable it cannot be chaotic. When studying Hamiltonian systems, we usually first transform the system into coordinates of action-angle variables (e.g., Anorld, 1989). Consider an integral twist mapping in the neighborhood of an elliptic points in the Poincare section of a time-periodic Hamiltonian system, e.g., $r_{n+1} = r_n$, $\theta_{n+1} = \theta_n + 2\pi\sigma(r_n)$. If $\sigma(r_n)$ is irrational and the trajectories will wrap densely (in mathematical sense) around a torus, never intersecting themselves. If $\sigma(r_n)$ is rational the trajectory will return exactly to its original position after a finite number of turns around the torus. What happens to the map if the system is slightly perturbed by a periodic perturbation? The KAM theory tells us what happens to the irrational tori of the integrable system. It says that under such perturbation most of the invariant irrational tori (a set of positive Legesgue measures) are conserved and do not disappear under the certain conditions. This can create an impassable barrier for transport and mixing so that the points inside the invariant circle, or KAM curve, can never escape the interior. If this irrational number does not satisfy the hypotheses of the KAM theorem, then the closed curve in the unperturbed system may break up into an invariant Cantor set on which the dynamics is quasi-periodic. This structure is often called a cantorus. Typically, the Cantori are hyperbolic in stability type and remnants of KAM invariant circles when increasing the strength of the perturbation. They form a partial barrier to transport and mixing, and moreover their hyperbolic nature indicates the existence of stable and unstable manifolds which can be used to form lobes. (See Percival, 1979; MacKay et al., 1984). The Poincare-Birkhoff theorem tells us what happens to the rational orbits that disappear. It says under such a perturbation the rational orbits break up into a collection of k-hyperbolic and k-elliptic points. Moser theorem ensures the existence of the invariant irrational torus. In a system with one and a half degrees of freedom such as a two-dimensional, periodic flow system, the domain of phase space or the fluid domain decomposes into regular regions or islands immersed in a chaotic sea. The islands are bounded by invariant KAM tori. Therefore, the invariant KAM tori act as barriers for transport.

In the current system (2.3) when there is no meandering jet, for steady wind there are separatrices connected by the stagnation points at $x = 0, 2, y = 0$ for instance, and two centers for two gyres. These were clearly shown in Fig.4 by solid lines. In the dynamical systems theory, the stagnation points are called the hyperbolic points or heteroclinic points in the present case and the separatrices are called the heteroclinic orbits, or heteroclinic cycles. The central point at the center of each gyre is called elliptic point. For each heteroclinic point there are stable manifolds and unstable manifolds. The stable manifold moves toward the heteroclinic point whereas the unstable manifold moves away from the point. Under a small periodic perturbation such as migration in the wind pattern or a meandering jet, the distance between the stable and unstable manifold can be measured by the Melnikov function and can be used to measure the transport. Samelson (1992) provided one such example (See Guckenheimer and Holmes, 1983; Lichetenberg and Lieberman, 1983; Wiggins, 1992 for general information). If these stable and unstable manifolds intersect once they will intersect infinitely many times to produce the boundary of a complicated two-dimensional lobe structure that is invariant. They are the cause of chaos and play a important role in the transport and mixing. The mechanism of chaotic transport and mixing is stretching and folding, which are equivalent to the horseshoe mapping topologically, by these stable and unstable manifolds. Thus, there are regions of chaos near the heteroclinic orbits, such as near the western boundary and intergyre boundary. The KAM theory says that most of the invariant irrational tori are conserved and do not disappear. Thus these KAM curves formed the impassable barriers to the transport if the perturbation is small and certain conditions are met. Otherwise the KAM closed curves may break up and form a partial barrier to transport and mixing. In Fig.4a and Fig.5a the streamlines near two centers of the subtropical gyre and subpolar gyre are conserved and fluid parcles there are completely isolated from the outside due to these KAM curves. They are impassbale barriers. However, there are partial barriers away from the two centers of gyres. The fluid parcles there do exchange with outside. In fact, these features of chaotic regions, partial barriers and impassable barriers have been observed in the conserved passive tracer and water temperature experiments here (Fig.4a and Fig.5a). (See Ottino, 1989 or Yang, 1993a for a general perspective and examples on the evolution of the heteroclinic orbits.).

When there is no wind forcing, the current model becomes a system with two traveling waves in shear flow. Such system has similar heteroclinic orbits or heteroclinic cycles (e.g., Knobloch and Weiss, 1987; Pierrehumbert, 1991a,b; Samelson, 1992; Lozier and Bercovici, 1992; del-Castillo-Negrete and Marrison, 1993; Yang, 1993a,b; Pratt et al., 1995). In traveling waves, the heteroclinic structure may appear in different terms such as critical layers or steering levels, closed streamlines, cat's eyes pattern,

Chirikov resonance overlap, or even wave breaking (McIntyre and Palmer, 1983, 1985; Haynes and McIntyre, 1987). These heteroclinic cycles lead to chaos when perturbed by transients. Conversely, there is region near the intergyre boundary where the stream lines are open. This region tends to form a barrier to transport and mixing. Therefore the jet core acts like barrier to transport and mixing. Many these features are observed in the experiments (Fig.4 and Fig.5). The nature of the Hamiltonian chaos is the reason behind chaotic transport regions and barriers to the transport. However, experiments here are more complicated since there are both the gyre mode circulation and the meandering jet is not weak. Near the intergyre boundary the meandering jet is more strong than the gyre circulation mode so that the jet acts as a barrier, especially in the western basin. One useful method to directly detect the chaos and barrier regions is the finite-time Lyapunov exponent.

5.2 The Finite-Time Lyapunov Exponent. The chaotic behavior in the transport is the inherent nature of the dynamical system governing the Lagrangian trajectory of fluid particle. The present dynamical system (2.3) is a two dimensional, time dependent system. Such system is capable of producing chaos (two dimensional nonautonomous dynamical system or three dimensional autonomous dynamical system is capable of producing chaos).

To quantify how chaotic the transport is and where the chaotic transport takes place, we introduced a very unique and important concept in the dynamical systems theory. It is called the Lyapunov exponent. The Lyapunov exponent is a measure of rate of exponential divergence of nearby trajectories in the phase space of the underlying dynamical system (Guckenheimer and Homes, 1983). In the current system, the phase space coincides with the physical space (x,y), as shown in the system (2.3). However, we would approximate the Lyapunov exponent by finite- time Lyapunov exponent since we are interested in the finite-time behavior of our system. Suppose that $\boldsymbol{X}(0)$ is an initial point of any trajectory in the phase space and $\boldsymbol{\delta}(X(0))$ is the initial variation vector; then the exponential divergence of nearby trajectories is quantified by the following finite-time Lyapunov exponent (Grassberger et al, 1989; or see Yang, 1993a,b; 1995a,b; Pierrehumbert and Yang, 1993 and references wherein)

$$\Lambda(t) = \frac{1}{t} \ln \frac{|\boldsymbol{\delta} X(t)|}{|\boldsymbol{\delta} X(0)|} \tag{5.1}$$

The chaotic region is indicated by positive Lyapunov exponent and regular or no chaotic region is represented by zero or small Lyapunov exponent. In terms of transport, the region with large positive Lyapunov exponent suggests a chaotic transport region or chaotic transport channel, whereas a region with zero or very small Lyapunov exponent represents a barrier to transport. The finite-time Lyapunov exponent analysis has been successfully applied in the transport and mixing by the geophysical flows to identify the chaotic transport channels and the barriers to the transport in the works cited above.

5.3 The Exponent Distributions. The Lyapunov exponent distribution was calculated and shown in Fig.6. It was found that there was clearly a wavy structure in the distribution near the intergyre boundary where the meandering jet was located. The large values of the Lyapunov exponent were along both sides of the jet center. The western boundary was also found to have the large values of the exponent. Other regions with large value were in eastern parts of the double gyre when the wind was steady (Fig.6a). The northern and southern boundary regions also had large values. These regions with large values of the exponent represent areas with strong chaotic transport, i.e., transport channels. The results suggest that the transport took place mainly near the intergyre boundary, especially within the western boundary and then penetrated further north in the eastern part of the subpolar gyre.

However, there were two large areas with zero or very small Lyapunov exponent, located at the center of each gyre. They were recirculation centers. These two regions were barriers to the transport. Because of the existence of such a barrier, thus, there was no transport between the center and its outside. The center was isolated from the outside. The subtropical water mass in the center of the subtropical gyre could not easily be transferred into the subpolar gyre. Likewise, the subpolar water mass in the subpolar gyre center was isolated from its outside and would not be transferred into the subtropical gyre. The value was also small in the eastern boundary near the intergyre boundary. More importantly, there was a wavy narrow region around the jet core, where the exponent was small, whereas on both sides of this narrow wavy belt the exponent was large. This exemplifies the central barrier effect of the jet as discussed in recent studies (e.g., Sommeria et al., 1988, 1991; Behringer et al., 1991; del-Castillo-Neregret and Marrison, 1993; Meyers, 1994; Yang, 1995a). Large values were found in the western boundary near the intergyre boundary. Therefore, there was a channel, through which, most subtropical water mass enters the subpolar gyre. However, because of the existence of the subpolar barrier (subpolar recirculation) and the barrier across the jet core, the water could not penetrate into the north in the western subpolar gyre. In stead, it flowed eastward near the intergyre boundary and then gradually penetrated into the northern subpolar gyre in the eastern part of the

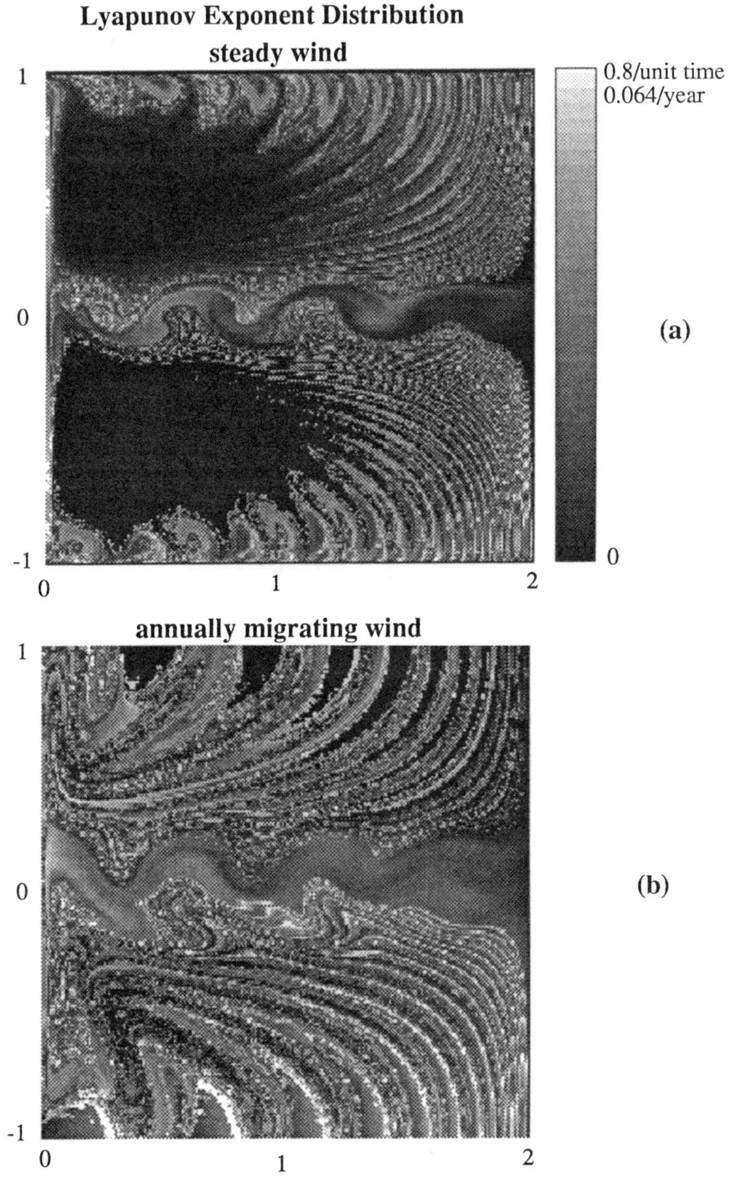

Figure 6. The Lyapunov exponent distribution, (a) in steady wind and (b) in annually migrating wind. The chaotic mixing and transport regions were indicated by large value of the exponent whereas the barriers to transport and mixing were represented by zero or small values of the exponent. From Yang (1995b).

gyre. These processes were clearly captured in the Lyapunov exponent distribution and were consistent with the Lagrangian modeling experiments performed above and observations. The migration of wind would enhance the chaotic transport and mixing as shown in Fig.6b. These Lyapunov exponent distributions exemplify the very nature of the chaotic transport and mixing, and barriers to transport and mixing discussed in 5.1 from a general perspective and quantify the chaotic nature observed in these direct Lagrangian modeling results.

5.5 The Chaotic Mixing and Diffusion. The relationship between the chaotic mixing and diffusion is of great importance. A recent study (Yang,1994) showed the relative importance between the chaotic mixing and diffusion in terms of finite-time Lyapunov exponent and the diffusivity of the system, concluding that in the chaotic regions, the mixing and transport was dominated by the chaotic mixing process, and in the weak chaotic region or regular region the mixing and transport was dominated by the diffusion process. The relative importance between the chaotic mixing and diffusion in the varying gyre ocean model was discussed in Yang and Liu (1994) and Liu and Yang (1994). It was found that the equivalent diffusivity due to the chaotic mixing was at the order of that due to the strong mesoscale eddies observed near the Gulf Stream. However, as we pointed out, the chaotic mixing cannot realistically be modeled by the diffusion process due to the nature of the large geophysical flows.

6. Concluding Remarks

The results of the Lagrangian modeling of the chaotic transport and mixing by the gyre-scale circulation motion are consistent with observations. The water mass transport, the conserved tracer and water temperature experiments, and the Lyapunov exponent distributions capture many important features observed near the Gulf Stream and its roles in the cross-jet transport and mixing (e.g.,Robinson, 1983 and references wherein; Bower et al., 1985; Bower and Rossby, 1989; Bower, 1991), including the water property front across the jet core and the ring formation and detachment. The results have implications in the transport of materials of biogeochemical importance as well as in the potential vorticity homogenization (see Yang and Liu, 1994 on this perspective). The Lagrangian trajectories indeed are chaotic in the current system and the results strongly suggest that the gyre scale chaotic mixing could also be the mechanism for the chaotic or fractal behavior of the quasi Lagrangian drifter trajectories observed in oceans by Osborne et al. (1986, 1989) and Brown and Smith

(1990, 1991).

The current model is very simple even though still dynamically consistent under some approximation. There are other forces at work in natural oceans. For example, the stochastic forcing in the ocean dynamics will induce the stochastic flow field in the circulation. This will further complicate the transport and mixing problem. The current knowledge of the stochastic forcing and the resulting stochastic behavior in the ocean motion and tracer is well described in this volume by other authors. The problem on how well these stochastic forcings represent the real physics in oceans still remains. The question that how the stochastic forcing in the flow field or the dynamics will alter the Lagrangian transport and mixing in the gyre ocean circulation is of great interest to oceanography in general. This needs further study. The current approach has definite advantage on investigating the mass transport and water property exchange. Nevertheless, this approach does have some drawbacks. First of all, the realistic and meaningful analytic flow field is hard to obtain in general. That means we may be forced to numerically solve the ocean dynamics. The present model employs the possible simplest model in which both gyre and jet are both present. For the potential vorticity, a dynamically consistent model within which the potential vorticity is conserved when the fluid has no direct contact with the sea surface may require. However, it should be noted that the potential vorticity is not strictly conserved in nature. Thus the first step of the Lagrangian modeling is a increasingly difficult task and hopeless for a realistic model. New trend is to incorporate with the general circulation model (GCM) in which full dynamics and/or physics is taken into account (e.g., Lozier and Riser, 1990; Liu et al., 1994; Bower and Lozier, 1994). Such approach has also been adapted in the atmosphere recently (e.g., Pierrehumbert and Yang, 1993; Yang and Pierrehubert, 1994; Yang, 1995a).

There are other methods that have been used to investigate the chaotic, Lagrangian transport and mixing in physical oceanography. For example, for the weak periodic perturbation Melnikov function can be employed to analytically investigate the transport since it can be shown that this function is proportional to the transport. Samelson (1992) used such method in cross-jet transport study (see Samelson in this volume for a review). It is well known in Hamiltonian chaos theory that there is a qualitative description of how the presence of two modes results in the resonance and generates chaotic trajectories. That is given in the form of the Chirikov resonance overlap criterion (Chirikov, 1979). According to the overlap criterion, when the sum of the half-widths of the two resonance islands equals the distance between the resonances the central barrier is destroyed, resulting in the global chaos. Meyers (1994) used the Chirokov criterion to determine the existence of global chaos, which is indicative of cross-jet transport, in a meandering jet, like the Gulf Stream. Also see del-Castillo-Negrete and Morrison (1993). However, both analytic methods are limited to the weak perturbation problem, requiring the knowledge of the mean flow and the perturbation.

Due to the increasing use of the Lagrangian floats and drifters in the oceanographic field research programs and possible in the ocean circulation prediction operation in near future, and the fact the only very limited floats or drifters would be deployed and tracked for a limited time, more Lagrangian modeling is urgently needed. A better understanding of the Lagrangian behaviors of the floats and drifters becomes necessary to understand the ocean circulation and ultimately to predict it. The information of the Lagrangian transport of the water mass is also needed in order to understand the water properties, their distributions, and variabilities. As shown here that the Lagrangian perspective may also provide additional information and mechanism to interpret observations. For example, the current study demonstrated the vital importance of transient or temporal variation of the gyre circulation mode on the regional and global climate study. It is the Lagrangian transport due to the transients or temporal variations in the gyre circulation mode, not the Eulerian transport due to the mean circulation that dominates the intergyre mass transport and result in long term (decadal and beyond) water property exchange, modification and distribution. It has potential in general oceanographic studies when the knowledge of a long term transport is needed. It is expected to see the increasing use of the Lagrangian modeling in the oceanographic research.

Acknowledgement: The work was made possible in part by the United States Geological Survey and the National Oceanic and Atmospheric Administration.

References

[1] Arnold, V.I., 1989: *Mathematical Methods of Classical Mechanics.*, Second Edition. Springer-Verlag, New York.

[2] Aref, H., 1984: Stirring by chaotic advection. *Journal of Fluid Mechanics.*, **143**, 1-21.

[3] Behringer, R.P., S.D. Meyers and H.L. Swinney, 1991: Chaos and mixing in geostrophic flows. *Physics of Fluids.*, **A 3**, 1243-1249.

[4] Bower, A.S., 1991: A simple kinematic mechanics for mixing fluid particles across a meandering jet. *Journal of Physical Oceanography.*, **20**, 173-180.

[5] Bower, A.S. and M.S. Lozier, 1994: A close look at particle exchange in the Gulf Stream. *Journal of Physical Oceanography.*, **24**, 1399-1418.

[6] Bower, A.S. and T. Rossby, 1989: Evidence of cross-frontal exchange processes in the Gulf Stream based on isopycnal RAFOS float data. *Journal of Physical Oceanography.*, **19**, 1177-1190.

[7] Bower, A.S. and T. Rossby and J.L. Lillibridge, 1985: The Gulf Stream – Barrier or blender. *Journal of Physical Oceanography.*, **15**, 24-32.

[8] Brown, M.G. and K.B. Smith, 1991: Stirring the oceans by chaotic advection. *Phys. Fluids A.*, **3**, 1186-1192.

[9] Brown, M.G. and K.B. Smith, 1990: Are SOFAR float trajectories chaotic?, *Journal of Physical Oceanography.*, **20**, 139-149.

[10] Chien, W.-L., H. Rising and J.M. Ottino, 1986: Laminar mixing and chaotic mixing in several cavity flows. *Journal of Fluid Mechanics.*, **170**, 355-377.

[11] Chirikov, B.V., 1979: A universal instability of many dimensional oscillator systems. *Physical Reports.*, **52**, 263-310.

[12] Cushman-Roisin, B., 1993: Trajectories in Gulf Stream meanders. *Journal Geophysical Research.*, **98**, 2543-2554.

[13] del-Castillo-Negrete, D. and P.J. Morrison, 1993: Chaotic transport by Rossby waves in shear flow. *Physics of Fluids.*, **A 5(4)**, 948-965.

[14] Eckart, C., 1948: An analysis of the stirring and mixing processes in incompressible fluids. *Journal of Marine Research.*, **7**, 265-275.

[15] Grassberger, P. and I. Procaccia, 1983: Measuring the strangeness of strange attractors. *Physica D.*, **9**, 189-208.

[16] Grassberger, P., R. Badii and A. Politi, 1989: Scaling laws for invariant measures on hyperbolic and non-hyperbolic attractors. *J. Stat. Phys.*, **51**, 135-178.

[17] Guckenheimer, J. and P. Holmes, 1993: *Nonlinear Oscillations, Dynamical Systems, and Bifurcations of Vector Fields.*, Applied Math. Sci. **42**, Springer-Verlag, New York.

[18] Haynes, P.H. and M.E. McIntrye, 1987: On the representation of Rossby-wave critical layers and wave breaking in zonally truncated models. *Journal of Atmospheric Sciences.*, **44**, 2359-2382.

[19] Huang, R.X., 1990: On the three-dimensional structure of the wind-driven circulation in the North Atlantic. *Dynamics of Atmosphere and Oceans.*, **15**, 117-159.

[20] Khakhar, D.V. , H. Rising and J.M. Ottino, 1986: Analysis of chaotic mixing in two model systems. *Journal of Fluid Mechanics.*, **172**, 419-

451.

[21] Knobloch, E. and J.B. Weiss, 1987: Chaotic advection by modulated traveling waves. *Physics Review.*, **A36**, 1522-1524.

[22] Lichetenberg, A.J. and M.A. Lieberman, 1983: *Regular and Stochastic Motion.*, Applied Math. Sciences, **38**, Springer-Verlag, New York.

[23] Liu, Z. and H. Yang, 1994: The intergyre chaotic transport. *Journal of Physical Oceanography.*, **24**, 1768-1782.

[24] Liu, Z. S.G.H. Philander and P.C. Pacanowski, 1994: A GCM study of tropical-subtropical upper-ocean water exchange. *Journal of Physical Oceanography.*, **24**, 2606-2623.

[25] Lozier, M.S. and D. Bercovici, 1992: Particle exchange in an unstable jet. *Journal of Physical Oceanography.*, **22**, 1506-1516.

[26] Lozier, M.S. and S.C. Riser, 1990: Potential vorticity sources and sinks in a quasi-geostrophic ocean: Beyond western boundary currents. *Journal of Physical Oceanography.*, **20**, 1608-1627.

[27] MacKay, R.S., J.D. Meiss and I.C. Percival, 1984: Transport in Hamiltonian systems. *Physica D.*, **13**, 55-81.

[28] McIntrye, M.E. and T.N. Palmer, 1983: Breaking planetary waves in the stratosphere. *Nature.*, **305**, 593-600.

[29] McIntrye, M.E. and T.N. Palmer, 1985: A note on the general concept of wave breaking for Rossby and gravity waves. *Pageoph.*, **123**, 964-975.

[30] Meyers, S.D., 1994: Cross-frontal mixing in a meandering jet. *Journal of Physical Oceanography.*, **24**, 1641-1646.

[31] Molinari, R.L., D. Battisti, K. Bryan and J. Walsh, 1994: The Atlantic Climate Change Program. *Bull. Amer. Met. Soc.*, **75**, 1191-1199.

[32] Osborne, A.R., A.D. Kirwan, A. Provenzale and L. Bergamasco, 1986: A search for chaotic behavior in large and mesoscale motions in the Pacific Ocean. *Physica D.*, **23**, 75-83.

[33] Osborne, A.R., A.D. Kirwan, A. Provenzale and L. Bergamasco, 1989: Fractal drifter trajectories in the Kuroshio extension. *Tellus.*, **41 A**, 416-435.

[34] Ottino, J.M., 1989: *The Kinematics of Mixing: Stretching, Chaos and Transport.* Cambridge University Press, Cambridge.

[35] Pedlosky, J., 1984: Cross-gyre ventilation of the subtropical gyre: An internal mode in the ventilated thermocline. *Journal of Physical Oceanography.*, **14**, 1172-1178.

[36] Percival, I.C., 1979: Variational principles for invariant tori and cantori. In *Nonlinear Dynamcs and the Beam-Beam Interaction*, Ed. Month and Herrera, Amer. Inst. Phys. Conf. Proc., **57**, 302-310.

[37] Pierrehumbert, R.T., 1991a: Large-scale horizontal mixing in plane-

tary atmospheres. *Physics of Fluids.*, **3**, 1250-1260.

[38] Pierrehumbert, R.T., 1991b: Chaotic mixing of tracer and vorticity by modulated travelling Rossby waves. *Geophys. Astrophys. Fluid Dynamics.*, **58**, 285-319.

[39] Pierrehumbert, R.T. and H. Yang, 1993: Global chaotic mixing on isentropic surfaces. *Journal of Atmospheric Sciences.*, **50**, 2462-2480.

[40] Pratt, L.J., M.S. Lozier and N. Beliakova, 1995: Parcel trajectories in quasi-geostrophic jets: Neutral modes. *Journal of Physical Oceanography.*, **25**, 1451-1466.

[41] Provenzale, A., A.R. Osbrone, A.D. Kirwan and L. Bergamasco, 1991: The study of fluid particle trajectories in large-scale ocean flow. *Nonlinear Topics in Ocean Physics*, Ed. by A.R. Osborne, North-Holland, Amsterdam, 367-402.

[42] Ridderinkhof, H. and J.W. Loder, 1994: Lagrangian characterization of circulation over submarine banks with application to the outer Gulf of Maine. *Journal of Physical Oceanography.*, **24**, 1184-1200.

[43] Ridderinkhof, H. and J.T.F. Zimmerman, 1992: Chaotic stirring in a tidal system. *Science.*, **258**, 1107-111.

[44] Rhines, P.B., and R. Schopf, 1991: The wind-driven circulation: Quasi-geostropic simulations and theory for nonsymmetric winds. *Journal of Physical Oceanography.*, **21**, 1438-1468.

[45] Robinson, A.R., 1983: *Eddies in Marine Science.* Springer-Verlag, New York.

[46] Samelson, R.M., 1992: Fluid exchange across a meandering jet. *Journal of Physical Oceanography.*, **22**, 431-440.

[47] Sanderson, B.G. and D.A. Booth, 1991: The fractal dimension of drifter trajectories and estimates of horizontal eddy-diffusivity. *Tellus.*, **43**, 334-349.

[48] Schmitz, W.J. and M.S. McCartney, 1993: On the North Atlantic circulation. *Reviews of Geophysics.*, **31**, 29-49.

[49] Sommeria, J., S.D. Meyers and H.L. Swinney, 1989: Laboratory model of planetary easterly jet. *Nature.*, **337**, 58-61.

[50] Sommeria, J., S.D. Meyers and H.L. Swinney, 1991: Experiments on vortices and Rossby waves in eastward and westward jets. *Nonlinear Topics in Ocean Physics.*, Ed. by A.R. Osborn, North-Holland, Amsterdam, 227-269.

[51] Wiggins, S., 1992: *Chaotic Transport in Dynamical Systems.* Springer-Verlag, New York.

[52] Yang, H., 1993a: Chaotic mixing and transport in wave systems and the atmosphere. *International Journal of Bifurcation and Chaos.*, **3**, 1423-1445.

[53] Yang, H., 1993b: Chaotic wave packet mixing and transport. *Canadian Applied Mathematics Quarterly.*, **1**, 569-602.

[54] Yang, H., 1994: On the relative importance between chaotic mixing and diffusion. *Physics Letters.*, **A, 185**, 191-195.

[55] Yang, H., 1995a: Three-dimensional transport of the Ertel potential vorticity and N_2O in the GFDL SKYHI model. *Journal of Atmospheric Sciences.*, **52**, 1513-1528

[56] Yang, H., 1995b: The subtropical/subpolar gyre exchange in the presence of annually migrating wind and a meandering jet: Water mass exchange. *Journal of Physical Oceanography.*, **25**, in press.

[57] Yang, H. and Z. Liu, 1994: Chaotic transport in a double gyre ocean. *Geophysical Research Letters.*, **21**, 545-548.

[58] Yang, H. and R.T. Pierrehumbert, 1994: Production of dry air by isentropic mixing. *Journal of Atmospheric Sciences.*, **51**, 3437-3454.

Department of Marine Science
University of South Florida
St. Petersburg, Florida 33701

yang@marine.usf.edu

Permissions

[1] Figure 6, page 39 is reprinted from *J. Geophys. Res.* **90**, C3, pp. 4756–4722, ©American Geophysical Union 1985.

[2] Figure 7, page 40 is reprinted from *J. Fluid Mech.* **117**, pp. 1–26, ©Cambridge University Press 1982.

[3] Figure 2, page 314 is reprinted from *J. Phys. Oceanogr.* **15**, pp. 1444, 1446, and 1447, ©American Meteorological Society 1985.

[4] Figure 3, page 317 is reprinted from *J. Atmos. Oceanogr. Tech.* **7**(2), pp. 285–295, ©American Meteorological Society 1990.

[5] Figure 4, page 321 is reprinted from *J. Atmos. Oceanogr. Tech.* **7**(2), pp. 285–295, ©American Meteorological Society 1990.

[6] Figure 5, page 322 is reprinted from *J. Atmos. Oceanogr. Tech.* **7**(2), pp. 285–295, ©American Meteorological Society 1990.

[7] Figure 6, page 324 is reprinted from *J. Atmos. Oceanogr. Tech.* **7**(2), pp. 285–295, ©American Meteorological Society 1990.

[8] Figure 7, page 325 is reprinted from *J. Atmos. Oceanogr. Tech.* **7**(2), pp. 285–295, ©American Meteorological Society 1990.

[9] Figure 10, page 333 is reprinted from *J. Phys. Oceanogr.* **18**, pp.1811–1853, ©American Meteorological Society 1988.

Progress in Probability

Editors

Progress in Probability is designed for the publication of workshops, seminars and conference proceedings on all aspects of probability theory and stochastic processes, as well as their connections with and applications to other areas such as mathematical statistics and statistical physics. It acts as a companion series to *Probability and Its Applications*, a context for research level monographs and advanced graduate texts.

We encourage preparation of manuscripts in some form of TeX for delivery in camera-ready copy, which leads to rapid publications, or in electronic form for interfacing with laser printers or typesetters.

Proposals should be sent directly to the editors or to:
Birkhäuser Boston, 675 Massachusetts Avenue, Cambridge, MA 02139, U.S.A.

12 ÇINLAR/CHUNG/GETOOR. Seminar on Stochastic Processes, 1985
13 ÇINLAR/CHUNG/GETOOR/GLOVER. Seminar on Stochastic Processes, 1986
14 DEVROYE. A Course in Density Estimation
15 ÇINLAR/CHUNG/GETOOR/GLOVER. Seminar on Stochastic Processes, 1987
16 KIPER. Random Perturbations of Dynamical Systems
17 ÇINLAR/CHUNG/GETOOR/GLOVER. Seminar on Stochastic Processes, 1988
18 ÇINLAR/CHUNG/GETOOR/FITZSIMMONS/WILLIAMS. Seminar on Stochastic Processes, 1989
19 ALEXANDER/WATKINS. Spatial Stochastic Processes: A Festschrift in Honor of Ted Harris on his 70th Birthday
20 HAAGERUP/HOFFMANN-JØRGENSEN/NIELSEN. Probability in Banach Spaces 6: Proceedings of the Sixth International Conference, Sandbjerg, Denmark 1986
21 EBERLEIN/KUELBS/MARCUS. Probability in Banach Spaces 7: Proceedings of the Seventh International Congress
22 PINSKY. Diffusion Processes and Related Problems in Analysis, Vol. I: Diffusions in Analysis and Geometry
23 HAHN/MASON/WEINER. Sums, Trimmed Sums and Extremes
24 ÇINLAR/FITZSIMMONS/WILLIAMS. Seminar on Stochastic Processes, 1990
25 CAMBANIS/SAMORODNITSKY/TAQQU. Stable Processes and Related Topics
26 CRUZEIRO/ZAMBRINI. Stochastic Analysis and Applications: Proceedings of the 1989 Lisbon Conference
27 PINSKY/WIHSTUTZ. Diffusion Processes and Related Problems in Analysis, Vol. II. Stochastic Flows
28 DURRETT/KESTEN. Random Walks, Brownian Motion and Interacting Particle Systems

29 ÇINLAR/CHUNG/SHARPE. Seminar on Stochastic Processes, 1991

30 DUDLEY/HAHN/KUELBS. Probability in Banach Spaces, 8, 1992

31 KÖREZLIOĞLU /ÜSTÜNEL. Stochastic Analysis and Related Topics, 1992

32 NUALART/SOLE. Barcelona Seminar on Stochastic Analysis: St. Feliu de Guixols, 1991

33 ÇINLAR/CHUNG/SHARPE. Seminar on Stochastic Processes, 1992

34 FREIDLIN. The Dynkin Festschrift. Markov Processes and their Applications, 1994

35 HOFFMANN-JØRGENSEN/KUELBS/ MARCUS. Probability in Banach Spaces, 9, 1993

36 BOLTHAUSEN/DOZZI/RUSSO. Seminar on Stochastic Analysis, Random Fields and Applications, 1993

37 BANDT/GRAF/ZÄHLE. Fractal Geometry and Stochastics, 1995

38 KÖREZLIOGLU/ØKSENDAL/ÜSTÜNEL. Stochastic Analysis and Related Topics V. The Silivri Workshop, 1994

39 ADLER/MÜLLER/ROZOVSKII. Stochastic Modelling in Physical Oceanography, 1996

F